T0293145

LONDON MATHEMATICAL SOCIETY LECTURE NOTE SERIES

The titles below are available from booksellers, or from Cambridge University Press at www.cambridge.org/mathematics

353 Trends in stochastic analysis, J. BLATH, P. MÖRTERS & M. SCHEUTZOW (eds)
354 Groups and analysis, K. TENT (ed)
355 Non-equilibrium statistical mechanics and turbulence, J. CARDY, G. FALKOVICH & K. GAWEDZKI
356 Elliptic curves and big Galois representations, D. DELBOURGO
357 Algebraic theory of differential equations, M.A.H. MACCALLUM & A.V. MIKHAILOV (eds)
358 Geometric and cohomological methods in group theory, M.R. BRIDSON, P.H. KROPHOLLER & I.J. LEARY (eds)
359 Moduli spaces and vector bundles, L. BRAMBILA-PAZ, S.B. BRADLOW, O. GARCÍA-PRADA & S. RAMANAN (eds)
360 Zariski geometries, B. ZILBER
361 Words: Notes on verbal width in groups, D. SEGAL
362 Differential tensor algebras and their module categories, R. BAUTISTA, L. SALMERÓN & R. ZUAZUA
363 Foundations of computational mathematics, Hong Kong 2008, F. CUCKER, A. PINKUS & M.J. TODD (eds)
364 Partial differential equations and fluid mechanics, J.C. ROBINSON & J.L. RODRIGO (eds)
365 Surveys in combinatorics 2009, S. HUCZYNSKA, J.D. MITCHELL & C.M. RONEY-DOUGAL (eds)
366 Highly oscillatory problems, B. ENGQUIST, A. FOKAS, E. HAIRER & A. ISERLES (eds)
367 Random matrices: High dimensional phenomena, G. BLOWER
368 Geometry of Riemann surfaces, F.P. GARDINER, G. GONZÁLEZ-DIEZ & C. KOUROUNIOTIS (eds)
369 Epidemics and rumours in complex networks, M. DRAIEF & L. MASSOULIÉ
370 Theory of p-adic distributions, S. ALBEVERIO, A.YU. KHRENNIKOV & V.M. SHELKOVICH
371 Conformal fractals, F. PRZYTYCKI & M. URBAŃSKI
372 Moonshine: The first quarter century and beyond, J. LEPOWSKY, J. MCKAY & M.P. TUITE (eds)
373 Smoothness, regularity and complete intersection, J. MAJADAS & A. G. RODICIO
374 Geometric analysis of hyperbolic differential equations: An introduction, S. ALINHAC
375 Triangulated categories, T. HOLM, P. JØRGENSEN & R. ROUQUIER (eds)
376 Permutation patterns, S. LINTON, N. RUŠKUC & V. VATTER (eds)
377 An introduction to Galois cohomology and its applications, G. BERHUY
378 Probability and mathematical genetics, N. H. BINGHAM & C. M. GOLDIE (eds)
379 Finite and algorithmic model theory, J. ESPARZA, C. MICHAUX & C. STEINHORN (eds)
380 Real and complex singularities, M. MANOEL, M.C. ROMERO FUSTER & C.T.C WALL (eds)
381 Symmetries and integrability of difference equations, D. LEVI, P. OLVER, Z. THOMOVA & P. WINTERNITZ (eds)
382 Forcing with random variables and proof complexity, J. KRAJÍČEK
383 Motivic integration and its interactions with model theory and non-Archimedean geometry I, R. CLUCKERS, J. NICAISE & J. SEBAG (eds)
384 Motivic integration and its interactions with model theory and non-Archimedean geometry II, R. CLUCKERS, J. NICAISE & J. SEBAG (eds)
385 Entropy of hidden Markov processes and connections to dynamical systems, B. MARCUS, K. PETERSEN & T. WEISSMAN (eds)
386 Independence-friendly logic, A.L. MANN, G. SANDU & M. SEVENSTER
387 Groups St Andrews 2009 in Bath I, C.M. CAMPBELL *et al* (eds)
388 Groups St Andrews 2009 in Bath II, C.M. CAMPBELL *et al* (eds)
389 Random fields on the sphere, D. MARINUCCI & G. PECCATI
390 Localization in periodic potentials, D.E. PELINOVSKY
391 Fusion systems in algebra and topology, M. ASCHBACHER, R. KESSAR & B. OLIVER
392 Surveys in combinatorics 2011, R. CHAPMAN (ed)
393 Non-abelian fundamental groups and Iwasawa theory, J. COATES *et al* (eds)
394 Variational problems in differential geometry, R. BIELAWSKI, K. HOUSTON & M. SPEIGHT (eds)
395 How groups grow, A. MANN
396 Arithmetic differential operators over the p-adic integers, C.C. RALPH & S.R. SIMANCA
397 Hyperbolic geometry and applications in quantum chaos and cosmology, J. BOLTE & F. STEINER (eds)
398 Mathematical models in contact mechanics, M. SOFONEA & A. MATEI
399 Circuit double cover of graphs, C.-Q. ZHANG
400 Dense sphere packings: a blueprint for formal proofs, T. HALES
401 A double Hall algebra approach to affine quantum Schur–Weyl theory, B. DENG, J. DU & Q. FU
402 Mathematical aspects of fluid mechanics, J.C. ROBINSON, J.L. RODRIGO & W. SADOWSKI (eds)

403 Foundations of computational mathematics, Budapest 2011, F. CUCKER, T. KRICK, A. PINKUS & A. SZANTO (eds)
404 Operator methods for boundary value problems, S. HASSI, H.S.V. DE SNOO & F.H. SZAFRANIEC (eds)
405 Torsors, étale homotopy and applications to rational points, A.N. SKOROBOGATOV (ed)
406 Appalachian set theory, J. CUMMINGS & E. SCHIMMERLING (eds)
407 The maximal subgroups of the low-dimensional finite classical groups, J.N. BRAY, D.F. HOLT & C.M. RONEY-DOUGAL
408 Complexity science: the Warwick master's course, R. BALL, V. KOLOKOLTSOV & R.S. MACKAY (eds)
409 Surveys in combinatorics 2013, S.R. BLACKBURN, S. GERKE & M. WILDON (eds)
410 Representation theory and harmonic analysis of wreath products of finite groups, T. CECCHERINI-SILBERSTEIN, F. SCARABOTTI & F. TOLLI
411 Moduli spaces, L. BRAMBILA-PAZ, O. GARCÍA-PRADA, P. NEWSTEAD & R.P. THOMAS (eds)
412 Automorphisms and equivalence relations in topological dynamics, D.B. ELLIS & R. ELLIS
413 Optimal transportation, Y. OLLIVIER, H. PAJOT & C. VILLANI (eds)
414 Automorphic forms and Galois representations I, F. DIAMOND, P.L. KASSAEI & M. KIM (eds)
415 Automorphic forms and Galois representations II, F. DIAMOND, P.L. KASSAEI & M. KIM (eds)
416 Reversibility in dynamics and group theory, A.G. O'FARRELL & I. SHORT
417 Recent advances in algebraic geometry, C.D. HACON, M. MUSTAŢĂ & M. POPA (eds)
418 The Bloch–Kato conjecture for the Riemann zeta function, J. COATES, A. RAGHURAM, A. SAIKIA & R. SUJATHA (eds)
419 The Cauchy problem for non-Lipschitz semi-linear parabolic partial differential equations, J.C. MEYER & D.J. NEEDHAM
420 Arithmetic and geometry, L. DIEULEFAIT *et al* (eds)
421 O-minimality and Diophantine geometry, G.O. JONES & A.J. WILKIE (eds)
422 Groups St Andrews 2013, C.M. CAMPBELL *et al* (eds)
423 Inequalities for graph eigenvalues, Z. STANIĆ
424 Surveys in combinatorics 2015, A. CZUMAJ *et al* (eds)
425 Geometry, topology and dynamics in negative curvature, C.S. ARAVINDA, F.T. FARRELL & J.-F. LAFONT (eds)
426 Lectures on the theory of water waves, T. BRIDGES, M. GROVES & D. NICHOLLS (eds)
427 Recent advances in Hodge theory, M. KERR & G. PEARLSTEIN (eds)
428 Geometry in a Fréchet context, C.T.J. DODSON, G. GALANIS & E. VASSILIOU
429 Sheaves and functions modulo *p*, L. TAELMAN
430 Recent progress in the theory of the Euler and Navier–Stokes equations, J.C. ROBINSON, J.L. RODRIGO, W. SADOWSKI & A. VIDAL-LÓPEZ (eds)
431 Harmonic and subharmonic function theory on the real hyperbolic ball, M. STOLL
432 Topics in graph automorphisms and reconstruction (2nd Edition), J. LAURI & R. SCAPELLATO
433 Regular and irregular holonomic D-modules, M. KASHIWARA & P. SCHAPIRA
434 Analytic semigroups and semilinear initial boundary value problems (2nd Edition), K. TAIRA
435 Graded rings and graded Grothendieck groups, R. HAZRAT
436 Groups, graphs and random walks, T. CECCHERINI-SILBERSTEIN, M. SALVATORI & E. SAVA-HUSS (eds)
437 Dynamics and analytic number theory, D. BADZIAHIN, A. GORODNIK & N. PEYERIMHOFF (eds)
438 Random walks and heat kernels on graphs, M.T. BARLOW
439 Evolution equations, K. AMMARI & S. GERBI (eds)
440 Surveys in combinatorics 2017, A. CLAESSON *et al* (eds)
441 Polynomials and the mod 2 Steenrod algebra I, G. WALKER & R.M.W. WOOD
442 Polynomials and the mod 2 Steenrod algebra II, G. WALKER & R.M.W. WOOD
443 Asymptotic analysis in general relativity, T. DAUDÉ, D. HÄFNER & J.-P. NICOLAS (eds)
444 Geometric and cohomological group theory, P.H. KROPHOLLER, I.J. LEARY, C. MARTÍNEZ-PÉREZ & B.E.A. NUCINKIS (eds)
445 Introduction to hidden semi-Markov models, J. VAN DER HOEK & R.J. ELLIOTT
446 Advances in two-dimensional homotopy and combinatorial group theory, W. METZLER & S. ROSEBROCK (eds)
447 New directions in locally compact groups, P.-E. CAPRACE & N. MONOD (eds)
448 Synthetic differential topology, M.C. BUNGE, F. GAGO & A.M. SAN LUIS
449 Permutation groups and cartesian decompositions, C.E. PRAEGER & C. SCHNEIDER
450 Partial differential equations arising from physics and geometry, M. BEN AYED *et al* (eds)
451 Topological methods in group theory, N. BROADDUS, M. DAVIS, J.-F. LAFONT & I. ORTIZ (eds)
452 Partial differential equations in fluid mechanics, C.L. FEFFERMAN, J.C. ROBINSON & J.L. RODRIGO (eds)
453 Stochastic stability of differential equations in abstract spaces, K. LIU
454 Beyond hyperbolicity, M. HAGEN, R. WEBB & H. WILTON (eds)
455 Groups St Andrews 2017 in Birmingham, C.M. CAMPBELL *et al* (eds)
456 Surveys in combinatorics 2019, A. LO, R. MYCROFT, G. PERARNAU & A. TREGLOWN (eds)
457 Shimura varieties, T. HAINES & M. HARRIS (eds)
458 Integrable systems and algebraic geometry I, R. DONAGI & T. SHASKA (eds)
459 Integrable systems and algebraic geometry II, R. DONAGI & T. SHASKA (eds)
460 Wigner-type theorems for Hilbert Grassmannians, M. PANKOV
461 Analysis and Geometry on Graphs and Manifolds M. KELLER, D. LENZ & R.K. WOJCIECHOWSKI

London Mathematical Society Lecture Note Series: 461

Analysis and Geometry on Graphs and Manifolds

MATTHIAS KELLER
The University of Potsdam

DANIEL LENZ
Friedrich-Schiller University of Jena

RADOSLAW K. WOJCIECHOWSKI
Graduate Center and York College of the City University of New York

CAMBRIDGE
UNIVERSITY PRESS

University Printing House, Cambridge CB2 8BS, United Kingdom

One Liberty Plaza, 20th Floor, New York, NY 10006, USA

477 Williamstown Road, Port Melbourne, VIC 3207, Australia

314–321, 3rd Floor, Plot 3, Splendor Forum, Jasola District Centre, New Delhi – 110025, India

79 Anson Road, #06–04/06, Singapore 079906

Cambridge University Press is part of the University of Cambridge.

It furthers the University's mission by disseminating knowledge in the pursuit of education, learning, and research at the highest international levels of excellence.

www.cambridge.org
Information on this title: www.cambridge.org/9781108713184
DOI: 10.1017/9781108615259

First published 2020

A catalogue record for this publication is available from the British Library.

Library of Congress Cataloging-in-Publication Data
Names: Analysis and Geometry on Graphs and Manifolds (2017 : Potsdam, Germany), author. | Keller, Matthias, editor. | Lenz, Daniel, editor. | Wojciechowski, Radoslaw K., editor.
Title: Analysis and geometry on graphs and manifolds / edited by Matthias Keller, Daniel Lenz, Radoslaw K. Wojciechowski.
Description: Cambridge ; New York, NY : Cambridge University Press, 2020. | Series: London Mathematical Society lecture note series | Includes bibliographical references.
Identifiers: LCCN 2020005418 | ISBN 9781108713184 (paperback) | ISBN 9781108615259 (epub)
Subjects: LCSH: Geometric analysis – Congresses. | Graph theory – Congresses. | Manifolds (Mathematics) – Congresses.
Classification: LCC QA360 .A63 2017 | DDC 515/.15–dc23
LC record available at https://lccn.loc.gov/2020005418

ISBN 978-1-108-71318-4 Paperback

Contents

List of Contributors *page* vii
Preface xi

1 Infinite Planar Graphs with Non-negative Combinatorial Curvature 1
Bobo Hua and Yanhui Su

2 Curvature Calculations for Antitrees 21
David Cushing, Shiping Liu, Florentin Münch, and Norbert Peyerimhoff

3 Gromov–Lawson Tunnels with Estimates 55
Józef Dodziuk

4 Norm Convergence of the Resolvent for Wild Perturbations 66
Colette Anné and Olaf Post

5 Manifolds with Ricci Curvature in the Kato Class: Heat Kernel Bounds and Applications 76
Christian Rose and Peter Stollmann

6 Multiple Boundary Representations of λ-Harmonic Functions on Trees 95
Massimo A. Picardello and Wolfgang Woess

7 Internal DLA on Sierpinski Gasket Graphs 126
Joe P. Chen, Wilfried Huss, Ecaterina Sava-Huss, and Alexander Teplyaev

8 Universal Lower Bounds for Laplacians on Weighted Graphs 156
D. Lenz and P. Stollmann

9 Critical Hardy Inequalities on Manifolds and Graphs 172
Matthias Keller, Yehuda Pinchover, and Felix Pogorzelski

10 Neumann Domains on Graphs and Manifolds 203
Lior Alon, Ram Band, Michael Bersudsky, and Sebastian Egger

11 On the Existence and Uniqueness of Self-Adjoint Realizations of Discrete (Magnetic) Schrödinger Operators 250
Marcel Schmidt

12 Box Spaces: Geometry of Finite Quotients 328
Ana Khukhro and Alain Valette

13 Ramanujan Graphs and Digraphs 344
Ori Parzanchevski

14 From Partial Differential Equations to Groups 368
Andrzej Zuk

15 Spectral Properties of Limit-Periodic Operators 382
David Damanik and Jake Fillman

16 Uniform Existence of the IDS on Lattices and Groups 445
C. Schumacher, F. Schwarzenberger, and I. Veselić

Contributors

Lior Alon
Department of Mathematics, Technion–Israel Institute of Technology, Haifa 32000, Israel

Colette Anné
Laboratoire de Mathématiques Jean Leray, CNRS–Université de Nantes, Faculté des Sciences, BP 92208, 44322 Nantes, France

Ram Band
Department of Mathematics, Technion–Israel Institute of Technology, Haifa 32000, Israel

Michael Bersudsky
Department of Mathematics, Technion–Israel Institute of Technology, Haifa 32000, Israel

Joe P. Chen
Department of Mathematics, Colgate University, 13 Oak Drive, Hamilton NY 13346, USA

David Cushing
Department of Mathematical Sciences, Durham University, Science Laboratories, South Road, Durham, England

David Damanik
Department of Mathematics, Rice University, Houston, TX 77005, USA

Józef Dodziuk
CUNY Graduate Center, 365 Fifth Avenue, New York, NY, 10016, USA

Sebastian Egger
Department of Mathematics, Technion–Israel Institute of Technology, Haifa 32000, Israel

Jake Fillman
Department of Mathematics, Virginia Polytechnic Institute and State University, 225 Stanger Street, Blacksburg, VA 24061, USA

Bobo Hua
School of Mathematical Sciences, LMNS, Fudan University, Shanghai 200433, China
Shanghai Center for Mathematical Sciences, Fudan University, Shanghai 200433, China

Wilfried Huss
Graz University of Technology, Department of Mathematical Structure Theory, Steyrergasse 30, 8010, Graz, Austria

Matthias Keller
Institut für Mathematik, Universität Potsdam, 14476 Potsdam, Germany

Ana Khukhro
Centre for Mathematical Sciences, University of Cambridge, Wilberforce Rd, Cambridge CB3 0WB, United Kingdom

Daniel Lenz
Mathematisches Institut, Friedrich Schiller Universität Jena, 07743 Jena, Germany

Shiping Liu
School of Mathematical Sciences, University of Science and Technology of China, 96 Jinzhai Road, Hefei 230026, Anhui Province, China

Florentin Münch
Universität Potsdam, Institut für Mathematik, Campus Golm, Haus 9, Karl-Liebknecht-Straße 24-25, 14476 Potsdam, Germany

Ori Parzanchevski
The Hebrew University of Jerusalem, Givat Ram. Jerusalem, 9190401, Israel

Norbert Peyerimhoff
Department of Mathematical Sciences, Durham University, Science Laboratories, South Road, Durham, England

Massimo A. Picardello
Dipartimento di Matematica, Università di Roma 'Tor Vergata', I-00133 Rome, Italy

Yehuda Pinchover
Department of Mathematics, Technion-Israel Institute of Technology, 3200003 Haifa, Israel

Felix Pogorzelski
Institut für Mathematik, Universität Leipzig, 04109 Leipzig, Germany

Olaf Post
Fachbereich 4 – Mathematik, Universität Trier, 54286 Trier, Germany

Christian Rose
Max-Planck Institute for Mathematics in the Sciences, D-04103 Leipzig, Germany

Ecaterina Sava-Huss
Graz University of Technology, Austria, Institute of Discrete Mathematics, Steyrergasse 30/III, Office ST 03 228, 8010 Graz, Austria

Marcel Schmidt
Mathematisches Institut, Friedrich Schiller Universität Jena, 07743 Jena, Germany

Christoph Schumacher
Technische Universität Dortmund, Fakultät für Mathematik, Lehrstuhl LSIX, Vogelpothsweg 87, 44227 Dortmund, Germany

Fabian Schwarzenberger
Universität Leipzig, Institut für Medizinische Informatik, Statistik und Epidemiologie, Härtelstraße 16-18, 04107 Leipzig, Germany

Peter Stollmann
Technische Universität Chemnitz, Faculty of Mathematics, D - 09107 Chemnitz, Germany

Yanhui Su
College of Mathematics and Computer Science, Fuzhou University, Fuzhou 350116, China

Alexander Teplyaev
Department of Mathematics, University of Connecticut, Storrs, CT 06269-1009, USA

Alain Valette
Institut de Mathématiques, Unimail, 11 Rue Emile Argand, CH-2000 Neuchâtel, Switzerland

Ivan Veselić
Technische Universität Dortmund, Fakultät für Mathematik, Lehrstuhl LSIX, Vogelpothsweg 87, 44227 Dortmund, Germany

Wolfgang Woess
Institut für Diskrete Mathematik, Technische Universität Graz, Steyrergasse 30, A-8010 Graz, Austria

Andrzej Żuk
Institut de Mathematiques, Universite Paris 7, 13 rue Albert Einstein, 75013 Paris, France

Preface

This book brings together contributions for the conference 'Analysis and geometry on graphs and manifolds,' which took place at the University of Potsdam in Potsdam, Germany, from 31st July to 4th August 2017. The aim of the conference was to bring together leading experts in geometric analysis, in both the discrete and continuous settings. This included researchers working on such diverse models as manifolds, graphs, fractals, groups, and metric measure spaces. The goal was for these researchers to share their expertise and to explore common ground. Each day there was also an extensive afternoon session which provided time for young researchers to present partial results and early work.

The overall theme of the conference and of the contributions contained in this volume is the interplay of geometry, spectral theory, and stochastics. This interplay has a long and fruitful history and can be seen as a driving force behind many developments in modern mathematics. The present volume focuses on the global effects of local properties. This can be explored in both the discrete and continuous settings, and there has been a continual interest in contrasting what happens in these two cases. The main goal of this volume is to give an expository overview of these topics. This is achieved by presenting a mixture of survey chapters which examine the landscape of certain subjects and shorter chapters which focus on specific techniques and problems.

We will now briefly comment on the content of the chapters contained in this volume. In doing so, we will also point out some of the connections between the chapters.

Curvature is a natural local quantity arising when studying the geometry of a space. It has been thoroughly investigated in the case of Riemannian manifolds. Recent years have seen an explosion of research aimed at establishing curvature notions in the discrete setting. Here we present several contributions in this direction. Namely, the chapter by Bobo Hua and Yanhui Su gives an overview of results concerning combinatorial curvature in the case of planar tessellations. In particular, they analyse combinatorial, potential, and spectral theoretic consequences of global lower combinatorial curvature bounds. The chapter by David Cushing, Shiping Liu, Florentin Münch and Norbert Peyerimhoff calculates both the Bakry–Émery and Ollivier–Ricci curvatures of a class of graphs called 'antitrees'. Antitrees have recently come of interest as they provide surprising counterexamples to direct analogues of statements from the continuous setting. For general graphs, Bakry–Émery and Ollivier–Ricci are the most commonly appearing curvature notions. As for curvature in the continuous setting, the chapter by Józef Dodziuk describes a procedure for constructing tunnels connecting manifolds of arbitrary dimension and positive scalar curvature while preserving the positivity of the curvature. A general treatment of convergence of operators while glueing and removing subsets in the case of suitable curvature control assumptions for manifolds can be found in the chapter by Colette Anné and Olaf Post.

Global geometry is strongly reflected in the properties of a Markov process exploring the space. An analytic approach to this investigation is provided via the heat equation. The connections between curvature and properties of the heat kernel in the case of unbounded Ricci curvature on manifolds are explored in the chapter by Peter Stollmann and Christian Rose. Another geometric feature explored by the Markov process is that of a boundary at infinity. By construction, this boundary captures global geometric properties of the space. Representations of generalizations of harmonic functions on such boundaries at infinity is the topic of the chapter by Massimo A. Picardello and Wolfgang Woess. Here, the areas of random walks, geometry, and potential theory merge. The trajectories of simple random walks on Sierpinski gasket graphs is the subject of the chapter by Joe P. Chen, Wilfried Huss, Ecaterina Sava-Huss, and Alexander Teplyaev.

Spectral geometry deals with the interplay of spectral theory and both local and global geometric properties. In particular, the bottom of the spectrum and properties of ground states have been investigated in both the discrete and continuous settings. In this book, three chapters deal with these issues. Specifically, the chapter by Daniel Lenz and Peter Stollmann provides lower bounds for eigenvalues on graphs in terms of the inradius of subsets. Furthermore, the chapter by Matthias Keller, Yehuda Pinchover, and Felix Pogorzelski discusses the Hardy inequality, and the chapter by Lior Alon, Michael Bersudsky, Sebastian Egger, and Ram Band deals with Neumann domains of eigenfunctions. These two chapters deal with both graphs and manifolds. The even more fundamental question of the existence and uniqueness of self-adjoint realizations in the discrete setting is detailed in the chapter by Marcel Schmidt.

The bottom of the spectrum is also a prominent topic in models investigated in geometric group theory. In particular, the chapter by Ana Khukhro and Alain Valette discusses expanders in the context of box spaces of Cayley graphs. Furthermore, the article by Ori Parzanchevski deals with Ramanujan digraphs, a counterpart to Ramanujan graphs, which are a special class of expanders. Finally, the chapter by Andrzej Żuk deals with the discretization of partial differential equations and connections to automata groups.

The spectral theory of Schrödinger operators in the case of a simple geometrical setting strongly depends on global features of the potential. This is analogous to how global features for complicated geometries influence the spectral theory of the Laplace–Beltrami operator. The impact of the potential on spectral theory is particularly prominent in the case of potentials generated by random processes or dynamical systems. The chapter by David Damanik and Jake Fillman gives a thorough overview of the spectral theory of Schrödinger operators on the one-dimensional lattice with potentials which are periodic or limit-periodic. The convergence of the integrated density of states in the case of random Schrödinger operators over lattices and amenable groups is investigated in the chapter by Christoph Schumacher, Fabian Schwarzenberger, and Ivan Veselić.

Acknowledgements. We gratefully acknowledge financial support for the conference provided by the DFG priority programme 'Geometry at infinity' SPP 2026, and the National Science Foundation (NSF) grant no. 1707722. Furthermore, the third editor acknowledges financial support provided by PSC-CUNY Awards, jointly funded by the Professional Staff Congress and the City University of New York, and the Collaboration Grant for Mathematicians, funded by the Simons Foundation. Finally, we are grateful for the hospitality and the financial support of the University of Potsdam.

1

Infinite Planar Graphs with Non-negative Combinatorial Curvature

Bobo Hua and Yanhui Su

Abstract

In this chapter, we survey some results on infinite planar graphs with non-negative combinatorial curvature, related to the total curvature, the number of vertices with positive curvature and the automorphism group.

1.1 Introduction

The combinatorial curvature for planar graphs was introduced by Nevanlinna, Stone, Gromov, and Ishida [Nev70, Sto76, Gro87, Ish90] respectively, which resembles the Gaussian curvature for smooth surfaces. Many interesting geometric and combinatorial results have been obtained under such curvature conditions since then (see, e.g., [Ż97, Woe98, Hig01, BP01, HJL02, LPZ02, HS03, SY04, RBK05, BP06, DM07, CC08, Zha08, Che09, Kel10, KP11, Kel11, Oh17, Ghi17]).

Let (V, E) be a (possibly infinite) locally finite, undirected simple graph with the set of vertices V and the set of edges E. It is called *planar* if it can be topologically embedded into the sphere $\mathbb{S}^2$ or the plane $\mathbb{R}^2$, where we distinguish $\mathbb{S}^2$ with $\mathbb{R}^2$ while they are identified in the theory of finite planar graphs. We write $G = (V, E, F)$ for the cellular complex structure of a planar graph induced by the embedding where F is the set of faces, i.e., connected components of the complement of the embedding image of the graph (V, E) in $\mathbb{S}^2$ or $\mathbb{R}^2$. We say that a planar graph G is a *planar tessellation* if the following hold (see, e.g., [Kel11]):

(i) Every face is homeomorphic to a disc whose boundary consists of finitely many edges of the graph.

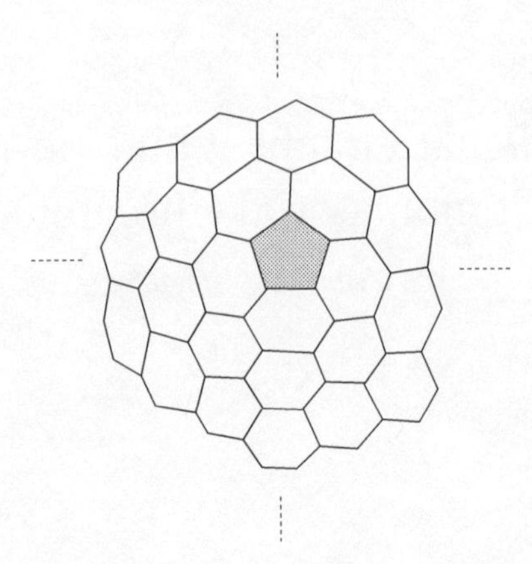

Figure 1.1 A planar graph G consists of a pentagon and infinitely many hexagons

(ii) Every edge is contained in exactly two different faces.
(iii) For any two faces whose closures have non-empty intersection, the intersection is either a vertex or an edge.

In this chapter, we only consider planar tessellations (see Figure 1.1 for an example) and call them *planar graphs* for the sake of simplicity. For a planar tessellation, it is finite (infinite resp.) if and only if it embeds into $\mathbb{S}^2$ ($\mathbb{R}^2$ resp.).

We say that a vertex x is incident to an edge e, denoted by $x \prec e$, (similarly, an edge e is incident to a face σ, denoted by $e \prec \sigma$; or a vertex x is incident to a face σ, denoted by $x \prec \sigma$) if the former is a subset of the closure of the latter. Two vertices x and y are called 'neighbours' if there is an edge e such that $x \prec e$ and $y \prec e$, in this case denoted by $x \sim y$. We denote by $\deg(x)$ the degree of a vertex x, i.e., the number of neighbours of a vertex x, and by $\deg(\sigma)$ the degree of a face σ, i.e., the number of edges incident to a face σ (equivalently, the number of vertices incident to σ). We always assume that for any vertex x and face σ,

$$\deg(x) \geq 3, \ \deg(\sigma) \geq 3.$$

We denote by

$$(\deg(\sigma_1), \deg(\sigma_2), \cdots, \deg(\sigma_N))$$

the pattern of a vertex x where $N = \deg(x)$, $\{\sigma_i\}_{i=1}^N$ are the faces which x is incident to, and $\deg(\sigma_1) \leq \deg(\sigma_2) \leq \cdots \leq \deg(\sigma_N)$.

Given a planar graph $G = (V, E, F)$, one may canonically endow its ambient space $\mathbb{S}^2$ or $\mathbb{R}^2$ with a piecewise flat metric as follows: assign each edge length one, replace each face by a regular Euclidean polygon of side length one with same facial degree, and glue these polygons along the common edges. The ambient space equipped with the induced metric constructed above is called the *regular polyhedral surface* of G, denoted by $S(G)$. In the following, we always call it the *polyhedral surface* for the sake of brevity. For a planar graph G, the *combinatorial curvature* at the vertex is defined as

$$\Phi(x) = 1 - \frac{\deg(x)}{2} + \sum_{\sigma \in F : x \prec \sigma} \frac{1}{\deg(\sigma)}, \quad x \in V. \tag{1.1.1}$$

In this chapter, we mean by the curvature of a planar graph the combinatorial curvature of it for simplicity. It turns out that the curvature of a planar graph is given by the generalized Gaussian curvature of the polyhedral surface $S(G)$ up to some normalization. Note that for the polyhedral surface $S(G)$ it is locally isometric to a flat domain in $\mathbb{R}^2$ near any interior point of an edge or a face, while it might be non-smooth near the vertices. As a metric surface, the generalized Gaussian curvature K of $S(G)$ vanishes at smooth points and can be regarded as a measure concentrated on the isolated singularities, i.e., on vertices. One can show that the mass of the generalized Gaussian curvature at each vertex x is given by $K(x) = 2\pi - \Sigma_x$, where Σ_x denotes the total angle at x in the metric space $S(G)$ (see [Ale05]). Moreover, by direct computation one has $K(x) = 2\pi\Phi(x)$, where the curvature $\Phi(x)$ is defined in (1.1.1). Hence, one can show that a planar graph G has non-negative curvature if and only if the polyhedral surface $S(G)$ is a generalized convex surface in the sense of Alexandrov (see [BGP92, BBI01, HJL15]). Furthermore, the polyhedral surface $S(G)$ can be isometrically embedded into $\mathbb{R}^3$ as a boundary of a compact or non-compact convex polyhedron by Alexandrov's embedding theorem ([Ale05]); see Figure 1.2 for an embedded image of $S(G)$ of the planar graph G in Figure 1.1.

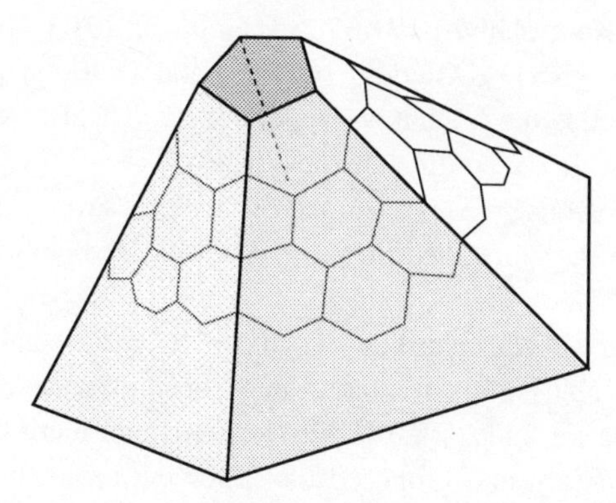

Figure 1.2 The isometric embedding of $S(G)$ of the planar graph G in Figure 1.1

In this chapter, we study planar graphs with non-negative curvature. We introduce two classes of planar graphs with positive or non-negative curvature as follows:

- $\mathcal{PC}_{>0} := \{G : \Phi(x) > 0, \forall x \in V\}$ is the class of planar graphs with positive curvature everywhere.
- $\mathcal{PC}_{\geq 0} := \{G : \Phi(x) \geq 0, \forall x \in V\}$ is the class of planar graphs with non-negative curvature everywhere.

We review some known results on the class $\mathcal{PC}_{>0}$. Stone [Sto76] first proved a Myers-type theorem: a planar graph with the curvature bounded below uniformly by a positive constant is a finite graph. Higuchi proposed a stronger conjecture that any $G \in \mathcal{PC}_{>0}$ is a finite graph (see [Hig01, Conjecture 3.2]). This is certainly wrong for smooth surfaces since there are many non-compact convex surfaces in $\mathbb{R}^3$, which have positive curvature everywhere. However, for a planar graph it is hopefully true by the combinatorial restriction of regular polygons as its faces. DeVos and Mohar [DM07] proved the conjecture by showing a generalized Gauss–Bonnet formula (see [SY04] for the case of cubic graphs).

For any finite planar graph $G \in \mathcal{PC}_{\geq 0}$, in particular any $G \in \mathcal{PC}_{>0}$, by Alexandrov's embedding theorem its polyhedral surface $S(G)$ can be isometrically embedded into $\mathbb{R}^3$ as a boundary of a convex polyhedron (see, e.g., [Ale05]). From this point of view, we obtain many examples for the class $\mathcal{PC}_{>0}$, e.g., the 1-skeletons of 5 Planotic solids, 13 Archimedean solids, and 92 Johnson solids. Any of them has regular Euclidean polygons as its faces in its embedded image in $\mathbb{R}^3$. Note that these are all examples of planar graphs in $\mathcal{PC}_{>0}$ whose faces of the embedded image in $\mathbb{R}^3$ are regular polygons (see [Joh66, Zal67]). Besides these, the class $\mathcal{PC}_{>0}$ contains many other examples, such as an example of 138 vertices constructed by Réti, Bitay, and Kosztolányi [RBK05], examples of 208 vertices by Nicholson and Sneddon [NS11], Ghidelli [Ghi17], and Oldridge [Old17], which cannot be realized as the boundary of a convex polyhedron whose faces are regular polygons. In fact, although any face of $G \in \mathcal{PC}_{>0}$ is isometric to a regular polygon in $S(G)$, it may split into several pieces of non-coplanar faces in the embedded image of $S(G)$ as the boundary of a convex polyhedron in $\mathbb{R}^3$.

There are two special families of graphs in $\mathcal{PC}_{>0}$ called *prisms* and *antiprisms*, both consisting of infinite many examples (see, e.g., [DM07]). Besides them, DeVos and Mohar [DM07] proved that there are only finitely many graphs in $\mathcal{PC}_{>0}$ and proposed the following problem to find out the largest graph among them.

Problem 1.1.1 ([DM07]) *What is the number*

$$C_{\mathbb{S}^2} := \max_{G=(V,E,F)} \sharp V,$$

where the maximum is taken over graphs in $\mathcal{PC}_{>0}$, which are not prisms or antiprisms, and $\sharp V$ denotes the cardinality of V?

On the one hand, as some examples of 208 vertices in $\mathcal{PC}_{>0}$ have been constructed in [NS11, Ghi17, Old17], we have the lower bound estimate that $C_{\mathbb{S}^2} \geq 208$. On the other hand, DeVos and Mohar [DM07] initiated to use the discharging methods to obtain the upper bound estimate $C_{\mathbb{S}^2} \leq 3444$. The discharging methods were adopted in the proof of the four-colour theorem in the literature (see [AH77, RSST97]). The upper bound was later improved to $C_{\mathbb{S}^2} \leq 380$ by Oh [Oh17]. By a delicate argument, Ghidelli [Ghi17] showed that $C_{\mathbb{S}^2} \leq 208$, which completely solves DeVos and Mohar's problem that $C_{\mathbb{S}^2} = 208$.

Next, we consider the class of planar graphs with non-negative curvature, i.e., $\mathcal{PC}_{\geq 0}$, which turns out to be much larger than $\mathcal{PC}_{>0}$ and contains many interesting examples. The class of $\mathcal{PC}_{>0}$ consists of essentially finite many examples, while the class $\mathcal{PC}_{\geq 0}$ contains infinitely many examples of different combinatorial types. A fullerene is a finite cubic planar graph whose faces are either pentagon or hexagon. There are plenty of examples of fullerenes which are important in the real-world applications, to cite a few examples [KHO+85, Thu98, BD97, BGM12, BE17a, BE17b]. Note that any fullerene is a planar graph with non-negative curvature. As shown by Thurston [Thu98], the number of combinatorial types of fullerenes with N hexagons grows as N^9 as $N \to \infty$. Besides these examples of finite graphs, there are plenty of examples of infinite graphs. Any planar tiling with regular polygons as tiles (see, e.g., [GS89, Gal09]) is in the class $\mathcal{PC}_{\geq 0}$. Note that there are infinitely many such planar tilings, for which only a few examples with symmetry can be classified. These motivate us to investigate the general structure of planar graphs in the class $\mathcal{PC}_{\geq 0}$.

1.2 Total Curvature of Planar Graphs with Non-negative Curvature

For a smooth non-compact surface with absolutely integrable Gaussian curvature, its total curvature encodes the global geometric information of the space, e.g., the boundary at infinity (see [SST03]). For example, the total curvature of a convex surface in $\mathbb{R}^3$ describes the apex angle of the cone at infinity of the surface, which is useful to study global geometric and analytic properties of the surface, such as harmonic functions and heat kernels, following [CM97b, Xu14]. For planar graphs with non-negative curvature G, we denote by

$$\Phi(G) := \sum_{x \in V} \Phi(x)$$

the total curvature of G whenever the summation converges absolutely. In case of finite graphs, the Gauss–Bonnet theorem reads as (see, e.g., [DM07])

$$\Phi(G) = 2. \tag{1.2.1}$$

For an infinite planar graph $G \in \mathcal{PC}_{\geq 0}$, the Cohn-Vossen type theorem, proven by [DM07, Theorem 1.3] or [Che09, Theorem 1.6], yields that

$$\Phi(G) \leq 1. \tag{1.2.2}$$

This means that for any infinite $G \in \mathcal{PC}_{\geq 0}$, the total curvature of G satisfies

$$0 \leq \sum_{x \in V} \Phi(x) \leq 1.$$

In this section, we study all possible values of total curvature of infinite planar graphs with non-negative curvature, i.e., the following set

$$\{\Phi(G) : \text{G infinite}, G \in \mathcal{PC}_{\geq 0}\}. \tag{1.2.3}$$

As is well known in Riemannian geometry that for any real number $0 \leq a \leq 2\pi$, there is a convex surface whose total curvature is given by a. Hence, the above set for non-compact convex surfaces turns out to be an interval in the continuous setting. However, combinatorial structure of planar graphs with non-negative curvature gives us more information and restrictions for the set (1.2.3).

For any $G = (V, E, F) \in \mathcal{PC}_{\geq 0}$, we denote by

$$T_G := \{v \in V : \Phi(x) > 0\} \tag{1.2.4}$$

the set of vertices with positive curvature, and by

$$D_G := \sup_{\sigma \in F} \deg(\sigma) \tag{1.2.5}$$

the maximal facial degree of G. Chen and Chen [CC08, Che09] proved an interesting result that the set of vertices with positive curvature in a planar graph with non-negative curvature is a finite set. Hence, the supremum in (1.2.5) is in fact the maximum.

Theorem 1.2.1 (Chen and Chen) *For any $G \in \mathcal{PC}_{\geq 0}$, T_G is a finite set.*

This result makes our combinatorial setting distinguished from the Riemannian setting. Note that there are many non-compact convex surfaces with positive curvature everywhere, e.g., the elliptic paraboloid, i.e., the revolution surface of the graph $y = x^2$ with respect to the z axis in $\mathbb{R}^3$.

Moreover, if the maximal facial degree D_G of $G \in \mathcal{PC}_{\geq 0}$ is at least 43, then G has rather special structure, analogous to the prisms or antiprisms in the finite case (see [HJL15] or Theorem 1.3.2 in this chapter). In that case, one gets $\Phi(G) = 1$. Hence, for our purposes to understand the set (1.2.3), it suffices to consider planar graphs G with $D_G \leq 42$. Note that there are finitely many vertex patterns, consisting of faces of degree at most 42, with positive curvature (see Table 1.1 in the Appendix). Then one is ready to see that the set (1.2.3) is a discrete subset in [0, 1] (see, e.g., [HS17b, Proposition 2.3]).

T. Réti [HL16, Conjecture 2.1] was motivated to determine the following value

$$\tau_1 := \inf\left\{\Phi(G) : G \in \mathcal{PC}_{\geq 0}, \Phi(G) > 0\right\},$$

which is called the *first gap of total curvature* for infinite planar graphs in the class of $\mathcal{PC}_{\geq 0}$. He suggested that $\tau_1 = \frac{1}{6}$ and the minimum is attained by the graph consisting of a pentagon and infinitely many hexagons, which is a kind of infinite fullerene (see Figure 1.1). In [HS17a], we give an answer to Réti's problem.

Theorem 1.2.2 (Theorem 1.3 in [HS17a])

$$\tau_1 = \frac{1}{12}.$$

A planar graph $G \in \mathcal{PC}_{\geq 0}$ satisfies $\Phi(G) = \frac{1}{12}$ if and only if the polyhedral surface $S(G)$ is isometric to either

(a) a cone with the apex angle $\theta = 2 \arcsin \frac{11}{12}$, or
(b) a 'frustum' with a hendecagon base (see Figure 1.3).

The proof strategy is straightforward and involves tedious case studies. For a vertex with positive curvature, if the curvature of the vertex is less than $\frac{1}{12}$, then we try to find some nearby vertices with positive curvature such that the sum of these curvatures is at least $\frac{1}{12}$ and prove the results case by case. Note that there are examples of graphs in $\mathcal{PC}_{\geq 0}$ whose total curvature attains the first gap $\frac{1}{12}$ (see Figure 1.4 and [HS17a] for more examples). Although graph structures of infinite planar graphs attaining the first gap of total curvature could be as complicated as planar tilings (see [HS17a]) we are able to classify

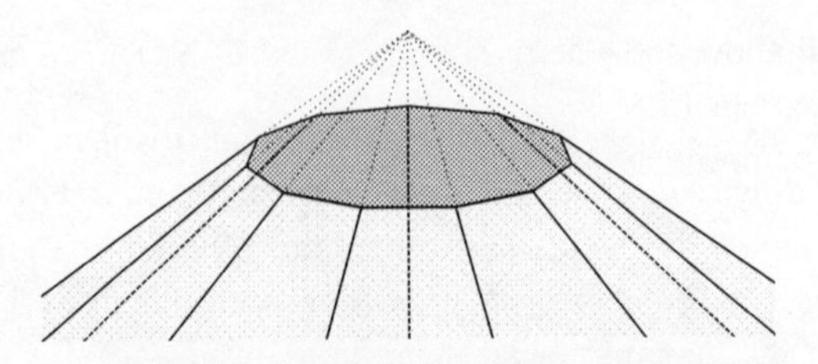

Figure 1.3 A 'frustum' with a hendecagon base

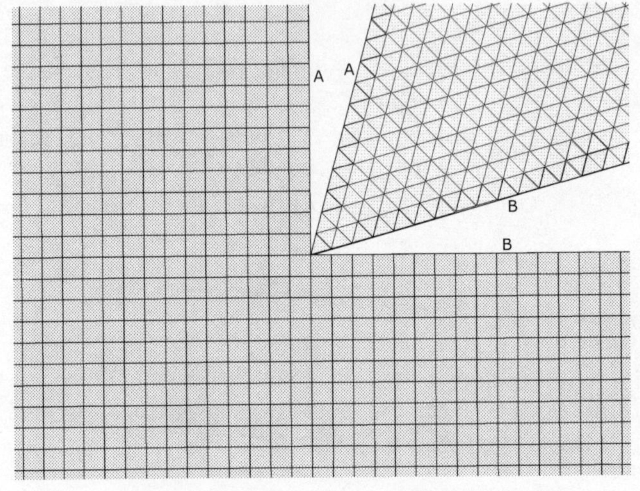

Figure 1.4 This is an example of total curvature $\frac{1}{12}$, where the half lines with same labels, A or B, are identified

metric structures of polyhedral surfaces for such planar graphs in the above theorem.

Inspired by Réti's question, it will be interesting to know other values in the set (1.2.3). Using Chen and Chen's result, Theorem 1.2.1, and the Gauss–Bonnet theorem for compact subsets with boundary, we are able to determine all possible total curvatures in the class $\mathcal{PC}_{\geq 0}$.

Theorem 1.2.3 (Theorem 1.1 in [HS17b]) *The set of all values of total curvature of infinite planar graphs with non-negative curvature* (1.2.3) *is given by*

$$\left\{\frac{i}{12} : 0 \leq i \leq 12, i \in \mathbb{Z}\right\}.$$

As a corollary, we also obtain that $\tau_1 = \frac{1}{12}$, which provides an alternative proof to Réti's problem. Moreover, as the part of the theorem, one may construct planar graphs with non-negative curvature whose total curvatures attain all values listed above (see [HS17b]). We sketch the proof of the theorem as follows: by Theorem 1.2.1, we know that T_G is a finite set. We choose a sufficiently large compact subset $K \subset S(G)$, homeomorphic to a closed disc, such that it contains T_G and consists of faces in F. Note that the vertices on the boundary of K have vanishing curvature, so that their patterns appear in the list of 17 possible patterns in Table 1.2 in the Appendix. By some combinatorial restrictions, one can further exclude several patterns from the list and conclude that any vertex on the boundary is incident to a triangle, a square, a hexagon, an octagon, or a dodecagon. Then using the Gauss–Bonnet formula on K, we may prove the theorem. Similar proof strategies apply to the problems on the total curvature of a planar graph with boundary, i.e., a graph embedded into the disc or a half plane (see [HS17b]).

Although we crucially use the finiteness structure of T_G in the proof of Theorem 1.2.3, we don't know much about the structure of the subset T_G which still lies in a black box. By a byproduct of the proof of Theorem 1.2.2, we can show that for $G \in \mathcal{PC}_{\geq 0}$, the induced subgraph on T_G has at most 14 connected components. It was conjecturally at most 12 (see [HS17a, Conjecture 5.2]).

1.3 The Vertices of Positive Curvature in Planar Graphs with Non-negative Curvature

In this section, we survey some results on the set of vertices with positive curvature in planar graphs with non-negative curvature. For any finite (infinite resp.) $G \in \mathcal{PC}_{\geq 0}$, Alexandrov's embedding theorem [Ale05] yields that an isometric embedding of the polyhedral surface $S(G)$ into $\mathbb{R}^3$ as a boundary of a compact (non-compact resp.) convex polyhedron. The set T_G serves as the set of the vertices/corners of the convex polyhedron, so that much geometric information of the polyhedron is contained in T_G. We are interested in the structure of the set T_G.

By the solution to DeVos and Mohar's problem [Ghi17], besides the prisms and antiprisms the largest number of vertices in a finite graph in $\mathcal{PC}_{>0}$ is 208. We would like to study analogous problems for planar graphs in $\mathcal{PC}_{\geq 0}$. We define some analogues to prisms and antiprisms in the class $\mathcal{PC}_{\geq 0}$.

Definition 1.3.1 *We call a planar graph $G = (V, E, F) \in \mathcal{PC}_{\geq 0}$ a prism-like graph if either*

(1) G is an infinite graph and $D_G \geq 43$, where D_G is defined in (1.2.5), *or*
(2) G is a finite graph and there are at least two faces with facial degree at least 43.

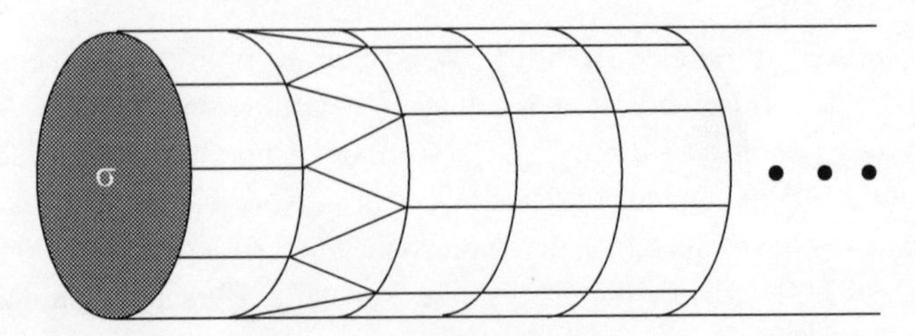

Figure 1.5 A half flat-cylinder in $\mathbb{R}^3$

By dividing hexagons into triangles, one may assume that there is no hexagon in G. Note that 'prism-like' graphs have rather special structures which can be completely determined by the following theorems. For any face σ, we denote by

$$\partial\sigma := \{x \in V : x \prec \sigma\}$$

the vertex boundary of σ.

Theorem 1.3.2 ([HJL15]) *Let $G = (V, E, F)$ be an infinite planar graph with non-negative curvature and $D_G \geq 43$. Then there is only one face σ of degree at least 43. Suppose that there is no hexagonal face. Then the set of faces F consists of σ, triangles or squares. Moreover,*

$$F = \sigma \cup (\cup_{i=1}^{\infty} L_i),$$

where L_i, $i \geq 1$, are sets of faces of the same type (triangle or square) which composite a band, i.e., an annulus, and is defined inductively: L_i is the next layer attaching to the previous layer L_{i-1} with $L_0 = \{\sigma\}$. $S(G)$ is isometric to the boundary of a half flat-cylinder in $\mathbb{R}^3$ (see Figure 1.5). Moreover, $\Phi(G) = 1$.

Theorem 1.3.3 ([HS18]) *Let $G = (V, E, F)$ be a finite prism-like graph. Then there are exactly two disjoint faces σ_1 and σ_2 of same facial degree at least 43. Suppose that there is no hexagonal face. Then the set of faces F consists of σ_1 and σ_2, triangles, or squares. Moreover,*

$$F = \sigma_1 \cup (\cup_{i=1}^{M} L_i) \cup \sigma_2,$$

where $M \geq 1$, and L_i, $1 \leq i \leq M$, are defined similarly as in Theorem 1.3.2. $S(G)$ is isometric to the boundary of a cylinder barrel in $\mathbb{R}^3$ (see Figure 1.6).

The following problem was proposed in [HL16] as an analogue to DeVos and Mohar's problem.

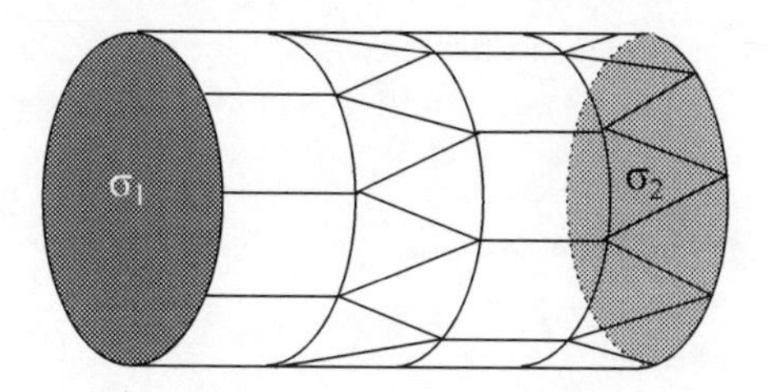

Figure 1.6 A cylinder barrel in $\mathbb{R}^3$

Problem 1.3.4 *What are the numbers*

$$K_{\mathbb{S}^2} := \max_{\text{finite } G} \sharp T_G, \quad K_{\mathbb{R}^2} := \max_{\text{infinite } G} \sharp T_G,$$

where the maxima are taken over finite and infinite graphs in $\mathcal{PC}_{\geq 0}$ which are not prism-like graphs respectively?

In [HS18], we prove the following theorem.

Theorem 1.3.5 ([HS18])

$$K_{\mathbb{R}^2} = 132.$$

Moreover, a graph in this class attains the maximum if and only if its polyhedral surface contains 12 *disjoint hendecagons.*

On the one hand, we give the upper bound $K_{\mathbb{R}^2} \leq 132$ by the discharging methods initiated by [DM07, Oh17, Ghi17] for the case $\mathcal{PC}_{>0}$. The curvature at vertices of a planar graph G can be regarded as the charge concentrated on vertices. The discharging method is to re-distribute the charge on vertices, via transferring the charge on vertices with large curvature to vertices with curvature less than $\frac{1}{132}$, such that the terminal charge on involved vertices in T_G after the distribution process is uniformly bounded below by $\frac{1}{132}$. Then the estimate of vertices in T_G follows from the Cohn–Vossen-type theorem (1.2.2). On the other hand, we can construct an example possessing 132 vertices of positive curvature (see [HS18]). This completely answers the second part of Problem 1.3.4.

For the first part of Problem 1.3.4, one can construct a family of infinitely many examples of finite graphs in $\mathcal{PC}_{\geq 0}$ with arbitrarily large number of vertices of positive curvature, see Example 3.3 in [HS18], which are not prism-like graphs. Hence

$$K_{\mathbb{S}^2} = \infty.$$

However, since the argument in the proof of Theorem 1.3.5 is local, we can prove the similar results for the modified quantity, $\widetilde{K}_{\mathbb{S}^2}$, for finite graphs in $\mathcal{PC}_{\geq 0}$.

Theorem 1.3.6 ([HS18]) *Let*

$$\widetilde{K}_{\mathbb{S}^2} := \max_{\text{finite } G} \sharp T_G,$$

where the maximum is taken over finite graphs in $\mathcal{PC}_{\geq 0}$ whose maximal facial degree are less than 132. *Then*

$$\widetilde{K}_{\mathbb{S}^2} = 264.$$

Moreover, a graph in this class attains the maximum if and only if its polyhedral surface contains 24 *disjoint hendecagons, i.e.,* 11*-gons.*

1.4 Automorphism Groups of Planar Graphs with Non-negative Curvature

In this section, we study automorphism groups of planar graphs with non-negative curvature. The automorphism groups of planar graphs have been extensively studied in the literature (see, e.g., [Man71, Bab75, CBGS08, SS98]). First, we introduce several definitions of isomorphisms on planar graphs.

Definition 1.4.1 *Let $G_1 = (V_1, E_1, F_1)$ and $G_2 = (V_2, E_2, F_2)$ be two planar graphs.*

(1) G_1 and G_2 are said to be graph-isomorphic *if there is a graph isomorphism between (V_1, E_1) and (V_2, E_2), i.e., $R : V_1 \to V_2$ such that for any $v, w \in V$, $v \sim w$ if and only if $R(v) \sim R(w)$.*

(2) G_1 and G_2 are said to be cell-isomorphic *if there is a cellular isomorphism $H = (H_V, H_E, H_F)$ between (V_1, E_1, F_1) and (V_2, E_2, F_2) in the sense of cell complexes, i.e., three bijections $H_V : V_1 \to V_2$, $H_E : E_1 \to E_2$ and $H_F : F_1 \to F_2$ preserving the incidence relations, that is, for any $v \in V, e \in E, \sigma \in F$, $v \prec e$ if and only if $H_V(v) \prec H_E(e)$ and $e \prec \sigma$ if and only if $H_E(e) \prec H_F(\sigma)$.*

(3) G_1 and G_2 are said to be metric-isomorphic *if there is an isometric map in the sense of metric spaces $L : S(G_1) \to S(G_2)$, such that the restriction map L is cell-isomorphic between (V_1, E_1, F_1) and (V_2, E_2, F_2).*

For a planar graph G, a graph (cellular, metric resp.) isomorphism from G to itself is called a *graph* (cellular, metric resp.) *automorphism of* G. We denote by $\mathrm{Aut}(G)$, ($\widetilde{\mathrm{Aut}}(G)$, $\mathcal{L}(G)$ resp.) the group of graph (cellular, metric resp.) automorphisms of a planar graph G. By the standard identification,

$$\mathcal{L}(G) \leq \widetilde{\mathrm{Aut}}(G) \leq \mathrm{Aut}(G),$$

where $\leq$ indicates that the former can be embedded as a sub-group of the latter. By our definition of polyhedral surfaces, it is easy to see that

$$\mathcal{L}(G) \cong \widetilde{\mathrm{Aut}}(G).$$

Moreover, by the results in [Whi33, Moh88] for a 3-connected planar graph G, any graph automorphism R of G can be uniquely realized as a cellular automorphism H such that $H_V = R$, which is called the *associated cellular automorphism of* R. This implies that for a 3-connected planar graph G,

$$\mathcal{L}(G) \cong \widetilde{\mathrm{Aut}}(G) \cong \mathrm{Aut}(G).$$

In [HS18], we prove that the cellular automorphism group of a planar graph in $\mathcal{PC}_{\geq 0}$ with positive total curvature is a finite group.

Theorem 1.4.2 *Let $G = (V, E, F)$ be a planar graph with non-negative curvature and positive total curvature. Then the automorphism group of G is finite. Set $a := \sharp T_G$ and $b := \max_{v \in T_G} \deg(v)$. We have the following:*

(1) If $D_G \leq 42$,

$$\sharp\widetilde{\mathrm{Aut}}(G) \text{ divides } a!b!.$$

(2) If $D_G > 42$, then

$$\sharp\widetilde{\mathrm{Aut}}(G) \text{ divides } \begin{cases} 2D_G, & G \text{ is infinite,} \\ 4D_G, & G \text{ is finite.} \end{cases}$$

By combining the estimates of the size of T_G in Theorems 1.3.5 and 1.3.6 with the above result, we obtain the estimates for the orders of cellular automorphism groups:

(1) If G is infinite, then

$$\sharp\widetilde{\mathrm{Aut}}(G) \leq \begin{cases} 132! \times 5!, & \text{for } D_G \leq 42, \\ 2D_G, & \text{for } D_G > 42. \end{cases}$$

(2) If G is finite, then

$$\sharp\widetilde{\mathrm{Aut}}(G) \leq \begin{cases} 264! \times 5!, & \text{for } D_G \leq 42, \\ 4D_G, & \text{for } D_G > 42. \end{cases}$$

1.5 Analysis on Planar Graphs with Non-negative Curvature

In this section, we study analysis problems on planar graphs with non-negative curvature. For any $G \in \mathcal{PC}_{\geq 0}$, it is easy to see that for any $x \in V$,

$$3 \leq \deg(x) \leq 6.$$

For any $x \in V$ and $r > 0$, we denote by

$$B_r(x) := \{y \in V : d(y, x) \leq r\}$$

the ball of radius r centred at x. We introduce the definitions of the volume-doubling property and the Poincaré inequality on graphs.

Definition 1.5.1

(DV) *A graph $G = (V, E)$ is called satisfying the* volume-doubling *property $DV(C)$ for constant $C > 0$ if for all $x \in V$ and all $r > 0$:*

$$\sharp B_{2r}(x) \leq C \sharp B_r(x).$$

(P) *A graph G is called satisfying the* Poincaré inequality *$P(C)$ for a constant $C > 0$ if for any functions f on V, $x_0 \in V$ and $r > 0$,*

$$\sum_{x \in B_r(x_0)} |f(x) - f_B|^2 \leq Cr^2 \sum_{x,y \in B_{2r}(x_0), y \sim x} (f(y) - f(x))^2,$$

where

$$f_B := \frac{1}{\sharp B_r(x_0)} \sum_{x \in B_r(x_0)} f(x).$$

The first author, Jost and Liu [HJL15], proved these properties for graphs in the class $\mathcal{PC}_{\geq 0}$.

Theorem 1.5.2 ([HJL15]) *For any $G \in \mathcal{PC}_{\geq 0}$, the volume-doubling property $DV(C_1)$ and the Poincaré inequality $P(C_2)$ hold for some $C_1, C_2 > 0$. Moreover, there is a constant C such that*

$$\sharp B_r(x) \leq Cr^2, \quad \forall x \in V, r > 0.$$

The general principle hidden in this result dates back to [CSC95], in which they showed that the volume-doubling property and the Poincaré inequality are quasi-isometric invariants (see [SC04, Woe00] for definitions). Since the planar graph G with bounded facial degree is properly embedded into the regular polygonal surface $S(G)$, they are in fact quasi-isometric to each other. For convex surfaces, even more general Alexandrov spaces with non-negative curvature, the volume-doubling property follows from the Bishop–Gromov volume comparison [BBI01] and the Poincaré inequality is obtained by [KMS01, Hua09].

For a graph $G = (V, E)$, the Laplace operator Δ, called *Laplacian* for short, is defined as follows for any function f on V:

$$\Delta f(x) = \frac{1}{\deg(x)} \sum_{y \in V : y \sim x} (f(y) - f(x)), \quad \forall x \in V.$$

A function f is called harmonic (subharmonic, superharmonic resp.) on $\Omega \subset V$ if $\Delta f \equiv 0$ ($\Delta f \geq 0$, $\Delta f \leq 0$ resp.) on Ω. As a corollary of Theorem 1.5.2, the simple random walk on a planar graph with non-negative curvature is recurrent, i.e., any positive superharmonic function is constant, which follows from the assertion that the volume growth is at most quadratic. Moreover, by Moser iteration, Delmotte [Del97] proved the elliptic Harnack inequality on graphs under the assumptions of the volume-doubling property and the Poincaré inequality. In particular, Theorem 1.5.2 implies that for $G \in \mathcal{PC}_{\geq 0}$ and any positive harmonic function f on $B_{2r}(p) \subset V$, $r > 0$, we have

$$\max_{B_r(p)} f \leq C \min_{B_r(p)} f, \tag{1.5.1}$$

where C is a constant independent of r. For any $k > 0$, we denote by

$$H^k(G) := \{f : V \to \mathbb{R} | \ \Delta f \equiv 0, \ |f(x)| \leq C(1 + d(x, p))^k, \\ \text{for some } p \in V, C > 0\}$$

the space of harmonic functions on V of polynomial growth whose growth order are less than or equal to k. The Harnack inequality (1.5.1) yields that

$$\dim H^k(G) = 1, \quad \text{for some } k < 1.$$

In fact, the combination of volume-doubling property and the Poincaré inequality turns out to be equivalent to the parabolic Harnack inequalities [Del99]. Furthermore, one can prove the finite-dimensional property of the space of harmonic functions of polynomial growth with growth rate bounded above, following Colding, Minicozzi, and Li [CM97a, CM98a, CM98b, Li97, STW00].

Theorem 1.5.3 ([HJL15, HJ15]) *For any $G \in \mathcal{PC}_{\geq 0}$,*

$$\dim H^k(G) \leq Ck, \quad \forall k \geq 1,$$

where C is a universal constant.

Table 1.1 *The patterns of a vertex with positive curvature*

Patterns		$\Phi(x)$
$(3,3,k)$	$3 \le k$	$1/6 + 1/k$
$(3,4,k)$	$4 \le k$	$1/12 + 1/k$
$(3,5,k)$	$5 \le k$	$1/30 + 1/k$
$(3,6,k)$	$6 \le k$	$1/k$
$(3,7,k)$	$7 \le k \le 41$	$1/k - 1/42$
$(3,8,k)$	$8 \le k \le 23$	$1/k - 1/24$
$(3,9,k)$	$9 \le k \le 17$	$1/k - 1/18$
$(3,10,k)$	$10 \le k \le 14$	$1/k - 1/15$
$(3,11,k)$	$11 \le k \le 13$	$1/k - 5/66$
$(4,4,k)$	$4 \le k$	$1/k$
$(4,5,k)$	$5 \le k \le 19$	$1/k - 1/20$
$(4,6,k)$	$6 \le k \le 11$	$1/k - 1/12$
$(4,7,k)$	$7 \le k \le 9$	$1/k - 3/28$
$(5,5,k)$	$5 \le k \le 9$	$1/k - 1/10$
$(5,6,k)$	$6 \le k \le 7$	$1/k - 2/15$
$(3,3,3,k)$	$3 \le k$	$1/k$
$(3,3,4,k)$	$4 \le k \le 11$	$1/k - 1/12$
$(3,3,5,k)$	$5 \le k \le 7$	$1/k - 2/15$
$(3,4,4,k)$	$4 \le k \le 5$	$1/k - 1/6$
$(3,3,3,3,k)$	$3 \le k \le 5$	$1/k - 1/6$

Table 1.2 *The patterns of a vertex with vanishing curvature*

(3, 7, 42),	(3, 8, 24),	(3, 9, 18),	(3, 10, 15),	(3, 12, 12),
(4, 5, 20),	(4, 6, 12),	(4, 8, 8),	(5, 5, 10),	(6, 6, 6),
(3, 3, 4, 12),	(3, 3, 6, 6),	(3, 4, 4, 6),	(4, 4, 4, 4),	(3, 3, 3, 3, 6),
(3, 3, 3, 4, 4),	(3, 3, 3, 3, 3, 3).			

1.6 Appendix

Table 1.1 lists all possible patterns of a vertex with positive curvature (see [DM07, CC08]); Table 1.2 lists all possible patterns of a vertex with vanishing curvature (see [GS89, CC08]).

Acknowledgements

B. H. is supported by NSFC (China) under grant nos. 11831004 and 11826031. Y. S. is supported by NSFC (China) under grant no. 11771083 and NSF of Fujian Province through grants 2017J01556 and 2016J01013.

References

[AH77] K. Appel and W. Haken. Every planar map is four colorable. I. Discharging. *Illinois J. Math.*, 21(3):429–490, 1977.

[Ale05] A. D. Alexandrov. *Convex polyhedra*. Springer Monographs in Mathematics. Springer-Verlag, Berlin, 2005. Translated from the 1950 Russian edition by N. S. Dairbekov, S. S. Kutateladze and A. B. Sossinsky, With comments and bibliography by V. A. Zalgaller and appendices by L. A. Shor and Yu. A. Volkov.

[Bab75] L. Babai. Automorphism groups of planar graphs. II. *Colloq. Math. Soc. János Bolyai*, 10:29–84, 1975.

[BBI01] D. Burago, Yu. Burago, and S. Ivanov. *A course in metric geometry*, volume 33 of *Graduate Studies in Mathematics*. American Mathematical Society, Providence, RI, 2001.

[BD97] G. Brinkmann and A. W. M. Dress. A constructive enumeration of fullerenes. *J. Algorithms*, 23(2):345–358, 1997.

[BE17a] V. Buchstaber and N. Erokhovets. Constructions of families of three-dimensional polytopes, characteristic patches of fullerenes, and Pogorelov polytopes. *Izv. Ross. Akad. Nauk Ser. Mat.*, 81(5):15–91, 2017.

[BE17b] V. Buchstaber and N. Erokhovets. Finite sets of operations sufficient to construct any fullerene from C_{20}. *Structural Chemistry*, 28(1):225–234, 2017.

[BGM12] G. Brinkmann, J. Goedgebeur, and B. D. McKay. The generation of fullerenes. *J. Chem. Inf. Model.*, 52(11):2910–2918, 2012.

[BGP92] Yu. Burago, M. Gromov, and G. Perelman. A. D. Aleksandrov spaces with curvatures bounded below. *Russian Math. Surveys*, 47(2):1–58, 1992.

[BP01] O. Baues and N. Peyerimhoff. Curvature and geometry of tessellating plane graphs. *Discrete Comput. Geom.*, 25(1):141–159, 2001.

[BP06] O. Baues and N. Peyerimhoff. Geodesics in non-positively curved plane tessellations. *Adv. Geom.*, 6(2):243–263, 2006.

[CBGS08] J. H. Conway, H. Burgiel, and C. Goodman-Strauss. *The symmetries of things*. A K Peters, Ltd., Wellesley, MA, 2008.

[CC08] B. Chen and G. Chen. Gauss-Bonnet formula, finiteness condition, and characterizations of graphs embedded in surfaces. *Graphs Combin.*, 24(3):159–183, 2008.

[Che09] B. Chen. The Gauss-Bonnet formula of polytopal manifolds and the characterization of embedded graphs with nonnegative curvature. *Proc. Amer. Math. Soc.*, 137(5):1601–1611, 2009.

[CM97a] T. Colding and W. Minicozzi. Harmonic functions on manifolds. *Ann. of Math. (2)*, 146(3):725–747, 1997.

[CM97b] T. Colding and W. Minicozzi. Harmonic functions with polynomial growth. *J. Differential Geom.*, 46(1):1–77, 1997.

[CM98a] T. Colding and W. Minicozzi. Liouville theorems for harmonic sections and applications. *Comm. Pure Appl. Math.*, 51(2), 1998.

[CM98b] T. Colding and W. Minicozzi. Weyl type bounds for harmonic functions. *Invent. Math.*, 131(2), 1998.

[CSC95] T. Coulhon and L. Saloff-Coste. Variétés Riemanniennes isométriques à l'infini. *Rev. Mat. Iberoamericana*, 11(3):687–726, 1995.

[Del97] T. Delmotte. Inégalité de Harnack elliptique sur les graphes. *Colloq. Math.*, 72(1):19–37, 1997.

[Del99] T. Delmotte. Parabolic Harnack inequalities and estimates of Markov chains on graphs. *Rev. Mat. Iberoamericana*, 15:181–232, 1999.

[DM07] M. DeVos and B. Mohar. An analogue of the Descartes-Euler formula for infinite graphs and Higuchi's conjecture. *Trans. Amer. Math. Soc.*, 359(7):3287–3300, 2007.

[Gal09] B. Galebach. n-Uniform tilings. *Available online at http://probability sports.com/tilings.html*, 2009.

[Ghi17] L. Ghidelli. On the largest planar graphs with everywhere positive combinatorial curvature. *arXiv:1708.08502*, 2017.

[Gro87] M. Gromov. Hyperbolic groups. In *Essays in group theory*, volume 8 of *Math. Sci. Res. Inst. Publ.*, pages 75–263. Springer, New York, 1987.

[GS89] B. Grünbaum and G. C. Shephard. *Tilings and patterns*. A Series of Books in the Mathematical Sciences. W. H. Freeman and Company, New York, 1989.

[Hig01] Y. Higuchi. Combinatorial curvature for planar graphs. *J. Graph Theory*, 38(4):220–229, 2001.

[HJ15] B. Hua and J. Jost. Geometric analysis aspects of infinite semiplanar graphs with nonnegative curvature II. *Trans. Amer. Math. Soc.*, 367(4):2509–2526, 2015.

[HJL02] O. Häggström, J. Jonasson, and R. Lyons. Explicit isoperimetric constants and phase transitions in the random-cluster model. *Ann. Probab.*, 30(1):443–473, 2002.

[HJL15] B. Hua, J. Jost, and S. Liu. Geometric analysis aspects of infinite semiplanar graphs with nonnegative curvature. *J. Reine Angew. Math.*, 700:1–36, 2015.

[HL16] B. Hua and Y. Lin. Curvature notions on graphs. *Front. Math. China*, 11(5):1275–1290, 2016.

[HS03] Y. Higuchi and T. Shirai. Isoperimetric constants of (d, f)-regular planar graphs. *Interdiscip. Inform. Sci.*, 9(2):221–228, 2003.

[HS17a] B. Hua and Y. Su. The first gap for total curvatures of planar graphs with nonnegative curvature. *arXiv:1709.05309*, 2017.

[HS17b] B. Hua and Y. Su. Total curvature of planar graphs with nonnegative curvature. *arXiv:1703.04119*, 2017.

[HS18] B. Hua and Y. Su. The set of vertices with positive curvature in a planar graph with nonnegative curvature. *arXiv:1801.02968*, 2018.

[Hua09] B. Hua. Generalized Liouville theorem in nonnegatively curved Alexandrov spaces. *Chin. Ann. Math. Ser. B*, 30(2):111–128, 2009.

[Ish90] M. Ishida. Pseudo-curvature of a graph. In *lecture at Workshop on topological graph theory*. Yokohama National University, 1990.

[Joh66] Norman W. Johnson. Convex solids with regular faces. *Canadian Journal of Mathematics*, 18:169–200, 1966.

[Kel10] M. Keller. The essential spectrum of the Laplacian on rapidly branching tessellations. *Math. Ann.*, 346(1):51–66, 2010.

[Kel11] M. Keller. Curvature, geometry and spectral properties of planar graphs. *Discrete Comput. Geom.*, 46(3):500–525, 2011.

[KHO+85] H. W. Kroto, J. R. Heath, S. C. O'Brien, R. F. Curl, and R. E. Smalley. C_{60} : Buckminsterfullerene. *Nature*, 318:162–163, 1985.

[KMS01] K. Kuwae, Y. Machigashira, and T. Shioya. Sobolev spaces, Laplacian, and heat kernel on Alexandrov spaces. *Math. Z.*, 238(2):269–316, 2001.

[KP11] M. Keller and N. Peyerimhoff. Cheeger constants, growth and spectrum of locally tessellating planar graphs. *Math. Z.*, 268(3–4):871–886, 2011.

[Li97] P. Li. Harmonic sections of polynomial growth. *Math. Res. Lett.*, 4(1), 1997.

[LPZ02] S. Lawrencenko, M. Plummer, and X. Zha. Isoperimetric constants of infinite plane graphs. *Discrete Comput. Geom.*, 28(3):313–330, 2002.

[Man71] P. Mani. Automorphismen von polyedrischen Graphen. *Math. Ann.*, 192:279–303, 1971.

[Moh88] B. Mohar. Embeddings of infinite graphs. *J. Combin. Theory Ser. B*, 44(1):29–43, 1988.

[Nev70] R. Nevanlinna. *Analytic functions*. Translated from the second German edition by Phillip Emig. Die Grundlehren der mathematischen Wissenschaften, Band 162. Springer-Verlag, New York-Berlin, 1970.

[NS11] R. Nicholson and J. Sneddon. New graphs with thinly spread positive combinatorial curvature. *New Zealand J. Math.*, 41:39–43, 2011.

[Oh17] B.-G. Oh. On the number of vertices of positively curved planar graphs. *Discrete Math.*, 340(6):1300–1310, 2017.

[Old17] P. R. Oldridge. Characterizing the polyhedral graphs with positive combinatorial curvature. *thesis, available at https://dspace.library.uvic.ca/handle/1828/8030*, 2017.

[RBK05] T. Réti, E. Bitay, and Z. Kosztolányi. On the polyhedral graphs with positive combinatorial curvature. *Acta Polytechnica Hungarica*, 2(2):19–37, 2005.

[RSST97] N. Robertson, D. Sanders, P. Seymour, and R. Thomas. The four-colour theorem. *J. Combin. Theory Ser. B*, 70(1):2–44, 1997.

[SC04] L. Saloff-Coste. Analysis on Riemannian co-compact covers. In *Surveys in differential geometry. Vol. IX*, volume 9 of *Surv. Differ. Geom.*, pages 351–384. Int. Press, Somerville, MA, 2004.

[SS98] B. Servatius and H. Servatius. Symmetry, automorphisms, and self-duality of infinite planar graphs and tilings. In *International Scientific Conference on Mathematics. Proceedings (Žilina, 1998)*, pages 83–116. Univ. Žilina, Žilina, 1998.

[SST03] K. Shiohama, T. Shioya, and M. Tanaka. *The geometry of total curvature on complete open surfaces*. Number 159 in Cambridge Tracts in Mathematics. Cambridge University Press, Cambridge, 2003.

[Sto76] D. A. Stone. A combinatorial analogue of a theorem of Myers. *Illinois J. Math.*, 20(1):12–21, 1976.

[STW00] C.-J. Sung, L.-F. Tam, and J. Wang. Spaces of harmonic functions. *J. London Math. Soc. (2)*, 61(3):789–806, 2000.

[SY04] L. Sun and X. Yu. Positively curved cubic plane graphs are finite. *J. Graph Theory*, 47(4):241–274, 2004.

[Thu98] W. P. Thurston. Shapes of polyhedra and triangulations of the sphere. In *The Epstein birthday schrift*, volume 1 of *Geom. Topol. Monogr.*, pages 511–549. Geom. Topol. Publ., Coventry, 1998.

[Whi33] H. Whitney. 2-Isomorphic Graphs. *Amer. J. Math.*, 55(1–4):245–254, 1933.

[Woe98] W. Woess. A note on tilings and strong isoperimetric inequality. *Math. Proc. Camb. Phil. Soc.*, 124:385–393, 1998.

[Woe00] W. Woess. *Random walks on infinite graphs and groups*. Number 138 in Cambridge Tracts in Mathematics. Cambridge University Press, Cambridge, 2000.

[Xu14] G. Xu. Large time behavior of the heat kernel. *J. Differential Geom.*, 98(3):467–528, 2014.

[Ż97] A. Żuk. On the norms of the random walks on planar graphs. *Ann. Inst. Fourier (Grenoble)*, 47(5):1463–1490, 1997.

[Zal67] V. Zalgaller. *Convex Polyhedra with Regular Faces*, volume 2 of *Zap. Nauchn. Semin. Leningr. Otd. Mat. Inst. Steklova (in Russian)*. 1967.

[Zha08] L. Zhang. A result on combinatorial curvature for embedded graphs on a surface. *Discrete Math.*, 308(24):6588–6595, 2008.

2

Curvature Calculations for Antitrees

David Cushing, Shiping Liu, Florentin Münch, and Norbert Peyerimhoff

Abstract

In this chapter we prove that antitrees with suitable growth properties are examples of infinite graphs exhibiting strictly positive curvature in various contexts: in the normalized and non-normalized Bakry–Émery setting as well as in the Ollivier–Ricci curvature case. We also show that these graphs do not have global positive lower curvature bounds, which one would expect in view of discrete analogues of the Bonnet–Myers theorem. The proofs in the different settings require different techniques.

2.1 Introduction and Results

The main protagonists in this chapter are *antitrees*. While these examples had been studied already in 1988, they were given the name *antitree* in talks by Radoslaw Wojciechowsi around 2010. A proper definition of antitrees, in their most general form, appeared first in [19]. Like in the case of a tree, the vertices of an antitree are partitioned in generations V_i, with the first generation V_1 called its *root set*. While trees are connected graphs with as few connections as possible between subsequent generations, antitrees have the maximal number of connections. More precisely, antitrees are simple (i.e., no loops and no multiple edges), connected graphs such that

(i) any root vertex $x \in V_1$ is connected to all vertices in V_2, and no vertices in $V_k, k \geq 3$,
(ii) any vertex $x \in V_k$, $k \geq 2$, is connected to all vertices in V_{k-1} and V_{k+1}, and no vertices in V_l, $|k - l| \geq 2$.

Note that this definition allows for the possibility of edges between vertices of the same generation. We will refer to such edges as *spherical edges*. Edges between vertices of different generations are called *radial edges*. Any radial or spherical edge incident to a vertex in V_1 is called *radial* or *spherical root-edge*, respectively. All other edges are called *inner edges*.

Antitrees are particularly interesting examples with regard to stochastic completeness. Section 2.2, provided by Radoslaw Wojciechowki, gives a more in-depth look at the history of antitrees. In this chapter, we investigate curvature properties of antitrees. Relations between curvature asymptotics and stochastic completeness were investigated recently in [17] in the Bakry–Émery setting and in [22] in the Ollivier–Ricci curvature setting.

For our curvature considerations, we consider only antitrees where the induced subgraph of any one generation V_k is complete, i.e., any two vertices in the same generation are neighbours. For any given finite or infinite sequence $(a_k)_{1 \leq k \leq N}$, $N \in \mathbb{N} \cup \{\infty\}$, the corresponding unique such antitree with $|V_k| = a_k$ for all $1 \leq k \leq N$ is denoted by $\mathcal{AT}((a_k))$. Note that in the case of a finite antitree, that is $N < \infty$, (ii) has to be understood in the case $k = N$ that any vertex $x \in V_N$ is connected to all vertices in V_{N-1}. Later in this introduction, we will only present results for infinite antitrees, but, since curvature is a local notion, we need only investigate curvatures of suitable finite antitrees for the proofs.

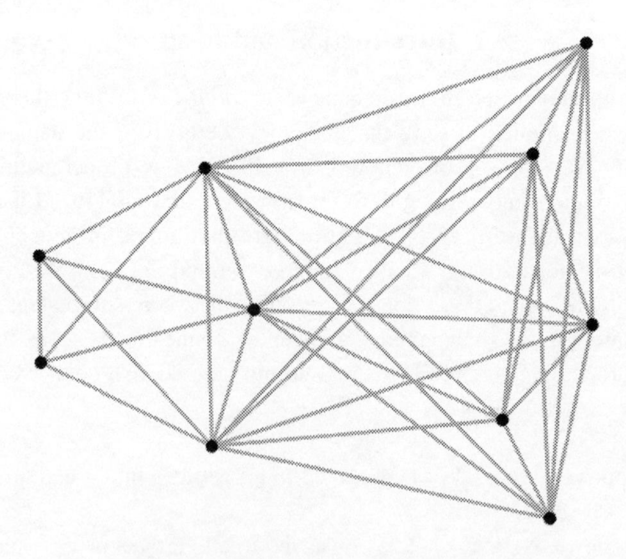

Figure 2.1 The antitree $\mathcal{AT}((2, 3, 5))$

Two particular curvature notions on graphs have been studied actively in recent years:

- *Bakry–Émery curvature* taking values on the vertices and based on Bochner's formula with respect to a suitable graph Laplacian,
- *Ollivier–Ricci curvature* taking values on the edges and based on optimal transport of lazy random walks.

Basic graph theoretical notions are introduced in Section 2.3.1 and precise definitions of these curvature concepts are given in Sections 2.3.2 and 2.3.3, respectively.

For both curvature notions there are graph theoretical analogues of the fundamental Bonnet–Myers theorem for Riemannian manifolds with strictly positive Ricci curvature bounded away from zero.

Let us first consider Bakry–Émery curvature. Generally, on a combinatorial graph $G = (V, E)$ with vertex set V and edge set E, the graph Laplacian on functions $f : V \to \mathbb{R}$ is of the form

$$\Delta f(x) = \frac{1}{\mu(x)} \sum_{y \sim x} (f(y) - f(x)), \tag{2.1.1}$$

with a vertex measure $\mu : V \to (0, \infty)$. In this chapter, we consider two specific choices of vertex measures:

- $\mu \equiv 1$, which we refer to as the *non-normalized case*,
- $\mu(x) = d_x$ (the vertex degree of $x \in V$), which we refer to as the *normalized case*.

The corresponding discrete Bonnet–Myers theorems in both settings are as follows.

Theorem 2.1.1 (see [21]) *Let $G = (V, E)$ be a connected graph satisfying $CD(K, \infty)$ for some $K > 0$ in the* non-normalized case *and $d_x \leq D$ for all $x \in V$ and some finite D. Then G is a finite graph and, furthermore,*

$$\operatorname{diam}(G) \leq \frac{2D}{K}.$$

Theorem 2.1.2 (see [21]) *Let $G = (V, E)$ be a connected graph satisfying $CD(K, \infty)$ for some $K > 0$ in the* normalized case *(possibly of unbounded vertex degree). Then G is a finite graph and, furthermore,*

$$\operatorname{diam}(G) \leq \frac{2}{K}.$$

Ollivier–Ricci curvature depends upon an idleness parameter $p \in [0, 1]$ describing the laziness of the associated random walk. Here, the discrete Bonnet–Myers theorem takes the following form.

Theorem 2.1.3 (see [23]) *Let $G = (V, E)$ be a connected graph satisfying $\kappa_p(x, y) \geq K > 0$ for all $x \sim y$ and a fixed idleness $p \in [0, 1]$. Then G is a finite graph and, furthermore,*

$$\mathrm{diam}(G) \leq \frac{2(1-p)}{K}. \tag{2.1.2}$$

These results give rise to the following natural questions:

- Do there exist examples of infinite connected graphs with strictly positive curvature? (That is, relaxing the condition of a uniform strictly positive lower curvature bound.)
- In the non-normalized case, does there exist an infinite connected graphs satisfying $CD(K, \infty)$ for $K > 0$ of unbounded vertex degree?

This chapter provides a positive answer to the first question. In fact, we show that antitrees $\mathcal{AT}((a_k))$ with suitable growth properties of the infinite sequence (a_k) have strictly positive curvature for all curvature notions mentioned above. More precisely, we have the following in the Bakry–Émery curvature case.

Theorem 2.1.4 *In both the normalized and non-normalized setting, the infinite antitree $\mathcal{AT}((k))$ satisfies $CD(K_x, \infty, x)$ for all vertices x with a family of constants $K_x > 0$ depending only on the generation of x. Furthermore,*

$$\liminf_{k \to \infty,\, x \in V_k} K_x = 0.$$

Remark 2.1.5 In fact, the method of proof relies on some Maple calculations which can be extended to also provide the following results (without going into the details):

(i) *Linear growth:* The same curvature results hold true for the infinite antitrees $\mathcal{AT}((1 + (k-1)t))$ with arbitrary $t \in \mathbb{N}$.
(ii) *Exponential growth:* The same curvature results hold true for the infinite antitree $\mathcal{AT}((2^{k-1}))$ in the normalized case and fail to satisfy $CD(0, \infty)$ in the non-normalized case.

Due to Bakry–Émery curvature being a local property, in order to calculate the curvatures $\mathcal{K}_{G,x}(\infty)$ of vertices x *in the first two generations* of $G = \mathcal{AT}((2^{k-1}))$ as defined later in (2.3.1), it is sufficient to consider the

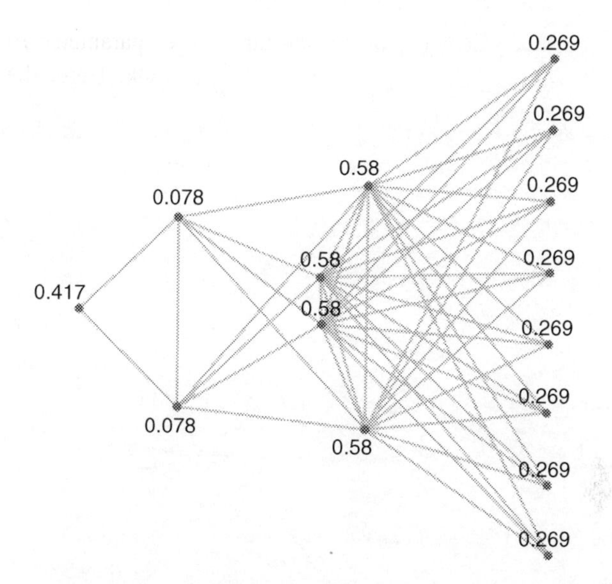

Figure 2.2 Normalized curvature $\mathcal{K}_{G,x}(\infty)$

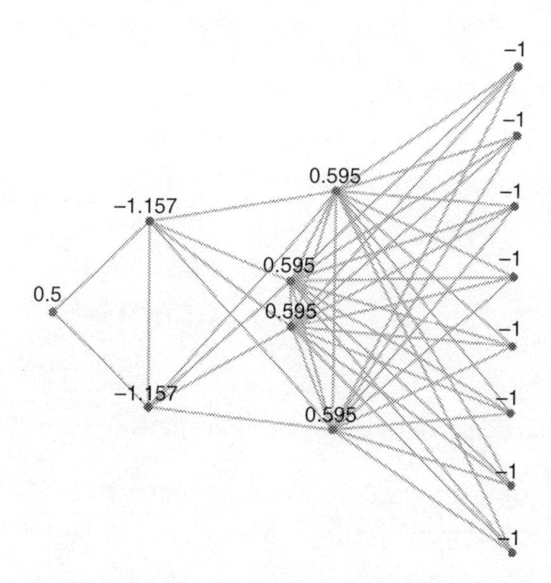

Figure 2.3 Non-normalized curvature $\mathcal{K}_{G,x}(\infty)$

graph presented in Figures 2.2 and 2.3 (spherical edges of 2-spheres around a vertex do not contribute to the curvature, see [7]). These figures are in agreement with the statements in Remark 2.1.5(ii).

Now we consider Ollivier–Ricci curvature. Here our main result is the following.

Theorem 2.1.6 *Let $G = \mathcal{AT}((a_k))$ be an infinite antitree with $1 = a_1$ and $a_{k+1} \geq a_k$ for all $k \in \mathbb{N}$ and x, y be neighbouring vertices in G.*

- Radial root edges: *If $x \in V_1$ and $y \in V_2$:*

$$\kappa_p(x, y) = \begin{cases} \frac{a_2-1}{a_2+a_3} + \frac{a_2+2a_3+1}{a_2+a-3}p, & \text{if } p \in \left[0, \frac{1}{a_2+a_3+1}\right], \\ \frac{a_2+1}{a_2+a_3}(1-p), & \text{if } p \in \left[\frac{1}{a_2+a_3+1}, 1\right]. \end{cases}$$

- Radial edges: *If $x \in V_k$ and $y \in V_{k+1}$, $k \geq 2$, $p \in [0, 1]$:*

$$\kappa_p(x, y) = \left(\frac{2a_k + a_{k+1} - 1}{a_k + a_{k+1} + a_{k+2} - 1} - \frac{2a_{k-1} + a_k - 1}{a_{k-1} + a_k + a_{k+1} - 1} \right)(1-p).$$

- Spherical edges: *If $x, y \in V_k$, $x \neq y$, $k \geq 2$:*

$$\kappa_p(x, y) = \begin{cases} \frac{a_{k-1}+a_k+a_{k+1}-2}{a_{k-1}+a_k+a_{k+1}-1} + \frac{a_{k-1}+a_k+a_{k+1}}{a_{k-1}+a_k+a_{k+1}-1}p, & \text{if } p \in \left[0, \frac{1}{a_{k-1}+a_k+a_{k+1}}\right], \\ \frac{a_{k-1}+a_k+a_{k+1}}{a_{k-1}+a_k+a_{k+1}-1}(1-p), & \text{if } p \in \left[\frac{1}{a_{k-1}+a_k+a_{k+1}}, 1\right]. \end{cases}$$

Let us consider special cases.

Corollary 2.1.7 (Linear growth) *Let $G = \mathcal{AT}((1+(k-1)t))$, $t \in \mathbb{N}$ arbitrary. Then*

$$\kappa_0(x, y) = \begin{cases} \frac{t}{3t+2} & \text{for } x \in V_1,\ y \in V_2, \\ \frac{6t^2}{(3kt+2)(3kt+2-3t)} & \text{for } x \in V_k,\ y \in V_{k+1}, \\ 1 - \frac{1}{3kt+2-3t} & \text{for } x, y \in V_k,\ x \neq y,\ k \geq 2. \end{cases}$$

In particular, κ_0 of radial edges decays asymptotically like $\frac{2}{3k^2}$ as $k \to \infty$.

Corollary 2.1.8 (Exponential growth) *We have for $G = \mathcal{AT}((r^{k-1})$, $r \in \mathbb{N}$:*

$$\kappa_0(x, y) = \begin{cases} \frac{r-1}{r(r+1)} & \text{for } x \in V_1,\ y \in V_2, \\ \frac{(r-1)^2(r+1)r^{k-2}}{(r^k+r^{k-1}+r^{k-2}-1)(r^{k+1}+r^k+r^{k-1}-1)} & \text{for } x \in V_k,\ y \in V_{k+1}, \\ 1 - \frac{1}{r^k+r^{k-1}+r^{k-2}-1} & \text{for } x, y \in V_k,\ x \neq y,\ k \geq 2. \end{cases}$$

In particular, κ_0 of radial edges decays asymptotically like $\frac{1}{r^k}$ as $k \to \infty$.

Remark 2.1.9 Note that for any finite sequence $(a_k)_{1 \leq k \leq N}$, $N \geq 2$, with $1 = a_1$ and $a_{k+1} \geq a_k$ for all $1 \leq k \leq N$, we can find a large enough $a_{N+1} \geq a_N$ such that $\kappa_0(x, y) < 0$ for $x \in V_{N-1}$ and $y \in V_N$.

The chapter is organised as follows: We start with some historical comments on antitrees in Section 2.2 which was provided by Radosław Wojciechowski. Section 2.3 introduces the readers into Bakry–Émery curvature and Ollivier–Ricci curvature. The following two Sections 2.4 and 2.5 present the concrete curvature investigations in both settings. Appendices A, B, and C provide the Maple code used for the results in Section 2.4.

2.2 A (Partial) History of Antitrees

To our knowledge, the first known appearance of an antitree is the case of $|S_r| = r + 1$ in the article of Dodziuk and Karp [8]. They study the normalized Laplacian Δ and give conditions for transience of the simple random walk in terms of $r\Delta r$ where r is the distance to a vertex. It appears in [8, Example 2.5] as a case of a transient graph with bottom of the spectrum 0 whose Green's function decays like $1/r$. The same antitree appears in the article of Weber [24]. Weber extends the result of Dodziuk and Mathai [9] concerning the stochastic completeness of the semigroup associated with the non-normalized Laplacian Δ. Indeed, Dodziuk/Mathai prove stochastic completeness in the case of bounded vertex degree. Weber improves this result to give stochastic completeness in the case of $\Delta r \geq K$ for some constant K. The antitree mentioned above is then given as an example of a graph whose vertex degree is unbounded but which satisfies $\Delta r \geq K$ (see [24, Figure 1, p. 156]). The general case of antitrees with arbitrary spherical growth $|S_r| = f(r)$ where f is any natural number-valued function is considered in [25, Example 4.11]. There it is shown that antitrees are stochastically complete if and only if

$$\sum_r \frac{\sum_{k=0}^r f(k)}{f(r)f(r+1)} = \infty.$$

This is used to give a counter-example to a direct analogue to Grigor'yan's result for stochastic completeness of manifolds (see [13]). Indeed, Grigor'yan's result says that any stochastically incomplete manifold must have super-exponential volume growth while the result above gives stochastically incomplete graphs which have only polynomial volume growth when the combinatorial graph metric is used. These examples give the smallest such examples in the combinatorial graph metric by a result of Huang, Grigor'yan and Masamune [12, Theorem 1.4], where the example (and name) of antitrees also appears. This might be the first time in print that the name is used and they refer to them as the '*antitree* of Wojciechowski'. A proper definition with the name of antitree first appears in [19, Definition 6.3]. Here the result on stochastic completeness is generalized to all weakly spherically symmetric graphs of which the antitrees are but an example. Furthermore, it is shown that the non-normalized Laplacian Δ on any such stochastically incomplete antitree has positive bottom of the spectrum (see [19, Corollary 6.6]). This gives a counter-example to a direct analogue to a theorem of Brooks [5] which states that the bottom of the spectrum of the Laplacian on any manifold with sub-exponential volume growth is zero. This sparked an interest in applying intrinsic metrics as defined by Frank, Lenz, and Wingert in [10] to study the question involving volume growth on graphs of unbounded vertex degree. In particular, the analogue to Grigor'yan's theorem was first proven in [11] (see also [18] for an analytic proof) while the analogue to Brooks' theorem was shown in [16]. Since then, antitrees appear in a variety of places. Their spectral theory is thoroughly analysed by Breuer and Keller in [4]. Here it should be noted that the spectrum consists mainly of eigenvalues with compactly supported eigenfunctions and a further spectral component which can be singular continuous in certain cases. Antitrees are also used as a counterexample to a conjecture presented by Golenia and Schumacher in [14] concerning the deficiency indices of the adjacency matrix (see [15]). They are also used to show the utility of the new bottom of the spectrum estimate for a Cheeger constant involving intrinsic metrics in [1].

2.3 Definitions and Notations

2.3.1 Basic Graph Theoretical Notations

Let $G = (V, E)$ be a locally finite connected *simple* combinatorial graph (that is, no loops and no multiple edges) with vertex set V and edge set E. For any $x, y \in V$ we write $x \sim y$ if $\{x, y\} \in E$. The *degree of a vertex* $x \in V$ is denoted by d_x. Let $d : V \times V \to \mathbb{N} \cup \{0\}$ be the *combinatorial distance function*, i.e., $d(x, y)$ is the length of the shortest path from x to y. For $x \in V$, the *combinatorial spheres* and *balls* of radius $r \geq 0$ around x are denoted by

$$S_r(x) = \{y \in V \mid d(x, y) = r\},$$

$$B_r(x) = \{y \in V \mid d(x, y) \leq r\},$$

respectively. The *diameter* of G is defined as

$$\operatorname{diam}(G) = \sup\{d(x, y) \mid x, y \in V\} \in \mathbb{N} \cup \{0, \infty\}.$$

2.3.2 Bakry–Émery Curvature

As mentioned before, this curvature notion is rooted on Bochner's formula using a Laplacian operator leading to the curvature-dimension inequality (CD-inequality for short). This approach was pursued by Bakry–Émery [2] via an elegant Γ-calculus and leads to a substitute of the lower Ricci curvature bound of the underlying space for much more general settings. (Some further information on the Bochner approach can be found, e.g., in [7, Remark 1.3].)

Recall definition (2.1.1) of the normalized ($\mu(x) = d_x$) and non-normalized Laplacian ($\mu \equiv 1$) from the Introduction. Such a choice of Laplacian leads to the following operator Γ for all $f, g : V \to \mathbb{R}$:

$$\begin{aligned}\Gamma(f,g)(x) &= \frac{1}{2}(\Delta(fg) - f\Delta g - g\Delta f)(x)\\ &= \frac{1}{2\mu(x)}\sum_{y\sim x}(f(y)-f(x))(g(y)-g(x)).\end{aligned}$$

For simplicity, we always write $\Gamma(f) := \Gamma(f,f)$. Iterating Γ, we can define another operator Γ_2, given by

$$\Gamma_2(f,g)(x) = \frac{1}{2}(\Delta\Gamma(f,g) - \Gamma(f,\Delta g) - \Gamma(g,\Delta f))(x).$$

Again, we abbreviate $\Gamma_2(f) = \Gamma_2(f,f)$. The Bakry–Émery curvature is defined via these operators in the following way.

Definition 2.3.1 Let $K \in \mathbb{R}$ and $N \in (0,\infty]$.

(i) The pointwise curvature dimension condition $CD(K,N,x)$ for $x \in V$ is defined by

$$\Gamma_2(f)(x) \geq K\Gamma(f)(x) + \frac{1}{N}(\Delta f)^2(x), \quad \text{for any} \quad f : V \to \mathbb{R}.$$

(ii) The global curvature dimension condition $CD(K,N)$ holds if and only if $CD(K,N,x)$ holds for any $x \in V$.

(iii) For any $x \in V$, we define

$$\mathcal{K}_{G,x}(N) := \sup\{K \in \mathbb{R} \mid CD(K,N,x)\}. \tag{2.3.1}$$

In this chapter, we are only concerned with ∞-curvature, that is, $N = \infty$. Following [7, Prop. 2.1], the condition $CD(K,\infty,x)$ is equivalent to

$$\Gamma_2(x) \geq K\Gamma(x), \tag{2.3.2}$$

where $\Gamma_2(x)$ and $\Gamma(x)$ are symmetric matrices of the corresponding quadratic forms evaluated at $x \in V$. Since only local information needs to be taken into account, they are of size $|B_2(x)| \times |B_2(x)|$ and $|B_1(x)| \times |B_1(x)|$, respectively, and to make sense of (2.3.2) the smaller size matrix must be padded with 0 entries. For more information in the non-normalized case, see [7, Sections 2.1–2.3]. The entries of these matrices in the general weighted case are explicitly given in [7, Section 12]. (Note that for the context of this chapter, the edge weights $w : E \to [0, \infty)$ take only values 0, 1 and reflect adjacency of vertices and the vertex measure $\mu : V \to (0, \infty)$ will only correspond to the normalized and non-normalized cases.)

The main tool to prove strictly positive curvature is [7, Corollary 2.7], that is, the following properties are equivalent:

- $\Gamma_2(x)$ is positive semi-definite with one-dimensional kernel,
- $\mathcal{K}_{G,x}(\infty) > 0$.

[7, Corollary 2.7] covers only the non-normalized case, but one can easily check that the equivalence holds also in the setting of general vertex measures.

2.3.3 Ollivier–Ricci Curvature

As mentioned before, Ollivier–Ricci curvature is based on optimal transport. Ollivier–Ricci curvature was introduced in [23]. A fundamental concept in optimal transport is the Wasserstein distance between probability measures.

Definition 2.3.2 Let $G = (V, E)$ be a locally finite graph. Let μ_1, μ_2 be two probability measures on V. The *Wasserstein distance* $W_1(\mu_1, \mu_2)$ between μ_1 and μ_2 is defined as

$$W_1(\mu_1, \mu_2) = \inf_{\pi} \sum_{y \in V} \sum_{x \in V} d(x, y)\pi(x, y), \tag{2.3.3}$$

where the infimum runs over all transportation plans $\pi : V \times V \to [0, 1]$ satisfying

$$\mu_1(x) = \sum_{y \in V} \pi(x, y), \quad \mu_2(y) = \sum_{x \in V} \pi(x, y).$$

The transportation plan π moves a mass distribution given by μ_1 into a mass distribution given by μ_2, and $W_1(\mu_1, \mu_2)$ is a measure for the minimal effort which is required for such a transition.

If π attains the infimum in (2.3.3) we call it an *optimal transport plan* transporting μ_1 to μ_2.

We define the following probability distributions μ_x for any $x \in V,\ p \in [0,1]$:

$$\mu_x^p(z) = \begin{cases} p, & \text{if } z = x, \\ \frac{1-p}{d_x}, & \text{if } z \sim x, \\ 0, & \text{otherwise.} \end{cases}$$

Definition 2.3.3 The p-Ollivier–Ricci curvature on an edge $x \sim y$ in $G = (V, E)$ is

$$\kappa_p(x, y) = 1 - W_1(\mu_x^p, \mu_y^p),$$

where $p \in [0, 1]$ is called the *idleness*.

The Ollivier–Ricci curvature introduced by Lin–Lu–Yau in [20] is defined as

$$\kappa_{LLY}(x, y) = \lim_{p \to 1} \frac{\kappa_p(x, y)}{1 - p}.$$

A fundamental concept in the optimal transport theory and vital to our work is Kantorovich duality. First we recall the notion of 1-Lipschitz functions and then state Kantorovich duality.

Definition 2.3.4 Let $G = (V, E)$ be a locally finite graph, $\phi : V \to \mathbb{R}$. We say that ϕ is 1-Lipschitz if

$$|\phi(x) - \phi(y)| \leq d(x, y)$$

for all $x, y \in V$. Let 1–Lip denote the set of all 1–Lipschitz functions.

Note that, by triangle inequality, ϕ is 1-Lipschitz iff $|\phi(x) - \phi(y)| \leq 1$ for all pairs $x \sim y$.

Theorem 2.3.5 (Kantorovich duality) *Let $G = (V, E)$ be a locally finite graph. Let μ_1, μ_2 be two probability measures on V. Then*

$$W_1(\mu_1, \mu_2) = \sup_{\substack{\phi: V \to \mathbb{R} \\ \phi \in 1\text{–Lip}}} \sum_{x \in V} \phi(x)(\mu_1(x) - \mu_2(x)).$$

If $\phi \in$ 1-Lip attains the supremum we call it an optimal Kantorovich potential *transporting μ_1 to μ_2.*

The following result on some properties of $p \mapsto \kappa_p(x, y)$ for $x \sim y$ and its consequences was useful in our curvature considerations.

Theorem 2.3.6 (see [3]) *Let $G = (V, E)$ be a locally finite graph. Let $x, y \in V$ with $x \sim y$. Then the function $p \mapsto \kappa_p(x, y)$ is concave and piecewise* *linear over* $[0, 1]$ *with at most 3 linear parts. Furthermore* $\kappa_p(x, y)$ *is linear on the intervals*

$$\left[0, \frac{1}{\mathrm{lcm}(d_x, d_y) + 1}\right] \quad \text{and} \quad \left[\frac{1}{\max(d_x, d_y) + 1}, 1\right].$$

Thus, if we have the further condition $d_x = d_y$, *then* $\kappa_p(x, y)$ *has at most two linear parts.*

2.4 Bakry–Émery Curvature of Antitrees

Let us first introduce some notation and a useful general fact (Lemma 2.4.1). The identity matrix of size d is denoted by Id_d and the all-zero and all-one matrix of size $d_1 \times d_2$ is denoted by $0_{d_1,d_2}$ and J_{d_1,d_2}, respectively. Moreover, if $d_1 = d_2$, we use the notation $J_{d_1} = J_{d_1,d_1}$, and if $d_2 = 1$, we use the notation $\mathbf{1}_{d_1}$ for the all-one column vector of size d_1. Moreover, the standard base of column vectors in $\mathbb{R}^N$ is denoted by $e_1, \ldots, e_N$.

Lemma 2.4.1 *Let* $d_1, \ldots, d_r \in \mathbb{N}$ *and* $A = (A_{ij})_{1 \le i,j \le r}$ *be a symmetric matrix, where the* A_{ij} *are block matrices of size* $d_i \times d_j$ *with* $A_{ji} = A_{ij}^\top$. *Assume that there exist constants* $\alpha_i, \beta_i \in \mathbb{R}$ *and* $\gamma_{ij} = \gamma_{ji} \in \mathbb{R}$ *such that, for* $1 \le i, j \le r$, $j \ne i$,

$$A_{ii} = \alpha_i \mathrm{Id}_{d_i} + \beta_i J_{d_i}$$

and

$$A_{ij} = \gamma_{ij} J_{d_i,d_j}.$$

Let $A_{\mathrm{red}} = (a_{ij})_{1 \le i,j \le r}$ *be the* $r \times r$*-matrix given by* $a_{ij} = \boldsymbol{1}_{d_i}^\top A_{ij} \boldsymbol{1}_{d_j}$, *i.e., for* $i \ne j$,

$$a_{ii} = \alpha_i d_i + \beta_i d_i^2,$$
$$a_{ij} = \gamma_{ij} d_i d_j.$$

For any vector $w = (w_1, \ldots, w_r)^\top \in \mathbb{R}^r$ *let*

$$\widehat{w} := (w_1 \boldsymbol{1}_{d_1}^\top, \ldots, w_r \boldsymbol{1}_{d_r}^\top)^\top \in \mathbb{R}^d$$

with $d = \sum_{j=1}^r d_j$. *Then we have the following two facts:*

(a) For every $d_i \ge 2$, *the* $(d_i - 1)$*-dimensional space*

$$E_i = \left\{ \sum_{j=1}^{d_i} c_j e_{j+d} \mid \sum_{j=1}^{d_i} c_j = 0 \right\}$$

with $d = \sum_{j=1}^{i-1} d_j$ *consists of eigenvectors to the eigenvalue* α_i.

(b) *For any $w \in \mathbb{R}^r$, the corresponding vector $\widehat{w}$ is orthogonal to all spaces E_i in (a) and we have*

$$\widehat{w}^\top A \widehat{w} = w^\top A_{\text{red}} w.$$

Proof. Choose a vector $\hat{u} = (u_1, \ldots, u_r)^\top \in \mathbb{R}^d$ with $u_j \in \mathbb{R}^{d_j}$ for $1 \leq j \leq r$. Then we have

$$Au = \left(\sum_{j=1}^{r} A_{1j} u_j, \ldots, \sum_{j=1}^{r} A_{rj} u_j \right)^\top.$$

Assume now that $u_j = (c_{j1}, \ldots, c_{jd_j})^\top \in \mathbb{R}^{d_j}$ satisfies

$$\sum_{k=1}^{d_j} c_{jk} = 0 \quad \text{for all } 1 \leq j \leq r. \tag{2.4.1}$$

This implies that $J_{d_i,d_j} u_j = 0$ for all $1 \leq i, j \leq r$ and, therefore,

$$\sum_{j=1}^{r} A_{ij} u_j = A_{ii} u_i = \alpha_i u_i,$$

which proves (a).

For the proof of (b), we assume that $w \in \mathbb{R}^r$ and $\widehat{w} \in \mathbb{R}^d$ are related as described in the lemma. It is then easy to see that $\widehat{w}$ is orthogonal to any vector u with components satisfying (2.4.1) and, therefore, to all eigenspaces E_i. Moreover, we have

$$\widehat{w}^\top A \widehat{w} = \sum_{i,j=1}^{r} \left(w_i \mathbf{1}_{d_i}^\top \right) A_{ij} \left(w_j \mathbf{1}_{d_j} \right) = \sum_{i,j=1}^{r} w_i w_j a_{ij} = w^\top A_{\text{red}} w.$$

This finishes the proof of (b). □

Now we start with our Bakry–Émery curvature considerations for antitrees. Due to localness of the Bakry–Émery curvature notion, we only need to consider $\mathcal{K}_{G,x}(\infty)$ for

(i) a vertex $x \in V_3$ in the finite antitree $\mathcal{AT}((a, b, c, d, e))$,
(ii) a vertex $x \in V_2$ in the finite antitree $\mathcal{AT}((b, c, d, e))$, and
(iii) a vertex $x \in V_1$ in the finite antitree $\mathcal{AT}((c, d, e))$.

The relevant results are given in the following theorems.

Theorem 2.4.2 *Let $x \in V_3$ be a vertex of the finite antitree $G = \mathcal{AT}((a, b, c, d, e))$. If*

$$a = n, \ b = n + 1, \ c = n + 2, \ d = n + 3, \ \textit{and } e = n + 4,$$

we have in both the normalized and non-normalized case:

$$\mathcal{K}_{G,x}(\infty) > 0. \tag{2.4.2}$$

Proof. In this proof, we will keep the values a, b, c, d, e general as long as possible and only specify them towards the end of the proof. Let $G = \mathcal{AT}((a, b, c, d, e))$, $1 \leq a \leq b < c \leq d \leq e$ and $x \in V_3$. To cover simultaneously both the normalized and non-normalized setting, we choose

$$\epsilon_- = \frac{\mu(x)}{\mu(y_-)} - 1, \quad \epsilon_+ = \frac{\mu(x)}{\mu(y_+)} - 1,$$

where $y_- \in V_2$ and $y_+ \in V_4$. (Note that $\mu(z)$ depends only the generation of z.) Using the results in [7, Section 12], a tedious but straightforward calculation shows the following: The matrix $A = 4\mu(x)^2\Gamma_2(x)$ is of the following block structure $A = (A_{ij})_{1\leq i,j\leq 6}$ where the blocks correspond to an ordering of $B_2(x)$ into the vertex sets $\{x\}$, $V_3\backslash\{x\}$, V_4, V_2, V_5, V_1:

$$A_{11} = d_x(d_x + 3) + 3b\epsilon_- + 3d\epsilon_+,$$
$$A_{12} = (-(d_x + 3) + b\epsilon_- + d\epsilon_+)J_{1,c-1},$$
$$A_{13} = (-(d_x + 3 + e) - (2 + c + e)\epsilon_+)J_{1,d},$$
$$A_{14} = (-(d_x + 3 + a) - (2 + a + c)\epsilon_-)J_{1,b},$$
$$A_{15} = (d + d\epsilon_+)J_{1,e},$$
$$A_{16} = (b + b\epsilon_-)J_{1,a},$$
$$A_{22} = (3(d_x + 1) + b\epsilon_- + d\epsilon_+)\mathrm{Id}_{c-1} - 2J_{c-1},$$
$$A_{23} = -(2 + 2\epsilon_+)J_{c-1,d},$$
$$A_{24} = -(2 + 2\epsilon_-)J_{c-1,b},$$
$$A_{25} = 0_{c-1,e},$$
$$A_{26} = 0_{c-1,a},$$
$$A_{33} = (-b + 3c + 3d + 3e + (3c + 4d + 3e)\epsilon_+)\mathrm{Id}_d - (2 + 4\epsilon_+)J_d,$$
$$A_{34} = 2J_{d,b},$$
$$A_{35} = -(2 + 2\epsilon_+)J_{d,e},$$
$$A_{36} = 0_{d,a},$$
$$A_{44} = (3a + 3b + 3c - d + (3a + 4b + 3c)\epsilon_-)\mathrm{Id}_b - (2 + 4\epsilon_-)J_b,$$
$$A_{45} = 0_{b,e},$$
$$A_{46} = -(2 + 2\epsilon_-)J_{b,a},$$
$$A_{55} = (d + d\epsilon_+)\mathrm{Id}_e,$$
$$A_{56} = 0_{e,a},$$
$$A_{66} = (b + b\epsilon_-)\mathrm{Id}_a.$$

Let A_{red} be the corresponding reduced symmetric 6×6 matrix $A_{\text{red}} = (a_{ij})_{1 \le i,j \le 6}$, as defined in Lemma 2.4.1.

Recalling the equivalence at the end of Section 2.3.2, $\mathcal{K}_{G,x}(\infty) > 0$ is equivalent to A being positive semi-definite and having one-dimensional kernel. Lemma 2.4.1 provides the following eigenvalues and multiplicities of A.

- Since $\epsilon_-, \epsilon_+ > -1$ and $d_x = b + c + d - 1$,
$$\alpha_2 = 3(d_x + 1_+ b\epsilon_- + d\epsilon_+) > 0$$
is a positive eigenvalue of multiplicity $c - 2 \ge 0$.
- Note that in both normalized and non-normalized case we have $\epsilon_+ \ge \frac{b+c+d-1}{c+d+e-1} - 1$ and
$$\begin{aligned}\alpha_3 = -b + 3c + 3d + 3e + (3c + 4d + 3e)\epsilon_+ \ge \\ \ge -b - d + \frac{3c + 4d + 3e}{c + d + e - 1}(b + c + d - 1) > 0\end{aligned}$$
is a positive eigenvalue of multiplicity $d - 1 \ge 1$.
- Note that in both normalized and non-normalized case we have $\epsilon_- \ge 0$ and
$$\alpha_4 = 3a + 3b + 3c - d + (3a + 4b + 3c)\epsilon_- \ge 3a + 3b + 3c - d > 0$$
if $d < 3(a + b + c)$. This eigenvalue has multiplicity $b - 1 \ge 0$.
- Since $\epsilon_-, \epsilon_+ > -1$,
$$\alpha_5 = d + d\epsilon_+ > 0 \quad \text{and } \alpha_6 = b + b\epsilon_- > 0$$
are both positive eigenvalues of multiplicities $e - 1 \ge 1$ and $a - 1 \ge 0$, respectively.

Moreover, it is easily checked that $A\mathbf{1}_{a+b+c+d+e} = 0$. The orthogonal complement of the direct sum of the corresponding eigenspaces E_i and $\mathbb{R}\mathbf{1}_{a+b+c+d+e}$ is 5-dimensional and given by $\widehat{W} = \{\widehat{w} \mid w \in W\}$, where $(d_1, d_2, d_3, d_4, d_5, d_6) = (1, c - 1, d, b, e, a)$ and
$$W := \{w \in \mathbb{R}^6, \sum_{i=1}^{6} w_i d_i = 0\}.$$

Under the assumption $d < 3(a + b + c)$, $\mathcal{K}_{G,x}(\infty) > 0$ is then equivalent to $A|_{\widehat{W}}$ being positive definite, which is equivalent to
$$\widehat{w}^\top A \,\widehat{w} = w^\top A_{\text{red}}\, w > 0 \quad \text{for all } w \in W \setminus \{0\}. \qquad (2.4.3)$$

Now we choose $(a, b, c, d, e) = (n, n+1, n+2, n+3, n+4)$, $n \in \mathbb{N}$. Then we have $d < 3(a+b+c)$ and we consider the characteristic polynomial of A_{red}, which is of the form

$$\chi_n(t) = \det(t\,\text{Id}_6 - A_{\text{red}}) = t^6 - p_5(n)t^5 + p_4(n)t^4 - p_3(n)t^3 + p_2(n)t^2 - p_1(n)t,$$

where $p_i(n)$ are polynomials in the variable n. (We do not have a constant term since $\mathbb{R} \cdot \mathbf{1}_6$ lies in the kernel of A_{red}.) A Maple calculation shows that all the $p_i(n)$ are strictly positive for any value of $n \in \mathbb{N}$ (see Appendix A for more details). This shows that we have $\chi_n(t) > 0$ for all $t < 0$, so A_{red} is positive semi-definite. Since $p_1(n) > 0$, A_{red} has a one-dimensional kernel $\mathbb{R} \cdot \mathbf{1}_6$.

Now we can show (2.4.3): Let $w_0 = \mathbf{1}_6, w_1, \ldots, w_5 \in \mathbb{R}^6$ be a basis of eigenvectors of A_{red}, i.e., $A_{\text{red}} w_j = \lambda_j w_j$ with $\lambda_j > 0$ for $j \in \{1, \ldots, 5\}$. Any vector $w \in W \backslash \{0\}$ is of the form $w = \sum_{j=0}^{5} c_j w_j$ with some $c_{j_0} \neq 0$, $j_0 \in \{1, \ldots, 5\}$, since $w_0 \notin W$. This implies

$$w^\top A_{\text{red}}\, w = \sum_{j=1}^{5} \lambda_j c_j^2 \geq \lambda_{j_0} c_{j_0}^2 > 0. \qquad \square$$

Theorem 2.4.3 *Let $x \in V_2$ be a vertex of the finite antitree $G = \mathcal{AT}((b, c, d, e))$. If $(c, d, e) = (1, 2, 3)$, we have in both the normalized and non-normalized case:*

$$\mathcal{K}_{G,x}(\infty) > 0.$$

Proof. We consider again the matrix $A = 4\mu(x)^2 \Gamma_2(x)$ and choose right from the beginning $(b, c, d, e) = (1, 2, 3, 4)$. It can be checked that this time the matrix A is of the form $A = (A_{ij})_{1 \leq i,j \leq 5}$ with A_{ij} as in the previous proof and $a = 0$. As in the previous proof, we conclude that A has eigenvalues $\alpha_3 = 27 + 30\epsilon_+ > 0$ of multiplicity 2 and $\alpha_5 = 1 + \epsilon_+ > 0$ of multiplicity 3 and that $A\mathbf{1}_{10} = 0$. In this case, A_{red} is a symmetric 5×5 matrix and its characteristic polynomial of A_{red} is (see Maple calculations in Appendix B)

$$\chi(t) = \det(t\,\text{Id}_5 - A_{\text{red}}) = t^5 - \frac{471}{4}t^4 + \frac{118743}{32}t^3 - \frac{593811}{16}t^2 + \frac{3082725}{64}t$$

in the normalized case and

$$\chi(t) = t^5 - 132t^4 + 3684t^3 - 25632t^2 + 8640t$$

in the non-normalized case. The same arguments as in the previous proof show that A is positive semi-definite with one-dimensional kernel, that is, $\mathcal{K}_{G,x}(\infty) > 0$. $\square$

Theorem 2.4.4 *Let $x \in V_1$ be a vertex of the finite antitree $G = \mathcal{AT}(c, d, e)$. If $(c, d, e) = (1, 2, 3)$, we have in both the normalized and non-normalized case:*

$$\mathcal{K}_{G,x}(\infty) > 0.$$

Proof. As in the previous proof, we consider the matrix $A = 4\mu(x)^2\Gamma_2(x)$ and choose $(c, d, e) = (1, 2, 3)$. This time A is of the form $A = (A_{ij})_{i,j \in I}$ with $I = \{1, 3, 4\}$ and A_{ij} as in the proof of Theorem 2.4.2 with $a = b = 0$. As before, we conclude that A has a simple eigenvalue $\alpha_3 = 18 + 20\epsilon_+ > 0$ and a double eigenvalue $\alpha_5 = 2 + 2\epsilon_+ > 0$ and $A\mathbf{1}_6 = 0$. A_{red} is now a symmetric 3×3 matrix with characteristic polynomial (see Maple calculations in Appendix B)

$$\chi(t) = t^3 - \frac{112}{5}t^2 + \frac{144}{5}t$$

in the normalized case and

$$\chi(t) = t^3 - 44t^2 + 72t$$

in the non-normalized case. Similarly as before, this implies that A is positive semi-definite with one-dimensional kernel, that is, $\mathcal{K}_{G,x}(\infty) > 0$. □

Remark 2.4.5 Alternatively, Theorem 2.4.4 could be proved, in the non-normalized case, by employing the fact that the root of $\mathcal{AT}((1, 2, 3))$ is S^1-out regular. For the definition of this notion and the corresponding curvature calculation, see [7, Definition 1.5 and Theorem 5.7].

The above theorems imply that the infinite antitree $\mathcal{AT}((k))$ has strictly positive Bakry–Émery curvature in all vertices. We finally prove that there is no uniform positive lower curvature bound.

Theorem 2.4.6 *Let $G = \mathcal{AT}((k))$ be the infinite antitree with vertex set $V = \bigcup_{k=1}^{\infty} V_k$. Then we have both in the normalized and normalized setting*

$$\inf_{x \in V} \mathcal{K}_{G,x}(\infty) = 0.$$

Proof. Let us first consider the normalized setting. If we had $\inf_{x \in V} \mathcal{K}_{G,x}(\infty) = K > 0$, then the discrete Bonnet–Myers theorem (Theorem 2.1.2 of the Introduction) would imply that G has bounded diameter, which is a contradiction. This argument does not work in the non-normalized setting. Let us now show in the non-normalized setting that

$$\lim_{n \to \infty, x \in V_n} \mathcal{K}_{G,x}(\infty) = 0.$$

For $\delta > 0$, let $A(\delta, n) = 4(\Gamma_2(x) - \delta\Gamma(x))$ for an arbitrary vertex $x \in V_{n+2}$, $n \in \mathbb{N}$, with respect to the vertex order

$$B_2(x) = \{x\} \sqcup (V_{n+2}\backslash\{x\}) \sqcup V_{n+3} \sqcup V_{n+1} \sqcup V_{n+4} \sqcup V_n.$$

The entries of $2\Gamma(x)$ in the non-normalized setting are given in [7, (2.2)], and using this information, we see that that matrix $A(\delta, n)$ is of the following block structure $A(\delta, n) = (A_{ij}(\delta, n))_{1 \le i,j \le 6}$:

$$\begin{aligned}
A_{11}(\delta, n) &= (3n+5)(3n+8) - (6n+10)\delta, \\
A_{12}(\delta, n) &= (-3n - 8 + 2\delta)J_{1,n+1}, \\
A_{13}(\delta, n) &= (-4n - 12 + 2\delta)J_{1,n+3}, \\
A_{14}(\delta, n) &= (-4n - 8 + 2\delta)J_{1,n+1}, \\
A_{15}(\delta, n) &= (n+3)J_{1,n+4}, \\
A_{16}(\delta, n) &= (n+1)J_{1,n}, \\
A_{22}(\delta, n) &= (9n + 18 - 2\delta)\mathrm{Id}_{n+1} - 2J_{n+1}, \\
A_{23}(\delta, n) &= -2J_{n+1,n+3}, \\
A_{24}(\delta, n) &= -2J_{n+1,n+1}, \\
A_{25}(\delta, n) &= 0_{n+1,n+4}, \\
A_{26}(\delta, n) &= 0_{n+1,n}, \\
A_{33}(\delta, n) &= (8n + 26 - 2\delta)\mathrm{Id}_{n+3} - 2J_{n+3}, \\
A_{34}(\delta, n) &= 2J_{n+3,n+1}, \\
A_{35}(\delta, n) &= -2J_{n+3,n+4}, \\
A_{36}(\delta, n) &= 0_{n+3,n}, \\
A_{44}(\delta, n) &= (8n + 6 - 2\delta)\mathrm{Id}_{n+1} - 2J_{n+1}, \\
A_{45}(\delta, n) &= 0_{n+1,n+4}, \\
A_{46}(\delta, n) &= -2J_{n+1,n}, \\
A_{55}(\delta, n) &= (n+3)\mathrm{Id}_{n+4}, \\
A_{56}(\delta, n) &= 0_{n+4,n}, \\
A_{66}(\delta, n) &= (n+1)\mathrm{Id}_n.
\end{aligned}$$

Let $\delta > 0$. Let $\lambda_j(\delta, n)$, $j \in \{1, \ldots, 5\}$ be the eigenvalues of the 6×6 matrix $A(\delta, n)_{red}$. The characteristic polynomial of $A(\delta, n)_{red}$ is of the form

$$\chi_{\delta,n}(t) = t^6 - p_5(\delta, n)t^5 + p_4(\delta, n)t^4 - p_3(\delta, n)t^3 + p_2(\delta, n)t^2 - p_1(\delta, n)t,$$

with polynomials $p_1, p_2, \ldots, p_5$, and a Maple calculation shows that

$$p_1(\delta, n) = -240\delta n^9 + q_8(\delta)n^8 + \cdots + q_1(\delta)n + q_0(\delta), \tag{2.4.4}$$

with polynomials $q_0, q_1, \dots, q_8$ (see Appendix C). By Vieta's formulas, we have

$$p_1(\delta, n) = \left(\prod_{j=1}^{5} \lambda_j(\delta, n) \right),$$

where $\lambda_j(\delta, n)$, $j = 1, \dots, 5$ are the eigenvalues (in ascending order) of $A(\delta, n)_{red}$ restricted to the orthogonal complement to the eigenvector $\mathbf{1}_6$. We conclude from (2.4.4) that there exists $k_0 > 0$ with $p_1(\delta, n) < 0$ for all $n \geq n_0$, i.e., $\lambda_1(\delta, n) < 0$. Applying Lemma 2.4.1, we conclude

$$(\widetilde{w})^\top A(\delta, n) \widetilde{w} = w^\top A(\delta, n)_{red} w = \lambda_1(\delta, n) \|w\|^2 < 0.$$

This implies that $\mathcal{K}_{G,x}(\infty) \in (0, \delta)$ for every $x \in V_{n+2}$ with $n \geq n_0$. □

2.5 Ollivier–Ricci Curvature of Antitrees

In this section, we calculate Ollivier–Ricci curvature for all idleness $p \in [0, 1]$ and the Lin–Lu–Yau curvature of all types of edges in antitrees.

Theorem 2.5.1 (Radial root-edges of an antitree) *Let $1 \leq a \leq b \leq c$, $\{x, y\}$ a radial root edge of the antitree $\mathcal{AT}((a, b, c))$, that is, $x \in V_1$, $y \in V_2$. Then we have:*

(a) If $a = 1$,

$$\kappa_p(x, y) = \begin{cases} \frac{b-1}{b+c} + \frac{b+2c+1}{b+c} p & \text{if } p \in [0, \frac{1}{b+c+1}], \\ \frac{b+1}{b+c}(1-p), & \text{if } p \in [\frac{1}{b+c+1}, 1]. \end{cases}$$

Therefore,

$$\kappa_{LLY}(x, y) = \frac{b+1}{b+c}.$$

(b) If $a \geq 3$ or ($a = 2$ and $b < c$),

$$\kappa_p(x, y) = \frac{1}{(a+b-1)(a+b+c-1)} \begin{cases} ((a+b-1)^2 - c(a-1)) + c(b+2a-2)p & \text{if } p \in [0, \frac{1}{a+b+c}], \\ ((a+b)(a+b-1) - c(a-1))(1-p), & \text{if } p \in [\frac{1}{a+b+c}, 1]. \end{cases}$$

Therefore,

$$\kappa_{LLY}(x, y) = \frac{(a+b)(a+b-1) - c(a-1)}{(a+b-1)(a+b+c-1)}.$$

(c) If $a = 2, b = c$,

$$\kappa_p(x,y) = \begin{cases} \frac{b}{2b+1} + \frac{3b+2}{2b+1}p & \text{if } p \in [0, \frac{1}{(2b+1)(b+1)1}], \\ \frac{b^2+b+1}{(2b+1)(b+1)} + \frac{b^2+2b}{(2b+1)(b+1)}p, & \text{if } p \in [\frac{1}{(2b+1)(b+1)+1}, \frac{1}{2(b+1)}], \\ \frac{b^2+2b+2}{(2b+1)(b+1)}(1-p), & \text{if } p \in [\frac{1}{2(b+1)}, 1]. \end{cases}$$

Therefore,

$$\kappa_{LLY}(x,y) = \frac{b^2+2b+2}{(2b+1)(b+1)}.$$

Proof. (a) Consider the following graph

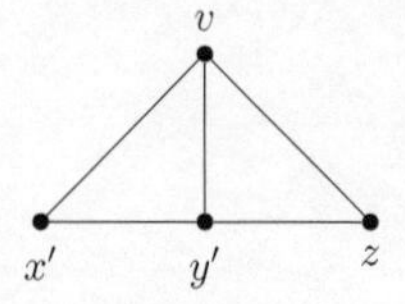

with associated probability measures μ_1^p, μ_2^p, defined as

$$\mu_1^p(x') = p, \ \mu_1^p(y') = \frac{1}{b}(1-p), \ \mu_1^p(v) = \frac{b-1}{b}(1-p), \ \mu_1^p(z) = 0,$$

$$\mu_2^p(x') = \frac{1}{b+c}(1-p), \ \mu_2^p(y') = p, \ \mu_2^p(v) = \frac{b-1}{b+c}(1-p),$$

$$\mu_2^p(z) = \frac{c}{b+c}(1-p).$$

One can verify that, due to the high connectivity of $\mathcal{AT}((a,b,c))$, we have $W_1(\mu_x^p, \mu_y^p) = W_1(\mu_1^p, \mu_2^p)$, where x' represents the root x, y' represents the vertex y, the vertex v represents all neighbours of y in V_2, and the vertex z represents all vertices in V_3.

Note that $\mu_1^p(x') < \mu_2^p(x')$ if and only if $p < \frac{1}{b+c+1}$. We will distinguish the cases.

Case $p < \frac{1}{b+c+1}$:
Note that

$$\mu_1^p(x') < \mu_2^p(x'), \quad \mu_1^p(z) < \mu_2^p(z),$$

$$\mu_1^p(y') > \mu_2^p(y'), \quad \mu_1^p(w) > \mu_2^p(w).$$

Thus when transporting μ_1^p to μ_2^p the only vertices that gain mass are x' and z. Note further all this mass can be transported over a distance of 1. Thus

$$\begin{aligned} W_1(\mu_x^p, \mu_y^p) &= W_1(\mu_1^p, \mu_2^p) \\ &\leq \mu_2^p(x') + \mu_2^p(z) - \mu_1^p(x') - \mu_1^p(z) \\ &= \frac{c+1}{b+c} - \frac{b+2c+1}{b+c} p. \end{aligned}$$

We verify that this is in fact equality by constructing the following $\phi \in 1-\mathrm{Lip}$,

$$\phi(x') = 0, \phi(y') = 1, \phi(w) = 1, \phi(z) = 0.$$

Then, by Theorem 2.3.5,

$$\begin{aligned} W_1(\mu_x^p, \mu_y^p) = W_1(\mu_1^p, \mu_2^p) &\geq \sum_v \phi(v)(\mu_1^p(v) - \mu_2^p(v)) \\ &= \frac{c+1}{b+c} - \frac{b+2c+1}{b+c} p. \end{aligned}$$

Therefore,

$$W_1(\mu_x^p, \mu_y^p) = \frac{c+1}{b+c} - \frac{b+2c+1}{b+c} p.$$

and

$$\kappa_p(x, y) = \frac{b-1}{b+c} + \frac{b+2c+1}{b+c} p, \tag{2.5.1}$$

for $p \in [0, \frac{1}{b+c+1})$. By continuity of $p \mapsto \kappa_p(x, y)$ this also holds for $p = \frac{1}{b+c+1}$.

Case $p \geq \frac{1}{b+c+1}$:
By [3, Theorem 4.4], $\kappa_p(x, y) = \frac{b+c+1}{b+c} \kappa_{\frac{1}{b+c+1}} (1-p)$ for $p \in [\frac{1}{b+c+1}, 1]$. Thus

$$\kappa_p(x, y) = \begin{cases} \frac{b-1}{b+c} + \frac{b+2c+1}{b+c} p & \text{if } p \in [0, \frac{1}{b+c+1}], \\ \frac{b+c+1}{b+c} \kappa_{\frac{1}{b+c+1}} (1-p), & \text{if } p \in [\frac{1}{b+c+1}, 1]. \end{cases}$$

Therefore it only remains to show that $\frac{b+c+1}{b+c} \kappa_{\frac{1}{b+c+1}} = \frac{b+1}{b+c}$.

We have, using (2.5.1),

$$\begin{aligned} \frac{b+c+1}{b+c} \kappa_{\frac{1}{b+c+1}} &= \frac{b+c+1}{b+c} \left(\frac{b-1}{b+c} + \frac{b+2c+1}{b+c} \frac{1}{b+c+1} \right) \\ &= \frac{b+1}{b+c}. \end{aligned}$$

(b) Similar to above we consider the simplified graph representing $\mathcal{AT}((a,b,c))$,

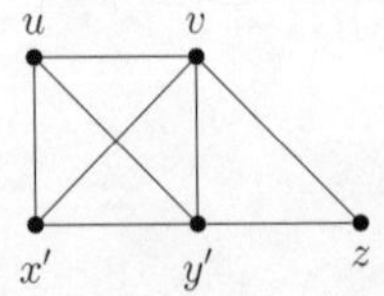

with associated probability measures μ_1^p, μ_2^p, defined as

$$\mu_1^p(x') = p,\ \mu_1^p(y') = \frac{1}{a+b-1}(1-p),\ \mu_1^p(u) = \frac{a-1}{a+b-1}(1-p),$$

$$\mu_1^p(v) = \frac{b-1}{a+b-1}(1-p),\ \mu_1^p(z) = 0,$$

$$\mu_2^p(x') = \frac{1}{a+b+c-1}(1-p), \mu_2^p(y') = p,\ \mu_2^p(u) = \frac{a-1}{a+b+c-1}(1-p),$$

$$\mu_2^p(v) = \frac{b-1}{a+b+c-1}(1-p),\ \mu_2^p(z) = \frac{c}{a+b+c-1}(1-p).$$

Again, one can verify that, due to the high connectivity of $\mathcal{AT}((a,b,c))$, we have $W_1(\mu_x^p, \mu_y^p) = W_1(\mu_1^p, \mu_2^p)$, where x' represents the root x, y' represents the vertex y, the vertex u represents all neighbours of x in V_1, the vertex v represents all neighbours of y in V_2, and the vertex z represents all vertices in V_3.

Let $p \in (0, \frac{1}{a+b+c})$. One can check that

$$\mu_1^p(x') < \mu_2^p(x'),\ \mu_1^p(z) < \mu_2^p(z),$$

$$\mu_1^p(y') > \mu_2^p(y'),\ \mu_1^p(u) > \mu_2^p(u),\ \mu_1^p(v) > \mu_2^p(v).$$

Thus the vertices x' and z must gain mass and the vertices u, v, and y must lose mass. We now show that some mass must be transported from u to z. Suppose that no mass is moved from u to v. Then the mass available to move from v and y' will be sufficient when moved to z. Therefore

$$\mu_1^p(y') + \mu_1^p(v) - \mu_2^p(y') - \mu_2^p(v) \geq \mu_2^p(z) - \mu_1^p(z).$$

Substituting in the values of the measures and rearranging gives $p \leq \frac{a+b+c-ac-1}{(a+b)(a+b-1)+bc} \leq 0$, a contradiction. Therefore some mass must be transported from u to z over a distance of 2 and all other mass can be transported over a distance of 1.

Thus,

$$\begin{aligned} W_1(\mu_x^p, \mu_y^p) &= W_1(\mu_1^p, \mu_2^p) \\ &\le (\mu_2^p(x) - \mu_1^p(x)) + 2(\mu_1^p(u) - \mu_2^p(u) - (\mu_2^p(x) - \mu_1^p(x))) \\ &\quad + (\mu_1^p(y') + \mu_1^p(v) - \mu_2^p(y') - \mu_2^p(v)) \\ &= (1-p)\left(\frac{a-1}{a+b-1} + \frac{c+1-a}{a+b+c-1}\right). \end{aligned}$$

We verify that this is in fact equality by constructing the following $\phi \in$ 1-Lip,

$$\phi(x') = 0, \phi(y') = 0, \phi(u) = 1, \phi(v) = 0, \phi(z) = -1.$$

Therefore,

$$\begin{aligned} \kappa_p(x, y) &= 1 - (1-p)\left(\frac{a-1}{a+b-1} + \frac{c+1-a}{a+b+c-1}\right) \\ &= \frac{((a+b-1)^2 - c(a-1)) + (bc + 2c(a-1))p}{(a+b-1)(a+b+c-1)}, \end{aligned}$$

for $p \in (0, \frac{1}{a+b+c})$.

As before, by [3, Theorem 4.4], $\kappa_p(x, y) = \frac{a+b+c}{a+b+c-1}\kappa_{\frac{1}{a+b+c}}(1-p)$ for $p \in [\frac{1}{a+b+c}, 1]$. Therefore,

$$\frac{a+b+c}{a+b+c-1}\kappa_{\frac{1}{a+b+c}} = \frac{(a+b)(a+b-1) - c(a-1)}{(a+b-1)(a+b+c-1)},$$

thus completing the proof.

(c) As in part (b) we consider the simplified graph representing $\mathcal{AT}((a, b, c))$,

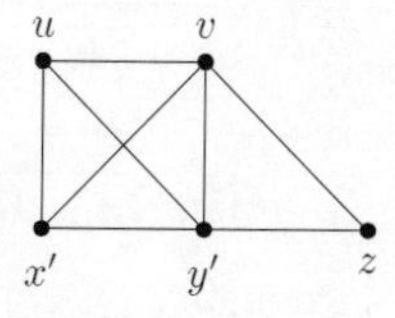

with the same associated probability measures μ_1^p, μ_2^p, defined as

$$\mu_1^p(x') = p, \;\; \mu_1^p(y') = \frac{1}{a+b-1}(1-p), \;\; \mu_1^p(u) = \frac{a-1}{a+b-1}(1-p),$$

$$\mu_1^p(v) = \frac{b-1}{a+b-1}(1-p), \;\; \mu_1^p(z) = 0,$$

$$\mu_2^p(x') = \frac{1}{a+b+c-1}(1-p),\ \ \mu_2^p(y') = p,$$
$$\mu_2^p(u) = \frac{a-1}{a+b+c-1}(1-p),$$

$$\mu_2^p(v) = \frac{b-1}{a+b+c-1}(1-p),\ \ \mu_2^p(z) = \frac{c}{a+b+c-1}(1-p).$$

Again, one can verify that, due to the high connectivity of $\mathcal{AT}((a,b,c))$, we have $W_1(\mu_x^p, \mu_y^p) = W_1(\mu_1^p, \mu_2^p)$, where x' represents the root x, y' represents the vertex y, the vertex u represents all neighbours of x in V_1, the vertex v represents all neighbours of y in V_2, and the vertex z represents all vertices in V_3.
We will distinguish the cases.

Case $p \in (0, \frac{1}{(2b+1)(b+1)1})$:
One can check that

$$\mu_1^p(x') < \mu_2^p(x'),\ \ \mu_1^p(z) < \mu_2^p(z),$$

$$\mu_1^p(y') > \mu_2^p(y'),\ \ \mu_1^p(u) > \mu_2^p(u),\ \ \mu_1^p(v) > \mu_2^p(v),$$

and

$$\mu_1^p(y') + \mu_1^p(v) - \mu_2^p(y') - \mu_2^p(v) \geq \mu_2^p(z) - \mu_1^p(z).$$

Thus the vertices x' and z must gain mass and the vertices u, v, and y must lose mass and it is possible for all mass to be moved over a distance of 1.

Thus,

$$\begin{aligned} W_1(\mu_x^p, \mu_y^p) &= W_1(\mu_1^p, \mu_2^p) \\ &\leq \mu_2^p(x') + \mu_2^p(z) - \mu_1^p(x') - \mu_1^p(z) \\ &= \frac{b+1}{2b+1} - \frac{3b+2}{2b+1}p. \end{aligned}$$

We verify that this is in fact equality by constructing the following $\phi \in$ 1–Lip,

$$\phi(x') = -1, \phi(y') = 0, \phi(u) = 0, \phi(v) = 0, \phi(z) = -1.$$

Therefore,

$$\kappa_p(x, y) = \frac{b}{2b+1} + \frac{3b+2}{2b+1}p.$$

Case $p \in (\frac{1}{(2b+1)(b+1)+1}, \frac{1}{2(b+1)})$:

One can check that we still have

$$\mu_1^p(x') < \mu_2^p(x'),\ \ \mu_1^p(z) < \mu_2^p(z),$$

$$\mu_1^p(y') > \mu_2^p(y'),\ \ \mu_1^p(u) > \mu_2^p(u),\ \ \mu_1^p(v) > \mu_2^p(v)$$

However we now have

$$\mu_1^p(y') + \mu_1^p(v) - \mu_2^p(y') - \mu_2^p(v) \leq \mu_2^p(z) - \mu_1^p(z).$$

Thus, as in part (b), some mass must be transported from u to z over a distance of 2 and all other mass can be transported over a distance of 1.

Therefore,

$$\begin{aligned} W_1(\mu_x^p, \mu_y^p) &= W_1(\mu_1^p, \mu_2^p) \\ &\leq (\mu_2^p(x) - \mu_1^p(x)) + 2(\mu_1^p(u) - \mu_2^p(u) - (\mu_2^p(x) - \mu_1^p(x))) \\ &\quad + (\mu_1^p(y') + \mu_1^p(v) - \mu_2^p(y') - \mu_2^p(v)) \\ &= (1-p)\left(\frac{1}{b+1} + \frac{b-1}{2b+1}\right). \end{aligned}$$

We verify that this is in fact equality by constructing the following $\phi \in$ 1−Lip,

$$\phi(x') = 0, \phi(y') = 0, \phi(u) = 1, \phi(v) = 0, \phi(z) = -1.$$

Therefore,

$$\kappa_p(x, y) = \frac{b^2+b+1}{(2b+1)(b+1)} + \frac{b^2+2b}{(2b+1)(b+1)}p.$$

Case $p \in (\frac{1}{2(b+1)}, 1)$: As before, by [3, Theorem 4.4], $\kappa_p(x, y) = \frac{2(b+1)}{2b+1}\kappa_{\frac{1}{2(b+1)}}(1-p)$ for $p \in [\frac{1}{2(b+1)}, 1]$. Thus

$$\frac{2(b+1)}{2b+1}\kappa_{\frac{1}{2(b+1)}} = \frac{b^2+2b+2}{(2b+1)(b+1)}, \qquad \square$$

thus completing the proof.

Theorem 2.5.2 (Inner radial edges of an antitree) *Let* $1 \leq a \leq b \leq c \leq d$, $\{x, y\}$ *an inner radial edge of the antitree* $\mathcal{AT}((a, b, c, d))$, *that is* $x \in V_2$, $y \in V_3$. *Then we have:*

$$\kappa_p(x, y) = \left(\frac{2b+c-1}{b+c+d-1} - \frac{2a+b-1}{a+b+c-1}\right)(1-p).$$

Proof. We first calculate $\kappa_0(x, y)$. We consider the simplified graph representing $\mathcal{AT}((a, b, c, d))$,

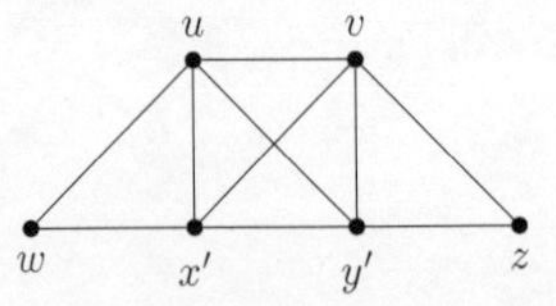

with the associated probability measures μ_1, μ_2, defined as

$$\mu_1(x') = 0,\ \ \mu_1(y') = \frac{1}{a+b+c-1},\ \mu_1(w) = \frac{a}{a+b+c-1},$$

$$\mu_1(u) = \frac{b-1}{a+b+c-1},\ \mu_1(v) = \frac{c-1}{a+b+c-1},\ \mu_1(z) = 0,$$

$$\mu_2(x') = \frac{1}{b+c+d-1},\ \mu_2(y') = 0,\ \ \mu_2(w) = 0,$$

$$\mu_2(u) = \frac{b-1}{b+c+d-1},\ \mu_2(v) = \frac{c-1}{b+c+d-1},\ \mu_2(z) = \frac{d}{b+c+d-1}.$$

Again, one can verify that, due to the high connectivity of $\mathcal{AT}((a, b, c, d))$, we have $W_1(\mu_x^0, \mu_y^0) = W_1(\mu_1, \mu_2)$, where x' represents the vertex x, y' represents the vertex y, the vertex w represents all the vertices in V_1, the vertex u represents all neighbours of x in V_2, the vertex v represents all neighbours of y in V_3, and the vertex z represents all vertices in V_4.

Observe that

$$\mu_1(x') < \mu_2(x'),\ \ \mu_1(z) < \mu_2(z),\ \ \mu_1(u) < \mu_2(u),\ \ \mu_1(v) < \mu_2(v),$$

$$\mu_1(y') > \mu_2(y'),\ \ \mu_1(w) > \mu_2(w).$$

Therefore the only vertices that gain mass are x' and z. Now, $\mu_1(w) - \mu_2(w) = \frac{a}{a+b+c-1} \geq \frac{1}{b+c+d-1} = \mu_2(x') - \mu_1(x')$, and so it is possible for x' to receive all of its needed mass from w. If we do this plan and send all other surplus mass to the vertex z we obtain

$$\begin{aligned} W_1(\mu_x^p, \mu_y^p) = W_1(\mu_1^p, \mu_2^p) \\ &\leq (\mu_2(x') - \mu_1(x')) + 3(\mu_1(w) - [\mu_2(x') \\ &\quad - \mu_1(x')] - \mu_2(w)) + 2(\mu_1(u) - \mu_2(u)) \\ &\quad + (\mu_1(v) - \mu_2(v)) + (\mu_1(y') - \mu_2(y')) \\ &= \frac{3a+2b+c-2}{a+b+c-1} - \frac{2b+c-1}{b+c+d-1}. \end{aligned}$$

We verify that this is in fact equality by constructing the following $\phi \in 1\text{-Lip}$,

$$\phi(w) = 3,\ \ \phi(x') = 2,\ \ \phi(u) = 2,\ \ \phi(y') = 1,\ \ \phi(v) = 1,\ \ \phi(z) = 0.$$

Thus,

$$\kappa_0(x, y) = \frac{2b + c - 1}{b + c + d - 1} - \frac{2a + b - 1}{a + b + c - 1}.$$

Observe that $\phi(x') - \phi(y') = 1$ and thus, by [3, Lemma 4.2], we have that $p \mapsto \kappa_p(x, y)$ is linear. Since $\kappa_1(x, y) = 0$, this gives

$$\kappa_p(x, y) \left(\frac{2b + c - 1}{b + c + d - 1} - \frac{2a + b - 1}{a + b + c - 1} \right) (1 - p) \qquad \square$$

Theorem 2.5.3 (Spherical root edges of an antitree) *Let* $2 \le a \le b$, $\{x, y\}$ *a spherical root edge of the antitree* $\mathcal{AT}((a, b))$, *that is* $x, y \in V_1$. *Then*

$$\kappa_p(x, y) = \begin{cases} \frac{a+b-2}{a+b-1} + \frac{a+b}{a+b-1} p & \text{if } p \in [0, \frac{1}{a+b}], \\ \frac{a+b}{a+b-1}(1 - p), & \text{if } p \in [\frac{1}{a+b}, 1]. \end{cases}$$

Proof. Since $d_x = d_y$, by [3, Theorem 5.3], we have

$$\kappa_p(x, y) = \begin{cases} ((a + b - 1)\kappa_{LLY}(x, y) - (a + b)\kappa_0(x, y))p + \kappa_0(x, y), & \text{if } p \in [0, \frac{1}{a+b}], \\ (1 - p)\kappa_{LLY}(x, y), & \text{if } p \in [\frac{1}{a+b}, 1]. \end{cases}$$

Therefore we will calculate $\kappa_p(x, y)$ for $p = 0$ and $p = \frac{1}{a+b}$.

Observe that $\mu_x^0(y) = \frac{1}{a+b-1}$ and 0 otherwise, and $\mu_y^0(x) = \frac{1}{a+b-1}$ and 0 otherwise. Thus we have

$$W_1(\mu_x^0, \mu_y^0) = \frac{1}{a + b - 1},$$

and so

$$\kappa_0(x, y) = \frac{a + b - 2}{a + b - 1}.$$

Note that

$$\mu_x^{\frac{1}{a+b}} \equiv \mu_y^{\frac{1}{a+b}},$$

so

$$\kappa_{LLY}(x, y) = \frac{a + b}{a + b - 1}\kappa_{\frac{1}{a+b}}(x, y) = \frac{a + b}{a + b - 1}.$$

Substituting these values into the above formula completes the proof. $\square$

Theorem 2.5.4 (Spherical inner edges of an antitree) *Let* $1 \le a \le b \le c$, $\{x, y\}$ *a spherical inner edge of the antitree* $\mathcal{AT}((a, b, c))$, *that is* $x, y \in V_2$. *Then*

$$\kappa_p(x, y) = \begin{cases} \frac{a+b+c-2}{a+b+c-1} + \frac{a+b+c}{a+b+c-1} p & \text{if } p \in [0, \frac{1}{a+b+c}], \\ \frac{a+b+c}{a+b+c-1}(1 - p), & \text{if } p \in [\frac{1}{a+b+c}, 1]. \end{cases}$$

Proof. The proofs follows in the same way as in the proof of Theorem 2.5.3. □

Appendices

A Maple Calculations for Theorem 2.4.2

In the *normalized case*, the Maple code to construct the matrix $A_{\text{red}} = 4\mu_x^2\Gamma_{2,\text{red}}(x)$ for $x \in V_3 \cong K_c$ of $\mathcal{AT}((a, b, c, d, e))$ is the following:

```
with(LinearAlgebra) : dx := b + c + d - 1 : eminus := (d - a)/(a + b + c - 1) : eplus := (e - b)/(c + d + e - 1) :
a11 := dx·(dx + 3) + 3·b·eminus + 3·d·eplus :
a12 := (c - 1)·(-(dx + 3) + b·eminus + d·eplus) : a21 := a12 :
a13 := d·(-(dx + 3 + e) - (2 + c + e)·eplus) : a31 := a13 :
a14 := b·(-(dx + 3 + a) - (2 + a + c)·eminus) : a41 := a14 :
a15 := e·(d + d·eplus) : a51 := a15 :
a16 := a·(b + b·eminus) : a61 := a16 :
a22 := (c - 1)·(3·(dx + 1) + b·eminus + d·eplus) - 2·(c - 1)^2 :
a23 := -(c - 1)·d·(2 + 2 eplus) : a32 := a23 :
a24 := -(c - 1)·b·(2 + 2·eminus) : a42 := a24 :
a25 := 0 : a52 := 0 :
a26 := 0 : a62 := 0 :
a33 := d·(-b + 3·c + 3·d + 3·e + (3·c + 4·d + 3·e)·eplus) - d^2·(2 + 4·eplus) :
a34 := 2·b·d : a43 := a34 :
a35 := -d·e·(2 + 2·eplus) : a53 := a35 :
a36 := 0 : a63 := 0 :
a44 := b·(3·a + 3·b + 3·c - d + (3·a + 4·b + 3·c)·eminus) - b^2·(2 + 4·eminus) :
a45 := 0 : a54 := 0 :
a46 := -a·b·(2 + 2·eminus) : a64 := a46 :
a55 := e·(d + d·eplus) :
a56 := 0 : a65 := 0 :
a66 := a·(b + b·eminus) :
Ared :=
Matrix([
[a11, a12, a13, a14, a15, a16],
[a21, a22, a23, a24, a25, a26],
[a31, a32, a33, a34, a35, a36],
[a41, a42, a43, a44, a45, a46],
[a51, a52, a53, a54, a55, a56],
[a61, a62, a63, a64, a65, a66]
]) :
```

Figure 2.4 Maple construction of A_{red} in the normalized case

For the generation of the coefficients of the characteristic polynomial $\chi_n(t)$ of A_{red} for $a = n, b = n + 1, c = n + 2, d = n + 3, e = n + 4$, see Figure 2.5. Note that there are no negative coefficients in the polynomials $p_1(n)$, $p_2(n)$, $p_3(n)$, $p_4(n)$, and $p_5(n)$.

The only modification of the above code in the *non-normalized case* is to set the variables `eminus` and `eplus` equal to 0. The coefficients of $\chi_n(t)$ for $a = n, b = n + 1, c = n + 2, d = n + 3, e = n + 4$ are given in Figure 2.6. Again, all coefficients of $p_j(n)$, $j = 1, 2, 3, 4, 5$, are non-negative.

```
p := CharacteristicPolynomial(Ared, t) :
p1 := -simplify(subs(a = n, b = n + 1, c = n + 2, d = n + 3, e = n + 4, coeff(p, t, 1))) :
p2 := simplify(subs(a = n, b = n + 1, c = n + 2, d = n + 3, e = n + 4, coeff(p, t, 2))) :
p3 := -simplify(subs(a = n, b = n + 1, c = n + 2, d = n + 3, e = n + 4, coeff(p, t, 3))) :
p4 := simplify(subs(a = n, b = n + 1, c = n + 2, d = n + 3, e = n + 4, coeff(p, t, 4))) :
p5 := -simplify(subs(a = n, b = n + 1, c = n + 2, d = n + 3, e = n + 4, coeff(p, t, 5))) :
simplify(coeff(p, t, 0)); p1; p2; p3; p4; p5; simplify(coeff(p, t, 6));
```

$$0$$

$$\frac{1}{(3n+2)^3(3n+8)^3}\big(72\,(45670 + 223332\,n + 446788\,n^2 + 476273\,n^3 + 293466\,n^4 + 104895\,n^5 + 20196\,n^6 + 1620\,n^7)\,n\,(n+1)^2\,(n+3)^2\,(n+4)\,(3n+5)\big)$$

$$\frac{1}{(3n+2)^3(3n+8)^3}\big(4\,(16441200 + 175285020\,n + 773219090\,n^2 + 1922518396\,n^3 + 3051173765\,n^4 + 3290645589\,n^5 + 2490470475\,n^6 + 1339378389\,n^7 + 509792130\,n^8 + 134257176\,n^9 + 2383830\,n^{11} + 23262390\,n^{10} + 109350\,n^{12})\,(n+1)\,(n+3)\big)$$

$$\frac{1}{(3n+2)^3(3n+8)^3}\big(4\,(38003904 + 340621920\,n + 1354291480\,n^2 + 3161194328\,n^3 + 4830046104\,n^4 + 5095272832\,n^5 + 3809906879\,n^6 + 2036918817\,n^7 + 773639964\,n^8 + 203785362\,n^9 + 3636981\,n^{11} + 35374239\,n^{10} + 167670\,n^{12})\big)$$

$$\frac{1}{(3n+2)^2(3n+8)^2}\big(949944 + 5548092\,n + 13510888\,n^2 + 17941498\,n^3 + 14250855\,n^4 + 6959448\,n^5 + 2048544\,n^6 + 333558\,n^7 + 23085\,n^8\big)$$

$$\frac{2\,(942 + 2684\,n + 2532\,n^2 + 981\,n^3 + 135\,n^4)}{(3n+2)\,(3n+8)}$$

$$1$$

Figure 2.5 Coefficients of $\chi_n(t) = \det(t\,\mathrm{Id}_6 - A_{\mathrm{red}})$, normalized case

$$0$$

$$10368\,n + 72648\,n^3 + 63432\,n^4 + 8496\,n^6 + 30960\,n^5 + 1224\,n^7 + 72\,n^8 + 43200\,n^2$$

$$8640 + 101376\,n + 509588\,n^3 + 434316\,n^4 + 61832\,n^6 + 215556\,n^5 + 9480\,n^7 + 600\,n^8 + 330612\,n^2$$

$$25632 + 97488\,n + 118508\,n^3 + 50100\,n^4 + 920\,n^6 + 10756\,n^5 + 150100\,n^2$$

$$3684 + 8100\,n + 2218\,n^3 + 285\,n^4 + 6421\,n^2$$

$$30\,n^2 + 118\,n + 132$$

$$1$$

Figure 2.6 Coefficients of $\chi_n(t) = \det(t\,\mathrm{Id}_6 - A_{\mathrm{red}})$, non-normalized case

B Maple Calculations for Theorems 2.4.3 and 2.4.4

For the Maple calculations needed for the proofs of these theorems, the code of Figure 2.4 is used again, followed by the code in Figure 2.7 (in the *normalized*

```
# Maple Calculations for Theorem 4.3
Ared := subs(a = 0, b = 1, c = 2, d = 3, e = 4,
Matrix([
 [a11, a12, a13, a14, a15],
 [a21, a22, a23, a24, a25],
 [a31, a32, a33, a34, a35],
 [a41, a42, a43, a44, a45],
 [a51, a52, a53, a54, a55]
 ])):
p := CharacteristicPolynomial(Ared, t);
```

$$t^5 - \frac{471}{4}t^4 + \frac{118743}{32}t^3 - \frac{593811}{16}t^2 + \frac{3082725}{64}t$$

```
# Maple Calculations for Theorem 4.4
Ared := subs(a = 0, b = 0, c = 1, d = 2, e = 3,
Matrix([
 [a11, a13, a15],
 [a31, a33, a35],
 [a51, a53, a55]
 ])):
p := CharacteristicPolynomial(Ared, t);
```

$$t^3 - \frac{112}{5}t^2 + \frac{144}{5}t$$

Figure 2.7 Calculation of $\chi(t) = \det(t\mathrm{Id} - A_{\mathrm{red}})$ for Theorems 2.4.3 and 2.4.4, normalized case

case). The reduced matrices A_{red} are here of dimension 5 and 3, respectively, and they can be extracted from the original 6×6 matrix as sub-matrices with specific choices for a, b, c, d, e. The crucial observation here is that the coefficients of the respective characteristic polynomials of degree 5 and 3 are alternating, guaranteeing that all non-zero roots are strictly positive. As before, the *non-normalized case* is treated analogously with the small modification to set the variables `eminus` and `eplus` equal to 0. This leads again to characteristic polynomials with alternating coefficients, given in the proofs of the theorems as

$$\chi(t) = t^5 - 132t^4 + 3684t^3 - 25632t^2 + 8640t$$

and

$$\chi(t) = t^3 - 44t^2 + 72t.$$

C Maple Calculations for Theorem 2.4.6

Using the information about $(A_{ij}(\delta, n))$ in the proof of Theorem 2.4.6, the Maple code to calculate the relevant polynomial $p_1(\delta, n)$ is given in Figure 2.8.

```
# Maple Calculations for Theorem 4.6
with(LinearAlgebra) :
a11 := (3·n + 5)·(3· n + 8) − (6·n + 10)·delta :
a12 := (n + 1)·( −3·n − 8 + 2·delta) : a21 := a12 :
a13 := (n + 3)·( −4·n − 12 + 2·delta) : a31 := a13 :
a14 := (n + 1)·( −4 n − 8 + 2·delta) : a41 := a14 :
a15 := (n + 4)·(n + 3) : a51 := a15 :
a16 := n·(n + 1) : a61 := a16 :
a22 := (n + 1)·(9·n + 18 − 2·delta) − 2·(n + 1)^2 :
a23 := −2·(n + 1)·(n + 3) : a32 := a23 :
a24 := −2·(n + 1)^2 : a42 := a24 :
a25 := 0 : a52 := 0 :
a26 := 0 : a62 := 0 :
a33 := (n + 3)·(8·n + 26 − 2·delta) − 2·(n + 3)^2 :
a34 := 2·(n + 3)·(n + 1) : a43 := a34 :
a35 := −2·(n + 3)·(n + 4) : a53 := a35 :
a36 := 0 : a63 := 0 :
a44 := (n + 1)·(8·n + 6 − 2·delta) − 2·(n + 1)^2 :
a45 := 0 : a54 := 0 :
a46 := −2·(n + 1)·n : a64 := a46 :
a55 := (n + 3)·(n + 4) :
a56 := 0 : a65 := 0 :
a66 := (n + 1)·n :
Ared :=
Matrix([
 [a11, a12, a13, a14, a15, a16],
 [a21, a22, a23, a24, a25, a26],
 [a31, a32, a33, a34, a35, a36],
 [a41, a42, a43, a44, a45, a46],
 [a51, a52, a53, a54, a55, a56],
 [a61, a62, a63, a64, a65, a66]
 ]) :
p := CharacteristicPolynomial(Ared, t) :
p1 := −simplify(coeff(p, t, 1)) :
simplify(coeff(p, t, 0)); sort( p1, order = plex(n, delta), descending);
```

$$0$$

$$-240 n^9 \delta + 264 n^8 \delta^2 - 4272 n^8 \delta + 72 n^8 - 48 n^7 \delta^3 + 4008 n^7 \delta^2 - 32208 n^7 \delta + 1224 n^7 - 624 n^6 \delta^3 + 24912 n^6 \delta^2 - 133968 n^6 \delta + 8496 n^6 - 3168 n^5 \delta^3 + 81840 n^5 \delta^2 - 335184 n^5 \delta + 30960 n^5 - 7968 n^4 \delta^3 + 152904 n^4 \delta^2 - 514896 n^4 \delta + 63432 n^4 - 10416 n^3 \delta^3 + 162216 n^3 \delta^2 - 473136 n^3 \delta + 72648 n^3 - 6768 n^2 \delta^3 + 90720 n^2 \delta^2 - 237744 n^2 \delta + 43200 n^2 - 1728 n \delta^3 + 20736 n \delta^2 - 50112 n \delta + 10368 n$$

Figure 2.8 Calculation of $p_1(\delta, n)$ in the proof of Theorem 2.4.6

Acknowledgements

We are grateful to Radoslaw Wojciechowski, Matthias Keller, and Jozef Dodziuk for providing useful information on antitrees. Some figures in this chapter are based on the curvature calculator by David Cushing and George Stagg (see [6]).

References

[1] F. BAUER, M. KELLER, AND R. K. WOJCIECHOWSKI, *Cheeger inequalities for unbounded graph Laplacians*, J. Eur. Math. Soc. (JEMS), 17(2):259–271, 2015.

[2] D. BAKRY AND M. ÉMERY, *Diffusions hypercontractives*, in Séminaire de probabilités, XIX, 1983/84, Lecture Notes in Math. 1123, 117–206, Springer, Berlin, 1985.

[3] D. BOURNE, D. CUSHING, S. LIU, F. MÜNCH, AND N. PEYERIMHOFF, *Ollivier-Ricci idleness functions of graphs*, SIAM J. Discrete Math., 32(2):1408–1424, 2018.

[4] J. BREUER AND M. KELLER, *Spectral analysis of certain spherically homogeneous graphs*, Oper. Matrices, 7(4):825–847, 2013.

[5] R. BROOKS, *A relation between growth and the spectrum of the Laplacian*, Math. Z., 178(4):501–508, 1981.

[6] D. CUSHING, R. KANGASLAMPI, V. LIPLÄINEN, S. LIU, AND G. W. STAGG, *The graph curvature calculator and the curvatures of cubic graphs*, arXiv:1712.03033, 2017, Experimental Mathematics, DOI:10.1080/10586458.2019.1660740.

[7] D. CUSHING, S. LIU, AND N. PEYERIMHOFF, *Bakry-Émery curvature functions of graphs*, arXiv:1606.01496, 2016, Canad. J. Math., 72(1):89–143, 2020.

[8] J. DODZIUK AND L. KARP, *Spectral and function theory for combinatorial Laplacians*, in Geometry of random motion (Ithaca, NY, 1987), Contemp. Math. 73, 25–40, Amer. Math. Soc., Providence, RI, 1988.

[9] J. DODZIUK AND V. MATHAI, *Kato's inequality and asymptotic spectral properties for discrete magnetic Laplacians In The ubiquitous heat kernel*, in The ubiquitous heat kernel, Contemp. Math. 398, 69–81, Amer. Math. Soc., Providence, RI, 2006.

[10] R. L. FRANK, D. LENZ, AND D. WINGERT, *Intrinsic metrics for non-local symmetric Dirichlet forms and applications to spectral theory*. J. Funct. Anal., 266(8):4765–4808, 2014.

[11] M. FOLZ, *Volume growth and stochastic completeness of graphs*, Trans. Amer. Math. Soc., 366(4):2089–2119, 2014.

[12] A. GRIGOR'YAN, X. HUANG, AND J. MASAMUNE' *On stochastic completeness of jump processes*, Math. Z., 271(3–4):1211–1239, 2012.

[13] A. GRIGOR'YAN, *Analytic and geometric background of re- currence and non-explosion of the Brownian motion on Riemannian manifolds*, Bull. Amer. Math. Soc. (N.S.), 36(2):135–249, 1999.

[14] S. GOLÉNIA AND CH. SCHUMACHER, *The problem of deficiency indices for discrete Schrödinger operators on locally finite graphs*, J. Math. Phys., 52(6), 2011.

[15] S. GOLÉNIA AND CH. SCHUMACHER, *Comment on 'The problem of deficiency indices for discrete Schrödinger operators on locally finite graphs'*, J. Math. Phys., 54(6), 2013.

[16] S. HAESELER, M. KELLER, AND R. K. WOJCIECHOWSKI, *Volume growth and bounds for the essential spectrum for Dirichlet forms*, J. Lond. Math. Soc. (2), 88(3):883–898, 2013.

[17] B. HUA AND F. MÜNCH, *Ricci curvature on birth-death processes*, arXiv:1712.01494, 2017.

[18] X. HUANG, *A note on the volume growth criterion for stochastic completeness of weighted graphs*, Potential Anal., 40(2):117–142, 2014.

[19] M. KELLER, D. LENZ, AND R. K. WOJCIECHOWSKI, *Volume growth, spectrum and stochastic completeness of infinite graphs*, Math. Z., 274(3–4):905–932, 2013.

[20] Y. LIN, L. LU, AND S.-T. YAU, *Ricci curvature of graphs*, Tohoku Math. J. (2), 63(4):605–627, 2011.

[21] S. LIU, F. MÜNCH, AND N. PEYERIMHOFF, *Bakry-Émery curvature and diameter bounds on graphs*, Calc. Var. Partial Differential Equations, 57(2):Art. 67, 9, 2018.

[22] F. MÜNCH AND R. K. WOJCIECHOWSKI, *Ollivier Ricci curvature for general graph Laplacians: Heat equation, Laplacian comparison, non-explosion and diameter bounds*, arXiv:1712.00875, 2017.

[23] Y. OLLIVIER, *Ricci curvature of Markov chains on metric spaces*, J. Funct. Anal., 256(3):810–864, 2009.

[24] A. WEBER, *Analysis of the physical Laplacian and the heat flow on a locally finite graph*, J. Math. Anal. Appl., 370(1):146–158, 2010.

[25] R. K. WOJCIECHOWSKI, *Stochastically incomplete manifolds and graphs*, in Random walks, boundaries and spectra, Progr. Probab. 64, 163–179, Birkhäuser/Springer Basel AG, Basel, 2011.

3

Gromov–Lawson Tunnels with Estimates

Józef Dodziuk

Abstract

In an appendix to an earlier paper [1] we showed how to construct tunnels of positive scalar curvature and of arbitrarily small length and volume connecting points in a *three-dimensional* manifold of *constant sectional curvature*. Here we generalize the construction to arbitrary dimensions and require only positivity of the scalar curvature.

3.1 Introduction

Suppose that X is a Riemannian manifold of positive scalar curvature of dimension $n \geq 3$. Gromov and Lawson [3, 4] and independently Schoen and Yau [9] proved that if a manifold M obtained from X by a surgery on a sphere of co-dimension greater than or equal to 3, then M carries a metric of positive scalar curvature. Obstructions to the existence of such metrics had been known previously cf. [5, 6, 7]. The breakthrough provided by Gromov–Lawson and Schoen–Yau constructions led to a great deal of understanding of which smooth manifolds carry metrics of positive scalar curvature.

More recently, Rosenberg and Stolz [8] re-visited the construction of Gromov and Lawson to correct a mistake in the proof. Our reason for doing this again is to obtain an estimate on the size of the part of the manifold where modification takes place. Before making this statement precise, we need to introduce some terminology and notation. As Gromov and Lawson point out, the case of connected sums is the most important since it generalizes easily to surgeries on spheres of positive dimensions. Thus we will consider only this case.

Suppose that $p_1 \neq p_2$ are two points of (possibly disconnected) X. Let $B(p_i, \delta) \subset X$, $i = 1, 2$, be disjoint balls in X with the radii δ smaller than the

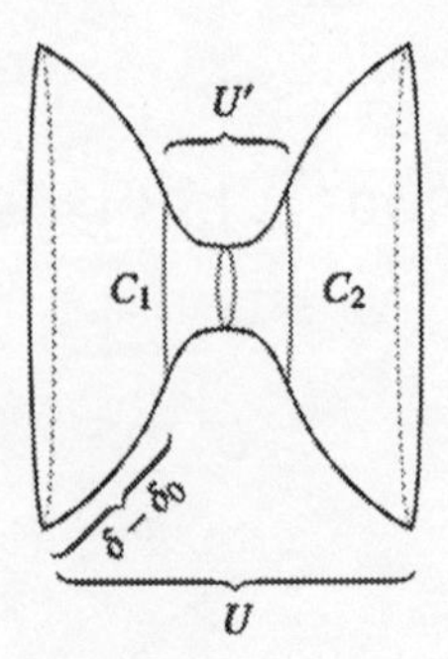

Figure 3.1 The tunnel

injectivity radii at p_1 and p_2. The connected sum M is obtained by removing the two balls from X and gluing in a cylindrical region U diffeormorphic to $S^{n-1} \times [0, 1]$. Thus

$$M = X \setminus (B(p_1, \delta) \cup B(p_2, \delta)) \sqcup U.$$

We fix $\delta_0 \in (0, \delta)$ sufficiently small to be specified later. The collars $C_i = B(p_i, \delta) \setminus B(p_i, \delta_0)$ are identified with subsets of M and the Riemannian metric g to be constructed on U will agree with the original metric h of X (see Figure 3.1). We will call the set U *the tunnel* and prove the following theorem.

Theorem 3.1.1 *There exists a constant $\delta_0 \in (0, \delta)$ and a Riemannian metric g on U with positive scalar curvature such that*

$$g \mid C_i = h \mid C_i \qquad \text{for} \quad i = 1, 2 \tag{3.1.1}$$

$$\operatorname{Diam} U = O(\delta) \quad \text{and} \quad \operatorname{Vol} U = O(\delta^n). \tag{3.1.2}$$

More precisely, the set $U' = U \setminus (C_1 \cup C_2)$ satisfies $\operatorname{Diam} U' = O(\delta_0)$ *and* $\operatorname{Vol} U' = O(\delta_0^n)$.

Note that the earlier constructions in [3, 4, 9], and [8] did not give information about the size of the tunnels. In [1], we proved the theorem above for manifolds of three dimensions and constant positive sectional curvature. This was sufficient for constructing examples of sequences of manifolds of positive scalar curvature whose limits (under any reasonable notions of convergence) did not have positive generalized scalar curvature in the sense explained in [1]. The main result of this chapter removes the restriction on dimension and allows variable sectional curvature.

3.2 Outline of the Proof

In this section, we establish the notation, describe the setup, and outline the construction of tunnels. Let X be a Riemannian manifold of positive scalar curvature and $D \subset X$ a geodesic ball centred at $p \in X$ of small radius δ. Using the normal geodesic coordinates $x^1, x^2, \ldots, x^n$, the metric on D is obtained by considering $D = \{x^1 e_1 + x^2 e_2 + \cdots x^n e_n \mid \|x\| \leq \delta\}$, where $e_1, e_2, \ldots, e_n$ is an orthonormal basis of $T_p X$, and pulling back the metric of X to D via the exponential map. We set $r(x) = \|x\|$ to be the distance of x to the origin of D and figure $S^{n-1}(\rho) = \{x \in D \mid r(x) = \rho\}$. We will construct a new metric on $D \setminus \{0\}$ of positive scalar curvature. Following Gromov and Lawson [3, Section 1] we consider the Riemannian product $D \times \mathbb{R}$ and a suitable curve γ (to be described and constructed below). We then define a hypersurface $M \subset D \times \mathbb{R}$ as $M = \{(x, t) \in D \times \mathbb{R} \mid (r(x), t) \in \gamma\}$. It is useful to think of M of as a hypersurface of revolution around the t-axis of a curve in $(x^1, 0, \ldots, 0, t)$-plane. This is a correct interpretation only if the metric of D has *constant* sectional curvature, but it is a very good approximation of the true picture if the radius of D is very small. The requirements on γ are that it begins along the positive r-axis and ends as a horizontal line segment $r = r_\infty$. Thus the metric on M extends the metric of D near ∂D and finishes as the product metric of the form $S^{n-1}(r_\infty) \times \mathbb{R}$. Of course, the main requirement on γ is that the resulting hypersurface M has positive scalar curvature and that the length of γ is $O(\delta)$. Construction of γ is the main difficulty of the proof. Note that $S^{n-1}(r_\infty)$ is not a round sphere. However, Lemma 1 of [3] (quoted below) allows us to modify the metric of M near the end of the tube so that the modified metric is a product of the round sphere of radius r_∞ with an interval. This will allow us to connect two such tubes to form a tunnel of very small length and volume and of positive scalar curvature.

The following lemma of [3] describes how the small geodesic spheres in X differ from round spheres of the same radius.

Lemma 3.2.1 *The principal curvatures of the hypersurfaces $S^{n-1}(\epsilon)$ in D are each of the form $-1/\epsilon + O(\epsilon)$ for small $\epsilon > 0$. Furthermore, let g_ϵ be the induced metric on $S^{n-1}(\epsilon)$ and let $g_{0,\epsilon}$ be the standard round metric of curvature $1/\epsilon^2$. Then as $\epsilon \to 0$, $(1/\epsilon^2) g_\epsilon \to (1/\epsilon^2) g_{0,\epsilon} = g_{0,1}$ in the C^2 topology. As a matter of fact, for an appropriate choice of the norm on the space of C^2 tensors, $\|(1/\epsilon^2) g_\epsilon - g_{0,1}\| = O(\epsilon^2)$.*

The estimate of principal curvatures above and the Gauss curvature equations lead to the following expression for the scalar curvature κ at the point $(x, t) \in M$, cf. formula (3.1.2) of [3]:

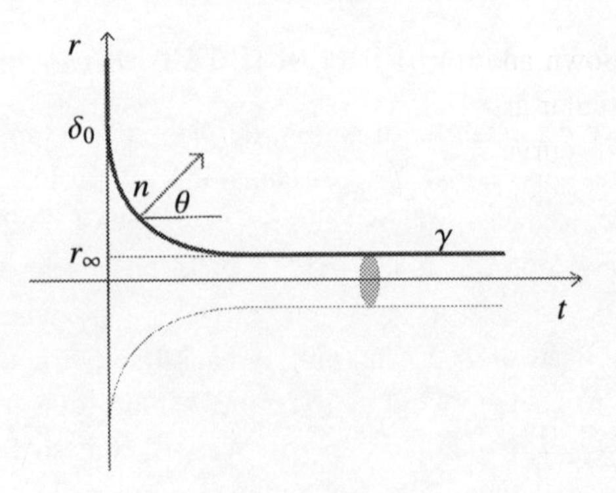

Figure 3.2 The curve γ

$$\begin{aligned}\kappa = \kappa^D &- 2\,\mathrm{Ric}^D\left(\frac{\partial}{\partial r}, \frac{\partial}{\partial r}\right)\sin^2\theta \\ &+ (n-1)(n-2)\left(\frac{1}{r^2} + O(1)\right)\sin^2\theta \\ &- (n-1)\left(\frac{1}{r} + O(r)\right)k\sin\theta\end{aligned} \tag{3.2.1}$$

where $\kappa^D(x,t) = \kappa^D(x)$ is the scalar curvature of D at x, Ric^D is the Ricci tensor of D at x, k is the geodesic curvature of γ at the point $(r(x), t) \in \gamma$, and θ is the angle between the normal to γ and the t-axis. We shall switch the order of variables r and t because we picture the t-axis as horizontal.

We write s for the arc length parameter along γ and define γ by specifying its geodesic curvature $k(s)$. Recall (cf. Theorem 6.7, [2]) how γ is determined by k. The unit tangent vector to γ and the curvature are given by

$$\frac{d\gamma}{ds} = (\sin\theta, -\cos\theta) \qquad \text{and} \qquad k = \frac{d\theta}{ds}.$$

Thus, if $\gamma(s)$ is defined for $s \le a$ and $k(s)$ is given for $s \ge a$, $\gamma(s) = (t(s), r(s))$ can be extended as follows:

$$\begin{aligned}\theta(s) &= \theta(a) + \int_a^s k(u)\,du \\ t(s) &= t(a) + \int_a^s \sin\theta(u)\,du \\ r(s) &= r(a) - \int_a^s \cos\theta(u)\,du.\end{aligned} \tag{3.2.2}$$

We remark that a segment of γ is a circular arc if and only if k is constant. During our construction, θ will increase from 0 to $\pi/2$ so that the point $\gamma(s)$

will be moving down and to the right as s increases. We will construct γ as a sequence of circular arcs, i.e., choosing $k(s)$ to be piecewise constant. This will only be a C^1 curve that will be smoothed out as discussed in Section 3.4. However, $\theta(s)$, $r(s)$, and $t(s)$ will have no discontinuities as the formulae above show. We will also show that the scalar curvature of M is positive at all points where κ is defined.

3.3 The Construction of a C^1 Curve

Begin by choosing a point $(0, \delta_0)$ in the (t, r)-plane with δ_0 small to be specified later. γ runs down the r-axis for $s \leq 0$ with $\gamma(0) = (0, \delta_0)$. As the initial segment of γ for positive s choose an arc of a circle of curvature $k = 1$, tangent to the r-axis at δ_0. We continue this arc for sufficiently small length $s_0 \leq \delta_0/2$ choosing s_0 to insure that that κ stays positive if $s \leq s_0$. Such a choice is possible by (3.2.1) since $\theta(s) = 0$ for $s \leq 0$ and κ^D has a positive lower bound on X. Let $\theta(s_0) = \theta_0$ and $\gamma(s_0) = (t_0, r_0)$. Observe that θ_0 is *positive*. The following lemma gives a sufficient condition for the scalar curvature κ of M to be positive.

Lemma 3.3.1 *For a sufficiently small δ_0 and $s \geq s_0$, κ will be positive provided*

$$\frac{\sin\theta}{4r} > k. \tag{3.3.1}$$

Proof. We rewrite the right-hand side of (3.2.1) as follows:

$$\begin{aligned}
&\frac{(n-1)(n-2)}{2r^2}\sin^2\theta + \left(\frac{(n-1)(n-2)}{2r^2} + O(1)\right)\sin^2\theta \\
&- 2(n-1)\frac{1}{r}k\sin\theta + \left((n-1)\frac{1}{r} - O(r)\right)k\sin\theta \\
&+ \kappa^D
\end{aligned}$$

Note that the term containing the Ricci tensor in (3.2.1) was absorbed in the term with $O(1)$ on the first line. $\kappa^D > 0$ by assumption and so are the second terms on the first and second lines above provided that $r \leq \delta_0$ is sufficiently small. Thus κ will be positive if

$$\frac{(n-1)(n-2)}{2r^2}\sin^2\theta - \frac{2(n-1)}{r}k\sin\theta > 0.$$

It is here that we have to make the choice of δ_0 so that $r(s)$ is sufficiently small to make the terms that we dropped positive; $r(s) \leq r_0 \leq \delta_0$. Now we cancel common factors and use the fact that $n \geq 3$ to obtain our sufficient condition:

$$\frac{\sin\theta}{4r} - k > 0. \qquad \square$$

Note that inequality (3.3.1) is, except for the constant factor in the denominator, the same as inequality (230) in [1]. It is therefore not surprising that we will carry out the construction in a very similar way to the construction in that paper. We now proceed to the main step in the construction. Suppose $\gamma(s) = (t(s), r(s))$ has been defined up to $s = a \geq s_0 > 0$. We extend it by a circular arc of constant curvature $k = \frac{\sin\theta(a)}{8r(a)}$ for length $\Delta s = r(a)/2$. Since $\frac{\sin\theta(s)}{r(s)}$ is an increasing function

$$\frac{\sin\theta(s)}{4r(s)} > \frac{\sin\theta(a)}{4r(a)} > \frac{\sin\theta(a)}{8r(a)} = k$$

on $[a, a + \Delta s]$ so that the curvature condition (3.3.1) is satisfied. Moreover, since $\Delta s = r(a)/2$, γ will not cross the t-axis. Most importantly, by (3.2.2), θ will increase by

$$\Delta\theta = k\Delta s = \frac{\sin\theta(a)}{8r(a)}\frac{r(a)}{2} = \frac{\sin\theta(a)}{16} \geq \frac{\sin\theta(s_0)}{16},$$

at least a fixed amount $\frac{\sin\theta_0}{16}$ *independent of the starting point* $\gamma(a)$.

We now proceed inductively beginning with $a = s_0$ by setting

$$s_i = s_{i-1} + \Delta s_i, \qquad \Delta s_i = \frac{r_{i-1}}{2}, \qquad k_i = \frac{\sin\theta_{i-1}}{8r_{i-1}}, \tag{3.3.2}$$

where r_j and θ_j denote $r(s_j)$ and $\theta(s_j)$, respectively. At every step, since $\theta(w)$ is increasing, the change in the angle $\Delta\theta_j = \theta_j - \theta_{j-1}$ is at least $\frac{\sin\theta_0}{16}$ so that

$$\theta_i \geq \theta_0 + i\frac{\sin\theta(s_0)}{16}. \tag{3.3.3}$$

We are trying to construct a curve γ for which the angle θ increases to $\pi/2$ when the tangent to γ becomes horizontal. In any case, for the argument above we need $\sin\theta$ to be increasing which will be the case only if $\theta \leq \pi/2$. Since the right-hand side of (3.3.3) becomes arbitrarily large, we can stop the construction if $\theta_{i-1} < \pi/2 \leq \theta_i$ at the value of s determined by $\theta(s) = \pi/2$. This would produce a curve with a 'full bend' but *without an estimate of its length.* Up to now our argument is a variation of [4, pp. 225–226]. To obtain a curve of controlled length, we break the induction off when the angle θ reaches the value $\overline{\theta}$ to be specified later but sufficiently close to $\pi/2$. We then complete the construction with a *single* circular arc. Thus define m so that $\theta_{m-1} < \overline{\theta} \leq \theta_m$ and re-define s_m so that $\theta(s_m) = \theta_m = \overline{\theta}$.

The lemma below will allow us to estimate the length s_m.

Lemma 3.3.2 *There exists a positive constant $C < 1$ that depends only on $\overline{\theta}$ such that for all i, $0 \leq i \leq m$,*

$$\frac{r_i}{r_{i-1}} \leq C.$$

Proof. We compute r_i explicitly using (3.2.2) and (3.3.2)

$$r_i = r_{i-1} - \int_{s_{i-1}}^{s_i} \cos\theta(u)\, du = r_{i-1} - \Delta s_i \cos\mu$$

for an angle $\mu \in [\theta_{i-1}, \theta_i]$. Since $\Delta s_i = r_{i-1}/2$ for $i \leq m-1$, $\Delta s_m \leq r_{m-1}/2$ by the definition of s_m, and $\theta_i \leq \overline{\theta}$

$$\frac{r_i}{r_{i-1}} = 1 - \frac{1}{2}\cos\mu \leq 1 - \frac{\cos\overline{\theta}}{2}$$

i.e., the lemma holds with $C = 1 - \frac{\cos\overline{\theta}}{2}$. □

From the lemma above using (3.3.2) and the inequality $\Delta s_m \leq r_{m-1}/2$

$$\frac{\Delta s_i}{\Delta s_{i-1}} \leq \frac{r_i}{r_{i-1}} \leq C.$$

Therefore

$$\begin{aligned} s_m &= s_0 + \Delta s_1 + \cdots \Delta s_m \\ &\leq s_0 + \frac{r_0}{2}\left(1 + C + \cdots + C^{m-1}\right) \\ &\leq s_0 + \frac{r_0}{2}\frac{1}{1-C} \end{aligned} \tag{3.3.4}$$

which is $O(\delta_0)$.

So now γ is defined on $[0, s_m]$ with $\theta(s_m) = \overline{\theta}$. We show that we can achieve the 'full bend' to $\theta = \pi/2$ by extending γ with a single circular arc. As above we need to define $k_{m+1} > 0$ and $s_{m+1} = s_m + \Delta s_{m+1}$. We continue the added circular segment until $\theta(s_{m+1}) = \theta_{m+1} = \pi/2$. By (3.2.2)

$$\begin{aligned} r_{m+1} = r(s_{m+1}) = r_m &- \int_{s_m}^{s_{m+1}} \cos\theta(u)\, du \\ &= r_m - \int_{s_m}^{s_{m+1}} \cos(s_m + k_{m+1}(u - s_m))\, du \\ &= r_m - \frac{1}{k_{m+1}}(\sin\theta_{m+1} - \sin\theta_m) \\ &= r_m - \frac{1}{k_{m+1}}\left(1 - \sin\overline{\theta}\right). \end{aligned}$$

So, for $r_{m+1} > 0$ we must have

$$k_{m+1} r_m > 1 - \sin \overline{\theta}.$$

On the other hand, condition (3.3.1) will be satisfied if

$$k_{m+1} r_m < \frac{\sin \overline{\theta}}{4}.$$

Thus we should choose k_{m+1} so that

$$1 - \sin \overline{\theta} < k_{m+1} r_m < \frac{\sin \overline{\theta}}{4}.$$

If $\sin \overline{\theta} > 4/5$ then $1 - \sin \overline{\theta} < \frac{\sin \overline{\theta}}{4}$ and we can choose k_{m+1} to satisfy both inequalities above. We fix $\overline{\theta} \in (\sin^{-1}(4/5), \pi/2)$. Note that for such a choice

$$\Delta s_{m+1} = \frac{1}{k_{m+1}} \left(\frac{\pi}{2} - \overline{\theta} \right)^{-1} < r_m (1 - \sin \overline{\theta})^{-1} \left(\frac{\pi}{2} - \overline{\theta} \right)$$

so that, for a fixed $\overline{\theta}$, $\Delta s_{m+1} = O(r_m) = O(\delta_0)$. Combining this with (3.3.4) we see that $s_{m+1} = O(\delta_0)$.

3.4 Smoothing

Let us recapitulate the result of our construction so far. $k(s)$ is a piecewise constant function, $k|(s_i, s_{i+1}] = k_{i+1}$ for $i = 1, 2, \ldots, m$ and $k_i < k_{i+1}$ for $i = 1, 2, \ldots, m-1$. The resulting curve γ is C^1 and piecewise C^∞. It will have to be smoothed out. Before smoothing extend γ to the interval $[0, s_{m+1} + 2\delta_0]$ by a horizontal line segment, i.e., by setting $k = k_{m+2} = 0$ on $[s_{m+1}, s_{m+1} + 2\delta_0]$. We set $S = s_{m+1} + 2\delta_0$ and smooth the function $k(s)$ on $[0, S]$.

The smoothing procedure is elementary and is described in some detail in the appendix to [1]. We recall it briefly. Observe first that $k(s)$ can be approximated with arbitrarily high accuracy in L^1 by C^∞ functions $\overline{k}(s)$ as, for example, pictured in Figure 3.3. We can construct a family of functions $k_\eta(s) = \overline{k}(s)$ for small positive η so that on each of the intervals (s_i, s_{i+1}) of continuity $k(s)$ and $\overline{k}(s)$ differ only near one or both end-points, $\overline{k}(s) \leq k(s)$ on $[0, s_{m+1}]$, and $\overline{k}(s)$ drops rapidly from $\overline{k}(s_{m+1}) = k_{m+1}$ to zero in $[s_{m+1}, S]$. We can achieve this together with the requirement that

$$\int_0^S \overline{k}(s)\, ds = \frac{\pi}{2}.$$

Since $k_\eta(s)$ converges to $k(s)$ in L^1, the components $t_\eta(s)$, $r_\eta(s)$, and the normal angle $\theta_\eta(s)$ of the curve γ_η determined by k_η will converge uniformly to

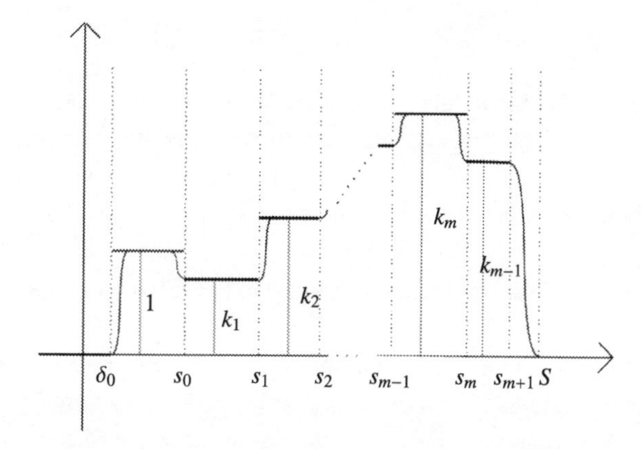

Figure 3.3 Smoothing of the curvature function

$t(s)$, $r(s)$, and $\theta(s)$, respectively. This is sufficient to conclude that for small η, the curve $\overline{\gamma} = \gamma_\eta$ will satisfy condition (3.3.1) and will give rise to a hypersurface of positive scalar curvature. For details of the smoothing, we refer to the appendix of [1]. We change the notation and from now on denote the smooth curve by γ.

By choosing η above sufficiently small, we can assume that γ is a horizontal line segment $r(s) = r(s_{m+1} + \delta_0) = r_\infty$ for $s \in [s_{m+1} + \delta_0, S]$ so that M is isometric to $S^{n-1}(r_\infty) \times [s_{m+1} + \delta_0, S]$ near the end of the tube.

The construction of γ and the resulting neck is now complete, but our aim is to construct a tunnel by connecting two tubes. For that we will need to modify the metric of M near its end. We first observe that the only place where the ambient manifold enters the construction is the choice of s_0 in the beginning of Section 3.3. This in turn is dictated by the bounds of the scalar curvature and the Ricci curvature on D. The same choice can be made for both discs that we are trying to connect. From then on the construction takes place entirely in the Euclidean (t, r)-plane and making the same choices as we go along we obtain equal values of r_∞ for both tubes.

We now modify the metric near the end of the tube to make it round. Let $a = t(s_{m+1} + \delta_0)$, $b = t(S)$, and $\epsilon = r_\infty$. The induced metric on the end of the tube $\{(x, t) \in M \mid a \leq t \leq b\}$ is $h_0 = g_\epsilon + dt^2$, where g_ϵ is the induced metric on $S^{n-1}(\epsilon)$. Recall that $\epsilon \leq \delta_0$. Let $h_1 = \epsilon^2 g_{0,1} + dt^2$ where g_ϵ and $g_{0,1}$ have the same meaning as in Lemma 3.2.1. Let $\phi(t) = \psi((t - a)/\delta_0$ where $\psi(u)$ is a smooth function on $[0, 1]$ vanishing near zero, increasing to 1 at $u = 3/4$ and equal to 1 for $u > 3/4$. Define the new metric h for $t \in [a, b]$ as

$$h(x,t) = g_\epsilon(x,t) + \phi(t)\left(\epsilon^2 g_{0,1} - g_\epsilon\right) + dt^2.$$

For t near a $h = h_0$, the induced metric on M, while for t near b, $h = h_1$, the round tube metric. The first and second derivatives of ϕ are of order $O(\delta_0^{-1})$ and $O(\delta_0^{-2})$, respectively. We consider h_0, h_1, and h as tensors on the product of the standard sphere S^{n-1} with the interval $[a,b]$ and see that

$$h(x,t) - h_0(x,t) = \phi(t)\left(\epsilon^2 g_{0,1}(x) - g_\epsilon(x)\right) = \phi(t)\epsilon^2\left(g_{0,1} - \frac{1}{\epsilon^2}g_\epsilon\right).$$

It follows from Lemma 3.2.1 that the metric h_0 has positive (and very large) scalar curvature. Moreover, since $\epsilon = r_\infty < \delta_0$, Lemma 3.2.1 and the bounds on derivatives of $\phi(t)$ show that all second-order derivatives of $h(x,t) - h_0(x,t)$ are of order $O(\delta_0^2)$. It follows that the scalar curvature of h is positive provided δ_0 is chosen sufficiently small. Our construction of the tunnel is now complete.

Clearly, the metric constructed on U has positive scalar curvature and agrees with the metric of X near the boundary. The estimates in (3.1.2) follows since the length of the constructed tube is $O(\delta_0)$ and each cross section $t = \text{const}$ is very close to a round sphere of radius $r(t) < \delta_0$. The proof of the theorem is now complete.

3.5 Tunnels of Prescribed Length

In view of possible applications we state the following consequence of the construction.

Proposition 3.5.1 *Let $p_1 \neq p_2 \in X$ be two points in a manifold X of positive scalar curvature. Then for every $L > 0$ there exists $\delta > 0$ and a Riemannian metric of positive scalar curvature on the tunnel U as in Figure 3.1 such that*

(a) *The new metric agrees with the original metric on the collars C_1 and C_2.*
(b) *The distance* $\operatorname{dist}(C_1, C_2) = L$ *and the diameter* $\operatorname{Diam}(U) = O(L)$.
(c) $\operatorname{Vol}(U) = O(L\delta^{n-1})$.
(d) *For every continuous curve α connecting the two components of the boundary of U, the tubular neighborhood of α of radius $2\pi\delta$ contains U, i.e.,*

$$\{x \in M \mid \operatorname{dist}(x,\alpha) \le 2\pi\delta\} \supset U.$$

Moreover, δ can be chosen arbitrarily small.

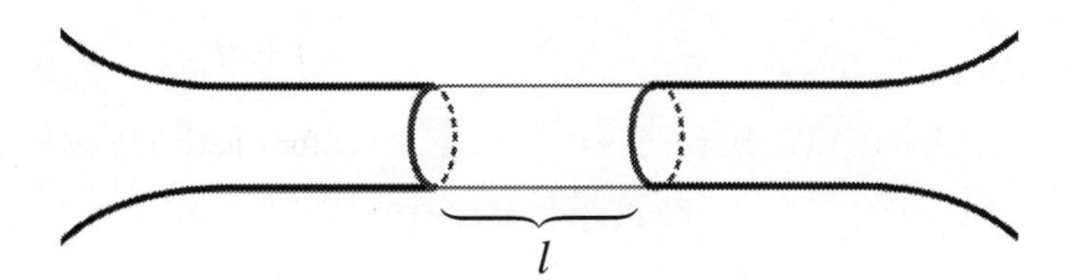

Figure 3.4 Telescoping the tunnel

Proof. We begin by choosing δ sufficiently small so that it is smaller than the injectivity radii at p_1 and p_2 and apply the theorem above to get a tunnel of diameter smaller than $L/2$. The central portion of the tunnel is isometric to the product of a round sphere of radius $r_\infty < \delta$ and an interval of length smaller than $L/2$. We simply replace this interval by one of appropriate length l to make the distance between the collars equal to L exactly (cf. Figure 3.4). Statements (c) and (d) follow since each cross section $t =$ const is very close to a round sphere of radius $r(t) < \delta$. □

References

[1] J. Basilio, J. Dodziuk and C. Sormani. Sewing Riemannian manifolds with positive scalar curvature. *J. Geom. Anal.*, 28(4):3553–3602, 2018.

[2] Gray, Alfred. *Modern differential geometry of curves and surfaces with Mathematica*. Second edition. CRC Press, Boca Raton, FL. pp. xxiv+1053, 1998.

[3] Mikhael Gromov and H. Blaine Lawson. The classification of simply connected manifolds of positive scalar curvature. *Ann. of Math.*, 111(3):423–434, 1980.

[4] Mikhael Gromov and H. Blaine Lawson, Jr. Spin and scalar curvature in the presence of a fundamental group. I. *Ann. of Math.*, 111(2):209–230, 1980.

[5] Nigel Hitchin. Harmonic spinors. *Advances in Math.*, 14:1–55, 1974.

[6] André Lichnerowicz. Laplacien sur une variété riemannienne et spineurs. *Atti Accad. Naz. Lincei Rend. Cl. Sci. Fis. Mat. Nat. (8)*, 33:187–191, 1962.

[7] André Lichnerowicz. Spineurs harmoniques. *C. R. Acad. Sci. Paris*, 257:7–9, 1963.

[8] Jonathan Rosenberg and Stephen Stolz. Metrics of positive scalar curvature and connections with surgery. In Andrew Ranicki Sylvain Cappell and Jonathan Rosenberg, editors, *Surveys on Surgery Theory*, number 149 in Annals of Mathematics Studies 2. Princeton University Press, 2001.

[9] R. Schoen and S. T. Yau. On the structure of manifolds with positive scalar curvature. *Manuscripta Math.*, 28(1–3):159–183, 1979.

4

Norm Convergence of the Resolvent for Wild Perturbations

Colette Anné and Olaf Post

Abstract

We present here recent progress in the convergence of the resolvent of Laplace operators under wild perturbations. In particular, we show convergence in *norm* in a generalized sense. We focus here on the excision of many small balls in a complete Riemannian manifold with bounded geometry.

4.1 Introduction

Rauch–Taylor's Contribution on Wild Perturbations What kind of convergence can we expect for the Laplace operator under wild perturbations such as removing many small holes or adding many thin handles? Such questions received already quite a lot of answers, following the seminal work of Rauch and Taylor [RT75]. We present here results from [AP18] and focus on convergence of the resolvents in *operator norm*. As the underlying spaces vary with the convergence parameter, we apply an abstract convergence result of the second author [P12] expressed in terms of quadratic forms acting in different Hilbert spaces (see Section 4.2).

The expression 'wild perturbation' goes back to [RT75]. Let us first recall the original result of Rauch and Taylor concerning the excision of small obstacles and convergence of the corresponding Dirichlet Laplacians. A typical result of their paper is as follows: let $\Omega \subset \mathbb{R}^m$ be an open and bounded set having some mild regularity, namely

$$H_0^1(\Omega) = \{u \in H^1(\mathbb{R}^m), \ \operatorname{supp} u \subset \overline{\Omega}\}.$$

Let K be a compact subset of Ω. We assume that $\Omega_n \to \Omega \setminus K$ as $n \to \infty$ *metrically*, i.e., every compact subset of $\Omega \setminus K$ is eventually in Ω_n, and every compact subset outside $\overline{\Omega \setminus K}$ is eventually outside $\overline{\Omega}_n$.

Let Δ_Ω and Δ_{Ω_n} be the (non-negative) Laplacians on Ω and Ω_n with Dirichlet boundary condition, respectively. Moreover, let $J_n f = \mathbb{1}_{\Omega \cap \Omega_n} f$ be the restriction of $f \in L_2(\Omega)$ onto $\Omega \cap \Omega_n$ extended by 0 on $\Omega_n \setminus \Omega$. Then $J_n^* u = \mathbb{1}_{\Omega \cap \Omega_n} u$ extended by 0 on $\Omega \setminus \Omega_n$.

Theorem 4.1.1 ([RT75, Theorem 2.3]) *If K has capacity zero then for any real-valued continuous and bounded function Φ and any $f \in L_2(\Omega)$ we have $J_n^* \Phi(\Delta_{\Omega_n}) J_n f \to \Phi(\Delta_\Omega) f$ in $L_2(\mathbb{R}^m)$ as $n \to \infty$.*

For a characterization of a set to have capacity zero, we refer, e.g., to [RT75, Lem. 2.1]. An example of a set of capacity zero is a finite set of points or more generally a subset of co-dimension 2.

We can think of this result as a (generalized) *strong* resolvent convergence (choose $\Phi(\lambda) = (1 + \lambda)^{-1}$, recall our convention $\Delta \geq 0$). Strong resolvent convergence implies the convergence of the discrete spectrum, as the limit spectrum cannot suddenly expand, but it can shrink suddenly in the limit (see the discussion after Theorem VIII.24 in [RS80]). This is probably the main disadvantage of *strong* resolvent convergence compared to *norm* resolvent convergence from a spectral viewpoint. Note that the sudden shrinkage leads to the so-called spectral pollution, i.e., spectral values in the approximation, which do not converge to a spectral value in the limit problem. The opposite effect is called *spectral exactness*, and holds for norm resolvent convergence in general (see [Bög17] and references therein for details).

Wild Perturbations and Norm Resolvent Convergence One of our main question in this chapter is as follows:

Question Can we show stronger convergence results for wild perturbations such as *norm* resolvent convergence and results which work also (without much modifications) for *unbounded* domains or manifolds?

As wild perturbation we focus here on the excision of many small balls as obstacle from a (not necessarily compact) Riemannian manifold of bounded geometry and the Dirichlet Laplacian on the manifold without the obstacles. Further results are shrinking *Neumann* obstacles (see [AP18]). Note that our perturbation result also works quite well when neither the perturbed space X_ε nor the limit space X_0 is subset of the other. This is, e.g., the case when adding many thin handles to a manifold; we treat this question in a forthcoming publication.

Domain Perturbations and Convergence Results Domain perturbation and (spectral) convergence results have a long history. We are not trying to give an exhaustive list of references here, but just highlight a few points.

Weidmann [W84] proved the continuous dependency of eigenvalues and eigenfunctions of elliptic differential operators and he also developed a general (strong resolvent) convergence theory for sequences of operators acting in different Hilbert spaces (which can be embedded in a larger common Hilbert space).

The asymptotic behaviour of Neumann eigenvalues was studied for a single hole for bounded domains or compact manifolds in [Oz83, Hem06, LdC12] and the Dirichlet eigenvalues in [CF78, Cou95] where we find precise estimates; it also applies to the ε-neighbourhood of compact subset (see also [CF88] for the calculation of the first correction term).

Daners [Dan03] considers the *norm* convergence of resolvents of Dirichlet Laplacians for perturbations of Euclidean *bounded* domains (or at least those with compact resolvent); the norm convergence follows from the strong one under the assumption of compactness of the limit resolvent (see also [Dan08] for a survey and the references therein). Our approach is more general as we do not assume a priori that the perturbed and unperturbed domains are embedded in a common space as in [Dan03, Dan08]. Moreover, we obtain explicit error estimates in terms of δ_ε. For an older survey about strong resolvent convergence and perturbations of Euclidean domains, we refer to [Hen94].

Finally, the work of Rauch and Taylor [RT75] is inspired with their *crushed ice problem* the study of *homogenization*. There is a critical density of balls removed under which the Dirichlet Laplacian converges to the original Laplacian with a shift in energy. Below that critical parameter, the limit is the original Laplacian itself; above, there are regions that 'become solid' in the sense that the limit Laplacian fulfils Dirichlet conditions there. The homogenization problem is usually also treated showing *strong* resolvent convergence (see [BN98, Bal88]) using Γ-convergence (see [DM93]). More recent works can be found in [Kh09] or [Kh13] and references therein. For a similar approach as in this chapter using the above-mentioned generalized *norm* resolvent convergence in the homogenization case, we refer to [KhP18] and the references cited therein. For an approach using the already shown strong resolvent convergence to upgrade to norm resolvent convergence (similarly as in [Dan03, Dan08], but even for general unbounded domains), we refer to [DCR17].

The notion of Γ- or Mosco convergence is another way of defining a convergence of quadratic forms acting in different Hilbert spaces (see, e.g., [KS03, Sec. 2]): note that this convergence is more or less equivalent with some sort of generalization of *strong* resolvent convergence; hence, our results are stronger.

4.2 Generalized Norm Resolvent Convergence

To achieve this goal we apply a rather general result of the second author [P06] (see also the monograph [P12]).

For each $\varepsilon \geq 0$, let $\mathscr{H}_\varepsilon$ be a separable Hilbert space together with a closed quadratic form $\mathfrak{q}_\varepsilon$ and domain $\mathscr{H}_\varepsilon^1$. We denote by $\Delta_\varepsilon \geq 0$ the corresponding self-adjoint operator. We define the generalized Sobolev spaces $\mathscr{H}_\varepsilon^k$ as $\mathscr{D}(\Delta^{k/2})$ together with the norms $\|u\|_k = \|(\Delta_\varepsilon + 1)^{k/2} f\|$, and choose the completion of $\mathscr{H}_\varepsilon$ with respect to the norm $\|\cdot\|_k$ if $k < 0$. Then all spaces $(\mathscr{H}_\varepsilon^k, \|\cdot\|_k)$ are complete. Note that

$$\|u\|_1^2 = \|u\|_{\mathscr{H}_\varepsilon}^2 + \mathfrak{q}_\varepsilon(u).$$

We suppose that there are transplantation or identification operators at the level of the Hilbert spaces and also at the level of the quadratic forms (we suppress here and in the following the dependency of ε in the notation):

$$J : \mathscr{H}_0, \to \mathscr{H}_\varepsilon \qquad J_1 : \mathscr{H}_0^1 \to \mathscr{H}_\varepsilon^1$$
$$J' : \mathscr{H}_\varepsilon \to \mathscr{H}_0 \qquad J_1' : \mathscr{H}_\varepsilon^1 \to \mathscr{H}_0^1.$$

We assume that these operators are bounded and need some compatibility, also called *δ_ε-quasi-unitary equivalence* of $\mathfrak{q}_\varepsilon$ and $\mathfrak{q}_0$, if $\delta_\varepsilon > 0$ and if

$$|\langle J'u, f\rangle - \langle u, Jf\rangle| \leq \delta_\varepsilon \|f\|_1 \, \|u\|_1, \tag{4.2.1}$$

$$\|f - J'Jf\| \leq \delta_\varepsilon \|f\|_1 \quad \text{and} \quad \|u - JJ'u\| \leq \delta_\varepsilon \|u\|_1, \tag{4.2.2}$$

$$\|(J_1 - J)f\| \leq \delta_\varepsilon \|f\|_1 \quad \text{and} \quad \|(J_1' - J')u\| \leq \delta_\varepsilon \|u\|_1, \tag{4.2.3}$$

$$|\mathfrak{q}_\varepsilon(J_1 f, u) - \mathfrak{q}_0(f, J_1' u)| \leq \delta_\varepsilon \|f\|_2 \, \|u\|_1 \tag{4.2.4}$$

for all f and u in the respective spaces. We have adopted the definition of quasi-unitary equivalence already to the situation here where the quadratic forms are estimated with respect to the *form* norm $\|\cdot\|_1$ on the perturbed space $\mathscr{H}_\varepsilon^1$ and the *graph* norm $\|\cdot\|_2$ on the unperturbed space $\mathscr{H}_0^2 = \operatorname{dom} \Delta_0$.

We have the following notion of *generalized norm resolvent convergence*:

Theorem 4.2.1 ([P12, Proposition 4.4.15]) *If the quadratic forms $\mathfrak{q}_\varepsilon$ and $\mathfrak{q}_0$ are δ_ε-quasi-unitary equivalent, then the resolvents $R_\varepsilon := (\Delta_\varepsilon + 1)^{-1}$ of the operators Δ_ε associated with $\mathfrak{q}_\varepsilon$ satisfy*

$$\|R_\varepsilon J - J R_0\| \leq 4\delta_\varepsilon.$$

Moreover, if $\delta_\varepsilon \to 0$, then we also have the convergence of (suitable) functions of the operators in norm, of the spectrum, and of the eigenfunctions also in energy norm.

4.3 Removing Many Small Balls: The Fading Case

Let (X, g) be a complete connected Riemannian manifold of dimension m with natural energy form defined by $\mathfrak{q}(f) = \int_X |df|^2 \, \mathrm{d}\,\mathrm{vol}_g$ for $f \in C_0^\infty(X)$. This form is closable (because the manifold is complete) and defines a non-negative self-adjoint operator $\Delta = \Delta_{(X,g)}$ (see, e.g., [RS80, Theorem VIII.15] for details), given in local coordinates $(y_1, \dots, y_m)$ by

$$\Delta(f) = - \sum_{1 \le i,j \le m} \frac{1}{\rho} \partial_{x_i} (\rho g^{ij} \partial_{x_j} f),$$

where (g^{ij}) is the inverse matrix of the metric tensor (g_{ij}), and where the Riemannian measure $\mathrm{d}\,\mathrm{vol}_g$ is locally given by $\rho \, \mathrm{d}y_1 \otimes \cdots \otimes \mathrm{d}y_m$ with $g_{ij} = g(\partial_{y_i}, \partial_{y_j})$.

As an example of application of the above generalized norm resolvent convergence, let us look at the problem of removing many small balls: Assume that (X, g) is a complete Riemannian manifold of dimension $m \ge 2$. Consider the following perturbation.

For any $\varepsilon > 0$, let $(x_j)_{j \in \mathscr{J}_\varepsilon}$ be a family of points in X such that $d(x_j, x_k) \ge 2\eta_\varepsilon$ for some $\eta_\varepsilon \gg \varepsilon$ (typically, we will choose $(\eta_\varepsilon = \varepsilon^\alpha$ for some $0 < \alpha < 1)$). Note that we do not assume any relation between points x_j for $j \in \mathscr{J}_\varepsilon$ for *different* values of ε.

We set

$$X_\varepsilon = X \setminus B_\varepsilon \qquad \text{with} \qquad B_\varepsilon = \bigcup_{j \in \mathscr{J}_\varepsilon} B(x_j, \varepsilon). \tag{4.3.1}$$

In this situation, let

$$\mathscr{H}_0 = L_2(X, g), \qquad \mathscr{H}_\varepsilon = L_2(X_\varepsilon, g)$$
$$\mathscr{H}_0^1 = H^1(X, g), \qquad \mathscr{H}_\varepsilon^1 = H_0^1(X_\varepsilon, g)$$

with the transplantation operators

$$J \colon \mathscr{H}_0 \to \mathscr{H}_\varepsilon \qquad J_1 \colon \mathscr{H}^1 \to \mathscr{H}_\varepsilon^1, \qquad Jf = f\restriction_{X_\varepsilon}, \quad J_1 f = \chi_\varepsilon f$$
$$J' \colon \mathscr{H}_\varepsilon \to \mathscr{H}_0 \qquad J_1' \colon \mathscr{H}_\varepsilon^1 \to \mathscr{H}_0^1 \qquad J'u = \overline{u}, \quad J_1'u = \overline{u},$$

where $\overline{u}$ is the extension of $u \colon X_\varepsilon \to \mathbb{C}$ onto X by 0, and where χ_ε is a cut-off function on X given by $\chi_\varepsilon(x) = \widetilde{\chi}(d(x, x_j))$ if $d(x, x_j) \in [0, \varepsilon^+)$, $\chi_\varepsilon(x) = 0$ if $d(x, x_j) \in [0, \varepsilon]$ and $\chi\varepsilon(x) = 1$ otherwise. Here,

$$\varepsilon \ll \varepsilon^+ \ll \eta_\varepsilon$$

and $\widetilde{\chi}_\varepsilon$ is given by $\widetilde{\chi}_\varepsilon(r) = 0$ if $r \in [0, \varepsilon]$ and

$$\widetilde{\chi}_\varepsilon(r) = \begin{cases} \dfrac{1/r^{m-2} - 1/\varepsilon^{m-2}}{1/(\varepsilon^+)^{(m-2)} - 1/\varepsilon^{m-2}}, & \text{for } m \geq 3 \\ \dfrac{\log(r/\varepsilon)}{\log(\varepsilon^+/\varepsilon)}, & \text{for } m = 2 \end{cases}$$

for $r \in (\varepsilon, \varepsilon^+)$. Note that $\widetilde{\chi}_\varepsilon$ and χ_ε are both Lipschitz continuous; hence, $\chi_\varepsilon f$ is in $H_0^1(X_\varepsilon)$ if $f \in H^1(X)$. In particular, J_1 is well defined.

The most difficult part to check in the assumptions of Theorem 4.2.1 is a control of assumption (4.2.4) on the forms (with $k = 2$). For this (and also the other assumptions (4.2.1)–(4.2.3)) we need the additional assumption of *bounded geometry* on the manifold.

Definition 4.3.1 The manifold (X, g) has *bounded geometry* if there exist $i_0 > 0$ and k_0 such that the injectivity radius and the Ricci curvature of X satisfy

$$\forall x \in X: \ \operatorname{Inj}(x) \geq i_0, \quad \operatorname{Ric}(x) \geq k_0\, g.$$

We assume throughout this chapter that (X, g) has bounded geometry. We know, for instance by the book of Hebey [Heb99], that these hypotheses assure the existence of a uniform harmonic radius r_0, i.e., a radius independent of the point such that inside the ball of this radius there exist *harmonic coordinates*. These coordinate assure a uniform control of the metric with respect to the Euclidean one: there exists $K > 0$ such that for all $x_0 \in X$ there are harmonic coordinates $(y_1, \ldots, y_m)$ in $B(x_0, r_0)$ such that

$$\forall x \in B(x_0, r_0): \ K^{-1}\delta_{ij} \leq g_x(\partial_{y_i}, \partial_{y_j}) \leq K\delta_{ij} \tag{4.3.2}$$

(see, e.g., [Heb99, Theorem 1.2]). These coordinates also assure that $C_0^\infty(X)$ is dense in $H^2(X, g) = \mathscr{H}_0^2$. As a consequence the Laplacian defined on $C_0^\infty(X)$ is essentially self-adjoint (see, e.g., [Heb99, Prop. 3.3].[1]

Let us now describe the first result in this context. By 'fading' we mean that there are not enough balls close to each other, so that one has no effect, i.e., the limit operator is the original Laplacian on X.

Theorem 4.3.2 (many small balls fading) *Let X be a complete Riemannian manifold with bounded geometry and X_ε as in* (4.3.1). *Moreover, let the centres of balls be separated by $2\eta_\varepsilon$ with*

$$\eta_\varepsilon = \begin{cases} \varepsilon^\alpha, & 0 < \alpha < \frac{m-2}{m} \text{ if } m \geq 3 \text{ and} \\ |\log \varepsilon|^{-\alpha}, & 0 < \alpha < \frac{1}{2} \quad \text{if } m = 2. \end{cases}$$

[1] The identification $H^2(X, g) = \mathscr{H}_0^2$ needs the Bochner–Weitzenböck formula.

Then the Laplacian with Dirichlet boundary conditions on X_ε converges in generalized norm resolvent sense to the Laplacian on X.

Remark The critical power $\varepsilon^{(m-2)/m}$ is related to the *capacity* of the obstacle (the balls of radius ε) being at a distance of order $\varepsilon^{(m-2)/2}$ away from other balls: This case needs more assumptions about the spacing of the points x_j; details about generalized norm resolvent convergence in this situation and capacity can be found in [KhP18]. In particular, the *capacity* determines about the limit behaviour of the crushed ice problem.

Proof. Let us sketch the proof of (4.2.4) for $k = 2$: For all $f \in \mathscr{H}_0^2$ and $u \in \mathscr{H}_\varepsilon^1$ we have

$$\begin{aligned}
&\left|\mathfrak{q}_0(f, J^{1\prime}u) - \mathfrak{q}_\varepsilon(J^1 f, u)\right| \\
&= \left|\langle df - d(\chi_\varepsilon f), du\rangle_{L_2(T^*B_{\varepsilon+},g)}\right| \\
&\le \left|\langle (1-\chi_\varepsilon) df, du\rangle_{L_2(T^*B_{\varepsilon+},g)}\right| + \left|\langle f\, d\chi_\varepsilon, du\rangle_{L_2(T^*B_{\varepsilon+},g)}\right| \\
&\le \big(\|df\|_{L_2(T^*B_{\varepsilon+},g)} + \|f\, d\chi_\varepsilon\|_{L_2(T^*B_{\varepsilon+},g)}\big)\|du\|_{L_2(T^*B_{\varepsilon+},g)}
\end{aligned}$$

To control $\|df\|^2_{L_2(T^*B_{\varepsilon+},g)}$ which is a sum of integrals on balls, we use the assumption of bounded geometry and (4.3.2). Hence, it suffices to control the estimate on Euclidean balls, namely

$$\forall \phi \in H^1(B_{\eta_\varepsilon}, \mathrm{eucl})\colon\ \|\phi\|_{L_2(B_{\varepsilon^+},\mathrm{eucl})} \le \tau_m\Big(\frac{\varepsilon^+}{\eta_\varepsilon}\Big)\|\phi\|_{H^1(B_{\eta_\varepsilon},\mathrm{eucl})} \tag{4.3.3}$$

where $\tau_m(r) = O(r)$ for $m \ge 3$ and $\tau_2(r) = O(r|\log r|^{1/2})$. This control, pulled back onto the balls of the manifold can be applied as well to $\phi = |df|$.

Now, for the second term, we conclude from the Hölder inequality that

$$\|f\, d\chi_\varepsilon\|^2_{L_2(T^*B_{\varepsilon+}(x),g)} \le \|f\|^2_{L_{2p}(B_{\varepsilon+}(x),g)}\|d\chi_\varepsilon\|^2_{L_{2q}(T^*B_{\varepsilon+}(x),g)}$$

for any $p \in (1,\infty)$ with $1/p + 1/q = 1$ and any $j \in \mathscr{J}_\varepsilon$.

In order to control $\|f\|_{L_{2p}(B_{\varepsilon+}(x),g)}$ we use a Sobolev embedding $H^2(B_2(0)) \hookrightarrow L_{2p}(B_1(0))$ for p small enough, rescaling gives a bad estimate in terms of ε, but this can be compensated by a rather good estimate of $\|d\chi_\varepsilon\|_{L_{2q}(T^*B_{\varepsilon+}(x),g)}$ if p is not too small. We assert that, for each dimension m, there exists good p_m, q_m, which do the job, and consequently there exists $\delta_\varepsilon = o(1)$ such that

$$\|f\, d\chi_\varepsilon\|_{L_2(T^*B_\varepsilon^+,g)} \le \delta_\varepsilon\|f\|_{\mathscr{H}_0^2}.$$

The details can be found in [AP18]. Note that $J' = J^*$, and that the remaining (non-trivial) assumptions of (4.2.2)–(4.2.3) also follow by (4.3.3) and the bounded geometry assumption. □

4.4 Removing Many Small Balls: The Solidifying Case

We also obtain results for the *solidifying* situation (named after [RT75]): here, the density of the removed balls is so high that it solidifies in the limit to an obstacle Ω_0. Again, (X, g) is a complete Riemannian manifold of dimension $m \geq 2$ and $X_\varepsilon = X \setminus B_\varepsilon$ with $B_\varepsilon = \bigcup_{j \in \mathscr{J}_\varepsilon} B(x_j, \varepsilon)$ for a set of points $(x_j)_{j \in \mathscr{J}_\varepsilon}$. We suppose now that there exists an open subset Ω_0 of X with regular boundary (see Definition 4.4.2) and that $B_\varepsilon \subset \Omega_0$. Moreover, we assume that there exists $N \in \mathbb{N}$, $\eta_\varepsilon \ggg \varepsilon$, and $\alpha_\varepsilon > 0$ such that

$$\Omega_{\alpha_\varepsilon} = \{x \in X;\ d(x, \Omega_0) < \alpha_\varepsilon\} \subset B_{\eta_\varepsilon} \tag{4.4.1}$$

$$\forall x \in X\ \forall \varepsilon > 0\colon\ \sharp\{j \in \mathscr{J}_\varepsilon;\ x \in B(x_j, \eta_\varepsilon)\} \leq N, \tag{4.4.2}$$

where $\sharp M$ is the cardinality of the set M. The first assertion assures that the family $(x_j)_{j \in \mathscr{J}_\varepsilon}$ is dense enough: at the scale η_ε it covers all Ω_0 and a bit more; it also implies that $\alpha_\varepsilon / \eta_\varepsilon$ is small or at least bounded. The second assertion assures that this cover is not too redundant. In particular, it follows from $B_\varepsilon \subset \Omega_0 \subset B_{\eta_\varepsilon}$ that $X_\varepsilon \underset{\varepsilon \to 0}{\longrightarrow} X \setminus \Omega_0 = X_0$.

We also need control of the first eigenvalue λ_ε of the Laplacian on $B_{\mathbb{R}^m}(0, \eta_\varepsilon) \setminus B_{\mathbb{R}^m}(0, \varepsilon)$ with Neumann boundary condition at $r = \eta_\varepsilon$ and Dirichlet boundary condition at $r = \varepsilon$. It is calculated in [RT75] that $\lambda_\varepsilon \geq C\varepsilon^{(m-2)}/\eta_\varepsilon^m$ (respectively $\lambda_\varepsilon \geq C/(\eta_\varepsilon^2 |\log \varepsilon|)$ for $m = 2$), where C depends only on the dimension m; this estimate carries over to balls on the manifold (by our assumption of bounded geometry).

Theorem 4.4.1 (many small balls solidifying) *In the situation just described, assume that* $\lim_{\varepsilon \to 0} \alpha_\varepsilon \lambda_\varepsilon = +\infty$, *then the Laplacian* Δ_ε *with Dirichlet boundary conditions on* $X_\varepsilon = X \setminus B_\varepsilon$ *converges in generalized norm resolvent sense to the Laplacian* Δ_0 *with Dirichlet boundary conditions on* $X \setminus \Omega_0$.

We check again the conditions of quasi-unitary equivalence in (4.2.1)–(4.2.4). We define here

$$\begin{aligned}
J\colon \mathscr{H}_0 &:= L_2(X_0, g) \longrightarrow \mathscr{H}_\varepsilon := L_2(X_\varepsilon, g), && f \mapsto \bar{f},\\
J_1\colon \mathscr{H}_0^1 &:= H_0^1(X_0, g) \longrightarrow \mathscr{H}_\varepsilon^1 := H_0^1(X_\varepsilon, g), && f \mapsto \bar{f},\\
J'\colon \mathscr{H}_\varepsilon &:= L_2(X_\varepsilon, g) \longrightarrow \mathscr{H}_0 = L_2(X_0, g), && u \mapsto u_{|X_\varepsilon},\\
J^{1\prime}\colon \mathscr{H}_\varepsilon^1 &:= H_0^1(X_\varepsilon, g) \longrightarrow \mathscr{H}_0^1 = H_0^1(X_0, g), && u \mapsto \chi_\varepsilon u,
\end{aligned}$$

where $\bar{f}$ is the extension of f by 0 onto X_ε, as $X_0 \subset X_\varepsilon$ and χ_ε is now a cut-off function depending on the distance to $\bar{\Omega}_0$. In particular, we need some control of the boundary of Ω_0.

Definition 4.4.2 We say that the open set $\Omega \subset X$ has a *regular boundary* $Y = \overline{\Omega} \setminus \Omega$ if Y is a smooth sub-manifold of X, which admits a *uniform tubular neighbourhood*, i.e., we assume that Y admits a global normal unitary vector field $\vec{N}$ (so that Y is orientable) and that there exists $r_0 > 0$ such that

$$\exp_\nu \colon Y \times [0, r_0) \to X, \qquad (y, t) \mapsto \exp_y(t\vec{N}(y)) \tag{4.4.3}$$

is a diffeomorphism.

Remark 4.4.3 This regularity assumption (together with the bounded geometry) implies that the principal curvatures of the hypersurface Y are bounded by a constant depending on $1/r_0$ and k_0 (see [HK78, Cor. 3.3.2]). But it is stronger: we also need that Y does not admit arbitrarily close points which are far away with respect to the inner distance.

References

[AP18] C. Anné and O. Post, *Wildly perturbed manifolds: Norm resolvent and spectral convergence*, arXiv:1802.01124 to appear in: J. Spectr. Theory.

[Bal88] M. Balzano, *Random relaxed Dirichlet problems*, Ann. Mat. Pura Appl. (4) **153** (1988), 133–174 (1989).

[BN98] M. Balzano and L. Notarantonio, *On the asymptotic behavior of Dirichlet problems in a Riemannian manifold less small random holes*, Rend. Sem. Mat. Univ. Padova **100** (1998), 249–282.

[Bög17] S. Bögli, *Convergence of sequences of linear operators and their spectra*, Integral Equations Operator Theory **88** (2017), 559–599.

[CF78] I. Chavel and E. A. Feldman, *Spectra of domains in compact manifolds*, J. Funct. Anal. **30** (1978), 198–222.

[CF88] ———, *Spectra of manifolds less a small domain*, Duke Math. J. **56** (1988), 339–414.

[Cou95] G. Courtois, *Spectrum of manifolds with holes*, J. Funct. Anal. **134** (1995), 194–221.

[Dan03] D. Daners, *Dirichlet problems on varying domains*, J. Differential Equations **188** (2003), 591–624.

[Dan08] ———, *Domain perturbation for linear and semi-linear boundary value problems*, Handbook of differential equations: stationary partial differential equations. Vol. VI, Handb. Differ. Equ., Elsevier/North-Holland, Amsterdam, 2008, pp. 1–81.

[DCR17] P. Dondl, K. Cherednichenko, and F. Rösler, *Norm-resolvent convergence in perforated domains*, arXiv:1706.05859 (2017).

[DM93] G. Dal Maso, *An introduction to Γ-convergence*, Progress in Nonlinear Differential Equations and their Applications, vol. 8, Birkhäuser Boston, Inc., Boston, MA, 1993.

[Heb99] E. Hebey, *Nonlinear analysis on manifolds: Sobolev spaces and inequalities*, Courant Lecture Notes in Mathematics, vol. 5, New York

University Courant Institute of Mathematical Sciences, New York, 1999.

[Hem06] R. Hempel, *On the lowest eigenvalue of the Laplacian with Neumann boundary condition at a small obstacle*, J. Comput. Appl. Math. **194** (2006), 54–74.

[Hen94] A. Henrot, *Continuity with respect to the domain for the Laplacian: a survey*, Control Cybernet. **23** (1994), 427–443, Shape design and optimization.

[HK78] E. Heintze and H. Karcher, *A general comparison theorem with applications to volume estimates for submanifolds*, Ann. Sci. École Norm. Sup. (4) **11** (1978), 451–470.

[Kh09] A. Khrabustovskyi, *On the spectrum of Riemannian manifolds with attached thin handles*, Zh. Mat. Fiz. Anal. Geom. **5** (2009), 145–169, 214.

[Kh13] A. Khrabustovskyi, *Homogenization of the spectral problem on the Riemannian manifold consisting of two domains connected by many tubes*, Proc. Roy. Soc. Edinburgh Sect. A **143** (2013), 1255–1289.

[KhP18] A. Khrabustovskyi and O. Post, *Operator estimates for the crushed ice problem*, Asymptot. Anal. **110** (2018), 137–161.

[KS03] K. Kuwae and T. Shioya, *Convergence of spectral structures: A functional analytic theory and its applications to spectral geometry*, Comm. Anal. Geom. **11** (2003), 599–673.

[LdC12] M. Lanza de Cristoforis, *Simple Neumann eigenvalues for the Laplace operator in a domain with a small hole. A functional analytic approach*, Rev. Mat. Complut. **25** (2012), 369–412.

[Oz83] S. Ozawa, *Point interaction potential approximation for $(-\Delta + U)^{-1}$ and eigenvalues of the Laplacian on wildly perturbed domain*, Osaka J. Math. **20** (1983), 923–937.

[P06] O. Post, *Spectral convergence of quasi-one-dimensional spaces*, Ann. Henri Poincaré **7** (2006), 933–973.

[P12] ———, *Spectral analysis on graph-like spaces*, Lecture Notes in Mathematics, vol. 2039, Springer, Heidelberg, 2012.

[RS80] M. Reed and B. Simon, *Methods of modern mathematical physics I: Functional analysis*, Academic Press, New York, 1980.

[RT75] J. Rauch and M. Taylor, *Potential and scattering theory on wildly perturbed domains*, J. Funct. Anal. **18** (1975), 27–59.

[W84] J. Weidmann, *Stetige Abhängigkeit der Eigenwerte und Eigenfunktionen elliptischer Differentialoperatoren vom Gebiet*, Math. Scand. **54** (1984), 51–69.

5

Manifolds with Ricci Curvature in the Kato Class: Heat Kernel Bounds and Applications

Christian Rose and Peter Stollmann

Abstract

The Kato condition is a tool from perturbation theory that gives a sufficient criterion for potentials to be semi-bounded with respect to a generator of a Dirichlet form. Moreover, it allows to transfer mapping properties of unperturbed semigroups to semigroups generated by Schrödinger operators. We review general properties of Kato potentials and discuss recent results in the context of Riemannian geometry where the negative part of the Ricci curvature is assumed to be such a potential.

5.1 Introduction

The 1973 paper by Tosio Kato entitled 'Schrödinger operators with singular potentials', published in the *Israel Journal of Mathematics* [23], was meant to establish essential self-adjointness for Schrödinger operators under very mild restrictions on the potential term. Along the way, the author introduced two concepts that bear his name and turned out to be useful in different contexts. Actually, those two concepts, *Kato's inequality* and the *Kato class* of potentials, can be combined to give new insights in analysis and geometry of Riemannian manifolds, and this is what the present survey is about.

We concentrate on the latter and record some of the implications that arise when the negative part of Ricci curvature obeys a Kato-type condition. Put very roughly, this is an application of methods from mathematical physics, more precisely, operator theory and Schrödinger operators, to questions about manifolds, namely, geometric properties that are related to the heat kernel. At the time being, papers concerning that topic are relatively recent and we have tried to record them all. If it should turn out that we missed a relevant paper, we would be grateful for references and include them in the future. Since ideas from two different communities are involved, we have decided to include some basics before stating the results.

In Section 5.2 we start by introducing the Kato class or Kato condition in a general set-up and explain its use in analysis and probability. The Kato condition can be seen as a condition of relative boundedness of a function (potential) V with respect to some reference operator H_0 on the space in question. This reference operator was the usual Laplacian in $\mathbb{R}^n$ in the case of Kato's original paper and it will be the Laplace–Beltrami operator on a Riemannian manifold in the application that we have in mind. The Kato condition means that V is, in a certain sense, small with respect to H_0 and that leads to the comforting fact that $H_0 + V$ will inherit some of the 'good' properties of H_0. In particular, mapping properties of the semigroup $\left(\mathrm{e}^{-tH_0}\right)_{t\geq 0}$ carry over to the perturbed semigroup $\left(\mathrm{e}^{-t(H_0+V)}\right)_{t\geq 0}$.

The next issue, also treated in the following section, is the connection between heat kernel bounds for the Laplace–Beltrami operator and the validity of the Kato condition for functions in appropriate L^p-spaces.

In Section 5.3 we give a short introductory account on domination of semigroups, a notion that is intimately connected with Kato's inequality and with the defining properties of Dirichlet forms in terms of the associated semigroups, known as the *Beurling–Deny criteria*. They allow for a pointwise comparison of the heat semigroup of the Laplace–Beltrami operator acting on functions and the heat semigroup of the Hodge–Laplacian acting on forms.

Throughout we will be concerned with a central point that makes the state of affairs somewhat complex. It is so important that we try to sketch it here and refer to the later sections for the technical details that we omit; we denote by M a Riemannian manifold. The Hodge–Laplacian on 1-forms, denoted by $\Delta^{(1)}$ (here you see that our sign convention differs from the preferred one in mathematical physics), can be calculated by the Weitzenböck formula as

$$\Delta^{(1)} = \nabla^*\nabla + \mathrm{Ric},$$

where the latter summand is considered as a matrix-valued function. Therefore, $\Delta^{(1)}$ itself looks like a Schrödinger operator, acting on vector-valued functions, though. Using Kato's inequality (introduced in that context by Hess, Schrader, and Uhlenbrock in their paper [22]) the heat semigroup $\left(\mathrm{e}^{-t\Delta^{(1)}}\right)_{t\geq 0}$ of the Hodge–Laplacian is *dominated* by the semigroup of the Schrödinger operator

$$\Delta + \rho \quad \text{on } L^2(M),$$

where $\Delta = \Delta^{\mathrm{LB}}$ is the Laplace–Beltrami operator on functions and

$$\rho(x) := \min\{\sigma(\mathrm{Ric}_x)\}, \quad x \in M,$$

picks the smallest eigenvalue of the symmetric matrix Ric_x considered as an endomorphism of the space of 1–forms. Note that due to our sign convention, $\Delta \geq 0$.

Knowing that $\rho_- = \max\{-\rho, 0\}$, the negative part of ρ, is in some sense small with respect to Δ, e.g., in terms of a Kato condition, allows one to control $\mathrm{e}^{-t(\Delta+\rho)}$ which in turn gives useful information on $\mathrm{e}^{-t\Delta^{(1)}}$ that can be used to estimate the first Betti number $b_1(M)$ in certain cases.

Now that looks like an easy lay-up, but there is a very important drawback. The implications of the Kato condition are perturbation theoretic in spirit and require some smallness of ρ_- with respect to Δ. But here, we cannot view ρ as a perturbation of Δ in the usual sense, since we cannot vary the potential ρ independently of Δ: both depend on the Riemannian metric that defines the manifold!

The good news are that especially the Kato condition provides good quantitative estimates and so we can arrive at interesting consequences, provided that ρ_- satisfies a suitable Kato condition.

This work should be seen as part of a general program concerning geometric properties under curvature assumptions that are less restrictive than uniform bounds. Here, as well, an important comment is in order: in the compact case, all the quantities we consider depend quite regularly on the space variable. In particular, ρ_- is a pointwise minimum of smooth functions, hence continuous and, therefore, bounded. Hence, ρ_- is certainly in the Kato class, what would also be true if the L^p-norm of ρ_- for certain p would be small enough in this case. However, ρ_- not only is relatively bounded, it is 'really bounded'. So why would one care about integrability conditions imposed on ρ_- or even the Kato condition? Well, the answer lies in the quantitative nature of our question and in the uniform control that is possible by assuming that, e.g., the L^p-norm of ρ_- for a family of metrics on M obeys a suitable bound. We could, of course, use the infimum of ρ as well but that would give much weaker estimates.

This being understood, we want to mention here the work of Gallot and co-authors in particular, who made important contributions in establishing analytic and geometric properties of Riemannian manifolds under the condition that the L^p-norm of ρ_- is small enough [2, 4, 10, 11]. Of course, the aforementioned program on geometric consequences of integral bounds on the Ricci curvature includes much more than the papers listed above (see, e.g., the original literature as well as [28, 29, 30, 35] for more information).

A natural question that comes up is whether a Kato condition on ρ_- is sufficient to control the heat kernel. This was established by one of us in [31], building on an observation made in [40]. This is given in Section 5.4, where we also record similar results by Carron from [7].

We used heat kernel bounds in relating L^p-properties and the Kato condition in order to prove upper bounds on $b_1(M)$ in [33] and to present conditions under which $b_1(M) = 0$, generalizing earlier results by Elworthy and Rosenberg [9]. Actually, the latter reference was the starting point and main source of motivation for our above-mentioned paper. Related work by different authors is collected in Section 5.5, starting from Bochner's seminal work [6].

In Section 5.6 we mention some more consequences that arise from the control of the Kato property of Ricci curvature.

5.2 The Kato class and the Kato condition

As mentioned in the introduction, the original definition of the property that defines the Kato class goes back to the celebrated paper [23] and was phrased as follows: the potential $V\colon \mathbb{R}^n \to \mathbb{R}$ is required to satisfy

$$\lim_{r\to 0}\left[\sup_{x\in\mathbb{R}^n}\int_{|y|\le r}|V(x-y)||y|^{2-n}\mathrm{d}y\right] = 0, \tag{5.2.1}$$

where we assume, in all that follows, that the space dimension satisfies $n \geq 3$ in order to avoid notational technicalities. Actually, an additional growth condition at 0 is present in Kato's paper. The important fact to notice is that condition (5.2.1) limits possible singularities, and uniformly so, in that V is convolved with a singular kernel, in fact with the kernel of the fundamental solution of the Laplacian on $\mathbb{R}^n$ (up to a constant). We refer to [38], Section A2 in particular, for more details on the prehistory of the Kato condition; we wish to underline one important point and cite from the latter article, p. 453f., that 'the naturalness of the Kato condition for L^p–properties was first noticed by Aizenman and Simon [1] in the path integral context (i.e., using the Feynman-Kac formula and Brownian motion) and by Agmon (cited as private communication in [38]) in the PDE context'. Let us point out the following facts, which can be found in [38], where original references are given; we write $V \in \mathcal{K}_n$, and say that V is in the Kato class, provided (5.2.1) holds.

Proposition 5.2.1 *For $W \geq 0$, the following are equivalent:*

(i) $W \in \mathcal{K}_n$,
(ii) $\|(\Delta+\alpha)^{-1}W\|_\infty \to 0$ *as* $\alpha \to \infty$,
(iii) $\|\mathrm{e}^{-t(\Delta-W)}\|_{\infty,\infty} \to 1$ *as* $t \to 0$.

See Theorem A.2.1 and Proposition A.2.3 in [38, p. 454], which go back to [1]. The analytic properties in the latter proposition were the starting point of a generalization of the Kato class given in [39]. There, the Laplacian is generalized to a self-adjoint operator H_0 on some $L^2(X,m)$ that is associated with a Dirichlet form (under some mild assumptions concerning X, see local citations for details). The fact that H_0 is associated with a Dirichlet form is equivalent to the fact that its semigroup is positivity preserving and contractive in the L^∞–sense, in which case we speak of a *Markovian semigroup*.

In the latter article, a Kato class of measures has been introduced and it has been shown that for this class $\hat{S}_K$ and $\mu \in \hat{S}_K$, $H_0 - \mu$ can be defined by form methods and that the semigroup $\left(\mathrm{e}^{-t(H_0-\mu)}\right)_{t\geq 0}$ shares many of the 'good properties' of $\left(\mathrm{e}^{-tH_0}\right)_{t\geq 0}$. The main idea is that in order to control L^p-L^q-norms, e.g., of $\left(\mathrm{e}^{-t(H_0-\mu)}\right)_{t\geq 0}$ uniformly in $\mu = W\mathrm{d}x$ for nice, say bounded $W \geq 0$, the relevant quantities are the following functions; as in [33] we omit the dependence on H_0 in the notation and set:

$$c_{\mathrm{Kato}}(W,\alpha) := \|(H_0+\alpha)^{-1}W\|_\infty \quad \text{for } \alpha > 0$$

as well as

$$b_{\mathrm{Kato}}(W,\beta) := \int_0^\beta \|\mathrm{e}^{-tH_0}W\|_\infty \mathrm{d}t \quad \text{for } \beta > 0\,.$$

As mentioned above, in most cases we are interested in dealing with bounded functions. In this case, the fact that the heat semigroup $\left(\mathrm{e}^{-tH_0}\right)_{t\geq 0}$ is Markovian implies that $\|\mathrm{e}^{-tH_0}W\|_\infty < \infty$ for $t > 0$ as well as $\|(H_0+\alpha)^{-1}W\|_\infty < \infty$ for $\alpha > 0$.

For general measurable $W \geq 0$ we can define

$$c_{\mathrm{Kato}}(W,\alpha) := \sup_{n\in\mathbb{N}} \|(H_0+\alpha)^{-1}(W \wedge n)\|_\infty \in [0,\infty]$$

and say that W satisfies a *Kato condition*, provided the latter quantity is finite for some $\alpha > 0$.

The quantities above are related via functional calculus:

$$(1-\mathrm{e}^{-\alpha\beta})c_{\mathrm{Kato}}(W,\alpha) \leq b_{\mathrm{Kato}}(W,\beta) \leq \mathrm{e}^{\alpha\beta}c_{\mathrm{Kato}}(W,\alpha), \tag{5.2.2}$$

see [14], and we get that the behaviour of $c_{\mathrm{Kato}}(W,\alpha)$ for $\alpha \to \infty$ controls the behaviour of $b_{\mathrm{Kato}}(W,\beta)$ for $\beta \to 0$ and vice versa.

An important property is the stability of the boundedness of the semigroup in different L^p–spaces under Kato perturbations. It is implicit in the equivalence (i)⇔(iii) of Proposition 5.2.1. The following explicit estimate goes back to [39, Theorem 3.3], and can be found in [33] in an equivalent dual form.

Proposition 5.2.2 *Let H_0 be as above (the generator of a Markovian semigroup on $L^2(X,m)$) and $V \in L^1_{\mathrm{loc}}(X)$ such that $b_{\mathrm{Kato}}(V_-,\beta) =: b < 1$ for some $\beta > 0$. Then*

$$\|\mathrm{e}^{-t(H_0+V)}\|_{\infty,\infty} \le \frac{1}{1-b}\mathrm{e}^{t\frac{1}{\beta}\log\frac{1}{1-b}}.$$

Consequently, if $(\mathrm{e}^{-tH_0}; t \ge 0)$ is ultra-contractive, i.e., maps L^1 to L^∞, then so is $(\mathrm{e}^{-t(H_0+V)}; t \ge 0)$, a fact that can be deduced from the above and an interpolation argument (see [39, Theorem 5.1]).

While we used an analytic set-up in the latter paper, one of the useful features of the Kato condition is that it is well suited for probabilistic techniques. So it is equally well possible to start from a given Markov process and define respective perturbations in a probabilistic manner. We refer to [1, 38] for the start and to [24, 25] for more recent contributions along these lines, as well as to [14] and the literature cited in these works.

One main point of interest here is to study the question whether the abstract version of a Kato condition as above, using quantities like b_{Kato}, c_{Kato}, can still be expressed in terms of kernels. In other words, whether a generalization of Proposition 5.2.1 holds true. The answer is yes, provided the *heat kernel* $p_t(\cdot,\cdot)$, the integral kernel of the heat semigroup, is controlled in some sense.

First of all, a very general condition for locally integrable functions to satisfy a Kato condition was given in [25] for general Markov processes associated with a Dirichlet form in $L^2(X,m)$ on a locally compact separable metric space (X,d). The authors defined the *Kato class relative to the Green kernel* of the generator. For fixed $\nu \ge \beta > 0$, a non-negative function $V \in L^1_{\mathrm{loc}}(X,m)$ belongs to this Kato class $K_{\nu,\beta}(X)$ if

$$\lim_{r\to 0}\sup_{x\in X}\int_{B(x,r)} G(x,y)V(y)\mathrm{d}m(y) = 0,$$

where $G(x,y) := G(d(x,y))$ with $G(r) = r^{\beta-\nu}$ and $B(x,r)$ denotes the metric ball around $x \in X$ of radius $r > 0$. Denote by $L^p_{\mathrm{unif}}(X)$ the set of functions f such that

$$\sup_{x\in X}\int_{B(x,1)} |f|^p\mathrm{d}m < \infty.$$

Theorem 5.2.3 ([25, Theorem 3.3]) *For all $\nu \ge \beta > 0$, $p > \nu/\beta$, we have*

$$f \in L^p_{\mathrm{unif}}(X) \quad \Rightarrow \quad |f| \in K_{\nu,\beta}(X)$$

provided there is a positive increasing function V on $(0,\infty)$ such that $r \mapsto V(r)/r^\nu$ is increasing or bounded and $\sup_{x\in X} m(B(x,r)) \le V(r)$ for all

$r > 0$, and the heat kernel satisfies upper and lower bounds in the following way: there exist two positive increasing functions $\varphi_1, \varphi_2 \colon (0, \infty) \to (0, \infty)$ such that

$$\int_1^\infty \frac{1}{t} \max\{V(t), t^\nu\} \varphi_2(t) \mathrm{d}t < \infty$$

and for any $x, y \in X$, $t \in (0, t_0]$, we have

$$\frac{1}{t^{\nu/\beta}} \varphi_1 \left(\frac{d(x, y)}{t^{1/\beta}} \right) \leq p_t(x, y) \leq \frac{1}{t^{\nu/\beta}} \varphi_2 \left(\frac{d(x, y)}{t^{1/\beta}} \right).$$

More specifically, in the case of a geodesically complete Riemannian manifold with bounded geometry, bounds on the heat kernel are explicit, leading to the following.

Theorem 5.2.4 ([14, Theorem 2.9]) *Let M be a geodesically complete Riemannian manifold of dimension $n \geq 2$ with Ricci curvature bounded below and assume that there are $C, R > 0$ such that for all $x \in M$ and $r \in (0, R]$, one has*

$$\mathrm{Vol}(B(x, r)) \geq C r^n.$$

Then we have

(i) $V \in K_{n,2}(M)$ if and only if

$$\lim_{t \to 0} \sup_{x \in M} \int_0^t p_t(x, y) |V(y)| \mathrm{dvol}(y) = 0.$$

(ii) for any $p > n/2$, we have $L^p_{\mathrm{unif}}(M) \subset K_{n,2}(M)$.

In particular, [14, Corollary 2.11] then gives $L^p(M) + L^\infty(M) \subset K_{n,2}(M)$ under the same assumptions. The non-collapsing of the volume of the balls seems strong, but can only be avoided by replacing it by a lower bound on the heat kernel, a condition that is stronger than the volume bound.

Theorem 5.2.5 ([19, Proposition 3.2]) *Let M be a Riemannian manifold of dimension $n \geq 2$ and $p > n/2$.*

(i) If there are $C, t_0 > 0$ such that for all $t \in (0, t_0]$ and all $x \in M$ we have $p_t(x, x) \leq C t^{-n/2}$, then, for any $V \in L^p(M) + L^\infty(M)$,

$$\lim_{t \to 0} \sup_{x \in M} \int_0^t p_t(x, y) |V(y)| \mathrm{dvol}(y) = 0.$$

(ii) Let M be geodesically complete and assume that there are positive constants $C_1, \ldots, C_6, t_0 > 0$ such that for all $t \in (0, t_0]$, $x, y \in M$, $r > 0$ one has $\mathrm{Vol}(B(x, r)) \leq C_1 r^n \mathrm{e}^{C_2 r}$ and

$$C_3 t^{-n/2} \mathrm{e}^{-C_4 \frac{d(x,y)^2}{t}} \leq p_t(x,y) \leq C_5 t^{-n/2} \mathrm{e}^{-C_6 \frac{d(x,y)^2}{t}}.$$

Then, one has

$$L^p_{\text{unif}}(M) + L^\infty(M) \subset K_{n,2}(M).$$

In the special case of compact manifolds, the potentials in the Kato class can be characterized with the help of uniform heat kernel estimates, i.e., there are some constants $C, k, t_0 > 0$ such that

$$\forall x, y \in M, t \in (0, t_0]: \quad p_t(x,y) \leq C t^{-k}. \tag{5.2.3}$$

Classically, such estimates follow from so-called isoperimetric inequalities under certain assumptions on the Ricci curvature. In [33], the authors exhibited the necessary analytic framework based on [10], leading to the following:

Proposition 5.2.6 ([33, Theorem 4.1]) *Let $D > 0$ and $q > n \geq 3$. There is an explicit constant $\varepsilon > 0$ such that for any compact Riemannian manifold M with $\dim M = n$, $\operatorname{diam}(M) \leq D$, and $\|\rho_-\|_{q/2} < \varepsilon$, for any $p > q/2$ there is $C > 0$ such that for any $0 \leq V \in L^p(M)$ we have*

$$c_{\text{Kato}}(V, \alpha) \leq C \|V\|_p \int_0^\infty \mathrm{e}^{-\alpha t} \max\{1, t^{-q/2p}\} \mathrm{d}t.$$

An analogous estimate also holds for $b_{\text{Kato}}(V, \beta)$ by a direct computation or by using relation (5.2.2). Due to the fact that the decay rate k of the heat kernel in (5.2.3) depends on the integrability of the negative part of the Ricci curvature, the integral on the right-hand side only converges for potentials with a higher integrability.

5.3 Domination of Semigroups, Kato's Inequality, and Comparison for the Heat Semigroup on Functions and on 1-forms

We start with some historical remarks and with the famous Kato's inequality that reads

$$\Delta|u| \leq \Re(\operatorname{sgn} \bar{u}) \Delta u \tag{5.3.1}$$

according to our sign convention for the Laplacian; here $\operatorname{sgn} \bar{u}(x) = \bar{u}(x)/|u(x)|$ if $u(x) \neq 0$ and 42 otherwise. Actually, in its original form as Lemma A in [23], magnetic fields were included on the right-hand side. We should note, in passing, that (5.3.1) is meant in the distributional sense and it is assumed that $\Delta u \in L^1_{\text{loc}}$. The interesting feature of (5.3.1) is that it can be expressed equivalently in terms of the following positivity property for the semigroup:

$$|e^{-t\Delta} f| \leq e^{-t\Delta} |f|, \quad (f \in L^2, t \geq 0)$$

which can be seen as the property that the heat semigroup *dominates itself.* To explain that, we follow the paper [21] in introducing the necessary concepts. See also [22, 36, 37] as well as [26] and the literature cited there for more recent contributions. The absolute value $|f|$ is replaced by a more general mapping, allowing vector-valued functions. We will use some terminology without explanation. All the necessary facts can be found in the articles above.

We start with a real Hilbert space $\mathcal{K}$ and a cone $\mathcal{K}^+$ that is compatible with the inner product $\langle \cdot, \cdot \rangle$. Given another, real or complex, Hilbert space $\mathcal{H}$, a map $|\cdot| \colon \mathcal{H} \to \mathcal{K}^+$ is called an absolutely pairing map, provided the following properties hold (we write $\langle \cdot, \cdot \rangle$ for both inner products in a slight abuse of notation):

- $\forall f_1, f_2 \in \mathcal{K}: \quad |\langle f_1, f_2 \rangle| \leq \langle |f_1|, |f_2| \rangle,$
- $\forall f \in \mathcal{H}: \quad \langle f, f \rangle = \langle |f|, |f| \rangle,$
- $\forall f_1 \in \mathcal{H} \forall g \in \mathcal{K}^+ \exists f_2 \in \mathcal{H}$ such that $|f_2| = g$ and $\langle f_1, f_2 \rangle = \langle |f_1|, g \rangle$.

Two functions f_1, f_2 satisfying the third condition are called *absolutely paired.* Given an absolutely pairing map, we can talk about domination of operators that act on $\mathcal{K}$ and $\mathcal{H}$, respectively. First, however, we give the example that is important for us here:

$$\mathcal{K} = L^2(M), \quad \mathcal{H} = L^2(M, \Omega^1),$$

the latter consisting of square integrable sections of the cotangent bundle, where the forms $\omega(x)$ are measured in terms of the inner product induced by the Riemannian metric, written as $|\omega(x)|$. It is not hard to see that

$$|\cdot| \colon L^2(M, \Omega^1) \to L^2(M), \quad \omega \mapsto |\omega(\cdot)|$$

is an absolutely pairing map; of course, both L^2-spaces are built upon the Riemannian volume.

Going back to the general case we can say that a bounded linear operator A on $\mathcal{H}$ is *dominated* by a bounded linear operator B on $\mathcal{K}$, provided

$$|Af| \leq B|f| \quad (f \in \mathcal{H}).$$

Clearly, this notion depends on the absolute map $|\cdot|$. The following powerful characterization of semigroup domination can be found in [21, Theorem 2.15]. Let H and K be (the negative of) generators of the strongly continuous semigroups $T_t = e^{-tH}$ and $S_t = e^{-tK}$.

Theorem 5.3.1 *In the situation above, the following statements are equivalent:*

(i) $(T_t)_{t\geq 0}$ *is dominated by* $(S_t)_{t\geq 0}$*, i.e.,*

$$|T_t f| \leq S_t |f| \quad (f \in \mathcal{H}).$$

(ii) H *and* K *satisfy a generalized Kato's inequality: For all* $f_1 \in \operatorname{dom}(H)$ *and* $f_2 \in \mathcal{H}$ *such that* $|f_2| \in \operatorname{dom}(K^*)$ *and* f_1 *and* f_2 *absolutely paired:*

$$\Re\langle Hf_1, f_2\rangle \geq \langle |f_1|, K^*|f_2|\rangle.$$

Let us mention that condition (ii) has a counterpart in a version of the first Beurling-Deny criterion (see [37] and [26]). The comforting fact for us is that $\left(\mathrm{e}^{-t\bar{\Delta}}\right)_{t\geq 0}$ and $\left(\mathrm{e}^{-t\Delta}\right)_{t\geq 0}$ enjoy the domination relation alluded to above, where $\bar{\Delta} = \nabla^*\nabla$ is the Bochner–Laplacian. This has been proven in terms of a Kato inequality by Hess, Schrader, and Uhlenbrock in [22], so that

$$|\mathrm{e}^{-t\bar{\Delta}}\omega| \leq \mathrm{e}^{-t\Delta}|\omega|$$

for all $t \geq 0$ and $\omega \in L^2(M, \Omega^1)$. The Weitzenböck formula and abstract results on sums of generators give

$$|\mathrm{e}^{-t\Delta^{(1)}}\omega| \leq \mathrm{e}^{-t(\Delta+\rho)}|\omega|$$

and, moreover,

$$\operatorname{Tr}\left(\mathrm{e}^{-t\Delta^{(1)}}\right) \leq n \operatorname{Tr}\left(\mathrm{e}^{-t(\Delta+\rho)}\right).$$

Both these inequalities have geometrical implications, as we already mentioned above and as we will see in more detail in Section 5.5.

We mention in passing that domination of semigroups can also be treated probabilistically (cf. [34]). This gives a nice pointwise estimate on the heat kernels, as shown in Theorem 3.5 of the latter paper by Rosenberg, which we state here in the special case under consideration. For this reason, denote by $p_t^{(1)}(x, y)$ the heat kernel of the Hodge–Laplacian $\Delta^{(1)}$ and $p_t^{(0)}(x, y)$ the heat kernel of $\Delta + \rho$ acting on functions.

Theorem 5.3.2 ([34, Theorem 3.5]) *For all* $t > 0$, $x, y \in M$*, we have*

$$|p_t^{(1)}(x, y)| \leq n\, p_t^{(0)}(x, y).$$

We also mention [5] where a Kato inequality was established for a class of submersions.

5.4 The Kato Condition Implies Heat Kernel Bounds

Proofs of the fact mentioned in the title of this section are based on the following result by Qi S. Zhang and M. Zhu.

Theorem 5.4.1 ([40, Theorem 1.1]) *Let M be a compact Riemannian manifold of dimension n, and u a positive solution to the heat equation*

$$\partial_t u = -\Delta u.$$

Suppose that either one of the following conditions holds:

(i) $\|\rho_-\|_p < \infty$ for $p > n/2$, and that there is a $c > 0$ such that for all $x \in M$ and $r \in (0, 1]$, we have $\mathrm{Vol}(B(x, r)) \geq c\, r^n$.

(ii) $\sup_{x \in M} \int_M \rho_-^2(y) d(x, y)^{2-n} \mathrm{dvol}(y) < \infty$ and the heat kernel is bounded from above.

Then, for any $\alpha \in (0, 1)$, there are an explicit continuous function $J\colon (0, \infty) \to (0, 1)$ and $c > 0$ such that

$$J(t)\frac{|\nabla u|^2}{u^2} - \frac{\partial_t u}{u} \leq \frac{c}{J(t)\,t}, \qquad (t \in (0, \infty)). \tag{5.4.1}$$

An inequality of the type (5.4.1) yields an explicit upper bound of the heat kernel by a nowadays standard technique introduced by Li and Yau [27]. A thorough inspection of the reasoning in [40] shows that the Kato condition on ρ_- indeed implies heat kernel estimates in the following sense.

Theorem 5.4.2 ([31]) *Let $n \geq 3$ and $\beta > 0$. For any closed Riemannian manifold M of dimension n satisfying* $\mathrm{diam}(M) \leq \sqrt{\beta}$, *and*

$$b := b_{\mathrm{Kato}}(\rho_-, \beta) < \frac{1}{16n},$$

there are explicit constants $C = C(n, b, \beta) > 0$ and $\kappa = \kappa(n, b, \beta) > 0$ such that we have

$$p_t(x, y) \leq \frac{C}{\mathrm{Vol}(M)} t^{-\kappa} \quad (t \in (0, \beta^2/4], x, y \in M). \tag{5.4.2}$$

A different version that is more explicit in the sense that it fits well with the Euclidean case was obtained independently by G. Carron. For its formulation we use the notation of this chapter.

Theorem 5.4.3 ([7]) *There is a constant c_n depending on n alone with the following property: Let $D := diam(M)$ and T the largest time such that*

$$b_{\mathrm{Kato}}(\rho_-, T) \leq \frac{1}{16n}.$$

Then

$$p_t(x,x) \leq \frac{c_n}{\mathrm{Vol}(B(x,\sqrt{t}))} \quad (t \in (0,T/2], x \in M). \tag{5.4.3}$$

See Corollary 3.9 in [7] and Section 3 of [7]. Note that the results above are closely related via the so-called volume doubling condition. [7, Proposition 3.8] also shows that the volume doubling condition is indeed satisfied under the curvature assumptions of the above theorems. See also [32] for the connection.

5.5 Bounds on the First Betti Number

The first Betti number, $b_1(M)$, is a tool for classifying compact Riemannian manifolds M of dimension n. By definition, $b_1(M)$ is the dimension of the first cohomology group, $b_1(M) := \dim \mathcal{H}^1(M)$, where $\mathcal{H}^1(M)$ is the real linear space quotient of the closed differential 1-forms on M by the exact forms. This group describes in some sense the $(n-1)$-dimensional holes of M and is therefore actually of topological nature, clarifying its relevance for the classification of manifolds. Bochner was the first to observe that it is quite easy to derive bounds on $b_1(M)$ if the Ricci tensor is non-negative everywhere and strictly positive in some point of M. More precisely, he showed in [6] the following theorem, although the result is not explicitly stated in the form below.

Theorem 5.5.1 ([6]) *Let M be a compact Riemannian manifold of dimension n. If the Ricci curvature is non-negative, then*

$$b_1(M) \leq n.$$

If, additionally, there is a point in M such that the Ricci curvature is strictly positive at that point, then

$$b_1(M) = 0.$$

Actually, the above theorem follows implicitly from the Weitzenböck formula

$$\Delta^{(1)} = \bar{\Delta} + \mathrm{Ric},$$

where $\Delta^{(1)}$ is the Hodge–Laplacian acting on one-forms, $\bar{\Delta} := \nabla^*\nabla$ the so-called rough or Bochner–Laplacian, and Ric denotes the Ricci tensor interpreted as a section of endomorphisms on the space of one forms as above. Any equivalence class in $\mathcal{H}^1(M)$ can be represented by a harmonic one-form, such that

$$\dim\ker(\Delta^{(1)}) = \dim\mathcal{H}^1(M).$$

Using the non-negativity of Ric in quadratic form sense implies directly that there are only parallel forms in $\ker(\Delta^{(1)})$, which vanish under the additional positivity assumption on Ric. This classical ansatz, known as the first application of the nowadays called Bochner method, seems not to lead to results allowing some negative Ricci curvature. However, instead of a geometric argument it is possible to use form methods to control the kernel of $\Delta^{(1)}$. The main observation here can be found in [9] by Elworthy and Rosenberg building on the domination property established by Hess, Schrader, and Uhlenbrock in [22], discussed in Section 5.3: namely, for any square integrable section of one-forms $\omega \in L^2(M;\Omega^1)$, we have:

$$|\mathrm{e}^{-t\Delta^{(1)}}\omega| \leq \mathrm{e}^{-t(\Delta+\rho)}|\omega|, \tag{5.5.1}$$

where the norms above are taken pointwise in the cotangent bundle of M. If ω is harmonic, the left-hand side equals $|\omega|$. If the semigroup on the right-hand side is generated by a positive operator $\Delta + \rho > 0$, we can let $t \to \infty$, so that $\mathrm{e}^{-t(\Delta+\rho)}|\omega| \to 0$ which gives $\omega = 0$, yielding a method to conclude the triviality of $\ker(\Delta^{(1)})$. The issue here is that we cannot easily treat ρ as a perturbation of Δ since both of them depend on the metric tensor of M. Therefore, it is not trivial to get positivity of the operator $\Delta + \rho$.

A general strategy is to control the part of the Ricci curvature lying below a certain positive threshold. Elworthy and Rosenberg derived the following theorem along these lines with a more complicated method of proof based on Sobolev embedding theorems and eigenfunction estimates.

Theorem 5.5.2 ([9]) *Let M be a compact manifold, $X \subset M$, $K, K_0 > 0$, $\mathrm{Ric} \geq -K_0$ on X, $\mathrm{Ric} \geq K$ on $M \setminus X$. There exists $a > 0$ depending on the quantities above such that if*

$$\mathrm{Vol}(X) < a\ \mathrm{Vol}(M),$$

then $b_1(M) = 0$.

Unfortunately, the constant a in the above theorem is far from being explicit and it also still depends on the lower bound K for the Ricci tensor.

When Elworthy and Rosenberg published their article there was already a result that implies the assertion in the latter theorem in Gallot's article [10] from 1988. In fact, Gallot proved an estimate of the first eigenvalue of $\Delta + V$ for some potential V in terms of its L^p-norm, so that (5.5.1) leads to the vanishing of $b_1(M)$; we also mention [3] in which the same basic idea is nicely explained in a little more restrictive context.

Rosenberg and Yang also recognized that integral bounds are the right thing to look for and arrived at the following result, Theorem 4 in [35].

Theorem 5.5.3 ([35]) *Let M be an n-dimensional complete Riemannian manifold. Assume that there exist constants $A, B > 0$ such that for any $f \in C_c^\infty(M)$*

$$\left(\int_M |f(x)|^{\frac{2n}{n-2}} \mathrm{dvol}(x)\right)^{\frac{n-2}{n}} \leq A \int_M |\nabla f(x)|^2 \mathrm{dvol}(x) + B \int_M |f(x)|^2 \mathrm{dvol}(x).$$

Then, whenever for some $\rho_0 > 0$,

$$\|(\rho - \rho_0)_-\|_{\frac{n}{2}} < \min\{A^{-1}, \rho_0 B^{-1}\},$$

it follows that there are no harmonic 1-forms on M. In particular, if M is compact, then $b_1(M) = 0$.

The main idea is to decompose

$$\Delta + \rho = \Delta + \rho_0 + (\rho - \rho_0) \geq \Delta + \rho_0 - (\rho - \rho_0)_-,$$

which is positive as soon as $(\rho - \rho_0)_-$ is relatively bounded w. r. t. Δ for some form-bound smaller than one and such an estimate can be deduced from a Sobolev embedding theorem. The nice fact about the latter result is that it allows a statement in the threshold case $\frac{n}{2}$ as far as integrability of $(\rho - \rho_0)_-$ is concerned. Moreover, the argument is quite direct and does not rely on explicit heat kernel estimates, an issue we turn to next.

Assuming that the semigroup generated by Δ is ultra-contractive, i.e., there are constants $C, k, t_0 > 0$ such that

$$\|\mathrm{e}^{-t\Delta}\|_{1,\infty} \leq C t^{-k}, \quad t \in (0, t_0),$$

perturbation techniques based on the Kato condition lead to quantitative results as well. With the decomposition of $\Delta + \rho$ as above, the assumption of ultra-contractivity allows to handle $(\rho - \rho_0)_-$ as a Kato perturbation, that means, we are looking for conditions that give

$$b_{\mathrm{Kato}}((\rho - \rho_0)_-, \rho_0^{-1}) := \int_0^{\rho_0^{-1}} \|\mathrm{e}^{-t\Delta}(\rho - \rho_0)_-\|_\infty \mathrm{d}t < 1. \tag{5.5.2}$$

Ultra-contractivity of the heat semigroup also implies its continuity from $L^p(M)$ to $L^\infty(M)$ for all $p \in (0, \infty]$, so that

$$b_{\mathrm{Kato}}((\rho - \rho_0)_-, \rho_0^{-1}) \leq C \|(\rho - \rho_0)_-\|_p \int_0^{\rho_0^{-1}} t^{-p/k} \mathrm{d}t < 1$$

if $p < k$ and the L^p-norm on the right-hand side is small enough. This explicitly computable quantity led to the result below.

Theorem 5.5.4 ([33]) *Let $n \in \mathbb{N}$, $n \geq 3$, $p > n/2$, $D > 0$. There is an explicitly computable $\varepsilon > 0$ such that for all compact Riemannian manifolds M of dimension n, $\operatorname{diam}(M) \leq D$, and*

$$\frac{1}{\operatorname{Vol}(M)} \int_M \rho_-^p \,\mathrm{dvol} < \varepsilon,$$

we have $b_1(M) = 0$.

At the heart of the proof lies a deep isoperimetric estimate from [10] that holds under the assumption that the averaged L^p-norm of the Ricci curvature is small enough, implying ultra-contractivity of the heat semigroup.

We now turn to the question whether it is enough to assume smallness of the Kato constant b_{Kato} to derive bounds on $b_1(M)$. The main observation is

$$\dim\ker(\Delta^{(1)}) \leq \operatorname{Tr}(\mathrm{e}^{-t\Delta^{(1)}}) \leq n \operatorname{Tr}(\mathrm{e}^{-t(\Delta-\rho_-)}) \leq n \operatorname{Vol}(M)\|\mathrm{e}^{-t(\Delta-\rho_-)}\|_{1,\infty}, \tag{5.5.3}$$

so that we get bounds on $b_1(M)$ as soon as we can control $\|\mathrm{e}^{-t(\Delta-\rho_-)}\|_{1,\infty}$. The ultra-contractivity estimate is crucial here as well as the stability of ultra-contractivity under Kato-class perturbations, stated in Proposition 5.2.2.

A little work and putting everything together yields the following.

Theorem 5.5.5 ([33]) $3 \leq n < p < 2q$ *and $D > 0$. There is an $\varepsilon > 0$ and a constant $K(p) > 0$ depending only on p such that for all compact Riemannian manifolds M with* $\dim M = n$, *and* $\operatorname{diam}(M) \leq D$ *with $\|\rho_-\|_q \leq \varepsilon$, we have*

$$b_1(M) \leq n \cdot \left(\frac{2}{1-\varepsilon^{-1}\|\rho_-\|_p}\right)^{2\frac{1+\varepsilon^{-1}\|\rho_-\|_p}{1-\varepsilon^{-1}\|\rho_-\|_p}+\frac{p}{2}} \left(1 + K(p)D^{\frac{p}{2}}\right).$$

Even though the Kato condition on the part of Ricci curvature below a positive level is sufficient to obtain the triviality of $\mathcal{H}^1(M)$, we do not know yet whether a Kato-bound on the negative part of Ricci curvature implies a non-trivial bound on $b_1(M)$. The ultra-contractivity is a necessary assumption such that equation (5.5.3) can be applied and calculated. Fortunately, Theorem 5.4.2 shows that the smallness of $b_{\mathrm{Kato}}(\rho_-, \beta)$ for some $\beta > 0$ implies a heat kernel upper bound, giving the desired ultra-contractive bound and in turn the bound on the first Betti number.

Theorem 5.5.6 ([31, 32]) *Let $n \geq 2$ and $\beta > 0$. Any compact Riemannian manifold with* $\dim M = n$, $\operatorname{diam} M \leq \sqrt{\beta}$, *and*

$$b := b_{\mathrm{Kato}}(\rho_-, \beta) \leq \frac{1}{16n},$$

satisfies

$$b_1(M) \leq n \cdot \left(\frac{2}{1-b}\right)^{\left(1+\frac{1}{\beta}\right)\frac{1+b}{1-b}+\frac{1}{n-1}} \mathrm{e}^{\frac{3n}{n-1}}.$$

Additionally, Carron showed in [7] that a clever improvement of the upper bound of the heat kernel and Gromov's trick lead to the following estimate.

Theorem 5.5.7 ([7]) *Let $n \geq 2$ and $\beta > 0$. There is an $\varepsilon > 0$ such that any compact Riemannian manifold with* $\operatorname{diam} M \leq \sqrt{\beta}$ *and $b_{\mathrm{Kato}}(\rho_-, \beta) < \varepsilon$ satisfies $b_1(M) \leq n$.*

5.6 Concluding Remarks

Here we first briefly mention some other results that have been obtained under the assumption that the negative part of Ricci curvature satisfies a Kato condition.

We already heavily cited [7] above. Apart from what we already referred to, Carron shows, amongst other things and assumptions, that such a curvature condition allows to control the volume growth of balls from above, giving volume doubling and an upper bound on the volume of balls that coincides with the Euclidean case.

Recently, in [8] it was shown that the Kato condition for the Ricci curvature also implies isoperimetric, Buser and Lichnerowicz-type inequalities on compact manifolds. Besides that, diameter bounds for compact manifolds and finiteness of the fundamental group are obtained, too.

In [18], the authors extend the heat semigroup characterization of functions of bounded variation to manifolds whose Ricci curvature is not necessarily bounded below, again assuming that the negative part of Ricci curvature satisfies a Kato condition.

We would also like to mention the works by Güneysu, who extended the concepts of Kato class potentials to the context of vector-valued functions on manifolds (see [12, 13, 14, 15, 16, 17] and the cited literature therein).

In the quite recent paper [20], the Birman–Schwinger principle is used for an estimate on the first Betti number.

Let us end with a meta question: While it is by now quite well understandable that Kato conditions on the negative part of Ricci curvature can be used to find bounds on $b_1(M)$ as we hopefully convinced our readers above, there is still some kind of mystery in the fact that Kato class Ricci curvature actually leads to heat kernel bounds and other geometric features. In fact, for the former results, one uses domination and a Schrödinger operator that features ρ_- as a potential term. For the latter case, however, the operator in question is the Laplace–Beltrami operator itself that exhibits no potential term.

Apart from the obvious fact that the proofs work: why is it true? A better understanding is certainly needed, e.g., for an extension of some of the results we mentioned to the non-compact case.

Acknowledgements

The second named author expresses his thanks to Daniel, Matthias, and Radek for organizing such a wonderful conference and creating an atmosphere of open and respectful exchange of ideas.

References

[1] M. Aizenman and B. Simon. Brownian motion and Harnack inequality for Schrödinger operators. *Comm. Pure Appl. Math.*, 35(2):209–273, 1982.

[2] P. H. Bérard. From vanishing theorems to estimating theorems: the Bochner technique revisited. *Bull. Amer. Math. Soc. (N.S.)*, 19(2):371–406, 1988.

[3] P. H. Bérard. A lower bound for the least eigenvalue of $\Delta + V$. *Manuscripta Math.*, 69(3):255–259, 1990.

[4] P Bérard and G Besson. Number of bound states and estimates on some geometric invariants. *Journal of Functional Analysis*, 94(2):375–396, 1990.

[5] G Besson. A Kato type inequality for Riemannian submersions with totally geodesic fibers. *Ann. Global Anal. Geom.*, 4(3): 273 – 289, 1986.

[6] S. Bochner. Vector fields and Ricci curvature. *Bull. Amer. Math. Soc.*, 52:776–797, 1946.

[7] G. Carron. Geometric inequalities for manifolds with Ricci curvature in the Kato class. arXiv: 1612.03027, 2016.

[8] G. Carron and C. Rose. Geometric and spectral estimates based on spectral Ricci curvature assumptions. *Preprint. https://arxiv.org/abs/1808.06965 [math.DG]*, 2018.

[9] K. D. Elworthy and S. Rosenberg. Manifolds with wells of negative curvature. *Invent. Math.*, 103(3):471–495, 1991. With an appendix by Daniel Ruberman.

[10] S. Gallot. Inégalités isopérimétriques et analytiques sur les variétés riemanniennes. *Astérisque*, (163–164):5–6, 31–91, 281 (1989), 1988. On the geometry of differentiable manifolds (Rome, 1986).

[11] S. Gallot. Isoperimetric inequalities based on integral norms of Ricci curvature. *Astérisque*, (157–158):191–216, 1988. Colloque Paul Lévy sur les Processus Stochastiques (Palaiseau, 1987).

[12] B. Güneysu. *On the Feynman-Kac formula for Schrödinger semigroups on vector bundles*. Dissertation, Rheinische Friedrich-Wilhelms-Universität Bonn, 2010.

[13] B. Güneysu. On generalized Schrödinger semigroups. *J. Funct. Anal.*, 262(11):4639–4674, 2012.

[14] B. Güneysu. Kato's inequality and form boundedness of Kato potentials on arbitrary Riemannian manifolds. *Proc. Amer. Math. Soc.*, 142(4):1289–1300, 2014.

[15] B. Güneysu. *Covariant Schrödinger semigroups on noncompact Riemannian manifolds*. Habilitation, Humboldt-Universität zu Berlin, 2016.

[16] B. Güneysu. Heat kernels in the context of Kato potentials on arbitrary manifolds. *Potential Anal.*, 46(1):119–134, 2017.

[17] B. Güneysu. *Covariant Schrödinger semigroups on Riemannian manifolds*. Operator Theory: Advances and Applications, 264. Birkhäuser/Springer, Cham, 2017.

[18] B. Güneysu and D. Pallara. Functions with bounded variation on a class of Riemannian manifolds with Ricci curvature unbounded from below. *Math. Ann.*, 363(3-4):1307–1331, 2015.

[19] B. Güneysu and O. Post. Path integrals and the essential self-adjointness of differential operators on noncompact manifolds. *Math. Z.*, 275(1–2):331–348, 2013.

[20] M. Hansmann, C. Rose and P. Stollmann: Bounds on the first Betti number – an approach via Schatten norm estimates on semigroup differences. arXiv: 1810.12205, 2018.

[21] H. Hess, R. Schrader, and D. A. Uhlenbrock. Domination of semigroups and generalization of Kato's inequality. *Duke Math. J.*, 44(4):893–904, 1977.

[22] H. Hess, R. Schrader, and D. A. Uhlenbrock. Kato's inequality and the spectral distribution of Laplacians on compact Riemannian manifolds. *J. Differential Geom.*, 15(1):27–37 (1981), 1980.

[23] T. Kato. Schrödinger operators with singular potentials. In *Proceedings of the international symposium on partial differential equations and the geometry of normed linear spaces (Jerusalem, 1972)*, volume 13, pages 135–148 (1973), 1972.

[24] K. Kuwae and M. Takahashi. Kato class functions of Markov processes under ultracontractivity. In *Potential theory in Matsue*, volume 44 of *Adv. Stud. Pure Math.*, pages 193–202. Math. Soc. Japan, Tokyo, 2006.

[25] K. Kuwae and M. Takahashi. Kato class measures of symmetric Markov processes under heat kernel estimates. *J. Funct. Anal.*, 250(1):86–113, 2007.

[26] D. Lenz, M. Schmidt, and M. Wirth. Domination of quadratic forms. *ArXiv e-print*, November 2017.

[27] P. Li and S.-T. Yau. On the parabolic kernel of the Schrödinger operator. *Acta Math.*, 156(3-4):153–201, 1986.

[28] P. Petersen and C. Sprouse. Integral curvature bounds, distance estimates and applications. *J. Differential Geom.*, 50(2):269–298, 1998.

[29] P. Petersen and G. Wei. Relative volume comparison with integral curvature bounds. *Geom. Funct. Anal.*, 7(6):1031–1045, 1997.

[30] P. Petersen and G. Wei. Analysis and geometry on manifolds with integral Ricci curvature bounds. II. *Trans. Amer. Math. Soc.*, 353(2):457–478, 2001.

[31] C. Rose. Li-Yau gradient estimate for compact manifolds with negative part of Ricci curvature in the Kato class. *Ann. Glob. Geom. Anal.*, 55(3):443–449, Apr 2019.

[32] C. Rose. *Heat kernel estimates based on Ricci curvature integral bounds.* Dissertation, Technische Universität Chemnitz, 2017.

[33] C. Rose and P. Stollmann. The Kato class on compact manifolds with integral bounds of Ricci curvature. *Proc. Amer. Math. Soc.* 145(5), 2199–2210, 2017.

[34] S. Rosenberg. Semigroup domination and vanishing theorems. In *Geometry of random motion (Ithaca, N.Y., 1987)*, volume 73 of *Contemp. Math.*, pages 287–302. Amer. Math. Soc., Providence, RI, 1988.

[35] S. Rosenberg and D. Yang. Bounds on the fundamental group of a manifold with almost nonnegative Ricci curvature. *J. Math. Soc. Japan*, 46(2):267–287, 1994.

[36] B. Simon. An abstract Kato's inequality for generators of positivity preserving semigroups. *Indiana Univ. Math. J.*, 26(6):1067–1073, 1977.

[37] B. Simon. Kato's inequality and the comparison of semigroups. *J. Funct. Anal.*, 32(1):97–101, 1979.

[38] B. Simon. Schrödinger semigroups. *Bull. Amer. Math. Soc. (N.S.)*, 7(3):447–526, 1982.

[39] P. Stollmann and J. Voigt. Perturbation of Dirichlet forms by measures. *Potential Anal.*, 5(2):109–138, 1996.

[40] Q. Zhang and M. Zhu. Li-Yau gradient bounds on compact manifolds under nearly optimal curvature conditions. *J. Funct. Anal.*, 275 (2): 478–515, 2018.

6

Multiple Boundary Representations of λ-Harmonic Functions on Trees

Massimo A. Picardello and Wolfgang Woess

Abstract

We consider a countable tree T, possibly having vertices with infinite degree, and an arbitrary stochastic nearest neighbour transition operator P. We provide a boundary integral representation for general eigenfunctions of P with eigenvalue $\lambda \in \mathbb{C}$, under the condition that the oriented edges can be equipped with complex-valued weights satisfying three natural axioms. These axioms guarantee that one can construct a λ-Poisson kernel. The boundary integral is with respect to distributions, that is, elements in the dual of the space of locally constant functions. Distributions are interpreted as finitely additive complex measures. In general, they do not extend to σ-additive measures: for this extension, a summability condition over disjoint boundary arcs is required. Whenever λ is in the resolvent of P as a self-adjoint operator on a naturally associated ℓ^2-space and the diagonal elements of the resolvent ('Green function') do not vanish at λ, one can use the ordinary edge weights corresponding to the Green function and obtain the ordinary λ-Martin kernel.

We then consider the case when P is invariant under a transitive group action. In this situation, we study the phenomenon that in addition to the λ-Martin kernel, there may be further choices for the edge weights which give rise to another λ-Poisson kernel with associated integral representations. In particular, we compare the resulting distributions on the boundary.

The material presented here is closely related to the contents of our 'companion' paper [17].

The first-named author acknowledges support by MIUR Excellence Departments Project awarded to the Department of Mathematics, University of Rome Tor Vergata, CUP E83C18000100006. The second author was partially supported by Austrian Science Fund: project FWF W1230.

6.1 Introduction

Let T be a countable tree, i.e., a connected graph without cycles. We allow vertices with infinite degree, but for simplicity, we exclude leaves (vertices with degree 1). Here, the degree $\mathsf{deg}(x)$ of a vertex x is the number of its neighbours. We tacitly identify T with its vertex set.

On T, we consider the stochastic transition matrix $P = \big(p(x,y)\big)_{x,y \in T}$ of a nearest neighbour random walk. This means that $p(x,y) > 0$ if and only if $x \sim y$ (i.e., x and y are neighbours). P acts on functions $f : T \to \mathbb{C}$ by

$$Pf(x) = \sum_y p(x,y) f(y), \tag{6.1.1}$$

where in case when $\mathsf{deg}(x) = \infty$ we postulate that the sum converges absolutely. For $\lambda \in \mathbb{C}$, a *λ-harmonic function* is a function $h : T \to \mathbb{C}$ which satisfies $Ph = \lambda \cdot h$.

For 'good' values of λ, every λ-harmonic function has a boundary integral representation over the geometric boundary at infinity of the tree. This is analogous to the Poisson integral formula for classical harmonic functions on the open unit disc, where the boundary is the unit circle. The Poisson kernel of the disc has to be replaced by the *λ-Martin kernel,* and the integral is with respect to a *distribution* on the boundary. The good values include in particular $\lambda = 1$, when the random walk is transient. More generally, they comprise at least all $\lambda \in \mathbb{C}$ where $|\lambda| > \rho$ with $\rho = \rho(P)$, the *spectral radius* of the random walk (the definitions will be given in more detail further on). For *positive* λ-harmonic functions – whose existence necessitates that $\lambda \geqslant \rho$ is real – the representing distribution on the boundary is a finite (σ-additive) Borel measure.

The results that we have mentioned in this last paragraph are due to Cartier [5] for the case when $\lambda \geqslant \rho$ and the tree is locally finite, and the extension to the non-locally finite case can be found in the book of Woess [22, Ch. 9]. For general complex λ, these results are proved in our recent paper [17], when λ is in the resolvent set of P and the diagonal elements of the *Green kernel* (Green function) do not vanish at λ. This was preceded by a result of Figà-Talamanca and Steger [8] for the locally finite case, when P is the transition matrix of a group invariant random walk on a free group, or a close relative of that group.

All this comprises the long known example of the *simple random walk* on $T = \mathbb{T}_q$, the regular tree with degree $q + 1 \geqslant 3$, where $p(x, y) = 1/(q + 1)$ when $x \sim y$. In this case, it follows from the results of Mantero and Zappa [13] that, besides the ordinary λ-Martin kernel, there is a second kernel which gives rise to a boundary integral representation of λ-harmonic functions. Indeed, this plays an important role in the context of the representation theory of free groups. Since then, this phenomenon has remained the object of repeated discussions, in particular between the first author and David Singman (George Mason University, Fairfax).

The purpose of this chapter is to shed more light on these multiple boundary integral representation by approaching them from a wider viewpoint. Thereby, part of our presentation lays out in detail several proofs which take up and generalize previous work.

We first (Section 6.2) recall the construction of the boundary at infinity ∂T of T and the corresponding compactification. We introduce distributions on ∂T and explain how locally constant functions on ∂T are integrated against a distribution.

Then (Section 6.3) we start with an arbitrary $\lambda \in \mathbb{C}$ and put weights on the oriented edges of T. They are required to satisfy certain axioms (this might not be possible for all λ) and then they can be used to define a general λ-potential kernel and subsequently a λ-Poisson kernel $k(x, y)$, $x, y \in T$. This kernel extends in the second variable to a locally constant function on ∂T, and we use it to prove a general Poisson–Martin boundary integral representation theorem for λ-harmonic functions.

Let us write $\mathsf{res}^*(P)$ for the set of all elements in the resolvent set of P as a self-adjoint operator for which the diagonal matrix elements of the resolvent (λ-Green function) do not vanish. For $\lambda \in \mathsf{res}^*(P)$, the classical weights satisfying the needed axioms are suitable quotients of the λ-Green function, which we call the *Green weights.* This leads to the above-mentioned representation proved in [17] and the preceding work.

Later on (Section 6.4), we restrict attention to the case when P is invariant under a transitive group of automorphisms of T. In this situation, we discuss the cases where in addition to the classical ones, one can find different sets of weights which also lead to boundary integral representations for the same space of λ-harmonic functions. In this case, however, we show that the distribution which arises for a given λ-harmonic function does typically not extend to a (σ-additive) Borel measure on the boundary, even when this is true with respect to the Green weights.

6.2 Boundary and Distributions

The End Compactification

For two vertices $x, y \in T$, the *geodesic* or *geodesic path* from x to y is the unique shortest path $\pi(x, y)$ from x to y, and the distance $d(x, y)$ is the length (number of edges) of $\pi(x, y)$.

A *ray* or *geodesic ray* in T is a sequence $[x_0, x_1, x_2, \dots]$ such that $x_{i-1} \sim x_i$ and $x_{i+1} \neq x_{i-1}$ for all i. Two rays are *equivalent,* if they differ by finitely many initial vertices. An *end* of T is an equivalence class of rays. If x is a vertex and ξ an end, there is a unique geodesic ray $\pi(x, \xi)$ which starts at x and represents ξ. The boundary ∂T of T is the set of all ends of T. For $x, y \in T$ with $x \neq y$, the *branch* or *cone* $T_{x,y}$ is the subtree spanned by all vertices w with $y \in \pi(x, w)$, and the *boundary arc* $\partial T_{x,y}$ is the set of all ends which have a representative ray in $T_{x,y}$.

We set $\widehat{T} = T \cup \partial T$ and $\widehat{T}_{x,y} = T_{x,y} \cup \partial T_{x,y}$. We put the following topology on $\widehat{T}$: it is discrete on the vertex set, and a neighbourhood base of $\xi \in \partial T$ is given by the collection of all $\widehat{T}_{x,y}$ which contain a ray that represents ξ. (Here, we may fix x and vary only $y \neq x$.) The resulting space is metrizable. It is compact precisely when T is locally finite, but otherwise, it is not complete. This can be overcome as follows. For each vertex x with infinite degree – following an idea of Soardi [4] – we add a boundary point as follows: we introduce a new *improper vertex* x^*, the *shadow* of x, and we set $T^* = \{x^* : x \in T\,,\ \mathsf{deg}(x) = \infty\}$, as well as $\partial^* T = T^* \cup \partial T$. Analogously, $\partial^* T_{x,y} = T^*_{x,y} \cup \partial T_{x,y}$.

Let us write $\overline{T} = T^* \cup \widehat{T}$ and $\overline{T}_{x,y} = T^*_{x,y} \cup \widehat{T}_{x,y}$. Again, T is discrete in $\overline{T}$. A neighbourhood base of end $\xi \in \partial T$ is now provided by all $\overline{T}_{x,y}$ which contain a geodesic that represents ξ. A neighbourhood sub-base of $x^* \in T^*$ is given by all $\overline{T}_{v,x}$, where $v \sim x$.

We now describe convergence of sequences in $\overline{T}$ in this topology. We choose a *root* vertex $o \in T$ and write $T_x = T_{o,x}$ for any $x \in T$; in particular, $T_o = T$. Throughout everything which follows, it will be useful to define the *predecessor* x^- of a vertex $x \neq o$. This is the neighbour of x on the geodesic $\pi(o, x)$, and x is a called a *forward neighbour* of x^-. For $x \in T$, set $|x| = d(o, x)$, the graph distance. For $\xi \in \partial T$, set $|\xi| = \infty$.

For any pair of elements $v, w \in \widehat{T}$ (i.e., not in T^*), their *confluent* $v \wedge w$ with respect to o is the last common element on the geodesics $\pi(o, v)$ and $\pi(o, w)$. It is a vertex, unless $v = w \in \partial T$, in which case the confluent is that end. Now, if (w_n) is a sequence in $\widehat{T}$, then

- $w_n \to x \in T$ when $w_n = x$ for all but finitely n.
- $w_n \to \xi \in \partial T$ when $|w_n \wedge \xi| \to \infty$.
- $w_n \to x^* \in T^*$ when w_n 'rotates' around x, that is, any $y \sim x$ lies on at most finitely many geodesics $\pi(x, w_n)$.

Finally, if (x_n^*) is a sequence of improper vertices, then

- $x_n^* \to x^* \in T^*$ or $x_n^* \to \xi \in \partial T$ when $x_n \to x^*$, resp. $x_n \to \xi$ in the above sense.

Now $\overline{T}$ is compact, and T is an open-discrete subset, so that also $\partial^* T$ is compact. For the understanding of distributions, the next considerations will be useful. They follow Cartwright, Soardi, and Woess [4] (see also [22, Theorem 7.13]).

Let X be a countable set. By a *compactification* of X we mean a compact Hausdorff space into which X embeds as a discrete, open, dense subset. Now let $\mathcal{F}$ be a *countable* family of bounded functions $f : X \to \mathbb{R}$. Then there is a unique minimal compactification $\overline{X}_{\mathcal{F}}$ of X such that each $f \in \mathcal{F}$ extends to a continuous function on $\overline{X}_{\mathcal{F}}$. Here, 'minimal' refers to the partial order on compactifications where one is smaller than the other if the identity mapping on X extends to a continuous surjection from the bigger to the smaller one, and two compactifications are considered equal, if that extension is a homeomorphism.

Now let T be a countable tree (or any connected, countable graph) with edge set $E = \{(x, y) \in T^2 : x \sim y\}$. A function $f : T \to \mathbb{C}$ is called *locally constant,* if the set of edges along which f changes its value,

$$\{e = [x, y] \in E : f(x) \neq f(y)\},$$

is finite. The vector space $\mathcal{V}$ of all locally constant functions is spanned by the countable set $\mathcal{F}$ of all those functions in $\mathcal{V}$ which take values in $\{0, 1\}$. Therefore, in the corresponding compactification $\overline{T}_{\mathcal{F}}$, every locally constant function on T has a continuous extension. Now, as explained in [4], when the tree (graph) is locally finite, then one gets the well-known *end compactification.* When the tree T is not locally finite, we just get the compactification $\overline{T} = T^* \cup \widehat{T}$ described above.

For the purposes of this chapter, the improper vertices remain an artefact which provides compactness, but will not be used in a specific way, except to clarify the view on the subject.

Distributions on the Boundary

The following material is adapted and extended from [17]. Consider a function $f \in \mathcal{V}$. Let E_f be the finite set of edges along which f changes value. We can choose a finite subtree τ of T which contains all those edges as well as the chosen root o. For a vertex $x \in \tau$, write $S_x(\tau)$ for the set of forward neighbours y of x in τ (it may be empty). The *boundary* $\partial\tau$ of τ in T consists of those $x \in \tau$ which have a neighbour outside τ. For each $x \in \partial\tau$, the function f is constant on the part of the tree which branches off at x, which is

$$T_x(\tau) = T_x \setminus \bigcup \{T_y : y \in S_x(\tau)\}$$

Now let $\partial\mathcal{V}$ be the trace of the vector space $\mathcal{V}$ on ∂T, and define $\partial\mathcal{F}$ correspondingly. By the above considerations, each element of $\partial\mathcal{F}$ is the indicator function of a subset of ∂T which can be written as a finite disjoint union of sets of the form

$$\partial T_x \setminus \bigcup \{\partial T_y : y \in S_x\},$$

where $x \in T$ and S_x is a finite collection of forward neighbours of x (possibly empty). If ν is an element in the dual space of $\partial\mathcal{V}$, then it can be seen as a complex-valued set function on the collection of all those sets, and we call it a *distribution*. The following is now obvious.

Lemma 6.2.1 *Any distribution ν is characterized by the property that, for every $x \in T$ and every finite set S_x of forward neighbours of x,*

$$\nu(\partial T_x) = \sum_{y \in S_x} \nu(\partial T_y) + \nu\Big(\partial T_x \setminus \bigcup \{\partial T_y : y \in S_x\}\Big).$$

In particular, if T is locally finite, then ν can be described as a set function on all boundary arcs such that

$$\nu(\partial T_x) = \sum_{y : y^- = x} \nu(\partial T_y) \quad \text{for every } x \in T. \tag{6.2.1}$$

In [17], we have defined distributions analogously in the non-locally finite case, requiring in that case that the sum in (6.2.1) converges absolutely. In this case, let us call ν a *strong distribution* here. For all results of [17] as well as this chapter, the distributions actually involved are always strong.

In particular, when ν is non-negative real, then it is not only strong, but extends to a finite, σ-additive Borel measure on ∂T, as explained in [17, 3.10]. As mentioned there, when ν is a complex-valued distribution, a necessary and sufficient condition for its extendability to a σ-additive, signed measure on the

Borel σ-algebra of ∂T is that there is $M < \infty$ such that for any sequence of pairwise disjoint boundary arcs ∂T_{x_n}, one has

$$\sum_n |\nu(\partial T_{x_n})| \leqslant M. \tag{6.2.2}$$

This is an easy extension of the corresponding condition in the locally finite case of Cohen, Colonna, and Singman [6].

For any distribution ν, we now write

$$\nu(\varphi) = \int_{\partial T} \varphi \, d\nu \quad \text{for} \quad \varphi \in \partial\mathcal{V}.$$

By the above, given φ, there are a finite subtree τ of T containing o and constants φ_x, $x \in \partial\tau$, such that

$$\begin{aligned} &\varphi \equiv \varphi_x \quad \text{on} \quad \partial T_x(\tau) = \partial T_x \setminus \bigcup\{\partial T_y : y \in S_x(\tau)\}, \quad \text{and} \\ &\int_{\partial T} \varphi \, d\nu = \sum_{x \in \partial\tau} \varphi_x \, \nu\big(\partial T_x(\tau)\big). \end{aligned} \tag{6.2.3}$$

By construction, this does not depend on the specific choice of the finite tree τ associated with φ. If ν extends to a σ-additive complex Borel measure on ∂T, then the integral is an ordinary one in the sense of Lebesgue.

Self-Adjointness of the Transition Operator

With the action defined by (6.1.1), the transition operator P is self-adjoint on the Hilbert space $\ell^2(T, \mathsf{m})$ of all functions $f : T \to \mathbb{C}$ with $\langle f, f \rangle < \infty$, where

$$\langle f, g \rangle = \sum_x f(x)\overline{g(x)}\, \mathsf{m}(x),$$

with the measure m on T as follows:

$$\text{for } x \in T \text{ with } \pi(o, x) = [x_0, x_1, \dots, x_k],$$
$$\mathsf{m}(x) = \frac{p(x_0, x_1) \cdots p(x_{k-1}, x_k)}{p(x_1, x_0) \cdots p(x_k, x_{k-1})}.$$

In particular, $\mathsf{m}(o) = 1$. Self-adjointness is a consequence of *reversibility:* $\mathsf{m}(x)p(x, y) = \mathsf{m}(y)p(y, x)$ for all x, y. The norm (spectral radius) of P is

$$\rho = \rho(P) = \limsup_{n \to \infty} p^{(n)}(x, y)^{1/n}$$

(independent of x and y), where $p^{(n)}(x, y)$ is the (x, y)-element of the matrix power P^n. Since trees are bipartite, the spectrum $\mathsf{spec}(P) \subset [-\rho, \rho]$ is symmetric around the origin.

Positive λ-harmonic functions exist if and only if $\lambda \geqslant \rho$ (real). At this point, we state a warning: when viewing λ-harmonic functions as 'eigenfunctions' of P, they are not considered as eigenfunctions of the above self-adjoint operator on $\ell^2(T, \mathsf{m})$. As a matter of fact, besides possibly for $\lambda = \pm\rho$ in very specific situations, our λ-harmonic functions will usually not belong to $\ell^2(T, \mathsf{m})$. In a variety of known cases, $\mathsf{spec}(P)$ contains no eigenvalues, that is, there is no point spectrum on $\ell^2(T, \mathsf{m})$. In any case, our methods and results do not cover the case where $\lambda \in \mathsf{spec}(P) \setminus \{\pm\rho\}$.

6.3 The General Integral Representation

We now fix a candidate eigenvalue $\lambda \in \mathbb{C}$ and we suppose that we can equip the *oriented* edge set $E(T) = \{(x, y) \in T^2 : x \sim y\}$ of T with λ-*weights* $f(x, y) \in \mathbb{C}$ satisfying the following properties for every $x \in T$ and every y with $x \sim y$.

$$f(x, y) f(y, x) \neq 1 \quad \text{for all pairs of neighbours } x, y, \tag{6.3.1}$$

$$u(x, x) \neq \lambda, \quad \text{where} \quad u(x, x) = \sum_v p(x, v) f(v, x), \tag{6.3.2}$$

$$\lambda f(x, y) = p(x, y) + \Big(u(x, x) - p(x, y) f(y, x)\Big) f(x, y). \tag{6.3.3}$$

If $\mathsf{deg}(x) = \infty$ then we require that the sum in (6.3.2) converges absolutely. Note that it follows from (6.3.3) that $f(x, y) \neq 0$ for all pairs of neighbours. The above three axioms arise by mimicking the main properties of the natural Green weights, which will be discussed at the end of this section.

Using these weights, for arbitrary $x, y \in T$ we define

$$f(x, y) = f(x_0, x_1) f(x_1, x_2) \cdots f(x_{k-1}, x_k), \quad \text{if} \quad \pi(x, y) = [x_0, \dots, x_k]. \tag{6.3.4}$$

In particular, $f(x, x) = 1$.

Lemma 6.3.1 *For fixed y, the function* $x \mapsto f(x, y)$ *satisfies*

$$Pf(x, y) = \lambda f(x, y) \quad \textit{if } x \neq y, \quad \textit{and} \quad Pf(y, y) = u(y, y).$$

Proof. The second identity is definition (6.3.2) of u. For the first identity, let $\pi(x, y)$ be as in (6.3.4). Consider the neighbours $x = x_0$ and x_1. Then (6.3.2) and (6.3.3) yield

$$
\begin{aligned}
Pf(x,y) &= p(x,x_1)f(x_1,y) + \sum_{v\neq x_1} p(x,v)f(v,x)f(x,y) \\
&= p(x,x_1)f(x_1,y) + \Big(u(x,x) - p(x,x_1)f(x_1,x)\Big) \\
&\qquad\qquad \times f(x,x_1)f(x_1,y) \\
&= \lambda\, f(x,x_1)f(x_1,y) = \lambda\, f(x,y),
\end{aligned}
$$

as stated. □

Note that absolute convergence of the sum in (6.3.2) is crucial for the lemma. It is this property that further on will give us strong distributions. Thanks to (6.3.2) we can set

$$
g(x,y) = \frac{f(x,y)}{\lambda - u(y,y)}, \qquad x,y \in T. \tag{6.3.5}
$$

Then we see from Lemma 6.3.1 that the function $x \mapsto g(x,y)$ satisfies the resolvent equation

$$
Pg(x,y) = \lambda\, g(x,y) - \delta_x(y). \tag{6.3.6}
$$

The following lemma shows how the transition probabilities can be recovered from the weights $f(x,y)$, compare with [5] for the locally finite case with standard non-negative Green weights.

Lemma 6.3.2 *For $x \in T$ and $y \sim x$,*

$$
\begin{aligned}
g(x,x)p(x,y) &= \frac{f(x,y)}{1 - f(x,y)f(y,x)}, \\
g(x,x)g(y,y) &= g(x,y)\left(\frac{1}{p(x,y)} + g(y,x)\right), \quad \textit{and} \\
\lambda\, g(x,x) &= 1 + \sum_{y:\, y\sim x} \frac{f(x,y)f(y,x)}{1 - f(x,y)f(y,x)}.
\end{aligned}
$$

When $\mathsf{deg}(x) = \infty$, the last sum converges absolutely.

Proof. We can rewrite (6.3.3) as

$$
p(x,y)\big(1 - f(x,y)f(y,x)\big) = f(x,y)\big(\lambda - u(x,x)\big).
$$

Since $\lambda - u(x,x) = 1/g(x,x)$, the first identity follows, and with $g(x,y) = f(x,y)g(y,y)$ as well as $g(y,x) = f(y,x)g(x,x)$, we get

$$
\begin{aligned}
&g(x,y)\Big(\frac{1}{p(x,y)} + g(y,x)\Big)\\
&= g(x,x)g(y,y)\ \underbrace{f(x,y)\Big(\frac{1}{g(x,x)p(x,y)} + f(y,x)\Big)}_{=1}.
\end{aligned}
$$

This is the second identity. We now multiply the first identity with $f(y,x)$. The sum over all neighbours y of x is absolutely convergent by assumption, so that we have indeed abolute convergence of the right-hand side of the third identity, and

$$
\sum_{y:y\sim x} \frac{f(x,y)f(y,x)}{1-f(x,y)f(y,x)} = g(x,x)\,u(x,x) = \lambda\, g(x,x) - 1
$$

by the definition of $g(x,x)$. □

We define the λ-*Poisson kernel* associated with our weights by

$$
k(x,w) = \frac{f(x, x\wedge w)}{f(o, x\wedge w)}, \quad x \in T\,,\ w \in \widehat{T}.
$$

Thus,

$$
k(x,w) = \frac{f(x,v)}{f(o,v)} = \frac{g(x,v)}{g(o,v)} \quad \text{for every vertex } v \in \pi(x\wedge w, w). \tag{6.3.7}
$$

By our assumptions, $Pk(\cdot, w)$ is well defined as a function of the first variable. That is, even at vertices with infinite degree, the involved sum is absolutely convergent, and for $\xi \in \partial T$,

$$
\sum_{y\sim x} p(x,y)\,k(y,\xi) = \lambda\, k(x,\xi) \quad \text{for every } x \in T. \tag{6.3.8}
$$

Now let $x \in T$ and $\pi(o,x) = [o = x_0\,, x_1\,, \dots, x_k = x]$. Then $x \wedge \xi \in \{x_0\,, x_1\,, \dots, x_k\}$ for every $\xi \in \partial T$, and

$$
k(x,\xi) = k(x,x_i) \quad \text{when} \quad \begin{cases} \xi \in \partial T_{x_i} \setminus \partial T_{x_{i+1}}\,, & i \leqslant k-1,\\ \xi \in \partial T_{x_k}\,, & i = k. \end{cases}
$$

Thus, $\varphi = k(x,\cdot)$ is locally constant on ∂T, and we can use $\pi(o,x)$ as a tree τ to which (6.2.3) applies. Then we have the following.

Proposition 6.3.3 *If ν is a strong distribution ν on ∂T, its* Poisson transform

$$
h(x) = \int_{\partial T} k(x,\xi)\,d\nu(\xi)
$$

is a λ*-harmonic function, and*

$$\begin{aligned} h(x) &= \sum_{i=0}^{k-1} k(x, x_i)\Big(\nu(\partial T_{x_i}) - \nu(\partial T_{x_{i+1}})\Big) + k(x, x)\,\nu(\partial T_x) \\ &= k(x, o)\nu(\partial T) + \sum_{i=1}^{k}\Big(k(x, x_i) - k(x, x_{i-1})\Big)\nu(\partial T_{x_i}). \end{aligned} \tag{6.3.9}$$

Proof. The proof of λ-harmonicity of h is obvious when T is locally finite. Otherwise, some care is needed, and we go through the details in order to show the necessity of the assumption that the distribution ν be strong. First of all, we show that $Ph(o) = \lambda\, h(o)$. By (6.3.9), if $x \sim o$,

$$h(x) = f(x, o)\nu(\partial T) + \left(\frac{1}{f(o, x)} - f(x, o)\right)\nu(\partial T_x). \tag{6.3.10}$$

Therefore

$$\begin{aligned} Ph(o) &= u(o, o)\,\nu(\partial T) + \sum_{x\sim o} p(o, x)\,\frac{1 - f(o, x)f(x, o)}{f(o, x)}\,\nu(\partial T_x) \\ &= \sum_{x\sim o}\left(u(o, o) + p(o, x)\,\frac{1 - f(o, x)f(x, o)}{f(o, x)}\right)\nu(\partial T_x) \\ &= \lambda \sum_{x\sim o} \nu(\partial T_x) = \lambda\,\nu(\partial T) = \lambda\, h(o) \end{aligned}$$

by (6.3.3). In case $\deg(x) = \infty$, we needed absolute convergence of the involved series. Similarly, let $x \neq o$. Then (6.3.9) yields the formula

$$h(x) = f(x, x^-)h(x^-) + \frac{1 - f(x, x^-)f(x^-, x)}{f(o, x)}\,\nu(\partial T_x), \tag{6.3.11}$$

that will also be important further below. To prove (6.3.11), we first observe that it is the same as (6.3.10) when $x \sim o$. Now let $k \geqslant 2$ in (6.3.9), and note that for $i \leqslant k-1$ we have $k(x, x_i) = f(x, x_i)/f(o, x_i) = f(x, x^-)k(x^-, x_i)$, with $x^- = x_{k-1}$. Then, using the first of the two identities of (6.3.9),

$$\begin{aligned} h(x) &= f(x, x^-)\sum_{i=0}^{k-2} k(x^-, x_i)\Big(\nu(\partial T_{x_i}) - \nu(\partial T_{x_{i+1}})\Big) \\ &\quad + f(x, x^-)k(x^-, x^-)\Big(\nu(\partial T_{x^-}) - \nu(\partial T_x)\Big) + k(x, x)\,\nu(\partial T_x) \\ &= f(x, x^-)h(x^-) - \frac{f(x, x^-)}{f(o, x^-)}\nu(\partial T_x) + \frac{1}{f(o, x)}\nu(\partial T_x). \end{aligned}$$

Since $f(o, x) = f(o, x^-)f(x^-, x)$, this reduces to the desired formula. For the following, we also need (6.3.11) for y with $y^- = x$. Absolute convergence in the first of the following identities is justified a posteriori, and the first identity of Lemma 6.3.2 is used for the under-braced as well as for the over-braced term, and again in the very last step:

$$
\begin{aligned}
Ph(x) &= p(x,x^-)h(x^-) + \sum_{y^-=x} p(x,y)f(y,x)h(x) \\
&\quad + \frac{1}{f(o,x)} \underbrace{\sum_{y^-=x} p(x,y)\frac{1-f(x,y)f(y,x)}{f(x,y)}}_{1/g(x,x)} \nu(\partial T_y) \\
&= \frac{p(x,x^-)}{f(x,x^-)}h(x) - \frac{1}{f(o,x)}\, \overbrace{p(x,x^-)\frac{1-f(x,x^-)f(x^-,x)}{f(x,x^-)}} \nu(\partial T_x) \\
&\quad + u(x,x)h(x) - p(x,x^-)f(x^-,x)h(x) + \frac{1}{g(o,x)}\nu(\partial T_x) \\
&= \left(p(x,x^-)\frac{1-f(x,x^-)f(x^-,x)}{f(x,x^-)} + u(x,x)\right) h(x) = \lambda\, h(x).
\end{aligned}
$$

In the second identity we made use of the assumption that ν is strong, and in the last identity we have used (6.3.3). □

The proof of the following is very similar to [22, Theorem 9.37]: we rewrite its main part here to take care of absolute convergence in the non-locally finite case.

Theorem 6.3.4 *Suppose that we have edge weights $f(x,y)$ which satisfy (6.3.1)–(6.3.3). A function $h : T \to \mathbb{C}$ satisfies $Ph = \lambda \cdot h$ if and only if it is of the form*

$$h(x) = \int_{\partial T} k(x,\xi)\, d\nu(\xi),$$

where ν is a strong complex distribution on ∂T. The distribution ν is determined by h, that is, $\nu = \nu^h$, where

$$\nu^h(\partial T) = h(o) \quad \text{and} \quad \nu^h(\partial T_x) = f(o,x)\,\frac{h(x) - f(x,x^-)h(x^-)}{1 - f(x^-,x)f(x,x^-)}, \quad x \neq o.$$

Proof. We first show that if h is λ-harmonic, then ν^h as defined in the theorem is indeed a strong distribution, and h is its Poisson transform. We start with the identity

$$\lambda\, g(x,x)h(x) = \sum_{y:y\sim x} g(x,x)p(x,y)h(y),$$

and recall that the sum on the right-hand side is assumed to converge absolutely when $\mathsf{deg}(x) = \infty$. Using Lemma 6.3.2, we rewrite this as

$$\left(1 + \sum_{y:y\sim x} \frac{f(x,y)f(y,x)}{1 - f(x,y)f(y,x)}\right)h(x) = \sum_{y:y\sim x} \frac{f(x,y)}{1 - f(x,y)f(y,x)}\, h(y).$$

Since the involved sums converge absolutely, we can regroup the terms and get

$$h(x) = \sum_{y\,:\,y\sim x} f(x,y)\,\frac{h(y) - f(y,x)h(x)}{1 - f(x,y)f(y,x)}. \tag{6.3.12}$$

Convergence is again absolute when $\mathsf{deg}(x) = \infty$.

For $x = o$, the last identity says that $\nu^h(\partial T) = \sum_{y\sim o} \nu^h(\partial T_y)$. If $x \neq o$, then by (6.3.12),

$$\begin{aligned}
\sum_{y:y^-=x} \nu^h(\partial T_y) &= f(o,x) \sum_{y\,:\,y^-=x} f(x,y)\frac{h(y) - f(y,x)h(x)}{1 - f(x,y)f(y,x)}\\
&= f(o,x)\left(h(x) - f(x,x^-)\frac{h(x^-) - f(x^-,x)h(x)}{1 - f(x,x^-)f(x^-,x)}\right)\\
&= f(o,x)\frac{h(x) - f(x,x^-)h(x^-)}{1 - f(x,x^-)f(x^-,x)} = \nu^h(\partial T_x).
\end{aligned}$$

So ν^h is indeed a signed distribution on $\mathcal{F}_o$. We verify that $\int_{\partial T} k(x,\xi)\,d\nu^h(\xi) = h(x)$. For $x = o$ this is clear, so let $x \neq o$. With notation as in (6.3.9), we simplify

$$\Big(k(x,x_i) - k(x,x_{i-1})\Big)\,\nu^h(\partial T_{x_i}) = f(x,x_i)h(x_i) - f(x,x_{i-1})h(x_{i-1}),$$

whence we obtain

$$\begin{aligned}
\int_{\partial T} K(x,\xi)\,d\nu^h(\xi) &= k(x,o)h(o) + \sum_{i=1}^{k}\Big(f(x,x_i)h(x_j) - f(x,x_{i-1})h(x_{i-1})\Big)\\
&= f(x,x)h(x) = h(x),
\end{aligned}$$

as stated.

Second, we need to verify that given ν and its Poisson transform h, we have $\nu = \nu^h$. This part of the proof is nothing but identity (6.3.11) in the proof of Proposition 6.3.3. □

The Natural Green weights

We now 'reveal' the origin of the axioms (6.3.1)–(6.3.3) for the edge weights. Let $\mathsf{res}(P)$ be the resolvent set of the self-adjoint operator P acting on $\ell^2(T, \mathsf{m})$ according to Section 6.2.C. For $\lambda \in \mathsf{res}(P)$, we write $\mathfrak{G}(\lambda) = (\lambda \cdot I - P)^{-1}$ for resolvent operator. Its matrix element

$$G(x, y|\lambda) = \mathfrak{G}(\lambda)\mathbf{1}_y(x)\,, \quad x, y \in T, \tag{6.3.13}$$

is the *Green function.* It is an analytic function of $\lambda \in \mathsf{res}(P) \supset \mathbb{C} \setminus [-\rho\,,\,\rho]$, and for $|\lambda| > \rho$,

$$G(x, y|\lambda) = \sum_{n=0}^{\infty} p^{(n)}(x, y)\,\lambda^{-n-1}.$$

At $\lambda = \rho$, the latter series converge or diverge simultaneously for all x, y. If they converge, i.e., $G(x, y|\rho) < \infty$ for all x, y, then P, resp. the associated random walk, is called *ρ-transient,* and otherwise it is called *ρ-recurrent.* Set

$$\mathsf{res}^*(P) = \big\{\lambda \in \mathsf{res}(P) : G(x, x|\lambda) \neq 0 \text{ for all } x \in T\big\}. \tag{6.3.14}$$

For $\lambda \in \mathsf{res}^*(P)$,

$$F(x, y|\lambda) = G(x, y|\lambda)/G(y, y|\lambda)\,, \quad x, y \in T, \tag{6.3.15}$$

is an analytic function of λ. For $|\lambda| \geqslant \rho$,

$$F(x, y|\lambda) = \sum_{n=0}^{\infty} f^{(n)}(x, y)/\lambda^n, \tag{6.3.16}$$

where $f^{(n)}(x, y)$ is the probability that the random walk starting at x hits y at time $n \geqslant 0$ for the first time. Also,

$$U(x, x|\lambda) = \sum_{y \sim x} p(x, y) F(y, x|\lambda) = \sum_{n=1}^{\infty} u^{(n)}(x, x)/\lambda^n,$$

where $u^{(n)}(x, x)$ is the probability that the random walk starting at x returns to x at time $n \geqslant 1$ for the first time. Now it is well known, and also explained in [5], [22] as well as in [17], that the edge weights

$$f(x, y) = F(x, y|\lambda)\,, \quad x, y \in T,\ x \sim y$$

are λ-weights which fulfil the requirements (6.3.1)–(6.3.3) for $\lambda \in \mathsf{res}^*(P)$, and for arbitrary $x, y \in T$,

$$\begin{aligned} F(x, y|\lambda) &= F(x_0\,, x_1|\lambda) \cdots F(x_{k-1}\,, x_k|\lambda)\,, \quad \text{where} \\ \pi(x, y) &= [x = x_0\,, x_1\,, \ldots, x_k = y]. \end{aligned} \tag{6.3.17}$$

With notation as in Section 6.3, we also have $u(x, x) = U(x, x|\lambda)$ and $g(x, y) = G(x, y|\lambda)$. The associated kernel according to (6.3.7), called the *λ-Martin kernel,* is

$$k(x, w) = K(x, w|\lambda) = \frac{G(x, v|\lambda)}{G(o, v|\lambda)} \quad \text{for every vertex } v \in \pi(x \wedge w, w), \tag{6.3.18}$$

where $x \in T$ and $w \in \widehat{T}$. All this also works for $\lambda = \pm\rho$ in the ρ-transient case. Thus, Theorem 6.3.4 yields the following, which we restate here once again.

Corollary 6.3.5 *For $\lambda \in \mathsf{res}^*(P)$, as well as for $\lambda = \pm\rho$ in the ρ-transient case, every λ-harmonic function h has an integral representation*

$$h(x) = \int_{\partial T} K(x, \xi|\lambda)\, d\nu(\xi).$$

The strong complex distribution $\nu = \nu^h$ on ∂T is determined by h,

$$\nu^h(\partial T) = h(o) \quad \text{and} \quad \nu^h(\partial T_x) = F(o, x|\lambda)\, \frac{h(x) - F(x, x^-|\lambda)h(x^-)}{1 - F(x^-, x|\lambda)F(x, x^-|\lambda)}$$

for $x \neq o$.

As already mentioned, this general result of [17] was preceded by various earlier ones, starting with the seminal paper [5] (that deals with locally finite trees and positive $\lambda > \rho$, and also $\lambda = \rho$ in the ρ-transient case), and another proof in [16]. In [8], one finds the result for complex λ in the locally finite case corresponding to nearest neighbour group invariant random walks on free groups (resp. closely related groups freely generated by involutions): the special case of the simple random walk in this environment goes back to [13]. A first proof for the non-locally finite case and $\lambda = 1$ (transient case) is in [22, §9.D].

Remark 6.3.6 If $\lambda \geqslant \rho$, or if $\lambda = \rho$ in the ρ-transient case, it is a well-known fact that for any positive λ-harmonic function h, one has

$$F(x, y|\lambda)\, h(y) \leqslant h(x) \quad \text{for all } x, y.$$

(This holds for any irreducible Markov chain.) In particular, the distribution ν^h of Corollary 6.3.5 is non-negative, whence it extends to a σ-additive measure on ∂T, and Corollary 6.3.5 leads to the classical Poisson–Martin representation. Furthermore, in that case, the real λ-harmonic functions which are Poisson transforms of *σ-additive* signed Borel measures on ∂T are precisely the differences of non-negative λ-harmonic functions. For the complex-valued case, the situation is analogous. □

There are many analogies between the structure, group actions, harmonic analysis, and potential theory on trees (in particular, regular trees) and the Poincaré disc, that is, the open unit disc with the hyperbolic metric. The discrete Laplacian P–I arising from a random walk on T is an analogue of the hyperbolic Laplace–Beltrami operator on the disc. See, e.g., Boiko and

Woess [2] for a mostly potential theoretic 'dictionary' regarding the correspondences. In this sense, our representation Theorem 6.3.4 should be seen as a discrete analogue of a result of Helgason [11] for a Poisson-type integral representation of *all* harmonic functions on rank 1 symmetric spaces, and in particular, the hyperbolic disc: see the beautifully written exposition by Eymard [7]. There, the integral representation is with respect to *analytic functionals* on the boundary (the unit circle), of which our strong distributions are the analogues in the tree setting.

6.4 Twin Kernels for Affine and Simple Random Walks

As we have seen above, the natural version of Theorem 6.3.4 is the one where the λ-weights are $f(x,y) = F(x,y|\lambda)$, where $\lambda \in \mathsf{res}^*(P)$, resp. $\lambda = \pm\rho$ in the ρ-transient case.

Now, there are cases where one has another choice for the collection of λ-weights $f(x,y)$ satisfying (6.3.1)–(6.3.3), leading to another kernel which can also be used to describe the λ-harmonic functions of P. The main aim of this section is to obtain a better understanding of such twin kernels and the different integral representations for a class of random walks which includes the simple random walk on a homogeneous tree.

We consider $T = \mathbb{T}_q$, the homogeneous tree with degree $q+1$, where $q \geqslant 1$. In case $q = 1$, this is just the bi-infinite integer line $\mathbb{Z}$.

For any end ξ of T, we define the associated *horocycle index*

$$\mathfrak{h}(x,\xi) = d(x, x \wedge \xi) - d(o, x \wedge \xi) \in \mathbb{Z},$$

(we recall that $\wedge$ stands for taking the confluent with respect to o). In addition to the root vertex, we choose and fix a reference end ϖ and write $\mathfrak{h}(x) = \mathfrak{h}(x,\varpi)$. The *horocycles* are the resulting level sets: $\mathfrak{H}_k = \{x \in T : \mathfrak{h}(x) = k\}$, $k \in \mathbb{Z}$. Thus, (following Cartier) one can imagine the tree as an infinite genealogical tree, where ϖ is the mythical ancestor, and the horocycles are the successive generations. Each of them is infinite, and each $x \in \mathfrak{H}_k$ has precisely one neighbour (parent) in $\mathfrak{H}_{k-1}$ and q neighbours (children) in $\mathfrak{H}_{k+1}$ (see Figure 6.1). The subgroup of $\mathsf{Aut}(\mathbb{T}_q)$ which preserves this genealogical order, i.e., the group of automorphisms which fix ϖ, is called the *affine group* $\mathsf{Aff}(\mathbb{T}_q)$ of $\mathbb{T}_q$. It was shown to be amenable by Nebbia [15], but non-unimodular for $q \geqslant 2$ (see Trofimov [19]). We note that the indexing of the horocycles here is opposite to the one which is commonly used in the unit disc, resp. hyperbolic upper half plane. The reason lies in the opposite behaviour of

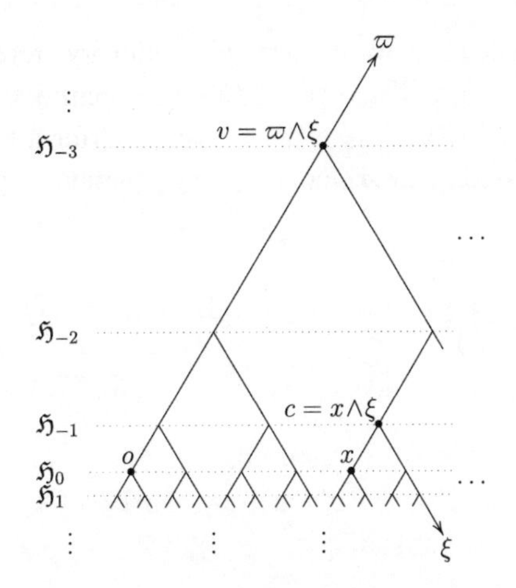

Figure 6.1 The regular tree in horocyclic layers

absolute values and q-adic norms. Very general random walks on $\mathsf{Aff}(\mathbb{T}_q)$ were studied in detail by Cartwright, Kaimanovich, and Woess [3].

Here we only consider nearest neighbour walks which are invariant under that group. Their transition probabilities are parametrized by an $\alpha \in (0\,,\ 1)$ as follows:

$$\text{for } x \sim y\,, \quad p(x,y) = \begin{cases} \alpha/q\,, & \text{if } \mathfrak{h}(y) = \mathfrak{h}(x) + 1, \\ 1 - \alpha\,, & \text{if } \mathfrak{h}(y) = \mathfrak{h}(x) - 1. \end{cases} \tag{6.4.1}$$

The simple random walk arises when $\alpha = q/(q+1)$. It is easy to see, and a consequence of the next computations, that the spectral radius is

$$\rho = \rho(P) = 2\sqrt{\alpha(1-\alpha)}.$$

Remark 6.4.1 In the group invariant case, $G(\lambda) = G(x,x|\lambda)$ is independent of x. (Do not confuse this with the resolvent operator $\mathfrak{G}(\lambda)$, of which $G(\lambda) = \mathfrak{G}(\lambda)\mathbf{1}_x(x)$ is the diagonal matrix element.) In the present example, we can use the argument at the end of [17, Remark 2.8] to see that $G(\lambda) \neq 0$ for any $\lambda \in \mathsf{res}(P)$. Indeed, as stated there, if $G(\lambda) = 0$ then some and thus every $x \in T$ would have a *unique* neighbour y such that $p(x,y)\,G(y,x|\lambda) = p(y,x)\,G(x,y|\lambda) = -1$. But since $\mathsf{Aff}(\mathbb{T}_q)$ acts transitively on the edges (preserving orientation, hence the 'parent relation'), this would hold for all pairs of neighbours, a contradiction. $\square$

We shall of course see this via explicit computation in a moment. By group invariance, there are only two types of functions $F(x,y|\lambda)$ for neighbours x, y. We set $F_+(\lambda) = F(x,y|\lambda)$ when $\mathfrak{h}(y) = \mathfrak{h}(x) + 1$, and $F_-(\lambda) = F(x,y|\lambda)$

when $\mathfrak{h}(y) = \mathfrak{h}(x) - 1$. As we have mentioned in Section 6.3.C, these functions, as λ-weights on the edges, satisfy (6.3.1)–(6.3.3): see [17, Lemma 2.3]. A priori, this is true for $|\lambda| > \rho$, and for other $\lambda \in \mathsf{res}(P)$, one uses analytic continuation. Now (6.3.3) yields the following equations.

$$\lambda\, F_-(\lambda) = (1-\alpha) + \alpha\, F_-(\lambda)^2\,, \quad \text{and} \tag{6.4.2}$$

$$\lambda\, F_+(\lambda) = \frac{\alpha}{q} + (1-\alpha) F_+(\lambda)^2 + \frac{q-1}{q}\, \alpha\, F_-(\lambda) F_+(\lambda). \tag{6.4.3}$$

Throughout this chapter we make the following habitual choice.

Convention 6.4.2 Our usual choice for the analytic continuation of the square root is the one on the slit plane without the negative half-axis, that is, $\sqrt{r\, e^{i\theta}} = \sqrt{r}\, e^{i\theta/2}$ for $r > 0$ and $-\pi < \theta < \pi$.

With this in mind, equation (6.4.2) has the two solutions

$$\begin{aligned} F_-(\lambda) &= \frac{\lambda}{2\alpha}\Big(1 - \sqrt{1 - 4\alpha(1-\alpha)/\lambda^2}\Big), \\ \widetilde{F}_-(\lambda) &= \frac{\lambda}{2\alpha}\Big(1 + \sqrt{1 - 4\alpha(1-\alpha)/\lambda^2}\Big). \end{aligned} \tag{6.4.4}$$

The solution that gives rise to the function defined in (6.3.16), and thus is associated with the resolvent $G(x, y|\lambda)$, is given by the convergent series (6.3.16) in powers of $1/\lambda$. It must be analytic for $\lambda \in \mathbb{C} \setminus [-\rho\,, \rho]$ and decreasing for real $\lambda > \rho$, so it is the former in (6.4.4). If we insert it into (6.4.3), then we get once more two solutions:

$$F_+(\lambda) = \frac{\alpha}{(1-\alpha)q}\, F_-(\lambda) \quad \text{and} \quad \widehat{F}_+(\lambda) = \frac{\alpha}{(1-\alpha)}\, \widetilde{F}_-(\lambda) \tag{6.4.5}$$

Again, the solution corresponding to the resolvent is $F_+(\lambda)$. On the other hand, if in (6.4.3) we insert $\widetilde{F}_-(\lambda)$ instead of $F_-(\lambda)$, then we get the following two other solutions of that equation:

$$\widetilde{F}_+(\lambda) = \frac{\alpha}{(1-\alpha)q}\, \widetilde{F}_-(\lambda) \quad \text{and} \quad \widehat{\widetilde{F}}_+(\lambda) = \frac{\alpha}{(1-\alpha)}\, F_-(\lambda). \tag{6.4.6}$$

A priori, we might consider to use any of the four pairs $(F_-\,, F_+)$, $(\widetilde{F}_-\,, \widetilde{F}_+)$, $(F_-\,, \widehat{F}_+)$, and $(\widetilde{F}_-\,, \widehat{\widetilde{F}}_+)$ for defining weights $f(x, y)$ on the edges in a way which remains invariant under $\mathsf{Aff}(\mathbb{T}_q)$. But $F_-(\lambda)\, \widehat{F}_+(\lambda) = \widetilde{F}_-(\lambda) \widehat{\widetilde{F}}_+(\lambda) = -1$, and this is not compatible with (6.3.1).

Thus, we have the natural choice $(F_-\,, F_+)$ and the 'twin' $(\widetilde{F}_-\,, \widetilde{F}_+)$. The weights provided by $\big(F_-(\lambda), F_+(\lambda)\big)$ in the sense of Section 6.3 are the Green weights, $f(x, y) = F_\pm(\lambda)$ for neighbours x, y with $\mathfrak{h}(y) = \mathfrak{h}(x) \pm 1$. An easy consequence of (6.3.5) is

$$G(x, x|\lambda) = G(\lambda) = \frac{2q/\lambda}{(q-1) + (q+1)\sqrt{1 - 4\alpha(1-\alpha)/\lambda^2}}.$$

We remark that from this one can deduce by classical spectral methods that $\mathsf{spec}(P) = [-\rho\,,\,\rho]$, where $\rho = 2\sqrt{\alpha(1-\alpha)}$. Namely, $G(x, x|\lambda)$ is the *Stieltjes transform* of the *Plancherel* or *spectral measure*, also called *KNS-measure* by Grigorchuk and Żuk [9]. That measure is the diagonal element of the resolution of the identity of the operator P; in the context of infinite graphs, see, e.g., Mohar and Woess [14]. Some more details will be considered in Section 6.5. The measure, and in this case, its density with respect to Lebesgue measure, can be computed via the inversion formula of Stieltjes–Perron (see Wall [20]). The spectrum is the support of that measure.

We also observe that our random walk is ρ-transient precisely when $q \geqslant 2$. We see that the Green weights fulfil the requirements (6.3.1)–(6.3.3) for any $\lambda \in \mathbb{C} \setminus [-\rho\,,\,\rho]$, as well as for $\lambda = \pm\rho$ when $q \geqslant 2$.

On the other hand, the only value of λ for which (6.3.1) does not hold, i.e., $\widetilde{F}_-(\lambda)\widetilde{F}_+(\lambda) = 1$, is

$$\lambda_0 = \frac{q+1}{2\sqrt{q}}\,\rho = \frac{\rho}{\rho(\mathrm{SRW})}, \tag{6.4.7}$$

where $\rho(\mathrm{SRW})$ is the spectral radius of the simple random walk on $\mathbb{T}_q$, that is, the random walk that arises for $\alpha = q/(q+1)$. It is also easy to check that

$$\begin{aligned}\widetilde{U}(\lambda) = \widetilde{U}(x, x|\lambda) &:= \sum_{y \sim x} p(x, y)\widetilde{F}(y, x|\lambda) \\ &= \frac{q+1}{2q}\lambda\Big(1 + \sqrt{1 - 4\alpha(1-\alpha)/\lambda^2}\Big)\end{aligned}$$

satisfies $\widetilde{U}(\lambda) = \lambda$ precisely when $\lambda = \lambda_0$. Thus, using $\big(\widetilde{F}_-(\lambda), \widetilde{F}_+(\lambda)\big)$, the weights $\tilde{f}(x, y) = \widetilde{F}_\pm(\lambda)$ for $x \sim y$ with $\mathfrak{h}(y) = \mathfrak{h}(x) \pm 1$ fulfil the requirements (6.3.1)–(6.3.3) for any $\lambda \in \mathbb{C} \setminus (-\rho\,,\,\rho)$, with the exception of λ_0.

According to (6.3.17), resp (6.3.4), for $\lambda \in \mathbb{C} \setminus [-\rho\,,\,\rho]$ and arbitrary $x, y \in T$ we have the extensions

$$\begin{aligned}F(x, y|\lambda) &= F_-(\lambda)^{d(x,v)}\, F_+(\lambda)^{d(v,y)} \quad \text{and} \\ \widetilde{F}(x, y|\lambda) &= \widetilde{F}_-(\lambda)^{d(x,v)}\, \widetilde{F}_+(\lambda)^{d(v,y)},\end{aligned}$$

where v is the unique point in $\pi(x, y)$ where $\mathfrak{h}(v)$ attains its minimum along that geodesic. The associated Poisson kernels are

$$K(x,\xi|\lambda) = F_-(\lambda)^{\mathfrak{h}(x,\xi)}\left(\frac{(1-\alpha)q}{\alpha}\right)^{\ell(x,\xi)} \quad \text{and}$$
$$\widetilde{K}(x,\xi|\lambda) = \widetilde{F}_-(\lambda)^{\mathfrak{h}(x,\xi)}\left(\frac{(1-\alpha)q}{\alpha}\right)^{\ell(x,\xi)}, \quad \text{where} \tag{6.4.8}$$
$$\ell(x,\xi) = d\big(x\wedge\xi, \pi(o,\varpi)\big).$$

This formula arises as follows: first, $\ell(x,\varpi) = 0$ so that $K(x,\varpi) = F_-(\lambda)^{\mathfrak{h}(x)}$. If $\xi \neq \varpi$ then let $v = \varpi \wedge \xi$ and $c = x \wedge \xi \in \pi(v,\varpi) \cup \pi(v,\xi)$; see Figure 6.1. If $c \in \pi(v,\varpi)$, then $F(x,c|\lambda) = F_-(\lambda)^{d(x,c)}$, $F(o,c|\lambda) = F_-(\lambda)^{d(o,c)}$, and $\ell(x,\xi) = 0$. On the other hand, if $c \in \pi(v,\xi)$ then still $F(x,c|\lambda) = F_-(\lambda)^{d(x,c)}$, but $F(o,c|\lambda) = F_-(\lambda)^{d(o,v)}F_+(\lambda)^{d(v,c)}$, and $d(v,c) = \ell(x,\xi)$. Now the first identity in (6.4.8) follows from (6.4.5). The same arguments apply to $\widetilde{K}$ and $\widetilde{F}$. Note that, for $\lambda = \pm\rho$, we have $\widetilde{K} = K$.

Remark 6.4.3 Consider the case when $q = 1$ and the random walk is on $\mathbb{T}_2 \equiv \mathbb{Z}$. Its non-zero transition probabilities are

$$p(x,x+1) = \alpha \quad \text{and} \quad p(x,x-1) = 1-\alpha\,, \quad x \in \mathbb{Z}.$$

Then it is natural to write $\partial\mathbb{T}_2 = \partial\mathbb{Z} = \{\pm\infty\}$, with $\varpi = -\infty$. Note that $\lambda_0 = \rho$. When $\lambda \in \mathbb{C} \setminus [-\rho\,,\,\rho]$, we have

$$\widetilde{K}(x,+\infty|\lambda) = K(x,-\infty|\lambda) \quad \text{and} \quad \widetilde{K}(x,-\infty) = K(x,+\infty|\lambda),$$

the kernels at $+\infty$ and $-\infty$ are distinct, and every λ-eigenfunction arises as a unique linear combination of those two kernels.

When $\lambda = \rho$, the function $K(x,+\infty|\rho) = K(x,-\infty|\rho)$ is the unique positive ρ-harmonic function with value 1 at the origin. □

This settles the special case $q = 1$. We are more interested in $q \geqslant 2$, where we get the following.

Corollary 6.4.4 *For $q \geqslant 2$, let $\lambda \in \mathbb{C} \setminus (-\rho\,,\,\rho)$, and let h be a λ-harmonic function. Then there is a unique strong distribution ν^h on ∂T such that*

$$h(x) = \int_{\partial T} K(x,\xi|\lambda)\,d\nu^h(\xi).$$

If in addition $\lambda \neq \lambda_0$ then there also is a unique strong distribution $\tilde{\nu}^h$ on ∂T such that

$$h(x) = \int_{\partial T} \widetilde{K}(x,\xi|\lambda)\,d\tilde{\nu}^h(\xi).$$

Of course, when $\lambda = \pm\rho$, we have $\widetilde{K} = K$ and $\tilde{\nu}^h = \nu^h$, but otherwise we shall see that the kernels and the representing distributions are distinct.

To our knowledge, this twin representation of λ-harmonic functions was first observed and used for the simple random walk in the context of the representation theory of free groups by Mantero and Zappa [13].

If $\lambda \geqslant \rho$, then it is well known that the functions $x \mapsto K(x, \xi|\lambda)$, $\xi \in \partial T$, are the *minimal* λ-harmonic functions, that is, the extremal elements of the convex set

$$\mathcal{H}_o(\lambda) = \{h : T \to (0\,,\,\infty) \mid h(o) = 1\,,\ Ph = \lambda \cdot h\}. \tag{6.4.9}$$

(When T is locally finite, this set is compact in the topology of pointwise convergence.) The index o stands for normalization at the reference point o.

Theorem 6.4.5 *Assume that $q \geqslant 2$. For $\lambda \in \mathbb{C} \setminus [-\rho\,,\,\rho]$, $\lambda \neq \lambda_0$, and for $\xi \in \partial T$, let ν^ξ and $\tilde{\nu}^\xi$ be the strong distributions on ∂T in the sense of Corollary 6.4.4 such that*

$$\widetilde{K}(x, \xi|\lambda) = \int_{\partial T} K(x, \cdot|\lambda)\, d\nu^\xi \quad \textit{and} \quad K(x, \xi|\lambda) = \int_{\partial T} \widetilde{K}(x, \cdot|\lambda)\, d\tilde{\nu}^\xi.$$

Then ν^ξ extends to a complex (σ-additive) Borel measure on ∂T, while this does not hold for $\tilde{\nu}^\xi$.

If, in particular, $\lambda > \rho$ is real, then the Borel probability measure ν^ξ is supported by all of ∂T, so that $\widetilde{K}(\cdot, \xi|\lambda)$ is not minimal in $\mathcal{H}_o(\lambda)$.

Proof. We start with an inequality that will be needed below:

$$\left|\frac{F_-(\lambda)}{\widetilde{F}_-(\lambda)}\right| = \left|\frac{1 - \sqrt{1 - \rho^2/\lambda^2}}{\rho/\lambda}\right|^2 < 1 \quad \text{for all } \lambda \in \mathbb{C} \setminus [-\rho\,,\,\rho]. \tag{6.4.10}$$

Recalling Convention 6.4.2, we obtain (6.4.10) by a few elementary computations.

Now let $x \in T \setminus \{o\}$. Noting that $\mathfrak{h}(x^-) = \mathfrak{h}(x) \pm 1$, we can use the first ones of the respective identities (6.4.5) and (6.4.6) plus (6.4.2) to compute

$$\begin{aligned} F(x, x^-|\lambda)\, \widetilde{F}(x^-, x|\lambda) &= F(x^-, x|\lambda)\, \widetilde{F}(x, x^-|\lambda) = F_-(\lambda)\widetilde{F}_+(\lambda) \\ &= F_+(\lambda)\widetilde{F}_-(\lambda) = F_-(\lambda)\, \widetilde{F}_-(\lambda)\, \frac{\alpha}{(1-\alpha)q} = \frac{1}{q}, \end{aligned} \tag{6.4.11}$$

because either $F(x, x^-|\lambda) = F_-(\lambda)$ and $\widetilde{F}(x^-, x|\lambda) = \widetilde{F}_+(\lambda)$, or $F(x, x^-|\lambda) = F_+(\lambda)$ and $\widetilde{F}(x^-, x|\lambda) = \widetilde{F}_-(\lambda)$. In particular,

$$|F_+(\lambda)F_-(\lambda)| = \left|\frac{F_-(\lambda)}{q\,\widetilde{F}_-(\lambda)}\right| < \frac{1}{q}. \tag{6.4.12}$$

By Theorem 6.3.4,

$$\nu^{\xi}(\partial T_x) = F(o, x|\lambda)\,\frac{\widetilde{K}(x, \xi|\lambda) - F(x, x^-|\lambda)\widetilde{K}(x^-, \xi|\lambda)}{1 - F_+(\lambda)F_-(\lambda)}$$

<u>*Case 1:*</u> $x \in \pi(o, \xi)$.

Then $\widetilde{K}(x, \xi|\lambda) = 1/\widetilde{F}(o, x|\lambda)$ and $\widetilde{K}(x^-, \xi|\lambda) = \widetilde{F}(x^-, x|\lambda)/\widetilde{F}(o, x|\lambda)$, and (6.4.11) yields

$$\nu^{\xi}(\partial T_x) = \frac{1 - 1/q}{1 - F_+(\lambda)F_-(\lambda)}\,\frac{F(o, x|\lambda)}{\widetilde{F}(o, x|\lambda)} = \frac{1 - 1/q}{1 - F_+(\lambda)F_-(\lambda)}\left(\frac{F_-(\lambda)}{\widetilde{F}_-(\lambda)}\right)^{d(o,x)}. \tag{6.4.13}$$

We note immediately that this is strictly positive when $\lambda > \rho$, because in view of (6.3.16), combined with (6.4.4) and (6.4.5), we then have $F_+(\lambda)F_-(\lambda) \leqslant F_+(\rho)F_-(\rho) = \left(\frac{\rho}{2\alpha}\right)^2 \frac{\alpha}{(1-\alpha)q} = \frac{1}{q} < 1$.

<u>*Case 2:*</u> $x \notin \pi(o, \xi)$.

Let $c = x \wedge \xi = x^- \wedge \xi$. Then $\widetilde{K}(x, \xi|\lambda) = \widetilde{F}(x, c|\lambda)/\widetilde{F}(o, c|\lambda)$ and $\widetilde{K}(x^-, \xi|\lambda) = \widetilde{K}(x, \xi|\lambda)/\widetilde{F}(x, x^-|\lambda)$. Now, (6.4.11) yields that $F(c, x|\lambda)\widetilde{F}(x, c|\lambda) = q^{-d(x,c)}$, because it is the product of $d(x, c)$ terms of the form $F(x_i^-, x_i|\lambda)\widetilde{F}(x_i, x_i^-|\lambda)$. Therefore, decomposing $F(o, x|\lambda) = F(o, c|\lambda)F(c, x|\lambda)$,

$$\begin{aligned}
\nu^{\xi}(\partial T_x) &= \overbrace{\frac{1 - F_-(\lambda)/\widetilde{F}_-(\lambda)}{1 - F_+(\lambda)F_-(\lambda)}}^{=:\,C(\lambda)}\, F(o, x|\lambda)\,\frac{\widetilde{F}(x, c|\lambda)}{\widetilde{F}(o, c|\lambda)} \\
&= C(\lambda)\,\frac{F(o, c|\lambda)}{\widetilde{F}(o, c|\lambda)}\, F(c, x|\lambda)\widetilde{F}(x, c|\lambda) \\
&= C(\lambda)\left(\frac{F_-(\lambda)}{\widetilde{F}_-(\lambda)}\right)^{d(o,c)}\left(\frac{1}{q}\right)^{d(x,c)}.
\end{aligned} \tag{6.4.14}$$

Again, this is strictly positive when $\lambda > \rho$, and we obtain that in this case the Borel probability measure ν^{ξ} is supported by all of ∂T.

We now prove that for any $\lambda \in \mathbb{C} \setminus (-\rho\,, \rho)$, the distribution ν^{ξ} extends to a σ-additive Borel measure on ∂T. Let $(x_n)_{n\geqslant 0}$ be a sequence of vertices such that the arcs ∂T_{x_n} are pairwise disjoint.

Write $\pi(o, \xi) = [o = v_0\,, v_1\,, \dots]$. There can be at most one x_n on that geodesic ray. In that case, suppose it is x_0, that is, $x_0 = v_k$ for some $k \geqslant 0$. By (6.4.10), (6.4.12) and (6.4.13),

$$|\nu^{\xi}(\partial T_{x_0})| < \left|\frac{1 - 1/q}{1 - F_+(\lambda)F_-(\lambda)}\right| < 1.$$

Next, let $A_k = \{n \geqslant 1 : x_n \wedge \xi = v_k\}$. We claim that, using (6.4.14), one has

$$\sum_{n\,:\,x_n \in A_k} \left|\nu^\xi(\partial T_{x_n})\right| \leqslant |C(\lambda)| \cdot \left|F_-(\lambda)/\widetilde{F}_-(\lambda)\right|^k.$$

Indeed, consider the equi-distribution $\overline{\nu}$ on ∂T, that is, $\overline{\nu}(\partial T_x) = 1/\big((q+1)\, q^{d(o,x)-1}\big)$ for $x \neq 0$. It extends to a Borel probability measure on ∂T, and for $k \geqslant 1$,

$$\sum_{n\,:\,x_n \in A_k} q^{-d(x_n, v_k)} = (q+1)q^{k-1} \sum_{n\,:\,x_n \in A_k} \overline{\nu}(\partial T_{x_n}) \leqslant (q+1)q^{k-1}\overline{\nu}(\partial T_{v_k} \setminus \partial T_{v_{k+1}})$$
$$= \frac{q-1}{q}.$$

For $k = 0$, the analogous computation yields the upper bound 1. By (6.4.10),

$$\sum_{n=0}^{\infty} \left|\nu^\xi(\partial T_{x_n})\right| \leqslant \left|\frac{1 - 1/q}{1 - F_+(\lambda)F_-(\lambda)}\right| + \sum_{k=0}^{\infty} \sum_{n\,:\,x_n \in A_k} \left|\nu^\xi(\partial T_{x_n})\right|$$
$$\leqslant 1 + \frac{|C(\lambda)|}{1 - |F_-(\lambda)/\widetilde{F}_-(\lambda)|}.$$

So condition (6.2.2) is satisfied, and ν^ξ has a σ-additive extension, as stated.

To obtain the analogous formulas to (6.4.13) and (6.4.14) for $\tilde{\nu}^\xi$, we just have to exchange F and $\widetilde{F}$ in each occurence. We write $\widetilde{C}(\lambda)$ for the resulting constant in the analogue of (6.4.14). In this case, let the sequence (x_n) consist of all the neighbours of the v_k, $k \geqslant 1$, which do not lie on $\pi(o, v)$. Thus, the set A_k defined above consists of the neighbours of v_k, and by the same computation we obtain

$$\sum_{n\,:\,x_n \in A_k} \left|\tilde{\nu}^\xi(\partial T_{x_n})\right| = |\widetilde{C}(\lambda)| \cdot \left|\widetilde{F}_-(\lambda)/F_-(\lambda)\right|^k.$$

The sum over all k diverges by (6.4.10), so that $\widetilde{\nu}^\xi$ does not satisfy the bounded variation condition (6.2.2). □

6.5 General Transitive Group Actions

After the detailed study of multiple integral representations in Section 6.4, we now turn to general transitive group actions in the place of $\mathsf{Aff}(\mathbb{T}_q)$. Once more, we take up material from our 'companion' paper [17, §4]: we assume that the transition probabilities are invariant under a general group Γ of automorphisms of the tree which acts transitively on the vertex set. That is,

$$p(\gamma x, \gamma y) = p(x, y) \quad \text{for all } x, y \in T \text{ and } \gamma \in \Gamma.$$

Let $I = \Gamma \backslash E(T)$ be the set of orbits of Γ on the set of oriented edges of T. If $j \in I$ is the orbit *(type)* of $(x, y) \in E(T)$, then we write $p_j = p(x, y)$ and $-j$ for the orbit of (y, x). Then $-j$ is independent of the representative (x, y), and $-(-j) = j$. In particular, $-j = j$ if and only if there is $\gamma \in \Gamma$ for which $\gamma x = y$ and $\gamma y = x$. For each $j \in I$ and fixed $x \in T$, we set $d_j = |\{y \sim x : (x, y) \in I\}|$. This is finite because $d_j \leqslant 1/p_j$, and independent of x by transitivity of Γ. For example, when $\Gamma = \mathsf{Aut}(\mathbb{T}_q)$ then $I = \{1\}$ with $d_1 = q+1$, while when $\Gamma = \mathsf{Aff}(\mathbb{T}_q)$ then $I = \{\pm 1\}$ with $d_{-1} = 1$ and $d_1 = q$. Thus, $\sum_{j \in I} d_j p_j = 1$, and $\mathsf{deg}(x) = \sum_{j \in I} d_j$.

As clarified in [17, Remark 4.4, second half], one can start with a finite or countable set I with an involution $j \mapsto -j$ and a collection $(d_j)_{j \in I}$ of natural numbers. Then for the regular tree T with degree $\sum_j d_j \leqslant \infty$, there is a group $\Gamma \leqslant \mathsf{Aut}(T)$ which acts transitively and such that I is in one-to-one correspondence with its set of orbits and the associated cardinalities are d_j.

For example, when $d_j = 1$ for all j, then we can choose Γ as the discrete group

$$\Gamma = \langle a_j, j \in I \mid a_j^{-1} = a_{-j} \text{ for all } j \in I \rangle. \tag{6.5.1}$$

Then, when $j \neq -j$, we can choose just one out of a_j and a_{-j} as a free generator. Instead, when $j = -j$, then a_j is a generator whose square is the group identity. In this example, Γ acts transitively with trivial stabilizers, and the fact that this provides all possible groups which act in this way on a countable tree is a well-known basic part of Bass–Serre theory (see Serre [18]). In all other cases, Γ will have non-discrete closure in $\mathsf{Aut}(T)$.

In the general situation of a transitive group action which leaves the transition probabilities invariant, it is shown in [17, Theorem 4.2] that $\mathsf{res}(P) \setminus \mathsf{res}^*(P) \subset \{0\}$. That is, $G(\lambda) \neq 0$ for all $\lambda \in \mathsf{res}(P) \setminus \{0\}$, where $G(\lambda) = G(x, x|\lambda)$, which is independent of x by transitivity. We remark that it may happen that 0 is part of the resolvent set of P [8].

Here, we shall always assume that the vertex degree is $\geqslant 3$, so that our random walk has to be ρ-transient by a result of Guivarc'h [10]. When I is infinite we make the additional assumption that

$$\sum_{j \in I} d_j \, p_{-j} < \infty. \tag{6.5.2}$$

Note that this is the sum over all neighbours of any vertex x of the incoming probabilities $p(y, x)$. The assumption is satisfied, for instance, if the quotients p_{-j}/p_j are bounded.

If (x, y) is an edge of type j, then $g(x, y) = G(x, y|\lambda) = G_j(\lambda)$ depends only on j. By reversibility, we have

$$p_j\, G_{-j}(\lambda) = p_{-j}\, G_j(\lambda),$$

and the second identity of Lemma 6.3.2 becomes

$$p_{-j}\, G_j(\lambda)^2 + G_j(\lambda) - p_j\, G(\lambda)^2 = 0. \tag{6.5.3}$$

When $\lambda > \rho$ is real, among the two solutions of this equation the meaningful one is

$$G_j(\lambda) = \frac{1}{2p_{-j}}\Big(\sqrt{1 + 4p_j p_{-j}\, G(\lambda)^2} - 1\Big), \tag{6.5.4}$$

because the functions $G(\lambda)$ and $G_j(\lambda)$ are decreasing in this range of λ. In other regions of the plane, there may be a minus sign in front of the root.

Proposition 6.5.1 *Let* $\kappa = \max\{2\sqrt{p_j p_{-j}} : j \in I\}$. *Then identity* (6.5.4) *holds for all* λ *in the set*

$$\mathcal{U} = \{\lambda \in \mathbb{C} : |\lambda| > \rho\} \setminus \{\pm i\, t : \rho < t \leqslant \kappa\}.$$

(When $\kappa \leqslant \rho$ *the last part is empty.)*

Proof. Each of the functions

$$\Phi_j(t) = \frac{1}{2}\Big(\sqrt{1 + 4p_j p_{-j}\, t^2} - 1\Big) \tag{6.5.5}$$

is analytic in the slit plane

$$\mathcal{W} = \mathbb{C} \setminus \{\pm i\, t : t \geqslant 1/\kappa\}. \tag{6.5.6}$$

We shall show that the function $G(\lambda)$ maps $\mathcal{U}$ into $\mathcal{W}$. This implies that the functions appearing in (6.5.4) are all analytic, so that the identity must hold on all of $\mathcal{U}$ by analytic continuation.

We use some well-known spectral theory. Let μ be the Plancherel measure of our random walk, introduced in Section 6.4. Recall that μ is a probability measure concentrated on $\mathsf{spec}(P)$, and is the diagonal matrix element at (x, x) (independent of $x \in T$ by group invariance) of the spectral resolution of the self-adjoint operator P on $\ell^2(T, \mathsf{m})$. In more classical terms, it is the measure on $[-\rho\,,\, \rho]$ whose nth moments are the return probabilities $p^{(n)}(x, x)$ for $n \geqslant 0$. Since in the present case, these probabilities are 0 when n is odd, μ is symmetric (invariant under the reflection $t \mapsto -t$). Thus

$$G(\lambda) = \int_{[-\rho\,,\,\rho]} \frac{1}{\lambda - t}\, d\mu(t)\,, \quad \lambda \in \mathbb{C} \setminus \mathsf{spec}(P).$$

Now let $|\lambda| > \rho$ be such that $\Re(\lambda) \neq 0$, and write $\bar{\lambda}$ for its complex conjugate. Then

$$\overline{G(\lambda)} = \int_{[-\rho\,,\,\rho]} \frac{1}{\bar{\lambda} - t}\, d\mu(t) = \int_{[-\rho\,,\,\rho]} \frac{1}{\bar{\lambda} + t}\, d\mu(t)\,, \quad \text{whence}$$

$$\begin{aligned}\Re\big(G(\lambda)\big) &= \frac{1}{2}\int_{[-\rho\,,\,\rho]} \left(\frac{1}{\lambda - t} + \frac{1}{\bar{\lambda} + t}\right) d\mu(t) \\ &= \Re(\lambda) \int_{[-\rho\,,\,\rho]} \frac{|\lambda|^2 - t^2}{(|\lambda|^2 - t^2)^2 + 4t^2\Im(\lambda)^2}\, d\mu(t).\end{aligned}$$

The last integral is > 0, so that also $\Re\big(G(\lambda)\big) \neq 0$. Therefore $G(\lambda) \in \mathcal{W}$.

Next, let $\lambda = i\,\beta$, where $\beta \in \mathbb{R}$ and $|\beta| > \max\{\rho, \kappa\}$. Then, using again that μ is symmetric (so that odd functions integrate to 0),

$$G(i\,\beta) = \int_{[-\rho\,,\,\rho]} \frac{-i\,\beta - t}{\beta^2 + t^2}\, d\mu(t) = -\frac{i}{\beta}\int_{[-\rho\,,\,\rho]} \frac{1}{1 + (t/\beta)^2}\, d\mu(t).$$

Therefore $|G(i\,\beta)| \leqslant 1/|\beta| < 1/\kappa$, and also $G(i\,\beta) \in \mathcal{W}$. □

We now obtain the following.

Theorem 6.5.2 *For* $\lambda \in \mathcal{U}$,

$$\lambda\, G(\lambda) = \Phi\big(G(\lambda)\big)\,, \quad \textit{where} \quad \Phi(t) = 1 + \sum_{j\in I} \frac{d_j}{2}\Big(\sqrt{1 + 4p_j\,p_{-j}\,t^2} - 1\Big).$$

The function $\Phi(t)$ *is analytic in the domain* $\mathcal{W}$ *of* (6.5.6). *Furthermore,*

$$\rho = \min\{\Phi(t)/t : t > 0\} = \Phi(\theta)/\theta,$$

where θ *is the unique positive real solution of the equation* $\Phi'(t) = \Phi(t)/t$.

Proof. First of all, observe that for $t \in \mathbb{C} \setminus \{i\,s : s \in \mathbb{R}\,,\ |s| \geqslant 1\}$,

$$\big|\sqrt{1 + t^2} - 1\big| < |t|.$$

Therefore, summing over all $j \in I$,

$$\sum \frac{d_j}{2}\Big|\sqrt{1 + 4p_j\,p_{-j}\,t^2} - 1\Big| < |t| \sum d_j \sqrt{p_j\,p_{-j}} \leqslant |t|\sqrt{\sum d_j\,p_{-j}},$$

which is finite by assumption (6.5.2). Consequently, even when I is infinite, the defining series of $\Phi(t)$ converges absolutely and locally uniformly on $\mathcal{W}$, so that $\Phi(t)$ is indeed analytic on that set. Now we can use (6.5.4) and Proposition 6.5.1: for $\lambda \in \mathcal{U}$,

$$\lambda\, G(\lambda) - 1 = \sum_y p(x, y) G(y, x|\lambda) = \sum_{j\in I} d_j\, p_j\, G_{-j}(\lambda) = \Phi\big(G(\lambda)\big) - 1.$$

The remaining statements of the theorem follow well-known lines, compare, e.g., with [22, Ex. 9.46], where the variable $z = 1/\lambda$ is used instead of λ, and see also below. □

Remarks 6.5.3 For the free group with (finitely or) infinitely many generators, the equation for $G(\lambda)$ of Theorem 6.5.2 was first deduced and used for finding the asymptotics of $p^{(n)}(x, x)$ by Woess [21]. Its validity was restricted to a complex neighbourhood of the real half-line $[\rho, +\infty)$. There, computations are performed in the variable $z = 1/\lambda$. A previous variant (for z, resp. λ positive real) is inherent in work of Levit and Molchanov [12]. Later on, Aomoto [1] considered equations of the same nature as (6.5.3) for the case of finitely generated free groups plus reasoning of algebraic geometry to study the nature of the involved functions and the spectrum of P. Similarly, Figà-Talamanca and Steger [8] considered the case when the group is discrete as in (6.5.1), I is finite, and $j = -j$ for all j. This served for an in-depth study of the associated harmonic analysis.

What is new here is

- the extension to the general group-invariant case, with I finite or infinite,
- the validity of the equation for $G(\lambda)$ in the large domain $\mathcal{U}$.

This domain can be further extended a bit by additional estimates, but for complex λ close to $\mathsf{spec}(P)$, the situation is more complicated. Indeed, in such regions, the correct solution of (6.5.3) may be the one where one has to use the negative branch of the square root in (6.5.4). The general formula instead of the one of Theorem 6.5.2 is then

$$\lambda\, G(\lambda) = 1 + \sum_{j \in I} \frac{d_j}{2}\Big(\pm\sqrt{1 + 4p_j p_{-j}\, G(\lambda)^2} - 1\Big),$$

where the signs may vary according to the region to which λ belongs. This requires some subtle algebraic geometry beyond the focus of this chapter [1], [8].

In the general group-invariant set-up, and even for non-locally finite T, we obtain the integral representation of Theorem 6.3.4 with respect to the Martin kernel $k(x, \xi) = K(\cdot, \cdot|\lambda)$ for any λ-harmonic function, whenever $0 \neq \lambda \in \mathbb{C} \setminus \mathsf{spec}(P)$, for $\lambda = \pm\rho$, and possibly also for $\lambda = 0$.

The study of twin kernels and the resulting integral representation of λ-harmonic functions becomes more delicate in view of the fact that $G(\lambda)$ and thus also the functions $F_j(\lambda) = G_j(\lambda)/G(\lambda)$ are only given via the implicit equation for $G(\lambda)$ of Theorem 6.5.2. Therefore we limit attention to the case when $\lambda \in (\rho, +\infty)$ is real. For real t, each function Φ_j of (6.5.5) describes the

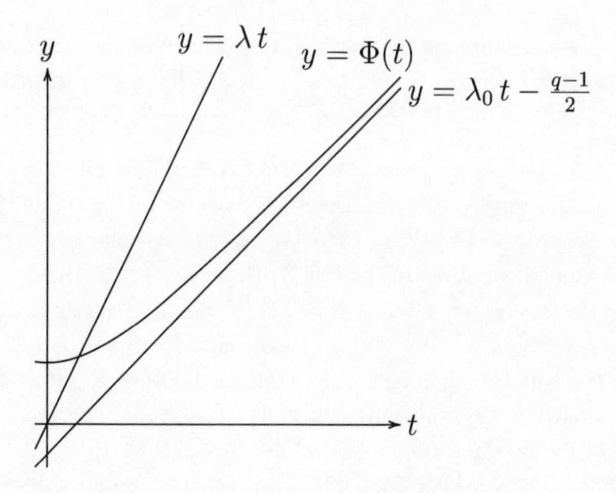

Figure 6.2 Illustration for the equation for $G\lambda$

upper branch of a hyperbola. Thus, the function Φ has the following properties: it is strictly increasing and strictly convex,

$$\Phi(0) = 1,\ \Phi'(0) = 0\,, \quad \text{and} \quad \lim_{t\to\infty} \Phi'(t) = \lambda_0\,, \ \text{where}\ \lambda_0 = \sum_{j\in I} d_j \sqrt{p_j p_{-j}}.$$

We have $\lambda_0 < \infty$ by assumption (6.5.2). Note that in the case of the affine random walks of Section 6.4, this is the same λ_0 as in (6.4.7). For $\lambda \geqslant \lambda_0$, the equation $\lambda t = \Phi(t)$ has a unique positive solution. This is $t = G(\lambda)$. See Figure 6.2, where we assume that $\mathsf{deg}(x) = q + 1$ is finite. With θ and ρ as in Theorem 6.5.2, it is clear from the shape of Φ that $\lambda_0 > \rho$, and for $\lambda_0 > \lambda > \rho$, there are precisely two solutions of the equation $\lambda t = \Phi(t)$. One is smaller than θ and the other is larger than θ. By continuity of $G(\cdot)$, the correct solution for $G(\lambda)$ is the one for which $G(\lambda) < \theta$: this is the solution that leads to the ordinary λ-Martin kernel $K(\cdot, \cdot|\lambda)$ and the resulting integral representation of any λ-harmonic function over ∂T. But we also have the second solution $\widetilde{G}(\lambda) > \theta$. Working with this one, we also find that for all $\lambda \in (\rho\,, \lambda_0)$ one has

$$\widetilde{G}(\lambda) \neq 0 \quad \text{and} \quad \widetilde{G}_j(\lambda) = \Phi_j\big(G(\lambda)\big)\big/p_{-j} \neq 0,$$

whence $\widetilde{F}_j(\lambda) = \widetilde{G}_j(\lambda)/\widetilde{G}(\lambda) \neq 0$. Also,

$$\widetilde{F}_j(\lambda)\,\widetilde{F}_{-j}(\lambda) = \frac{\Phi_j\big(\widetilde{G}(\lambda)\big)^2}{p_j p_{-j}\,\widetilde{G}(\lambda)^2} < 1,$$

since $\sqrt{1+t^2}-1 < t$ for $t > 0$. Thus, (6.3.1) holds for the weights $f(x,y) = \widetilde{F}_j(\lambda)$, when (x,y) is an oriented edge of type j. Let us verify (6.3.2):

$$\sum_v p(x,v)f(v,x) = \sum_{j\in I} d_j\, p_j\, \frac{\widetilde{G}_{-j}(\lambda)}{\widetilde{G}(\lambda)} = \frac{\Phi\big(\widetilde{G}(\lambda)\big)-1}{\widetilde{G}(\lambda)} = \frac{\lambda\,\widetilde{G}(\lambda)-1}{\widetilde{G}(\lambda)} < \lambda.$$

Finally, (6.3.3) reduces to equation (6.5.3), which holds for $\widetilde{G}_j(\lambda)$ as well as for $G_j(\lambda)$. We conclude that these edge weights lead to a second kernel

$$k(x,\xi) = \widetilde{K}(x,\xi|\lambda)\,, \quad x\in T\,,\ \xi\in\partial T\,, \quad \lambda\in(\rho\,,\lambda_0),$$

so that $x\mapsto \widetilde{K}(x,\xi|\lambda)$ is positive λ-harmonic. Thus, every λ-harmonic function has a second integral representation as in Theorem 6.3.4, in addition to the one with respect to the ordinary Martin kernel $K(\cdot,\cdot|\lambda)$.

Again, for any $\xi\in\partial T$, there is a positive (σ-additive !) Borel probability measure ν^ξ on ∂T such that

$$\widetilde{K}(x,\xi|\lambda) = \int_{\partial T} K(x,\cdot|\lambda)\,d\nu^\xi.$$

We omit the computation which shows that ν^ξ is supported by all of ∂T, which is a consequence of the fact that T has degree $\geqslant 3$. In particular, $\widetilde{K}(x,\xi|\lambda)$ cannot be a minimal λ-harmonic function, i.e., an extremal point of the set $\mathcal{H}_o$ of (6.4.9). Therefore the converse representing distribution $\tilde{\nu}^\xi$ that by Theorem 6.3.4 gives the integral representation

$$K(x,\xi|\lambda) = \int_{\partial T} \widetilde{K}(x,\cdot|\lambda)\,d\tilde{\nu}^\xi,$$

cannot have a σ-additive extension.

We may ask how to proceed for $\lambda > \lambda_0$, while we exclude the case $\lambda = \lambda_0$, since we have already seen in Section 6.4 that for affine random walks there is no natural choice for a second family of weights for λ_0. We choose to proceed as follows, *requiring here that I be finite* and $\sum_j d_j = q+1$.

The second solution of (6.5.3) is

$$\widetilde{G}_j(\lambda) = \frac{1}{p_{-j}}\,\widetilde{\Phi}_j\big(\widetilde{G}(\lambda)\big), \quad \text{where} \quad \widetilde{\Phi}_j(t) = \frac{1}{2}\Big(-\sqrt{1+4p_jp_{-j}t^2}-1\Big).$$

Then we set

$$\widetilde{\Phi}(t) = \sum_{j\in I} d_j\,\widetilde{\Phi}_j(t).$$

(When I is infinite, the series does not converge.) While $\Phi(t)$ is a sum of upper branches of hyperbolic functions, $\widetilde{\Phi}(t)$ it the sum of the associated lower branches. The two asymptotes of $\Phi(t)$ and $\widetilde{\Phi}(t)$ are $y = \pm\lambda_0 t - (q-1)/2$.

Thus, any line $y = \lambda t$ has exactly two intersection points with the 'twin curve' $\big(\Phi(t), \widetilde{\Phi}(t)\big)$, except for $\lambda = \pm\rho$, in which cases there is only one double solution, and $\lambda = \pm\lambda_0$, in which case there is only one simple solution. Thus, for $\lambda > \lambda_0$, we choose $\widetilde{G}(\lambda)$ as the unique solution of

$$\lambda\, \widetilde{G}(\lambda) = \widetilde{\Phi}\big(\widetilde{G}(\lambda)\big),$$

which is negative. The associated solution for $\widetilde{G}_j(\lambda)$ is

$$\widetilde{G}_j(\lambda) = \frac{1}{p_{-j}}\, \widetilde{\Phi}_j\big(\widetilde{G}(\lambda)\big),$$

so that indeed

$$\sum_{j\in I} d_j\, p_j\, \widetilde{G}_{-j}(\lambda) = \widetilde{\Phi}\big(\widetilde{G}(\lambda)\big) - 1 = \lambda\, \widetilde{G}(\lambda) - 1$$

Note that also $\widetilde{G}_j(\lambda) < 0$, so that $\widetilde{F}_j(\lambda) = \widetilde{G}_j(\lambda)/\widetilde{G}(\lambda) > 0$. The associated edge weights are again given by $f(x,y) = \widetilde{F}_j(\lambda)$, when (x,y) is an oriented edge of type j. It is straightforward to see that they also satisfy the requirements (6.3.1)–(6.3.3), so that we also obtain a positive kernel $k(x,\xi) = \widetilde{K}(x,\xi\,|\,\lambda)$ with the same properties as above.

By symmetry, analogous properties hold for negative $\lambda \in (-\infty\,,\,-\rho) \setminus \{-\lambda_0\}$.

References

[1] Aomoto, K.: *Spectral theory on a free group and algebraic curves.* J. Fac. Sci. Univ. Tokyo Sect. IA Math. **31** (1984) 297–318.

[2] Boiko, T., and Woess, W.: *Moments of Riesz measures on Poincaré disk and homogeneous tree – a comparative study.* Expo. Math. **33** (2015) 353–374.

[3] Cartwright, D. I., Kaimanovich, V. A., and Woess, W.: *Random walks on the affine group of local fields and of homogeneous trees.* Ann. Inst Fourier (Grenoble) **44** (1994) 1243–1288.

[4] Cartwright, D. I., Soardi, P. M., Woess, W.: *Martin and end compactifications of non locally finite graphs.* Trans. Amer. Math. Soc. **338** (1993) 679–693.

[5] Cartier, P.: *Fonctions harmoniques sur un arbre.* Symposia Math. **9** (1972) 203–270.

[6] Cohen, J. M., Colonna, F., and Singman, D.: *Distributions and measures on the boundary of a tree.* J. Math. Anal. and App. **293** (2004) 89–107.

[7] Eymard, P.: *Le noyau de Poisson et la théorie des groupes.* Symposia Math. **22** (1977) 107–132.

[8] Figà-Talamanca, A., and Steger, T.: *Harmonic analysis for anisotropic random walks on homogeneous trees.* Mem. Amer. Math. Soc. **110** (1994), no. 531.

[9] Grigorchuk, R. I., and Żuk, A.: *The Ihara zeta function of infinite graphs, the KNS spectral measure and integrable maps.* In: *Random Walks and Geometry* (Vienna, 2001), 141–180, de Gruyter, Berlin, 2004.

[10] Guivarc'h, Y.: *Sur la loi des grands nombres et le rayon spectral d'une marche aléatoire.* Astŕisque **74** (1980) 47–98.

[11] Helgason, S.: *Eigenspaces of the Laplacian; integral representations and irreducibility.* J. Functional Analysis **17** (1974), 328–353.

[12] Levit, B. Ja., and Molčanov, S. A.: *Invariant chains on a free group with a finite number of generators.* (Russian) Vestnik Moskov. Univ. Ser. I Mat. Meh. **26** (1971) 80–88.

[13] Mantero, A. M., and Zappa, A.: *The Poisson transform and representations of a free group.* J. Funct. Anal. **51** (1983) 372–399.

[14] Mohar, B., and Woess, W.: *A survey on spectra of infinite graphs.* Bull. London Math. Soc. **21** (1989) 209–234.

[15] Nebbia, C.: *On the amenability and the Kunze-Stein property for groups acting on a tree.* Pacific J. Math. **135** (1988) 371–380.

[16] Picardello, M.A., and Woess, W.: *Finite truncations of random walks on trees* (appendix to: Korànyi, A., Picardello, M. A., and Taibleson M: *Hardy-spaces on non-homogeneous trees*), Symposia Math. **29** (1988) 255–265.

[17] Picardello, M.A., and Woess, W.: *Boundary representations of λ-harmonic and polyharmonic functions on trees.* Potential Analysis **51** (2019) 541–561.

[18] Serre, J-P.: *Trees.* Springer-Verlag, Berlin-New York, 1980.

[19] Trofimov, V. I.: *Automorphism groups of graphs as topological groups.* Math. Notes **38** (1985) 717–720.

[20] H. S. Wall, H. S.: *Analytic Theory of Continued Fractions.* Van Nostrand, New York, 1948.

[21] Woess, W.: *Puissances de convolution sur les groupes libres ayant un nombre quelconque de générateurs.* In: *Random Walks and Stochastic Processes on Lie Groups* (Nancy, 1981), 181–190, Inst. Élie Cartan **7**, Univ. Nancy, 1983.

[22] Woess, W.: *Denumerable Markov Chains. Generating functions, Boundary Theory, Random Walks on Trees.* European Math. Soc. Publishing House, 2009.

7

Internal DLA on Sierpinski Gasket Graphs

Joe P. Chen, Wilfried Huss, Ecaterina Sava-Huss, and Alexander Teplyaev

Abstract

Internal diffusion-limited aggregation (IDLA) is a stochastic growth model on a graph G which describes the formation of a random set of vertices growing from the origin (some fixed vertex) of G. Particles start at the origin and perform simple random walks; each particle moves until it lands on a site which was not previously visited by other particles. This random set of occupied sites in G is called the IDLA cluster. In this chapter we consider IDLA on Sierpinski gasket graphs and show that the IDLA cluster fills balls (in the graph metric) with probability 1.

7.1 Introduction

The *internal diffusion limited aggregation* model (in short *IDLA* or *internal DLA*) is a stochastic growth model introduced by Diaconis and Fulton [9]. To formally define the process, let G be an infinite connected graph with a distinguished vertex o which will be called the *origin*. Then IDLA on G is defined as follows. For $i = 1, 2, \ldots$, let $\left(X^i(t)\right)_{t\geq 0}$ be a sequence of iid simple random walks on G starting at o, where $X^i(t)$ represents the random position of the ith random walk at time t. The IDLA cluster is built up one site at a time, by letting the ith particle walk until it exits the set of sites already occupied by the previous $i-1$ particles. Denote by $\mathscr{I}(i)$ the IDLA cluster with i particles. Set $\mathscr{I}(0) = \{o\}$, and for $i \geq 1$ define the sequence of stopping times $(\sigma^i)_{i\geq 0}$, with $\sigma^0 = o$ and

$$\sigma^i = \inf\left\{t > 0 : X^i(t) \notin \mathscr{I}(i-1)\right\}.$$

The IDLA cluster with i particles is defined inductively as

$$\mathscr{I}(i) = \mathscr{I}(i-1) \cup \{X^i(\sigma^i)\}. \tag{7.1.1}$$

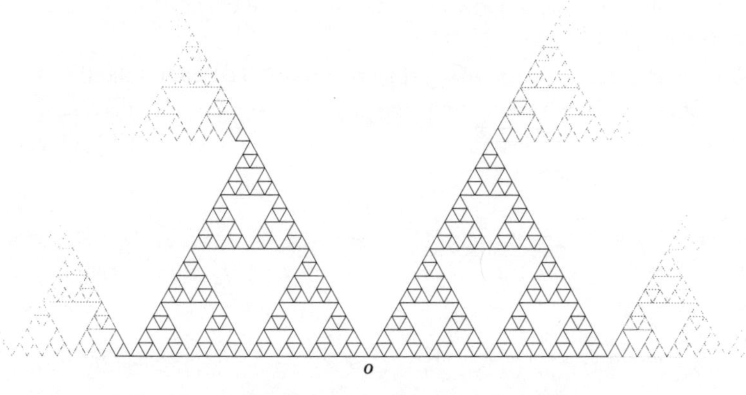

Figure 7.1 Doubly infinite Sierpinski gasket graph

For large i, we are interested in the shape of the IDLA cluster $\mathscr{I}(i)$ after the ith particle stops. Does the random set $\mathscr{I}(i)$ exhibit a regular shape once i is large enough? On $\mathbb{Z}^d$ Lawler, Bramson, and Griffeath [19] were the first to identify the limit shape as an Euclidean ball. If instead of simple random walks on $\mathbb{Z}^d$ one takes drifted random walks, then the limiting shape is shown to be a true heat ball in [21]; the proof is based on an unfair divisible sandpile model. On other state spaces, there are several results concerning the IDLA limit shape: on discrete groups with exponential growth [7], on non-amenable graphs [13], on supercritical percolation clusters on $\mathbb{Z}^d$ [10, 23], and on comb lattices [1, 15]. For a survey of these results see [22].

In this work we investigate IDLA on the pre-fractal Sierpinski gasket graph, specifically the doubly infinite Sierpinski gasket graph SG, shown in Figure 7.1. To construct SG, we consider in $\mathbb{R}^2$ the sets

$$V_0 = \left\{(0,0), (1,0), (1/2, \sqrt{3}/2)\right\}$$

and

$$E_0 = \left\{\big((0,0),(1,0)\big), \big((0,0),(1/2,\sqrt{3}/2)\big), \big((1,0),(1/2,\sqrt{3}/2)\big)\right\}.$$

Now recursively define $(V_1, E_1), (V_2, E_2), \ldots$ by

$$V_{n+1} = V_n \cup \{(2^n, 0) + V_n\} \bigcup \left\{\left(2^{n-1}, 2^{n-1}\sqrt{3}\right) + V_n\right\}$$

and

$$E_{n+1} = E_n \cup \{(2^n, 0) + E_n\} \bigcup \left\{\left(2^{n-1}, 2^{n-1}\sqrt{3}\right) + E_n\right\},$$

where $(x, y) + S := \{(x, y) + s : s \in S\}$. Let $V_\infty = \cup_{n=0}^{\infty} V_n$, $E_\infty = \cup_{n=0}^{\infty} E_n$, $V = V_\infty \cup \{-V_\infty\}$, and $E = E_\infty \cup \{-E_\infty\}$. Then the doubly infinite Sierpinski gasket graph SG is the graph with vertex set V and edge set E. Set the origin $o = (0, 0)$.

Our main result is the following spherical shape theorem for the IDLA cluster on SG, consisting of random walks launched successively from o. Denote by $B_o(n)$ the ball of radius n and centre o in the graph distance of SG, and by $b_n := \mid B_o(n) \mid$ its cardinality.

Theorem 7.1.1 (Shape theorem for internal DLA) *On* SG*, the IDLA cluster of* b_n *particles occupies a set of sites close to a ball of radius n. That is, for all* $\epsilon > 0$*, we have*

$$B_o\,(n(1-\epsilon)) \subset \mathcal{I}(b_n) \subset B_o\,(n(1+\epsilon))\,,\ \textit{for all n sufficiently large}$$

with probability 1.

Our proof of Theorem 7.1.1 combines relevant arguments from previous work on IDLA [10, 18, 19] and information about the geometry and the potential theory on SG [2, 6, 17]. A key reason why we can prove Theorem 7.1.1 is because SG is a finitely ramified fractal, and this makes the analysis of random walks [2, 6] and quantities related to random walks, such as Green functions and harmonic functions [17, 24], easier.

Let us remark that it is possible to prove Theorem 7.1.1 on the one-sided infinite Sierpinski gasket graph (where the origin o has degree 2) upon appropriate modification of the proofs.

The rest of the chapter is organized as follows. In Section 7.2 we give the basic notions on random walks and related quantities such as Green function and harmonic function. In Section 7.2.2 we recall some well-known properties and results on Sierpinski gasket graphs and random walks on them, which will be subsequently used in the proof of the limit shape for the IDLA cluster. Then in Sections 7.3.1 and 7.3.2 we prove, respectively, the inner bound and the outer bound of Theorem 7.1.1. We conclude with some remarks and open questions.

7.2 Preliminaries

7.2.1 Random Walks

Let G be any infinite, locally finite, connected graph. We shall identify G with its vertices and for $x, y \in G$ we write $x \sim y$ if (x, y) is an edge in G. We write $d(x, y)$ for the natural graph distance in G, i.e., the length of the shortest path between x and y in G. Given a subset $A \subset G$, we define its boundary

$$\partial A = \{y \in G \setminus A : y \sim x \text{ for some } x \in A\}.$$

For $x \in G$ and $n \geq 0$, denote by $B_x(n) = \{y \in G : d(x, y) < n\}$ the ball of centre x and radius n in G, and by $\partial B_x(n) = \{y \in B_x(n)^c\colon y \sim x \text{ for some } x \in B_x(n)\}$ the boundary of $B_x(n)$. Denote by $\deg(x)$ the degree of x in G, that is, the number of neighbours of x.

The (discrete-time) simple random walk $\big(X(t)\big)_{t\geq 0}$ on G is the (time homogeneous) Markov chain with one-step transition probabilities given by

$$p(x, y) := \mathbb{P}[X(t+1) = y \mid X(t) = x] = \frac{1}{\deg(x)}$$

if $y \sim x$, and 0 otherwise. The walk $X(t)$ is reversible with respect to deg, since

$$\deg(x) \cdot p(x, y) = \deg(y) \cdot p(y, x).$$

We denote by $\mathbb{P}_x$ and $\mathbb{E}_x$ the probability law and the expectation of the random walk $X(t)$ starting at $x \in G$, and omit the subscript if the random walk starts at the origin $o \in G$ (some fixed vertex to be chosen later). The t-step transition probabilities are then defined as

$$p_t(x, y) = \mathbb{P}[X(t) = y \mid X(0) = x]$$

and we have that

$$p_t(x, y) = \frac{\deg(y)}{\deg(x)} \cdot p_t(y, x).$$

Green function The Green function g is defined by

$$g(x, y) = \mathbb{E}_x\Big[\sum_{t=0}^{\infty} \mathbf{1}_{\{X(t)=y\}}\Big] = \sum_{t=0}^{\infty} p_t(x, y)$$

and represents the expected number of visits to y of the random walk $X(t)$ started at x. If $X(t)$ is a recurrent random walk, then g is not defined since every vertex is visited infinitely many times. A quantity of interest in the context of aggregation models is the *stopped Green function* g_n upon exiting a ball in the graph. For some vertex $x \in G$, if

$$\tau_n(x) = \inf\{t : X(t) \notin B_x(n)\} \tag{7.2.1}$$

is the first time when the random walk $X(t)$ exits $B_x(n)$, then

$$g_n(x, y) = \mathbb{E}_x\Big[\sum_{t=0}^{\tau_n(o)-1} \mathbf{1}_{\{X(t)=y\}}\Big] \tag{7.2.2}$$

and represents the expected number of visits to y before time $\tau_n(o)$, of the random walk $X(t)$ starting at $X(0) = x$. We write $\tau_n := \tau_n(o)$ if there is no risk for confusion. For $y \in G$ define

$$\tau_y = \inf\{t : X(t) = y\} \tag{7.2.3}$$

to be the first time the random walk $X(t)$ visits $y \in G$. By standard Markov chain theory, we have that

$$\mathbb{P}_y[\tau_x < \tau_n] = \frac{g_n(y,x)}{g_n(x,x)} \quad \text{and} \quad \mathbb{P}_o[\tau_x < \tau_n] = \frac{g_n(o,x)}{g_n(x,x)}, \tag{7.2.4}$$

For a function $h : G \to \mathbb{R}$, the *(probabilistic) graph Laplacian* of h is defined as

$$\Delta h(x) = \frac{1}{\deg(x)} \sum_{y \sim x} h(y) - h(x).$$

We say that h is *harmonic* on $S \subset G$ if $\Delta h = 0$ (that is, the discrete mean value property holds) on S.

Definition 7.2.1 *We say that the graph G satisfies an* elliptic Harnack inequality (EHI) *if there exists a positive constant C such that for all $x \in G$, $n > 0$, and functions $h \geq 0$ which are harmonic on $B_x(2n)$,*

$$\sup_{y \in B_x(n)} h(y) \leq C \inf_{y \in B_x(n)} h(y). \tag{EHI}$$

Definition 7.2.2 *A weighted graph (G, P) satisfies the condition (p_0) if there exists $p_0 > 0$ such that*

$$\frac{p(x,y)}{\sum_y p(x,y)} \geq p_0 \text{ for all } x \sim y, \tag{p_0}$$

where P is the transition matrix of the simple random walk $(X(t))$ on G.

We shall also use the connection between random walks and electrical networks. For a function $f : G \to \mathbb{R}$ define its *energy* by

$$\mathscr{E}(f) = \frac{1}{2} \sum_{x,y \in G, x \sim y} (f(x) - f(y))^2 \, p(x,y) \deg(x),$$

which represents the energy dissipation in the network G associated with the potential f.

Definition 7.2.3 *The effective resistance between two (disjoint) subsets $A, B \subset G$ is defined as*

$$R_{\text{eff}}(A,B) = \left[\inf\{\mathscr{E}(f) \mid f : G \to \mathbb{R},\ f|_A = 1,\ f|_B = 0\}\right]^{-1}, \tag{7.2.5}$$

with the convention that $\inf \emptyset = \infty$.

We write $R_{\text{eff}}(x, y)$ for $R_{\text{eff}}(\{x\}, \{y\})$. Moreover, $R_{\text{eff}}(x, y)$ defines a metric on G.

7.2.2 Sierpinski Gasket Graphs

For the rest of the chapter, the state space is the doubly infinite Sierpinski gasket graph denoted by SG (see Figure 7.1). Let $(X(t))_{t\geq 0}$ be a simple random walk on SG starting at o, which is recurrent. Actually, it is strongly recurrent both in the sense of [5, Definition 1.2]. We recall below some known facts about the growth of SG and the behaviour of random walks and Green functions on SG.

Notation: For two sequences a_n, b_n of real numbers, we write $a_n \asymp b_n$, if there exist a constant $C \geq 1$ such that for all $n \in \mathbb{N}$

$$\frac{1}{C} b_n \leq a_n \leq C b_n.$$

On SG there are three main quantities of interest: the spectral dimension d_s, the walk dimension β (sometimes also denoted as d_w), and fractal dimension α (sometimes also called *uniform volume growth* and denoted as d_f). Throughout this chapter we shall use α and β for the two quantities mentioned above. They are given by

$$d_s = 2\frac{\log 3}{\log 5}, \quad \beta = \frac{\log 5}{\log 2} \approx 2.32, \quad \alpha = \frac{\log 3}{\log 2} \approx 1.56, \tag{7.2.6}$$

and $\beta - \alpha \approx 0.76$.

Proposition 7.2.4 *The following holds on* SG*:*

1. ***Uniform volume growth:*** *For every $x \in$ SG and $n \in \mathbb{N}$, the balls $B_x(n)$ around x of radius n have growth of order α:*

$$| B_x(n) | \asymp n^{\alpha}, \tag{VG}$$

2. ***Elliptic Harnack inequality (EHI):*** SG *satisfies (EHI) (see [17, Corollary 2.1.8 and Proposition 3.2.7] for a proof).*

The uniform volume growth, called also *Ahlfors regularity condition*, can be easily deduced on SG; for more details, see [2, 3]. Moreover, equation (VG) implies also the *volume-doubling condition*: there exist $c > 0$ such that

$$| B_x(2n) | \leq c \, | B_x(n) |, \qquad \text{for all } x \in \text{SG}, n \geq 1. \tag{VD}$$

Remark 7.2.5 *If P is the transition matrix of the simple random walk $X(t)$ on* SG*, the (p_0) condition is satisfied with $p_0 = \frac{1}{4}$.*

The next result is well known in the study of random walks on fractal graphs (see, e.g., [2, Proposition 8.11] or [3, Corollary 2.3]).

Lemma 7.2.6 *Simple random walk on SG is sub-diffusive or sub-Gaussian: the expected exit time from balls in* SG *has order* $\beta > 2$, *that is, for every* $x \in \mathsf{SG}$ *and radius* $n \in \mathbb{N}$

$$\mathbb{E}_x[\tau_n(x)] \asymp n^\beta, \tag{E_β}$$

where $\tau_n(x)$ *is defined as in* (7.2.1).

In the proof of Theorem 7.1.1 we will need the expected exit time $\tau_n(o)$ from balls $B_o(n)$ when starting the random walk at an arbitrary point x inside $B_o(n)$. The following upper bound is easily deduced from existing results.

Lemma 7.2.7 (Uniform upper bound for the exit time) *There exists* $C > 0$ *such that for every* $n \geq 1$, $x \in \mathsf{SG}$ *and* $y \in B_x(n)$, *we have* $\mathbb{E}_y[\tau_n(x)] \leq Cn^\beta$.

Proof. Since all the conditions from [5, Proposition 3.4] are satisfied for SG with the function $\eta(n) = n^\beta$, the bound follows immediately. ☐

A corresponding lower bound holds, provided that x is not too close to the boundary of $B_o(n)$.

Lemma 7.2.8 (Lower bound for the exit time) *For every* $\epsilon \in (0, 1)$, *there exists* $c = c(\epsilon) > 0$ *such that for every* $n \geq 1$ *and every* $x \in B_o(n(1-\epsilon))$, *we have* $\mathbb{E}_x[\tau_n(o)] \geq cn^\beta$.

Proof. If $x \in B_o(n(1-\epsilon))$, then $B_x(\epsilon n) \subset B_o(n)$. Since a random walk started at x must exit $B_x(\epsilon n)$ before leaving $B_o(n)$, it follows that

$$\mathbb{E}_x[\tau_n(o)] \geq \mathbb{E}_x[\tau_{\epsilon n}(x)] \geq C^{-1}\epsilon^\beta n^\beta,$$

for a constant $C \geq 1$ independent of n and ϵ, where we used the lower bound in (E_β). ☐

Remark 7.2.9 *The function* $h(x) = \mathbb{E}_x[\tau_n]$ *solves the Dirichlet problem*

$$\begin{cases} \Delta h = 1 & \text{in } B_o(n), \\ h = 0 & \text{on } \partial B_0(n). \end{cases}$$

On SG an exact expression of h *can be obtained when* $n = 2^k$ *for* $k \in \mathbb{N}$. *Indeed, in the proof of Lemma 7.3.1, we will encounter a related Dirichlet problem with* $\Delta h = 1$ *replaced by* $\Delta h = 1 - |B_o(n)|\delta_o$, *whose solution is fully addressed in [16]. We believe it is possible to find sharper estimates of* h *for all radii* n *using harmonic splines [25], see also the recent work [12]. In any case, the estimates contained in Lemmas 7.2.7 and 7.2.8 will suffice for the purposes of this chapter.*

Lemma 7.2.10 (Upper bound for the stopped Green function) *There exists $C > 0$ such that for every $n \geq 1$ and every $x \in B_o(n)$,*

$$g_n(x, x) \leq Cn^{\beta-\alpha} \quad \text{and} \quad g_n(o, x) \leq Cn^{\beta-\alpha}$$

Proof. The graph SG is a strongly recurrent graph as defined in [5, Definition 1.2] and has volume growth of exponent $\alpha < \beta$. Therefore it satisfies condition $(VG(\beta-))$ from [5, Theorem 1.3]. Since $|B_x(d(x, y))| \asymp d(x, y)^\alpha$, [5, Theorem 1.3] gives

$$R_{\text{eff}}(x, y) \asymp \frac{d(x, y)^\beta}{|B_x(d(x, y))|} \asymp d(x, y)^{\beta-\alpha},$$

where $R_{\text{eff}}(x, y)$ represents the effective resistance between x and y. In view of the connection between random walks and electrical networks as in the proof of [5, Proposition 3.4], together with [5, Theorem 1.3], we obtain

$$g_n(x, x) = R_{\text{eff}}(x, B_o(n)^c) \leq R_{\text{eff}}(x, y) \asymp d(x, y)^{\beta-\alpha},$$
$$\text{for all } x \in SG,\ y \in B_o(n)^c.$$

The above inequality together with $d(x, y) < 2n$ yield $g_n(x, x) \leq Cn^{\beta-\alpha}$ for some constant $C > 0$, which proves the first part of the claim. Equation (7.2.4) implies that $g_n(o, x) \leq g_n(x, x)$, and this completes the proof. □

Lemma 7.2.11 *The stopped Green function $g_n(o, o)$ has growth of order $\beta - \alpha$:*

$$g_n(o, o) \asymp n^{\beta-\alpha}$$

Proof. The upper bound $Cn^{\beta-\alpha}$ follows from Lemma 7.2.10 by taking $x = o$. For the lower bound, from Equation (E_β), we have

$$\frac{1}{C}n^\beta \leq \mathbb{E}_o[\tau_n] = \sum_{x \in B_o(n)} g_n(o, x) = \sum_{x \in B_o(n)} g_n(x, o) = \sum_{x \in B_o(n)} \mathbb{P}_x[\tau_o < \tau_n] g_n(o, o)$$
$$\leq \sum_{x \in B_o(n)} g_n(o, o) = |B_o(n)| g_n(o, o) \leq C' n^\alpha g_n(o, o),$$

which gives

$$\frac{1}{C \cdot C'} n^{\beta-\alpha} \leq g_n(o, o) \leq C \cdot C' n^{\beta-\alpha},$$

and this finishes the proof. □

7.3 IDLA on Sierpinski Gasket Graphs

In this section we prove the shape result in Theorem 7.1.1. This will be done in two parts: in Section 7.3.1 we prove the inner bound $B_o(n(1-\epsilon)) \subset \mathscr{I}(b_n)$, and in Section 7.3.2 based on the inner bound, we prove the outer bound $\mathscr{I}(b_n) \subset B_o(n(1+\epsilon))$. Many of the existing proofs of the limiting shape for the IDLA cluster use the elegant idea of [19] for internal DLA on $\mathbb{Z}^d$. Of course, on other state spaces, for different estimates for Green functions, expected exit times from balls are needed, depending very much on the geometry of the underlying graph.

7.3.1 The Inner Bound for the IDLA Cluster

The proof of the inner bound in Theorem 7.1.1 is based on understanding of the divisible sandpile model and its shape on SG. This is done in [16], where the authors obtain the limit shape for the divisible sandpile cluster, by using a modified divisible sandpile process in waves, that behaves nicely due to the self-similarity of SG. The divisible sandpile model was introduced in [20], and it was also used on comb lattices as in [15] in order to prove the shape result for the IDLA cluster. The next result is similar to [19, Lemma 3], but the proof is based on the divisible sandpile model and the odometer function on SG, since no such fine estimates for the Green function and expected exit time from balls are available, like in the case of random walks on $\mathbb{Z}^d$.

Lemma 7.3.1 *Fix $\epsilon > 0$. For n sufficiently large and $z \in B_o(n(1-\epsilon))$, we have*

$$\sum_{y \in B_o(n)} g_n(y,z) \le |B_o(n)| g_n(o,z).$$

Proof. Define the function $h_n : B_o(n) \to \mathbb{R}$, by

$$h_n(z) = |B_o(n)| g_n(o,z) - \sum_{y \in B_o(n)} g_n(y,z).$$

By using the linearity of the Laplace operator together with

$$\Delta g_n(o,z) = -\delta_o(z) \text{ and } \Delta g_n(y,z) = -\delta_y(z),$$

we obtain that $h_n(z)$ solves the following Dirichlet problem on SG:

$$\begin{cases} \Delta h_n(z) &= 1 - |B_o(n)|\delta_o(z), \quad z \in B_o(n) \\ h_n(z) &= 0, \quad z \in B_o^c(n). \end{cases}$$

Recall now the definition of the divisible sandpile model and of the odometer function u on SG, whose Laplace satisfies equation (3) from [16], when starting with initial sand distribution μ_0. That is, if we start with mass $|B_o(n)|$ at the origin o, then the odometer function satisfies

$$\begin{cases} \Delta u(z) &= 1 - |B_o(n)|\delta_o(z), \quad z \in \mathscr{S} \\ u(z) &= 0, \quad z \in \mathscr{S}^c, \end{cases}$$

where $\mathscr{S}$ is the divisible sandpile cluster on SG, and the sandpile cluster is defined as the set of vertices z in SG where $u(z) > 0$. By [16, Theorem 1.1], the shape of the sandpile cluster $\mathscr{S}$ when starting with mass $|B_o(n)|$ at o is given by $B_o(n-1) \subseteq \mathscr{S} \subseteq B_o(n)$. The difference there is that the balls are closed instead of open, therefore translating [16, Theorem 1.1] in our setting of open balls, we have $B_o(n) \subseteq \mathscr{S} \subseteq B_o(n+1)$.

Define the function $k_n : B_o(n+1) \to \mathbb{R}$ by $k_n(z) := u(z) - h_n(z)$, which is harmonic on $B_o(n)$. We have

$$\begin{cases} \Delta k_n(z) &= 0, \quad z \in B_o(n) \\ k_n(z) &= u(z) \geq 0, \quad z \in \partial B_o(n). \end{cases}$$

Applying both the minimum and the maximum principle to k_n, if $\max_{z \in \partial B_o(n)} k_n(z) = c \geq 0$, then we obtain $u(z) - c < h_n(z) < u(z)$ for all $z \in B_o(n)$. The odometer function $u(z)$ is strictly positive on $\mathscr{S}$ and decreasing in the distance from z to the origin o. This means that there exists a constant c', such that $u(z) > c$ on $B_o(n - c')$, which implies that $h_n(z) > 0$ on $B_o(n - c')$. For n large enough $B_o(n(1-\epsilon)) \subset B_o(n - c')$, and this yields $h_n(z) > 0$ on $B_o(n(1-\epsilon))$ as well, which proves the claim. □

We first prove that for every $\epsilon > 0$,

$$B_o(n(1-\epsilon)) \subset \mathscr{I}(b_n(1+\epsilon)), \text{ for all sufficiently large } n, \tag{7.3.1}$$

with probability 1. Recall that $b_n = |B_o(n)|$. Taking the intersection on both sides over all $\epsilon' < \epsilon$, we get the inner bound $B_o(n(1-\epsilon)) \subset \mathscr{I}(b_n)$ in Theorem 7.1.1. By Borel–Cantelli, a sufficient condition for (7.3.1) is

$$\sum_n \sum_{z \in B_o(n(1-\epsilon))} \mathbb{P}[E_z(b_n(1+\epsilon))] < \infty, \tag{7.3.2}$$

where

$$E_z(b_n(1+\epsilon)) := \{z \notin \mathscr{I}(b_n(1+\epsilon))\}$$

is the event that z does not belong to the IDLA cluster $\mathscr{I}(b_n(1+\epsilon))$. We want to show that the probability of this event decreases exponentially in n.

Let us first fix $z \in B_o(n(1-\epsilon))$ and look at the first $b_n(1+\epsilon)$ random walks $(X^i(t))_{t\geq 0}$ and $i = 1, 2, \ldots, b_n(1+\epsilon)$ which build the IDLA cluster. We let these $b_n(1+\epsilon)$ walks evolve forever, even after they have left the IDLA cluster. We upper bound $\mathbb{P}[E_z(b_n(1+\epsilon))]$ as follows. Let M be the number of walks that visit z before exiting the ball $B_o(n)$. Furthermore, let L be the number of walks that visit z before exiting the ball $B_o(n)$ but after leaving the occupied IDLA cluster. We have

$$E_z(b_n(1+\epsilon)) \subset \{M = L\}.$$

Then for any given $a \geq 0$

$$\mathbb{P}[E_z(b_n(1+\epsilon))] < \mathbb{P}[M = L] \leq \mathbb{P}[M \leq a \text{ or } L \geq a] \leq \mathbb{P}[M \leq a] + \mathbb{P}[L \geq a]. \tag{7.3.3}$$

We choose a later so that the above two probabilities can be made sufficiently small. We have to show that M includes more walks on average while L includes fewer terms.

Consider the following stopping times:

$$\begin{aligned}
\sigma^i &= \inf\left\{t > 0 : X^i(t) \notin \mathcal{I}(i-1)\right\} \\
&= \text{ the time it takes the } i\text{th walk to leave the IDLA cluster} \\
\tau_n^i &= \inf\{t > 0 : X^i(t) \notin B_o(n)\} \\
&= \text{ the time it takes the } i\text{th walk to leave the ball } B_o(n) \\
\tau_z^i &= \inf\{t > 0 : X^i(t) = z\} \\
&= \text{ the time it takes the } i\text{th walk to hit } z.
\end{aligned}$$

Recall that all particles start their journey at the fixed origin $o \in \mathsf{SG}$. In terms of these stopping times we can write

$$M = \sum_{i=1}^{b_n(1+\epsilon)} \mathbf{1}_{\{\tau_z^i < \tau_n^i\}} \quad \text{and} \quad \mathbb{E}[M] = b_n(1+\epsilon)\mathbb{P}_o[\tau_z < \tau_n],$$

since the summands in M are iid. On the other hand,

$$L = \sum_{i=1}^{b_n(1+\epsilon)} \mathbf{1}_{\{\sigma^i < \tau_z^i < \tau_n^i\}},$$

but the summands in L are not identically distributed and not independent, since after each walk exits the IDLA cluster, the shape of the cluster is modified. Thus $\mathbb{E}[L]$ is hard to determine, but a good upper bound for it would suffice. Note that only those walks that exit the IDLA cluster inside the ball $B_o(n)$ contribute to L, and for each $y \in B_o(n)$ there is at most one index i for which $X^i(\sigma^i) = y$. Then the walks started at y that hit z after leaving the ball $B_o(n)$ are independent. So in order to get rid of the dependence of the summands in L, we enlarge the index to all $y \in B_o(n)$, start a random walk at y, and look if this walk visits z before leaving $B_o(n)$. That is, if we let

$$\tilde{L} = \sum_{y \in B_o(n)} \mathbf{1}^y_{\{\tau_z < \tau_n\}},$$

where the indicators $\mathbf{1}^y$ correspond to independent random walks starting at y, we have that $L \leq \tilde{L}$ and

$$\mathbb{E}[\tilde{L}] = \sum_{y \in B_o(n)} \mathbb{P}_y[\tau_z < \tau_n].$$

The next step is to use a large deviation result in order to bound the sum of a large number of independent random variables, but we need to know more about $\mathbb{E}[M]$, $\mathbb{E}[\tilde{L}]$, and the relationship between them.

Using equation (7.2.4) together with the symmetry of the stopped Green function, we have

$$\mathbb{E}[M] = \lfloor b_n(1+\epsilon) \rfloor \frac{g_n(o,z)}{g_n(z,z)}$$

$$\mathbb{E}\left[\tilde{L}\right] = \frac{1}{g_n(z,z)} \sum_{y \in B_o(n)} g_n(y,z) = \frac{1}{g_n(z,z)} \sum_{y \in B_o(n)} g_n(z,y) = \frac{1}{g_n(z,z)} \mathbb{E}_z[\tau_n]$$

Then, by Lemma 7.3.1, we can write

$$\mathbb{E}[M] \geq \left(1 + \frac{\epsilon}{2}\right) \frac{b_n g_n(o,z)}{g_n(z,z)} \geq \left(1 + \frac{\epsilon}{2}\right) \frac{\sum_{y \in B_o(n)} g_n(y,z)}{g_n(z,z)} = \left(1 + \frac{\epsilon}{2}\right) \mathbb{E}[\tilde{L}]. \tag{7.3.4}$$

Lemma 7.3.2 *The expectation of the random variable $\tilde{L}$ can be bounded from below by*

$$\mathbb{E}\left[\tilde{L}\right] \geq c' n^{\alpha},$$

where $c' > 0$ depends on nothing but ϵ.

Proof. Recall that $\mathbb{E}[\tilde{L}] = \dfrac{1}{g_n(z,z)} \mathbb{E}_z[\tau_n]$. From Lemma 7.2.8 we know that there exists $c = c(\epsilon) > 0$ such that $\mathbb{E}_z[\tau_n(o)] \geq cn^{\beta}$, and Lemma 7.2.10 gives the bound $g_n(z,z) \leq Cn^{\beta-\alpha}$, for c and C both being positive constants. Putting these two relations together, for $c' = \frac{c}{C} > 0$ we get the claim. □

On account of (7.3.4), one has the same lower bound for $\mathbb{E}[M]$. We shall use the following large deviation estimate for sums of independent indicator random variables. For a proof see [19, Lemma 4].

Lemma 7.3.3 *Let S be a finite sum of independent indicator random variables with mean μ. For any $0 < \gamma < 1/2$, and for all sufficiently large μ,*

$$\mathbb{P}\left[|S-\mu| \geq \mu^{1/2+\gamma}\right] \leq 2\exp\left\{-\frac{1}{4}\mu^{2\gamma}\right\}.$$

Since both M and $\tilde{L}$ are finite sums of indicator random variables, we can apply the above lemma to both of them. Recall that we want to choose a number a such that the probabilities $\mathbb{P}[M \leq a]$ and $\mathbb{P}[\tilde{L} \geq a]$ can be made sufficiently small.

Proof of the inner bound in Theorem 7.1.1. Recall that in order to prove that $B(n(1-\epsilon)) \subset \mathscr{I}(b_n(1+\epsilon))$, it is sufficient to upper bound the probabilities $\mathbb{P}[M \leq a]$ and $\mathbb{P}[\tilde{L} \geq a]$ and to show that they are summable over n and $z \in B_o(n(1-\epsilon))$.

Let us choose $a = \left(1+\frac{\epsilon}{4}\right)\mathbb{E}\left[\tilde{L}\right]$ and $\gamma = \frac{1}{4}$, and let us first show that for $\epsilon > 0$ and n large enough we have

$$\mathbb{E}\left[\tilde{L}\right] + \mathbb{E}\left[\tilde{L}\right]^{1/2+\gamma} \leq \underbrace{\left(1+\frac{\epsilon}{4}\right)\mathbb{E}\left[\tilde{L}\right]}_{=a} \leq \mathbb{E}[M] - \mathbb{E}[M]^{1/2+\gamma}. \tag{7.3.5}$$

The first inequality in (7.3.5) comes from Lemma 7.3.2:

$$\mathbb{E}\left[\tilde{L}\right]\left(1+\mathbb{E}\left[\tilde{L}\right]^{-1/4}\right) \leq \mathbb{E}\left[\tilde{L}\right]\left(1+\frac{1}{c'n^{\alpha/4}}\right) \leq \left(1+\frac{\epsilon}{4}\right)\mathbb{E}\left[\tilde{L}\right] = a,$$

for n large enough.

The second inequality in (7.3.5) is obtained using equation (7.3.4) and Lemma 7.3.2:

$$\begin{aligned}
\mathbb{E}[M]\left(1-\mathbb{E}[M]^{-1/4}\right) &\geq \left(1+\frac{\epsilon}{2}\right)\mathbb{E}\left[\tilde{L}\right]\left(1-\frac{1}{(1+\frac{\epsilon}{2})^{1/4}}\mathbb{E}\left[\tilde{L}\right]^{-1/4}\right) \\
&\geq \left(1+\frac{\epsilon}{2}\right)\mathbb{E}\left[\tilde{L}\right]\left(1-\mathbb{E}\left[\tilde{L}\right]^{-1/4}\right) \\
&= \left(1+\frac{\epsilon}{4}\right)\mathbb{E}\left[\tilde{L}\right] + \frac{\epsilon}{4}\mathbb{E}\left[\tilde{L}\right]\left(1-\frac{4+2\epsilon}{\epsilon}\mathbb{E}\left[\tilde{L}\right]^{-1/4}\right) \\
&\geq \left(1+\frac{\epsilon}{4}\right)\mathbb{E}\left[\tilde{L}\right] + \frac{\epsilon}{4}\mathbb{E}\left[\tilde{L}\right]\left(1-\frac{4+2\epsilon}{\epsilon}\frac{1}{c'n^{\alpha/4}}\right) \\
&\geq \left(1+\frac{\epsilon}{4}\right)\mathbb{E}\left[\tilde{L}\right], \text{ for } n \text{ large enough.}
\end{aligned}$$

The last inequality follows from the fact that for n sufficiently large the quantity $\left(1-\frac{4+2\epsilon}{\epsilon}\frac{1}{cn^{\alpha/4}}\right)$ is greater than 0 and $\mathbb{E}[\tilde{L}] > 0$. Therefore we have proved (7.3.5). Thus, recalling that we defined $a = \left(1+\frac{\epsilon}{4}\right)\mathbb{E}[\tilde{L}]$ and $1/2+\gamma = 3/4$

$$\mathbb{P}\Big[\tilde{L} \geq a\Big] \leq \mathbb{P}\Big[\tilde{L} \geq \mathbb{E}[\tilde{L}] + \mathbb{E}[\tilde{L}]^{3/4}\Big] \leq 2\exp\Big\{-\frac{1}{4}\mathbb{E}[\tilde{L}]^{1/2}\Big\} \leq \exp\Big\{-cn^{\frac{\alpha}{2}}\Big\},$$

where $c = c(\epsilon) > 0$ is independent of n. The last two inequalities above follow from Lemmas 7.3.3 and 7.3.2. Similarly for M, we use Lemma 7.3.3, (7.3.4), and Lemma 7.3.2 to obtain

$$\begin{aligned}\mathbb{P}[M \leq a] \leq \mathbb{P}\Big[M \leq \mathbb{E}[M] - \mathbb{E}[M]^{3/4}\Big] &\leq 2\exp\Big\{-\frac{1}{4}\mathbb{E}[M]^{1/2}\Big\} \\ &\leq 2\exp\Big\{-\frac{1}{4}\Big(1+\frac{\epsilon}{2}\Big)^{1/2}\mathbb{E}[\tilde{L}]^{1/2}\Big\} \\ &\leq 2\exp\Big\{-\frac{1}{4}\mathbb{E}[\tilde{L}]^{1/2}\Big\} \leq \exp\Big\{-cn^{\frac{\alpha}{2}}\Big\}.\end{aligned}$$

Putting together the previous two inequalities and using (VG), we have that for $\epsilon > 0$ there exists $n_\epsilon \in \mathbb{N}$ such that for all $n \geq n_\epsilon$,

$$\begin{aligned}&\sum_{n\geq n_\epsilon}\sum_{z\in B_o(n(1-\epsilon))}\mathbb{P}[z \notin \mathcal{I}(b_n(1+\epsilon))] \\ &\quad\leq \sum_{n\geq n_\epsilon}\sum_{z\in B_o(n(1-\epsilon))}\Big(\mathbb{P}[M \leq a] + \mathbb{P}[\tilde{L} \geq a]\Big) \\ &\quad\leq \sum_{n\geq n_\epsilon}\sum_{z\in B_o(n(1-\epsilon))} 2\exp\Big\{-cn^{\frac{\alpha}{2}}\Big\} \\ &\quad\leq \sum_{n\geq n_\epsilon} cn^\alpha \exp\big\{-n^{\frac{\alpha}{2}}\big\} < \infty,\end{aligned}$$

with $\alpha = \frac{\log 3}{\log 2}$. By the Borel–Cantelli lemma, we have proved that for $\epsilon > 0$

$$B_o(n(1-\epsilon)) \subset \mathcal{I}(b_n(1+\epsilon)), \text{ for } n \text{ large}$$

with probability 1, and this implies the inner bound in Theorem 7.1.1. □

7.3.2 The Outer Bound for IDLA Cluster

In order to prove the outer bound $\mathcal{I}(b_n) \subset B_o(n(1+\epsilon))$ in Theorem 7.1.1, we shall use parts of the main result of [10]. More precisely, in well-behaved environments, once a good inner bound is obtained, we can control the number of particles not contained in the inner bound and obtain a good upper bound, as a corollary of [10, Theorem 1.2].

In order to be able to use [10, Theorem 1.2] and [10, Corollary 1.3], one should check that the conditions required in these two results are fulfilled when we perform IDLA on SG. Unfortunately, not all the conditions required there are available on SG, but we have good estimates on the stopped Green function which allow us to overcome some of those conditions mentioned above. Before going into details about the conditions required, we shall first set the notation about stopped IDLA clusters. We use the same notation as in [10].

Let $S \subset \mathsf{SG}$ be a finite subset of SG. In order to define the stopped IDLA clusters, we first define the aggregate $\mathscr{I}(S; x)$ when we start with an existing finite cluster S, and run a simple random walk $X(t)$ starting at some vertex $x \in S$ until it exits S. Let σ_S be the first time when $X(t)$ exits S. Then define

$$\mathscr{I}(S; x) := S \cup \{X(\sigma_S)\}.$$

For the outer boundary, we will need a slightly more general process, where the growth of the IDLA cluster is stopped before exiting slightly bigger balls. For some radius $r > 0$ such that $S \subset B_o(r)$, denote by $\mathscr{I}(S; x \mapsto B_o(r))$ the cluster which is obtained as follows. For some $x \in S$ start a simple random walk $X(t)$ at x and let it run until it either exits S or reaches $B_o(r)^c$. If σ_S is as above, and $\tau_r = \tau_r(o)$ is the first time the random walk exits $B_o(r)$ as defined in (7.2.1),

$$\mathscr{I}\big(S; x \mapsto B_o(r)\big) := S \cup \{X(\sigma_S \wedge (\tau_r - 1))\}.$$

We need to keep track of the paused particles, and their positions on $B_o(r)^c$; the paused particles will all be at distance r from the origin. Define

$$P(S; x \mapsto B_o(r)) = \begin{cases} X(\tau_r), & \text{if } \tau_r \le \sigma_S \\ \perp, & \text{otherwise,} \end{cases}$$

where $\perp$ indicates that the random walk attached to the existing aggregate S before exiting the ball $B_o(r)$, so there is no particle to be paused. For vertices $x_1, x_2, \ldots, x_k$ in SG, a set $S \subset \mathsf{SG}$, and a ball $B_o(r)$, define $\mathscr{I}\big(S; x_1, x_2, \ldots, x_k \mapsto B_o(r)\big)$ to be the IDLA cluster when starting with the occupied set S and k random walks with starting points $x_1, x_2, \ldots, x_k$, respectively, and paused upon exiting $B_o(r)$. One can then inductively define $\mathscr{I}\big(S; x_1, x_2, \ldots, x_k \mapsto B_o(r)\big)$ by taking

$$S_0 = S, \quad S_j = \mathscr{I}\big(S_{j-1}; x_j \mapsto B_o(r)\big) \quad \text{for} \quad j \in \{1, \ldots, k\},$$

and $\mathscr{I}\big(S; x_1, x_2, \ldots, x_k \mapsto B_o(r)\big) = S_k$. Since some of the k particles may be stopped on $B_o(r)^c$ before attaching to the existing cluster, we will keep a track of these particles in the following way. Define $P(S; x_1, \ldots, x_k \mapsto B_o(r))$ to be the sequence of the paused particle in the process above. More precisely, if $p_j = P(S_{j-1}; x_j \mapsto B_o(r))$ for $j \in \{1, \ldots, k\}$, then $P(S; x_1, \ldots, x_k \mapsto B_o(r))$ is just the sequence $(p_j : p_j \neq \perp)$. If particles are not paused before exiting some ball, then the aggregate is simply denoted by $\mathscr{I}(S; x_1, \ldots, x_k)$. The reason for working with the paused IDLA process, before exiting bigger and bigger balls, is that the IDLA process possesses the *abelian property*: the unstopped cluster $\mathscr{I}(S; x_1, \ldots, x_k)$ has the same distribution as

$$\mathscr{I}\big(\mathscr{I}(S; x_1, \ldots, x_k \mapsto B_o(r)); P(S; x_1, \ldots, x_k \mapsto B_o(r))\big). \tag{7.3.6}$$

For details on this property see [9, 19]. As defined in Section 7.1, the IDLA cluster $\mathscr{I}(n)$ is built by starting n particles at the origin, and letting them run until exiting the previously occupied cluster. We are interested in the shape $\mathscr{I}(b_n)$ of IDLA cluster when we start $b_n = |B_o(n)|$ particles at the origin which is a special case of the stopped process defined above. Nevertheless, in proving the outer bound in Theorem 7.1.1, the general stopped process will be used. In terms of the stopped process defined above, we have

$$\mathscr{I}(n) = \mathscr{I}\big(\emptyset; \underbrace{o, \ldots, o}_{n \text{ times}}\big)$$

and set also

$$\mathscr{I}_n(x \mapsto r) := \mathscr{I}\big(\emptyset; \underbrace{x, \ldots, x}_{n \text{ times}} \mapsto B_x(r)\big)$$

and

$$P_n(x \mapsto r) := P\big(\emptyset; \underbrace{x, \ldots, x}_{n \text{ times}} \mapsto B_x(r)\big).$$

Having set the notations, our aim is to prove a similar result to [10, Theorem 1.2]. The condition *weak lower bound* (wLB) from [10], which says that when releasing $|B_x(n)|$ particles at x, the IDLA cluster contains $B_x(n)$ with noticeable probability, does not hold for every starting point x. We have such a lower bound on the IDLA cluster in Theorem 7.1.1, only when we start $b_n = |B_n(o)|$ particles at the origin $o \in \mathsf{SG}$. We believe that one should be able to adapt the proof of the inner bound for the IDLA cluster when releasing particles at vertices $x \in \mathsf{SG}$ other than o, and to get the same result with some additional technical difficulties on Green function estimates and expected exit time from balls for the random walks. Nevertheless, we are not going to do this here, but we will overcome the assumption (wLB) in [10], which is used only in [10, Lemma 3.2], by giving a different proof of this lemma. Our proof uses finer estimates on the stopped Green function, estimates which are not available in the general setting of [10]. Our approach to this result is similar to the one in [18, Lemma 11].

The *continuity assumption* (C) [10, Page 4/8] holds automatically, since the balls we work with are considered with respect to the graph metric, that is, ρ and d_G coincide in our case. Moreover, *the regular volume growth* (VG) condition holds for any radius r and centre x: $|B_x(r)|$ has growth of order α, as defined in (VG). The fractal growth α as defined in (7.2.6) will play the role of d in [10]. We next prove an estimate for the infimum of the stopped Green function.

Lemma 7.3.4 *Let $u \in (0, 1)$. There exists $c = c(u) > 0$ such that for all sufficiently large radii $r > 0$ and all $x \in \mathsf{SG}$*

$$\inf_{y \in B_x(ur)} g_r(x, y) \geq c r^{\beta-\alpha}, \tag{7.3.7}$$

where $g_r(x, y)$ is the stopped Green function as defined in (7.2.2).

Proof. We prove this lemma in two steps. First we show that (7.3.7) holds for $u = \frac{1}{2}$ (and hence for all $u \in \left(0, \frac{1}{2}\right)$). Then we use a chaining argument to extend the estimate to any $u \in \left(\frac{1}{2}, 1\right)$. The elliptic Harnack inequality (EHI) is used in both steps.

Step 1: To prove (7.3.7) for $u = \frac{1}{2}$, we need a related Harnack inequality for the stopped Green function stated in [11]. We say that the Green function satisfies the condition (HG) if there exist a constant $C_1 \geq 1$ such that for all $x \in \mathsf{SG}$, $r > 0$, and finite sets $D \supset B_x(r)$,

$$\sup_{y \in B_x(r/2)^c} g_D(x, y) \leq C_1 \inf_{y \in B_x(r/2)} g_D(x, y), \tag{HG}$$

where $g_D(x, y)$ represents the expected number of visits to y before leaving the set D, when starting the random walk at x. It is shown in [4, Theorem 2] that under the conditions (p_0) and (EHI), there exists a constant $C_2 \geq 1$ such that if $x_0 \in \mathsf{SG}$, $r > 0$, $d(x_0, x) = d(x_0, y) = r/2$, and $B_{x_0}(r) \subset D$, then

$$C_2^{-1} g_D(x_0, y) \leq g_D(x_0, x) \leq C_2 g_D(x_0, y). \tag{7.3.8}$$

In particular this implies that (EHI) and (HG) are equivalent. Based on this equivalence, it can be further shown (see [26, Lemma 3.7 and Proposition 3.7]) that under (p_0) and (EHI), there exists $C_3 > 0$ such that for any ball $B_x(ur)$ with $u \in \left(0, \frac{1}{2}\right]$,

$$\inf_{y \in B_x(ur)} g_r(x, y) \geq C_3 R_{\text{eff}}\left(B_x(ur), B_x(r)^c\right). \tag{7.3.9}$$

The conditions (p_0) and (EHI) hold on G (see Section 7.2.2). Furthermore the effective resistance estimate is known on Sierpinski gasket graphs SG; on account of [26, Proposition 2.3] there exist $C \geq c > 0$ such that for all $x \in \mathsf{SG}$ and $r > 0$,

$$c r^{\beta-\alpha} \leq R_{\text{eff}}(B_x(r/2), B_x(r)^c) \leq C r^{\beta-\alpha}. \tag{7.3.10}$$

Altogether (7.3.9) and (7.3.10) yield

$$\inf_{y \in B_x(r/2)} g_r(x, y) \geq cC_3 r^{\beta-\alpha} \tag{7.3.11}$$

for all $r > 0$.

Step 2: We now extend the estimate to the ball $B_x(ur)$, for $u \in \left(\frac{1}{2}, 1\right)$. Call $h := g_r(x, \cdot)$, which is a non-negative harmonic function on $B_x(r) \setminus \{x\}$. Let $x_1 \in \partial B_x(ur)$ be such that $h(x_1) = \inf_{y \in \partial B_x(ur)} h(y)$.

For $x, y \in \mathsf{SG}$, denote by $\gamma(x, y)$ the shortest path in SG connecting x and y. Let y_1 be the intersection point of $\gamma(x, x_1)$ and $\partial B_x(r/2 - 1)$. Along the path $\gamma(y_1, x_1)$ we construct a minimal chain of intersecting balls $\{B_{z_k}(r') : k \in \{1, 2, \cdots, K\}\}$, where $z_k \in \gamma(y_1, x_1)$ for every k, $y_1 \in B_{z_1}(r')$, and $x_1 \in B_{z_K}(r')$. In order to apply (EHI) to each ball $B_{z_k}(r') \subset B_{z_k}(2r')$, we choose the radius $r' > 0$ such that $B_{z_k}(2r') \subset (B_x(r) \setminus \{x\})$ for every k. An easy geometric reasoning shows that one can take $r' = \lfloor (1-u)r \rfloor$, and the chain consists of K balls with

$$K \leq \frac{(u - \frac{1}{2})r}{2(r' - 1)} \leq \frac{1}{2} \frac{u - \frac{1}{2}}{1 - u} + \mathcal{O}\left(\frac{1}{r}\right) \tag{7.3.12}$$

as $r \to \infty$. Applying (EHI) to the function h successively yields the comparison

$$h(y_1) \leq Ch(z_1) \leq C^2 h(z_2) \leq \cdots \leq C^{K+2} h(x_1), \tag{7.3.13}$$

In other words, there exists $C_4 = C_4(u) > 0$ such that for all sufficiently large r,

$$h(x_1) \geq C_4 h(y_1). \tag{7.3.14}$$

Recall the maximum (or minimum) principle for harmonic functions, e.g., [26, Proposition 3.1]: if h is harmonic on a finite set $A \subset \Gamma$, then

$$\max_{\overline{A}} h = \max_{\partial A} h, \quad \min_{\overline{A}} h = \min_{\partial A} h, \tag{7.3.15}$$

where $\overline{A} = A \cup \partial A$. By taking $A = B_x(ur) \setminus \{x\}$, we have $\partial A = \partial B_x(ur) \cup \{x\}$. By the minimum principle for the function $h := g_r(x, \cdot)$ on the set A

$$\inf_A h \geq \inf_{\overline{A}} h = \inf_{\partial A} h. \tag{7.3.16}$$

Since $g_r(x, y) \leq g_r(x, x)$ for all $y \in \mathsf{SG}$, $h := g_r(x, \cdot)$ attains a maximum at x, which together with (7.3.16) implies

$$\inf_{B_x(ur)} h \geq \inf_{\partial B_x(ur)} h.$$

Using (7.3.14) and (7.3.11), we deduce that for all sufficiently large r,

$$\inf_{B_x(ur)} h \geq \inf_{\partial B_x(ur)} h = h(x_1) \geq C_4 h(y_1) \geq C_4 \inf_{B_x(r/2)} h \geq C_4 c C_3 r^{\beta-\alpha}, \tag{7.3.17}$$

and this proves the claim. □

We now prove that the probability that a random walk on SG hits a set whose complement has size at most ϵr^{α} is bounded away from zero. The proof is similar to [18, Lemma 11].

Lemma 7.3.5 *Let $x \in$ SG and $(X(t))$ be a simple random walk on* SG *started at x. For every $\epsilon > 0$, there exists $\eta > 0$ such that if $A \subset B_x(r)$, with $|A| \geq \epsilon |B_x(r)|$, then*

$$\mathbb{P}_x\left[\sigma_A < \tau_r(x)\right] \geq \eta,$$

where σ_A is the first time $X(t)$ hits A.

Proof. Without loss of generality we prove the inequality for all sufficiently large $r > 0$. Let V be the number of visits to A of the random walk $\big(X(t)\big)$ started at x, and before leaving $B_x(r)$:

$$V = \sum_{t=0}^{\tau_r(x)-1} \mathbf{1}_{\{X(t)\in A\}}.$$

Then we have

$$\mathbb{P}_x[\sigma_A < \tau_r(x)] \geq \mathbb{P}_x[V \geq 1] = \frac{\mathbb{E}_x[V]}{\mathbb{E}_x[V|V \geq 1]}.$$

Observe that on the event $\{V \geq 1\}$, $\sigma_A < \tau_r(x)$ and $V \leq \tau_r(x) - \sigma_A$. Combine this with the Markov property and we get

$$\begin{aligned}
\mathbb{E}_x[V|V \geq 1] &\leq \mathbb{E}_x\left[\tau_r(x) - \sigma_A | V \geq 1\right] \\
&= \mathbb{E}_x\left[\mathbb{E}_{X(\sigma_A)}[\tau_r(x) - \sigma_A] | V \geq 1\right] \\
&= \mathbb{E}_x\left[\mathbb{E}_{X(\sigma_A)}[\tau_r(x)] | V \geq 1\right] \\
&\leq \mathbb{E}_x[C r^{\beta} | V \geq 1] = C r^{\beta},
\end{aligned} \tag{7.3.18}$$

where in the last line we used Lemma 7.2.7, which states that $\mathbb{E}_y[\tau_r(x)] \leq C r^{\beta}$ for all $y \in A \subset B_x(r)$.

Now we need a lower bound of order β for $\mathbb{E}_x[V]$. Recall that the balls in SG have growth of order α; see (VG). Let $\epsilon > 0$. Then we can find $u = u(\epsilon) < 1$ such that for all sufficiently large $r > 0$

$$|B_x(r) \setminus B_x(ur)| \leq \frac{\epsilon}{2} |B_x(r)|.$$

Since $|A \setminus B_x(ur)| \leq |B_x(r) \setminus B_x(ur)|$ and $A \subset B_x(r)$ with $|A| \geq \epsilon |B_x(r)|$, we have

$$|A \cap B_x(ur)| = |A| - |A \setminus B_x(ur)| \geq \frac{\epsilon}{2}|B_x(r)|.$$

On the other hand,

$$\mathbb{E}_x[V] = \sum_{y \in A} g_r(x,y) \geq \sum_{y \in B_x(ur) \cap A} g_r(x,y) \geq \frac{\epsilon}{2}|B_x(r)| \inf_{y \in B_x(ur)} g_r(x,y).$$

Lemma 7.3.4 together with equation (VG) imply that there exists a constant $c = c(u)$ such that

$$\mathbb{E}_x[V] \geq \frac{\epsilon}{2} c r^{\alpha} r^{\beta-\alpha} = c(\epsilon) r^{\beta},$$

which together with (7.3.18) yield

$$\mathbb{P}_x[\sigma_A < \tau_r(x)] \geq \frac{c(\epsilon)}{C} = \eta > 0,$$

whence the claim. □

The statement of the next result is similar to [10, Lemma 3.3]. Nevertheless, the proof there uses the (wLB) on IDLA which we do not have. We use instead Lemma 7.3.5.

Lemma 7.3.6 *There exist $\rho, \eta \in (0,1]$ such that for large enough n and $n^{\frac{1}{\alpha(\alpha+1)}} < r < n$, the following holds. Let $x \in B_o(n)$ and let $S \subset B_o(n+r)$ be such that $|S \setminus B_o(n)| \leq \rho r^{\alpha}$. Let $X(t)$ be a simple random walk started at x, σ_Q be the first time $(X(t))$ hits the set $Q := B_o(n+r) \setminus (S \cup B_o(n))$, and $\tau_{n+r}(o)$ be the first time $(X(t))$ exits the ball $B_o(n+r)$ of radius $n+r$ around the origin o. Then*

$$\mathbb{P}_x[\sigma_Q < \tau_{n+r}(o)] \geq \eta.$$

Proof. For every path γ from inside $B_o(n)$ to outside $B_o(n+r)$, let $y(\gamma)$ be the first vertex on this path for which $d(y(\gamma), B_o(n)) = \frac{r}{2}$. Let us denote by Y the set of all vertices $y(\gamma)$ for paths γ. Moreover, every path from $x \in B_o(n)$ to outside $B_o(n+r)$ must hit the set Y. Therefore by Markov's property, it suffices to prove the result for starting points $y \in Y$. Let us fix such a $y \in Y$, and consider the ball $B_y(r/3)$ of radius $r/3$ around y. By letting $A = B_y(r/3) \setminus S$ and using (VG), there exists $C \geq 1$ such that

$$\begin{aligned}|A| \geq |B_y(r/3)| - |S| &\geq \frac{1}{C}\left(\frac{r}{3}\right)^{\alpha} - \rho r^{\alpha} \geq \frac{1}{C}\left(\frac{r}{3}\right)^{\alpha}\left(1 - \left(\frac{3}{4}\right)^{\alpha}\right) \\ &\geq \frac{1}{C^2}\left(1 - \left(\frac{3}{4}\right)^{\alpha}\right) \cdot \left|B_y(r/3)\right|,\end{aligned}$$

for $\rho = \frac{4-\alpha}{C} \in (0, 1]$. Putting $A = B_y(r/3) \setminus S \subset B_y(r/3)$ and $\epsilon = \frac{1}{C^2}\left(1 - \left(\frac{3}{4}\right)^\alpha\right)$ in Lemma 7.3.5, we then deduce the existence of $\eta > 0$ (and, without loss of generality, it is understood that $\eta \leq 1$) such that

$$\mathbb{P}_y[\sigma_A < \tau_{r/3}(y)] \geq \eta.$$

Next, since $d(y, B_o(n)) = r/2$, we have that $d(y, \mathsf{SG} \setminus B_o(n + r)) > r/3$ and $B_y(r/3) \subset B_o(n + r) - B_o(n)$. Then

$$\mathbb{P}_x[\sigma_Q < \tau_{n+r}(o)] \geq \mathbb{P}_y[\sigma_A < \tau_{r/3}(y)] \geq \eta.$$ □

The previous lemma investigates the behaviour of a single particle attaching to the IDLA cluster. The next lemma, which claims that with high probability, a constant fraction of the IDLA aggregate is absorbed in a fine annulus of SG, resembles [10, Lemma 3.3], with α (the volume growth of SG) in place of d. For the reader's convenience, we state both the result and its proof adapted to our case and to our notation.

Lemma 7.3.7 *There exist $\delta > 0$ and $p < 1$ such that all n large enough, for all $n^{1/(\alpha+1)} < k < n^\alpha$ and $x_1, \ldots, x_k \in B_o(n)$, and for all $S \subset B_o(n)$, the following holds:*

$$\mathbb{P}\left[\left|\mathscr{I}\left(S; x_1, \ldots, x_k \mapsto B_o\left(n + k^{1/\alpha}\right)\right) \setminus S\right| \leq \delta k\right] \leq p^k.$$

Proof. Let $r = k^{1/\alpha}$ and fix $\rho, \eta \in (0, 1]$ as in Lemma 7.3.6. Moreover, let $X_1(t), \ldots, X_k(t)$ be simple random walks that start at $x_1, \ldots, x_k$ respectively and stop when exiting $B_o(n + r)$, and that generate the stopped IDLA cluster. Let $k' = \lfloor \rho k \rfloor \leq \rho r^\alpha$, and for $j \in \{1, \ldots k'\}$, denote

$$\mathscr{I}_j = \mathscr{I}\left(S; x_1, \ldots, x_j \mapsto B_o(n + r)\right).$$

By construction, since only j vertices can add to the IDLA cluster, we have $\left|\mathscr{I}_j \setminus B_o(n)\right| \leq j \leq \rho r^\alpha$, so we are in the setting of Lemma 7.3.6, with $\mathscr{I}_j \subset B_o(n + r)$ instead of the set S, for all $j \in \{1, \ldots k'\}$. This implies that for all $j \in \{1, \ldots k'\}$

$$\mathbb{P}[X_{j+1} \cap \left(B_o(n + r) \setminus \mathscr{I}_j\right) \neq \emptyset | \mathscr{I}_j] \geq \eta.$$

Thus $|\mathscr{I}(S; x_1, \ldots, x_k \mapsto B_o(n + r)) \setminus S|$ dominates a (k', η)-binomial random variable, which implies that there exist $\delta > 0$ and $p < 1$ depending only on ρ, η such that

$$\mathbb{P}\left[|\mathscr{I}(S; x_1, \ldots, x_k \mapsto B_o(n + r)) \setminus S| \leq \delta k\right] \leq p^k,$$

and this proves the desired. □

We next show that the condition *lower bound* (LB) from [10] holds on SG. To do this, we first need to estimate the growth of the annulus $B_o(n) \setminus B_o(n(1-\epsilon))$ in SG. Recall that $b_n = |B_o(n)|$.

Lemma 7.3.8 *For $1/n < \epsilon < 1$, the growth of the annulus $B_o(n) \setminus B_o(n(1-\epsilon))$ in SG satisfies the upper bound*

$$b_n - b_{n(1-\epsilon)} \leq 4\epsilon^{\alpha-1} b_n. \tag{7.3.19}$$

Proof. To motivate the proof, we first carry out the estimate using closed balls, even though the statement calls for open balls. Let $n = 2^k$ and $\epsilon = 2^{-m}$ for some positive integers k and m with $m < k$. Then $\overline{B}_o(n)$ is the union of two triangles of side 2^k joined at o. It is easily shown via induction that each triangle has cardinality $(3^{k+1}+3)/2$, so $|\overline{B}_o(n)| = 2[(3^{k+1}+3)/2] - 1 = 3^{k+1}+2$. Meanwhile one observes that the difference $\overline{B}_o(n) \setminus \overline{B}_o(n(1-\epsilon))$ consists of two copies of the union of 2^m identical triangles each of side 2^{k-m}. Therefore $|\overline{B}_o(n)| - |\overline{B}_o(n(1-\epsilon))|$ is less than

$$\begin{aligned} 2 \cdot 2^m \cdot \frac{3^{k-m+1}+3}{2} &= \frac{2^m(3^{k-m+1}+3)}{3^{k+1}+2} |\overline{B}_o(n)| \leq 2^m(3^{-m}+3^{-k}) |\overline{B}_o(n)| \\ &\leq 2\left(\frac{2}{3}\right)^m |\overline{B}_o(n)| = 2\epsilon^{\alpha-1} |\overline{B}_o(n)|. \end{aligned}$$

For the actual proof, we consider the case when $2^k < n \leq 2^{k+1}$ and $2^{-m-1} < \epsilon \leq 2^{-m}$ for positive integers m and k with $m < k$. Then the difference $B_o(n) \setminus B_o(n(1-\epsilon))$ can be covered by a union of at most $2 \cdot 2^{m+1}$ identical triangles each of side 2^{k-m}. Therefore, the left-hand side in (7.3.19) is less than

$$\begin{aligned} 2 \cdot 2^{m+1} \cdot \frac{3^{k-m+1}+3}{2} &\leq 2\frac{2^m(3^{k-m+1}+3)}{3^{k+1}+2} |B_o(n)| \leq 2^{m+2}3^{-m}|B_o(n)| \\ &\leq 4\epsilon^{\alpha-1}|B_o(n)|. \end{aligned}$$

□

Remark 7.3.9 *If the centre of the ball (or annulus) is an arbitrary vertex x of SG rather than o, then a similar argument shows that (7.3.19) holds with the constant 4 replaced by 8. This is due to the fact that for $2^k < n \leq 2^{k+1}$, the ball $B_x(n)$ can be covered by two joint triangles of side 2^{k+1}.*

Proposition 7.3.10 *Condition LB holds on SG, that is,*

$$\frac{|\mathcal{I}_{b_n}(o \mapsto n)|}{b_n} \to 1, \quad \textit{almost surely.}$$

Proof. By construction $\mathscr{I}_{b_n}(o \mapsto n)$ is a subset of $B_o(n)$, that is $\frac{|\mathscr{I}_{b_n}(o \mapsto n)|}{b_n} \leq 1$. On the other hand, from the inner boundary for IDLA cluster in Theorem 7.1.1 we have that for every $\epsilon > 0$, $B_o(n(1-\epsilon)) \subset \mathscr{I}(b_n)$, for n large with probability 1. Actually, the proof of the inner boundary implies the stronger result that $B_o(n(1-\epsilon)) \subset \mathscr{I}_{b_n}(o \mapsto n)$, since in the random variables M and L we count only particles that visit a point $z \in B_o(n(1-\epsilon))$ before exiting $B_o(n)$. Therefore, for every $\epsilon > 0$, we have

$$1 - \frac{b_n - b_{n(1-\epsilon)}}{b_n} = \frac{b_{n(1-\epsilon)}}{b_n} \leq \frac{|\mathscr{I}_{b_n}(o \mapsto n)|}{b_n} \quad \text{almost surely.}$$

On the other hand, Lemma 7.3.8 yields

$$1 - 4\epsilon^{\alpha-1} \leq 1 - \frac{b_n - b_{n(1-\epsilon)}}{b_n},$$

and the left-hand side goes to 1 as ϵ goes to 0, which together with the first claim of the proof gives $\frac{|\mathscr{I}_{b_n}(o \mapsto n)|}{b_n} \to 1$ almost surely. □

Now we have all ingredients needed for the proof of the outer boundary for IDLA cluster in Theorem 7.1.1. This would be an application of [10, Theorem 1.2 and Corollary 1.3]. Since there are some minor gaps in their proofs, for the sake of completeness, we give the whole proof, adapted to our case, here. We want to stress again that we can prove [10, Theorem 1.2] without the wLB condition, a condition which was used only in [10, Lemma 3.2]. We overcame this by having finer estimates on the stopped Green function, similar to the case of lattices $\mathbb{Z}^d$. The estimates are of course not as precise as on $\mathbb{Z}^d$, but fine enough for our purposes.

As in [10, Theorem 1.2], we construct inductively a sequence of IDLA aggregates $\mathscr{I}_j$, by stopping the particles at different distances n_j from the origin. If k_j is the number of stopped particles, we choose the next distance n_{j+1}, at which we pause particles again, in terms of k_j and n_j. We iterate this procedure until there are fewer than $n_j^{1/(\alpha+1)}$ particles, at which point there are too few particles to affect the limiting outer radius of the IDLA.

Proof of the outer bound in Theorem 7.1.1. To prove that for every $\epsilon > 0$, $\mathscr{I}(b_n) \subset B_o(n(1+\epsilon))$ for n large enough with probability 1, we bound the event $[\mathscr{I}(b_n) \not\subset B_o(n(1+\epsilon))]$ by another event whose probability is exponentially decreasing in n, and then apply Borel–Cantelli.

As mentioned just above, we define recursively the sequence of aggregates $\mathscr{I}_j$ and the quantities n_j, P_j, k_j, $j = 0, 1, \ldots$ as follows. Fix first $n > 0$, and let

$$\begin{cases} n_0 & = n \\ \mathscr{I}_0 & = \mathscr{I}_{b_n}(o \mapsto n) \\ P_0 & = P_{b_n}(o \mapsto n) \\ k_0 & = |P_0|. \end{cases}$$

In words, we start the general internal DLA process with $b_n = |B_o(n)|$ particles at o and build the cluster $\mathscr{I}_0$ by stopping particles either when they attach to the existing cluster, or pausing them on $B_o(n)^c$, that is, when they reach distance n from o. So $\mathscr{I}_0$ is a subset of the unstopped IDLA cluster $\mathscr{I}(b_n)$ as defined in the introduction. Then P_0 gives the positions of the paused particles, which will continue their journey (only if there are enough particles to contribute to the behaviour of the IDLA outer boundary) in order to build the next cluster $\mathscr{I}_1$. If they do not attach to $\mathscr{I}_1$ before reaching the distance n_1 (still to be defined) from the root, then they are paused again and used for the subsequent aggregate. Formally, for $j \geq 0$, let

$$\begin{cases} n_{j+1} & = \begin{cases} n_j + k_j^{1/\alpha} & \text{if } k_j > n_j^{1/(\alpha+1)} \\ \infty & \text{otherwise .} \end{cases} \\ \mathscr{I}_{j+1} & = \mathscr{I}\left(\mathscr{I}_j;\, P_j \mapsto B_o\left(n_{j+1}\right)\right) \\ P_{j+1} & = P\left(\mathscr{I}_j;\, P_j \mapsto B_o\left(n_{j+1}\right)\right) \\ k_{j+1} & = |P_{j+1}|. \end{cases}$$

We continue this iterative construction as long as we have enough particles. Let J be the minimum value of j for which $k_j \leq n_j^{1/(\alpha+1)}$. At this point, the particles left over ($|P_J|$ of them) evolve until attaching to the aggregate, without being paused anymore, and for all $j \geq J+1$, we have $\mathscr{I}_j = \mathscr{I}_{J+1}$. By the abelian property of the internal DLA process (7.3.6), we have that $\mathscr{I}_{J+1}$ and $\mathscr{I}(b_n)$ have the same distribution. Since the aggregate at time J is stopped before exiting $B_o(n_J)$, it holds $\mathscr{I}_J \subset B_o(n_J)$ and there are exactly $k_J \leq n_J^{1/(\alpha+1)}$ particles left for completing the aggregate $\mathscr{I}_{J+1}$, particles which cannot build too long tentacles. This means that after releasing the last k_J particles, the radius of the ball which contains $\mathscr{I}_{J+1}$ cannot increase with more than $c^* k_J \leq c^* n_J^{1/(\alpha+1)}$, for some $c^* \leq 1$, which implies that $\mathscr{I}_{J+1} \subset B_o\left(n_J + c^* n_J^{1/(\alpha+1)}\right)$. Then

$$\mathbb{P}\left[\mathscr{I}(b_n) \not\subset B_o(n(1+\epsilon))\right] \leq \mathbb{P}\left[n_J + c^* n_J^{1/(\alpha+1)} > n(1+\epsilon)\right],$$

and we will upper bound the probability on the right-hand side in the previous inequality. We have $k_J \leq n_J^{1/(\alpha+1)}$ and $n_{J-1}^{1/(\alpha+1)} < k_{J-1} < \cdots < k_o < n^\alpha$. The fact $k_0 < n^\alpha$ follows from $k_0 \leq b_n - b_{n(1-\epsilon)}$ together with Lemma 7.3.8, for n large enough. Moreover, for every $j = 1, \dots, J-1$, we can apply Lemma 7.3.7, by starting with the occupied cluster $\mathscr{I}_{j-1} \subset B_o(n_{j-1})$, and the paused particles $P_{j-1} \in \partial B_o(n_{j-1})$. The number of paused particles k_{j-1} used to build $\mathscr{I}_j$ fulfil the relation $n_j^{1/(\alpha+1)} < k_{j-1} < n^\alpha < n_j^\alpha$. Then there exist $\delta < 1$ and $p < 1$ such that

$$\begin{aligned}
\mathbb{P}\left[\left|\mathscr{I}_j - \mathscr{I}_{j-1}\right| \le \delta k_{j-1}\right] &= \mathbb{P}[k_j \ge (1-\delta)k_{j-1}] \\
&= \sum_{l=n_j^{1/(\alpha+1)}}^{n_j^\alpha} \mathbb{P}[k_j \ge (1-\delta)k_{j-1} | k_{j-1} = l] \cdot \mathbb{P}[k_{j-1} = l] \\
&= \sum_{l=n_j^{1/(\alpha+1)}}^{n_j^\alpha} \mathbb{P}[k_j \ge (1-\delta)l] \cdot \mathbb{P}[k_{j-1} = l] \\
&\le \sum_{l=n_j^{1/(\alpha+1)}}^{n_j^\alpha} p^l \cdot \mathbb{P}[k_{j-1} = l] \le p^{n_j^{1/(\alpha+1)}} \le p^{n^{1/(\alpha+1)}},
\end{aligned}$$

for all $j = 1, \dots, J$. Since $J \le n^\alpha$, together with the union bound we obtain

$$\mathbb{P}[\exists\, 1 \le j \le J : k_j \ge (1-\delta)k_{j-1}] \le \sum_{j=1}^{J} \mathbb{P}[k_j \ge (1-\delta)k_{j-1}] \le n^\alpha\, p^{n^{1/(\alpha+1)}}. \tag{7.3.20}$$

In view of the inclusion of the events $\left\{\forall\, 1 \le j \le J : k_j < (1-\delta)k_{j-1}\right\} \subseteq \left\{k_l < (1-\delta)^l k_0\right\}$ for any fixed $l \le J$, we get that for any $l \le J$

$$\mathbb{P}\left[k_l \ge (1-\delta)^l k_0\right] \le \mathbb{P}\left[\exists 1 \le j \le l : k_j \ge (1-\delta)k_{j-1}\right].$$

Thus

$$\begin{aligned}
\mathbb{P}\left[\exists 1 \le l \le J : k_l \ge (1-\delta)^l k_0\right] &\le \mathbb{P}\left[\bigcup_{l \le J} \{k_l \ge (1-\delta)^l k_0\}\right] \\
&\le \mathbb{P}\left[\bigcup_{l \le J} \left\{\exists 1 \le j \le l : k_j \ge (1-\delta)k_{j-1}\right\}\right] \\
&= \mathbb{P}\left[\exists\, 1 \le j \le J : k_j \ge (1-\delta)k_{j-1}\right].
\end{aligned}$$

Altogether equation (7.3.20) and the previous inequality imply that for some $\delta < 1$ and $p < 1$

$$\mathbb{P}\left[\exists 1 \le j \le J : k_j \ge (1-\delta)^j k_0\right] \le \mathbb{P}\left[\exists\, 1 \le j \le J : k_j \ge (1-\delta) k_{j-1}\right] \le n^\alpha p^{n^{1/(\alpha+1)}}, \tag{7.3.21}$$

that is

$$\mathbb{P}\left[\forall\, 1 \le j \le J : k_j < (1-\delta)^j k_0\right] \ge 1 - n^\alpha p^{n^{1/(\alpha+1)}}.$$

In other words, with probability at least $1 - n^\alpha p^{n^{1/(\alpha+1)}}$

$$n_J = n + \sum_{j=0}^{J-1} k_j^{1/\alpha} < n + k_0^{1/\alpha} \sum_{j=0}^{J-1} \left((1-\delta)^{1/\alpha}\right)^j < n + k_0^{1/\alpha} \frac{1}{1-(1-\delta)^{1/\alpha}}.$$

Meanwhile, if $n_J + c^* n_J^{1/(\alpha+1)} > n(1+\epsilon)$, then there exists c' such that for n big enough

$$n + k_0^{1/\alpha} \frac{1}{1-(1-\delta)^{1/\alpha}} + c' n^{1/(\alpha+1)} >$$

$$> n + k_0^{1/\alpha} \frac{1}{1-(1-\delta)^{1/\alpha}} + c^* \left(n + k_0^{1/\alpha} \frac{1}{1-(1-\delta)^{1/\alpha}}\right)^{\frac{1}{\alpha+1}} > n(1+\epsilon),$$

which implies that for some constant $c_1 > 0$ and n big enough

$$k_0^{1/\alpha} > c_1 \epsilon n \quad \Rightarrow \quad k_0 > c \epsilon^\alpha b_n.$$

So by conditioning on the event $n_J < n + k_0^{1/\alpha} \frac{1}{1-(1-\delta)^{1/\alpha}}$, we obtain

$$\begin{aligned}
&\mathbb{P}\left[\mathcal{I}(b_n) \not\subset B_o(n(1+\epsilon))\right] \le \mathbb{P}\left[n_J + c^* n_J^{1/(\alpha+1)} > n(1+\epsilon)\right] \\
&\quad = \mathbb{P}\left[n_J + c^* n_J^{1/(\alpha+1)} > n(1+\epsilon) \,\middle|\, n_J < n + k_0^{1/\alpha} \frac{1}{1-(1-\delta)^{1/\alpha}}\right] \\
&\quad \cdot \mathbb{P}\left[n_J < n + k_0^{1/\alpha} \frac{1}{1-(1-\delta)^{1/\alpha}}\right] \\
&\quad + \mathbb{P}\left[n_J + c^* n_J^{1/(\alpha+1)} > n(1+\epsilon) \,\middle|\, n_J \ge n + k_0^{1/\alpha} \frac{1}{1-(1-\delta)^{1/\alpha}}\right] \\
&\quad \cdot \mathbb{P}\left[n_J \ge n + k_0^{1/\alpha} \frac{1}{1-(1-\delta)^{1/\alpha}}\right] \le \mathbb{P}\left[k_0 > c\epsilon^\alpha b_n\right] \\
&\quad + \mathbb{P}\left[n_J \ge n + k_0^{1/\alpha} \frac{1}{1-(1-\delta)^{1/\alpha}}\right] \le \mathbb{P}\left[k_0 > c\epsilon^\alpha b_n\right] \\
&\quad + \mathbb{P}\left[\exists 1 \le j \le J : k_j \ge (1-\delta)^j k_0\right].
\end{aligned}$$

Finally, applying Borel–Cantelli to the events involved in the previous inequality, together with the bound in (7.3.21), and using that $k_0 = b_n - |\mathscr{I}_{b_n}(o \mapsto n)|$ give that for every $\epsilon > 0$,

$$\mathbb{P}\left[\mathscr{I}(b_n) \not\subset B_o(n(1+\epsilon)) \quad i.o.\right] \leq \mathbb{P}\left[\frac{|\mathscr{I}_{b_n}(o \mapsto n)|}{b_n} < (1 - c\epsilon^\alpha) \quad i.o.\right].$$

By Proposition 7.3.10, the event on the right-hand side above can happen only finite number of times. Therefore for every $\epsilon > 0$,

$$\mathscr{I}(b_n) \subset B_o(n(1+\epsilon)) \qquad \text{for all sufficiently large } n$$

with probability 1. This concludes the proof of the outer bound in Theorem 7.1.1. □

7.4 Open Questions

7.4.1 Rotor-Router Aggregation on Sierpinski Gasket Graphs

Rotor-router aggregation is a deterministic version of IDLA, which describes the growth of a cluster of particles, where the particles perform deterministic walks (called *rotor-router walks*) instead of random walks. In a rotor-router walk on a graph G, each vertex is equipped with an arrow (rotor) pointing to one of the neighbours. A particle performing a rotor-router walk first changes the rotor at the current position to point to the next neighbour, in a fixed order of neighbours chosen at the beginning, and then the particle moves to the neighbour the rotor is pointing towards. In rotor-router aggregation, for a fixed initial configuration of rotors, we start n particles at the origin of o, and let each of these particles perform rotor-router walk until reaching a site previously unvisited, where it stops. Then a new particle starts at the origin, without resetting the configuration of rotors. The resulting deterministic set $\mathscr{R}(n)$ of n occupied sites is called the *rotor-router cluster*. As in the case of IDLA, one of the questions here is to determine if the set $\mathscr{R}_n$ of occupied sites has a limiting shape regardless of the initial configuration of rotors. IDLA and rotor-router aggregation have similar behaviour on several state spaces, as shown in [20] on $\mathbb{Z}^d$, and in [14, 15] on comb lattices. On the Sierpinski gasket SG, it has been proven in [8] that the rotor cluster has the same limit shape from Theorem 7.1.1. Even more, a fourth growth model, the abelian sandpile model, has the same limit shape on SG; see again [8].

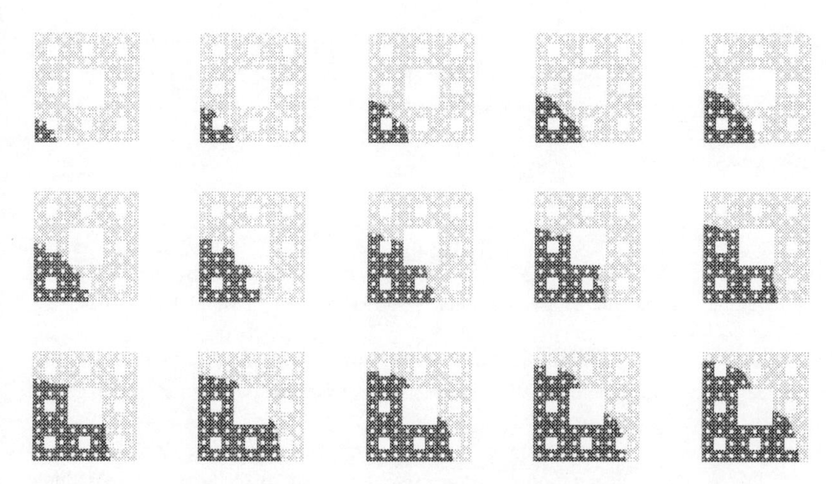

Figure 7.2 IDLA clusters on the Sierpinski carpet for 10000 up to 150000 particles.

7.4.2 IDLA on Other Fractal Graphs

Another fractal graph with interesting properties is the graphical Sierpinski carpet SC, where it may be very interesting to investigate the IDLA process. Most of the computations in this chapter can be carried over to Sierpinski carpet graphs, with the exception of Lemma 7.3.1 which is more delicate on SC. The proof of Lemma 7.3.1 uses the divisible sandpile model on SG, whereon explicit computations can be carried out, thanks to the finite ramification of SG. In contrast, SC is infinitely ramified, and fine estimates of the solution to the corresponding Dirichlet problem are not known at the moment. According to computer simulations, there does not seem to exist a unique scaling limit for the IDLA clusters. Actually, the simulations suggest that there is a whole family of scaling limits, and that these scaling limits seem to have a fractal boundary. Figure 7.2 shows IDLA clusters on the graphical Sierpinski carpet SC in dimension 2, for 10000 up to 150000 random walks starting at the origin.

Acknowledgements

We are grateful to Lionel Levine for providing useful comments on an earlier version of this chapter. The research of Joe P. Chen was supported by the National Science Foundation (NSF) grant DMS-1262929 and the Research Council of Colgate University. The research of Wilfried Huss was supported by the Austrian Science Fund (FWF): J3628-N26 and P25510-N26. The research of Ecaterina Sava-Huss was supported by the Austrian Science Fund (FWF): J3575-N26. The research of Alexander Teplyaev was supported by the National Science Foundation (NSF) grants DMS-1262929 and DMS-1613025.

References

[1] Amine Asselah and Houda Rahmani. Fluctuations for internal DLA on the comb. *Ann. Inst. Henri Poincaré Probab. Stat.*, 52(1):58–83, 2016.

[2] Martin T. Barlow. Diffusions on fractals. In *Lectures on probability theory and statistics (Saint-Flour, 1995)*, volume 1690 of *Lecture Notes in Math.*, pages 1–121. Springer, Berlin, 1998.

[3] Martin T. Barlow. Heat kernels and sets with fractal structure. In *Heat kernels and analysis on manifolds, graphs, and metric spaces (Paris, 2002)*, volume 338 of *Contemp. Math.*, pages 11–40. Amer. Math. Soc., Providence, RI, 2003.

[4] Martin T. Barlow. Some remarks on the elliptic harnack inequality. *Bull. London Math. Soc.*, 37(2):200–208, 2005.

[5] Martin T. Barlow, Thierry Coulhon, and Takashi Kumagai. Characterization of sub-Gaussian heat kernel estimates on strongly recurrent graphs. *Comm. Pure Appl. Math.*, 58(12):1642–1677, 2005.

[6] Martin T. Barlow and Edwin A. Perkins. Brownian motion on the Sierpiński gasket. *Probab. Theory Related Fields*, 79(4):543–623, 1988.

[7] Sébastien Blachère and Sara Brofferio. Internal diffusion limited aggregation on discrete groups having exponential growth. *Probab. Theory Related Fields*, 137(3-4):323–343, 2007.

[8] Joe P. Chen and Jonah Kudler-Flam. Laplacian growth & sandpiles on the Sierpinski gasket: limit shape universality and exact solutions. Ann. Inst. Henri Poincaré (D), to appear, 2020. Also available at https://arxiv.org/abs/1807.08748.

[9] P. Diaconis and W. Fulton. A growth model, a game, an algebra, Lagrange inversion, and characteristic classes. *Rend. Sem. Mat. Univ. Politec. Torino*, 49(1):95–119 (1993), 1991. Commutative algebra and algebraic geometry, II (Italian) (Turin, 1990).

[10] Hugo Duminil-Copin, Cyrille Lucas, Ariel Yadin, and Amir Yehudayoff. Containing internal diffusion limited aggregation. *Electron. Commun. Probab.*, 18:no. 50, 8, 2013.

[11] Alexander Grigor'yan and Andras Telcs. Harnack inequalities and sub-gaussian estimates for random walks. *Math. Ann.*, 324(3):521–556, 2002.

[12] Zijian Guo, Rachel Kogan, Hua Qiu, and Robert S. Strichartz. Boundary value problems for a family of domains in the Sierpinski gasket. *Illinois J. Math.*, 58(2):497–519, 2014.

[13] Wilfried Huss. Internal diffusion-limited aggregation on non-amenable graphs. *Electron. Commun. Probab.*, 13:272–279, 2008.

[14] Wilfried Huss and Ecaterina Sava. Rotor-router aggregation on the comb. *Electron. J. Combin.*, 18(1):Paper 224, 23, 2011.

[15] Wilfried Huss and Ecaterina Sava. Internal aggregation models on comb lattices. *Electron. J. Probab.*, 17:no. 30, 21, 2012.

[16] Wilfried Huss and Ecaterina Sava-Huss. Divisible sandpile on Sierpinski gasket graphs. Fractals, 27(03):1950032, 2019.

[17] Jun Kigami. *Analysis on fractals*, volume 143 of *Cambridge Tracts in Mathematics*. Cambridge University Press, Cambridge, 2001.

[18] G. Lawler. Subdiffusive fluctuations for internal diffusion limited aggregation. *Ann. Probab.*, 23:71–86, 1995.

[19] Gregory F. Lawler, Maury Bramson, and David Griffeath. Internal diffusion limited aggregation. *Ann. Probab.*, 20(4):2117–2140, 1992.

[20] Lionel Levine and Yuval Peres. Strong spherical asymptotics for rotor-router aggregation and the divisible sandpile. *Potential Anal.*, 30(1):1–27, 2009.

[21] Cyrille Lucas. The limiting shape for drifted internal diffusion limited aggregation is a true heat ball. *Probab. Theory Related Fields*, 159(1-2):197–235, 2014.

[22] Ecaterina Sava-Huss. From fractals in external DLA to internal DLA on fractals. Fractal Geometry and Stochastics VI, to appear, 2020.

[23] Eric Shellef. IDLA on the supercritical percolation cluster. *Electron. J. Probab.*, 15:no. 24, 723–740, 2010.

[24] Robert S. Strichartz. *Differential equations on fractals: A tutorial*. Princeton University Press, Princeton, NJ, 2006.

[25] Robert S. Strichartz and Michael Usher. Splines on fractals. *Math. Proc. Cambridge Philos. Soc.*, 129(2):331–360, 2000.

[26] András Telcs. *The art of random walks*, volume 1885 of *Lecture Notes in Mathematics*. Springer-Verlag, Berlin.

8

Universal Lower Bounds for Laplacians on Weighted Graphs

D. Lenz and P. Stollmann

Abstract

We discuss optimal lower bounds for eigenvalues of Laplacians on weighted graphs. These bounds are formulated in terms of the geometry and, more specifically, the inradius of subsets of the graph. In particular, we study the first non-zero eigenvalue in the finite volume case and the first eigenvalue of the Dirichlet Laplacian on subsets that satisfy natural geometric conditions.

8.1 Introduction

The first main objective of this chapter is to give an overview over recent results that deal with lower bounds on the first non-zero eigenvalue of Laplacians on graphs. These lower bounds involve the inradius of subsets. This topic certainly has a long history and key arguments have been re-discovered over and again. It seems that the classical paper by Beurling and Deny [4] is the first source for the key step, a fact that we just recently became aware of and that has gone unnoticed in the literature. The lower bounds we mentioned can be seen as discrete analogues of Poincaré or Payne–Weinberger inequalities [21], the latter both being of fundamental importance for analysis on continuum spaces.

A second point to be emphasized it that the bounds in question are *universal*. By this we mean inequalities that do not require additional assumptions on the graph, be it combinatorial assumptions, e.g., regularity or finiteness or geometrical assumptions, e.g., curvature restrictions. This is in particular motivated by applications in mathematical physics, where random (sub-)graphs have seen a great deal of interest. By the very nature of randomness these subgraphs tend to not satisfy further regularity assumptions.

Since we deal with Laplacians of general weighted graphs, our treatment includes several important classes considered in different mathematical fields, in particular combinatorial Laplacians and normalized Laplacians in the usual setting of combinatorial graphs. The estimates we derive in the finite volume case are quite easy to state and to prove and determine a lower bound that basically depends on the diameter (or the inradius) as well as on the volume. Simple examples as well as more involved ones show that these estimates are optimal. Yet, there is even a more conceptual way of showing optimality that was put forward in the recent paper [18] based on the following observation: in a first step towards a Poincaré inequality one estimates the variation of a function in terms of the energy and an appropriate metric. Such an estimate was called *topological Poincaré inequality* in the last named paper and might maybe better be called a *metric Poincaré inequality*. This crucial bound had already been proved in the above-mentioned paper [4] and does not involve the underlying measure. It gives rise to a Poincaré inequality with respect to a different metric, sometimes called *resistance metric*. Varying the measure one can show that the corresponding Poincaré inequality is optimal; see Section 8.5 for details.

The third main objective of this chapter is a generalization of the already quite general set-up considered in [19]. We briefly explain the situation here and refer to Section 8.4 for the full picture. A non-zero lower bound for the first Neumann eigenvalue can in general not be expected in the infinite volume case. However, the lowest Dirichlet eigenvalue $\lambda_0^D(\Omega)$ of the Laplacian on the open subset Ω can be strictly positive, even if the volume of Ω is infinite. The reader may think of a strip in Euclidean space for a continuum analogue. Such an estimate can be deduced with the help of a Voronoi decomposition for Ω that exhibit suitable relative boundedness properties, of which finite inradius is the property that replaces finite diameter and a second condition can be thought of as a relative finite volume property. In Euclidean space, a Voronoi decomposition can easily be written down. In general weighted graphs, however, the situation is much more complicated and certain topological properties of the graph are necessary (see Proposition 8.4.2 and Example 8.4.3 for details). Our new result (Theorem 8.4.4), now holds in full generality, in particular in situations where Voronoi decompositions are no longer available. The main new idea of the proof is really simple: for any particular given function, we work on a tailor-made finite subgraph to verify the appropriate energy estimate.

While it is not the topic of this chapter a comment on Cheeger inequalities may be in order. Cheeger inequalities provide lower bounds on the minimum of the spectrum of Laplacians in terms of geometry viz the isoperimetric constant. For graphs such inequalities were first discussed by Alon and Milman [1] (for finite graphs) and by Dodziuk [7] (for infinite graphs). The work [7] features a somewhat unsatisfactory additional dependence on the measure. This could later be removed in the situation of normalized Laplacians by Dodziuk and Kendall [8]. A satisfactory answer for general Laplacians on infinite graphs was only recently established by Bauer, Keller, and Wojciechowski [3]. It involves intrinsic metrics. The corresponding bounds do not require any finiteness condition on the inradius. Hence, they apply in situations where the bounds discussed below do not give any useful information. On the other hand in situations involving subsets of lattices, where our bounds apply, they tend to be better than bounds via Cheeger estimates (see [19], Section 8.6, for comparison).

8.2 Set-up and Main Results

In this section we present our set-up and the main results. These results connect geometric data and spectral data of a graph, where our concept of graph is a very general one. In the presentation of this section we take special care to introduce the spectral data with as little technical effort as possible. A thorough discussion of the operator and form theoretic background is given in the last section.

A weighted graph is a triple (X, b, m) satisfying the following properties:

- X is an arbitrary set, whose elements are referred to as *vertices*;
- $b : X \times X \to [0, \infty)$ is a symmetric function with $b(x, x) = 0$ for all $x \in X$;
- $m : X \to (0, \infty)$ is a function on the vertices.

An element $(x, y) \in X \times X$ with $b(x, y) > 0$ is then called an *edge* and b is denoted as *edge weight*. The positive function $m : X \to (0, \infty)$ gives a measure on X of which we think as a volume. In particular, we define

$$\mathrm{vol}(\Omega) := \sum_{x \in \Omega} m(x)$$

for $\Omega \subset X$.

A sequence of vertices $\gamma = (x_0, \ldots, x_k)$ is called a *path* from x to y if $x_0 = x$, $x_k = y$ and $b(x_l, x_{l+1}) > 0$ for $l = 0, \ldots, k - 1$. Throughout our graphs will be assumed to be *connected*, i.e., to allow for a path between arbitrary vertices.

A natural distance to consider is

$$d(x, y) := \inf\{L(\gamma) \mid \gamma \text{ a path from } x \text{ to } y\},$$

where the length $L(\gamma)$ of a path γ is given by

$$L(\gamma) := \sum_{j=0,\dots,k-1} \frac{1}{b(x_j, x_{j+1})}.$$

Setting $d(x,x) = 0$ we obtain a pseudo-metric, i.e., d is symmetric and satisfies the triangle inequality. Clearly, in this generality, d need not separate the points of X, a fact that is of no importance for what follows.

We call the graph *geodesic* if for any $x, y \in X$ there exists a path γ from x to y with $L(\gamma) = d(x,y)$.

We denote by

$$U_r(x) := \{y \in X \mid d(x,y) < r\} \text{ and } B_r(x) := \{y \in X \mid d(x,y) \leq r\}$$

the open and closed balls of radius r, respectively. The *diameter* of X is given by

$$\operatorname{diam}(X) := \sup\{d(x,y) : x, y \in X\}.$$

The positive function $m : X \to (0, \infty)$ together with the distance d gives the geometrical data of the space. For further details on weighted graphs and their geometry as expressed by d and related metrics we refer to the recent studies [9, 10] as well as the surveys [13, 15] and the upcoming monograph [16].

Spectral data are given in terms of the *energy* $\mathcal{E}$ associated with the graph defined by the quadratic form

$$\mathcal{E}(f) := \frac{1}{2} \sum_{x,y \in X} b(x,y)(f(x) - f(y))^2 \text{ for } f \in \mathbb{R}^X,$$

which may assume the value ∞ for the time being. The underlying Hilbert space is

$$\ell^2(X,m) := \{f \in \mathbb{R}^X \mid \sum_{x \in X} f(x)^2 m(x) < \infty\}$$

with inner product and norm given by

$$\langle f, g \rangle = \sum_{x \in X} f(x) g(x) m(x) \text{ and } \|f\| = \sqrt{\langle f, f \rangle}$$

respectively.

The Neumann Laplacian $H = H^N(X, b, m)$ is the self-adjoint operator associated with $\mathcal{E}$ on its maximal domain in $\ell^2(X,m)$. We refer the reader to Appendix 8.6 for a precise definition as well as further details concerning forms, operators, and all that. The quantity we want to estimate in the case of finite volume and finite diameter is the first non-zero eigenvalue λ_1^N of H. This can be written down in variational terms as

$$\lambda_1^N(X) = \inf\{\mathcal{E}(f) \mid f \in \ell^2(X,m) \text{ with } \mathcal{E}(f) < \infty, \langle f, 1\rangle = 0 \text{ and } \|f\| = 1\}.$$

Again, we refer to the Appendix for the fact that this is the first non-zero eigenvalue of H in the case of finite volume and finite diameter. Readers who do not want to bother with operators should just stick to the definition above. No operator or spectral theoretic technicalities enter the quite elementary proof of our first main result (Theorem 8.3.4), which says that

$$\lambda_1^N \geq \frac{4}{\operatorname{diam}(X)\operatorname{vol}(X)}.$$

Even though our proof partly uses known techniques, we did not find any reference in this generality (see the discussion in Section 8.3). Let us point out that the finite volume estimate above can be thought of as a universal Poincaré or spectral gap inequality that holds without any further restriction on the weights. In particular, it is valid for combinatorial and normalized Laplacians. Moreover, it is optimal. This supports the feeling that the distance d is rather well suited for spectral geometry in a general setting where no further information on the graph is available.

We also deduce a lower bound for the Dirichlet Laplacian H_Ω for $\Omega \subset X$. The latter is again defined in terms of forms and the relevant functions in the form domain are supposed to vanish on the complement $D := X \setminus \Omega$. The spectral quantity of interest is the first eigenvalue. It can be written down in variational terms as

$$\begin{aligned}\lambda_0^D(\Omega) = \inf\{&\mathcal{E}(f) \mid f \in \ell^2(X,m) \text{ with} \\ &\mathcal{E}(f) < \infty,\ f = 0 \text{ on } D \text{ and } \|f\| = 1\ \}.\end{aligned}$$

It is easily seen (see Appendix 8.6) that

$$\lambda_0^D(\Omega) = \min \sigma(H_\Omega).$$

(In the unlikely event that $\mathcal{E}(f) = \infty$ for all f with support contained in Ω, $\lambda_0^D(\Omega) = \infty$ in accordance with the usual conventions.)

Theorem 8.3.2 says

$$\lambda_0^D(\Omega) \geq \frac{1}{R_\Omega \operatorname{vol}(\Omega)};$$

with the *inradius* R_Ω given by

$$R_\Omega := \sup\{r > 0 \mid \text{ there exists } x \in \Omega \text{ such that } U_r(x) \subset \Omega\}.$$

Clearly, if both $\mathrm{vol}(\Omega)$ and R_Ω are finite, this gives a positive lower bound; otherwise the usual convention $\infty^{-1} = 0$ makes the statement valid but evident.

As will be seen below the proof of Theorem 8.3.4 and the proof of Theorem 8.3.2 are very similar in nature. In this context it may be elucidating to point out that diameter and inradius are strongly related concepts. In fact, it is not hard to see that

$$\mathrm{diam}(X) = \sup\{R_{X\setminus\{x\}} : x \in X\}.$$

So one may actually think of both Theorems 8.3.2 and 8.3.4 as giving bounds in terms of inradius. Indeed, it is possible to even derive the bound given in Theorem 8.3.4 (up to the factor 4) from Theorem 8.3.2, compare discussion at the end of Section 8.3.

Subsequently, in Section 8.4, we are able to give a lower bound on the Dirichlet Laplacian in the infinite volume case. Of course, this can only be hoped for if $D := X \setminus \Omega$ is relatively dense, i.e., that there exist an $R > 0$ such that any point in Ω has distance no more than R to a point of D. Note that this relative denseness can equivalently be seen as finiteness of the inradius of Ω. Indeed, the inradius of Ω is nothing but the infimum over all possible such $R \geq 0$ (see [19] for details). So, here again, the relevant geometry enters via the inradius.

Moreover a relative volume estimate enters our analysis measured in terms of

$$\mathrm{vol}^\sharp[r] := \inf_{s>r} \sup\{m(U_s(x)) \mid x \in X\}$$

and we get the following result:

$$\lambda_0^D(\Omega) \geq \frac{1}{R_\Omega \mathrm{vol}^\sharp[R_\Omega]}.$$

The proof combines a decomposition technique from [19] with a new approximation argument. In fact, the decomposition into Voronoi cells used in [19] needs some geometric properties of the underlying weighted graphs. Specifically, it was established in the latter reference for locally compact geodesic weighted graphs.

8.3 Lower Bounds in the Finite Volume Case

The proofs of both theorems in this section go along similar lines. One main idea is an estimate that relates the energy of a function f and the maximal growth of f over a certain distance. This estimate is given in the following lemma. It has been noted in various places and it seems [4], Remarque 3, p. 208, is the original source. We include the simple proof for completeness reasons and later also interpret the estimate below as an estimate between different metrics, as done in the latter reference.

Proposition 8.3.1 (Basic proposition) *Let γ be a path from x to y and $f \in \mathbb{R}^X$ with $\mathcal{E}(f) < \infty$ be given. Then,*

$$(f(x) - f(y))^2 \leq L(\gamma)\mathcal{E}(f).$$

Proof. Let $f, \gamma = (x_0, \ldots, x_k)$ be as above. Then,

$$\begin{aligned}(f(x) - f(y))^2 &= (f(x_0) - f(x_k))^2 \\ &= \left(\sum_{j=0}^{k-1} \sqrt{b(x_j, x_{j+1})}(f(x_j) - f(x_{j+1})) \frac{1}{\sqrt{b(x_j, x_{j+1})}} \right)^2 \\ &\leq \sum_{j=0}^{k-1} b(x_j, x_{j+1}) \left(f(x_j) - f(x_{j+1})\right)^2 \sum_{j=0}^{k-1} \frac{1}{b(x_j, x_{j+1})} \\ &\leq L(\gamma)\mathcal{E}(f).\end{aligned}$$

□

Now, we can directly derive the result on the Dirichlet case.

Theorem 8.3.2 *Let (X, b, m) be as above. Let a non-empty $\Omega \subsetneq X$ be given and assume $\mathrm{vol}(\Omega) < \infty$ and $\mathrm{Inr}(\Omega) < \infty$. Then*

$$\lambda_0^D(\Omega) \geq \frac{1}{R_\Omega \mathrm{vol}(\Omega)}.$$

Proof of Theorem 8.3.2. Let $R > R_\Omega$. Consider $x \in \Omega$ and $f \in \mathbb{R}^X$ with $\mathcal{E}(f) < \infty$ and $f = 0$ on $D = X \setminus \Omega$. By definition of the inradius there is $x_0 \in U_R(x) \setminus \Omega$. In particular, there is a path $\gamma = (x_0, \ldots, x_k)$ from x_0 to $x = x_k$ of length at most R and $f(x_0) = 0$. Using the above proposition, we get

$$f(x)^2 \leq R\mathcal{E}(f).$$

Summing over $x \in \Omega$ and using that $R > R_\Omega$ was arbitrary we conclude that

$$\|f\|_{\ell^2}^2 \leq R_\Omega \mathrm{vol}(\Omega)\mathcal{E}(f),$$

and hence, the assertion of the theorem follows. □

The proof of Theorem 8.3.4 requires one more ingredient borrowed from [18], where it was stated in the special situation at hand.

Proposition 8.3.3 *Let $(Y, \mathcal{B}, \mu)$ be a finite measure space. Then, for any bounded and measurable $f : Y \to \mathbb{R}$ with $f \perp 1$:*

$$\|f\|_2^2 \leq \frac{1}{4} \sup_{x,y \in Y} (f(x) - f(y))^2 \mu(X).$$

Theorem 8.3.4 *Let (X, b, m) be as above. Assume* $\mathrm{vol}(X) < \infty$ *and* $\mathrm{diam}(X) < \infty$. *Then*

$$\lambda_1^N(X) \geq \frac{4}{\mathrm{diam}(X)\mathrm{vol}(X)}.$$

Proof. This directly follows from Proposition 8.3.1 and the previous proposition. □

Remark 8.3.5 *(a) For finite combinatorial graphs, the lower bound is a familiar bound and our proof follows known lines, compare Lemma 1.9 in [5] and Lemma 2.4 in [2] for related estimates as well as [20], estimate (4.1) in Theorem 4.2, p. 62, where it is attributed to McKay.*

(b) Theorem 4.1 from [2] gives that the lower bound is optimal up to constants, i.e., it is achieved, up to constants by balls in certain graphs. As mentioned in the introduction, we will present another approach to optimality in terms of variation of the measure m in Section 8.5.

As it is instructive we conclude this section by showing how a (slightly weaker) version of Theorem 8.3.4 can easily be derived from Theorem 8.3.2: We consider the situation of Theorem 8.3.4 and let $f \in \ell^2(X, m)$ with $\mathcal{E}(f) < \infty$, $\|f\| = 1$, and $f \perp 1$ be given. Setting $f_+ := \max\{f, 0\}$ and $f_- := \max\{-f, 0\}$ we obtain the decomposition $f = f_+ - f_-$. Clearly $\Omega_+ := \{x \in X : f(x) > 0\}$ and $\Omega_- := \{x \in X : f(x) < 0\}$ are disjoint with f_+ vanishing outside Ω_+ and f_- vanishing outside of Ω_-. This easily gives

$$\begin{aligned}\mathcal{E}(f_+, f_-) &= \frac{1}{2} \sum_{x,y \in X} b(x, y)(f_+(x) - f_+(y))(f_-(x) - f_-(y)) \\ &= - \sum_{x,y \in X} b(x, y) f_+(x) f_-(y) \leq 0.\end{aligned}$$

Now, a direct computation involving Theorem 8.4.4 gives

$$\begin{aligned}\mathcal{E}(f) &= \mathcal{E}(f_+) + 2\mathcal{E}(f_+, f_-) + \mathcal{E}(f_-) \\ &\geq \mathcal{E}(f_+) + \mathcal{E}(f_-) \\ \text{(Theorem 8.3.2)} \quad &\geq \frac{\|f_+\|^2}{\mathrm{vol}(\Omega_+) R_{\Omega_+}} + \frac{\|f_-\|^2}{\mathrm{vol}(\Omega_-) R_{\Omega_-}}\end{aligned}$$

$$\geq \frac{\|f_+\|^2 + \|f_-\|^2}{\text{vol}(X)\text{diam}(X)}$$
$$= \frac{1}{\text{vol}(X)\text{diam}(X)}.$$

Here, we used the obvious bounds $R_{\Omega_+}, R_{\Omega_-} \leq \text{diam}(X)$ and $\text{vol}(\Omega_+), \text{vol}(\Omega_-) \leq \text{vol}(X)$ in the penultimate step

8.4 Lower Bounds for the Dirichlet Laplacian in the Infinite Volume Case

In this section we consider H_Ω again, this time without assuming that $\text{vol}(\Omega) < \infty$. Clearly, the estimate from Theorem 8.3.2 breaks down in this case. The basic idea, already employed in [19], is to decompose the large, infinite graph X into finite volume pieces on which the above estimate can be used. The resulting energy estimates can be summed up to give the Poincaré type inequality. The way this was implemented in [19] required strong assumptions on the underlying geometry. Here we show how, based on the result of [19], one can actually get rid of all these assumptions.

We start with a discussion of the relevant decompositions. These were introduced in [19] and had already been used in [24] in a more restrictive set–up.

Definition 8.4.1 *Let (X, b, m) be as above and $D \subset X$ non-empty. A* Voronoi decomposition *of X with centres from D is a pairwise disjoint family $(V_p)_{p\in D}$ such that following conditions hold:*

(V1) For each $p \in D$ the point p belongs to V_p and for all $x \in V_p$ there exists a path γ from p to x that lies in V_p and satisfies $L(\gamma) = d(p, x)$.
(V2) For each $p \in D$ and for all $x \in V_p$ the inequality $d(p, x) \leq d(q, x)$ holds for any $q \in D$.
(V3) $\bigcup_{p\in D} V_p = X$.

Given a Voronoi decomposition with centres from D one can obtain a lower bound on the values of $Q(f)$ for f vanishing on D as follows: Let $D \subset X$ be given and let V_p, $p \in D$, be an Voronoi decomposition with centres from D. Assume that there exists for each $p \in D$ a $c_p > 0$ with

$$\frac{1}{2} \sum_{x,y\in V_p} b(x, y)(f(x) - f(y))^2 \geq c_p \sum_{x\in V_p} f(x)^2 m(x)$$

for all f with $f(p) = 0$. Then, a direct summation gives

$$\mathcal{E}(f) \geq (\inf_{p \in D} c_p) \|f\|^2$$

for all f which vanish on D (compare [19] as well).

We now turn to existence of Voronoi decompositions. This was shown in [19] under the additional strong geometric condition of compactness of balls and this condition was shown to be necessary. Here is the precise result.

Proposition 8.4.2 *Let (X, b, m) be a connected graph such that all balls $B_r(x)$, $x \in X$ and $r > 0$, are finite. Assume that $D \subset X$ is non-empty and let $\Omega = X \setminus D$. Then there exists a Voronoi decomposition with centres from D. Moreover, whenever $R = R_\Omega$ is finite then any Voronoi decomposition $(V_p)_{p \in D}$ of X with centres from D has the property that $V_p \subset B_R(p)$ for all $p \in D$.*

Let us give a simple example of a geodesic weighted graph that does not allow a Voronoi decomposition.

Example 8.4.3 *Let $X := (\mathbb{N} \times \{0\}) \cup \{(1, 1)\}$ with weight $b((n, 0); (n + 1, 0)) = 2$ for $n \in \mathbb{N}$, $b((n, 0), (1, 1)) = 1 + \frac{1}{n}$ for $n \in \mathbb{N}$ and $b(x, y) = 0$ else. Since none of the points from $D := \mathbb{N} \times \{0\}$ is closest to the point $(1, 1)$, there is no Voronoi decomposition of X with centres in D in the sense of [19] in this case.*

In the general setting discussed in this chapter finiteness of balls does not hold in general. So we cannot directly appeal to the proposition to get a Voronoi decomposition (and subsequent lower bounds). Our method to circumvent this extra complication is quite easy. For given f vanishing on D, for which we want to estimate the ℓ^2-norm by the energy, we consider a finite subgraph of (X, b, m) on which a Voronoi decomposition exists by the result from [19]. We get the following estimate and underline the fact that we have assumed no geometric restrictions apart from connectedness!

Theorem 8.4.4 *Let (X, b, m) be a connected weighted graph and $\Omega \subsetneq X$ such that $R_\Omega < \infty$ and $\mathrm{vol}^\sharp[R_\Omega] < \infty$. Then*

$$\lambda_0^D(\Omega) \geq \frac{1}{R_\Omega \cdot \mathrm{vol}^\sharp[R_\Omega]}.$$

Proof. Let $\varepsilon > 0$, $R > R_\Omega$, and $f \in \ell^2(X)$ with $\|f\| = 1$. We have to show that

$$1 - \varepsilon \leq R \cdot \sup\{m(U_R(x)) \mid x \in X\} \cdot \mathcal{E}(f) \qquad (*).$$

To this end, first note that there is a finite subset $X_f^0 \subset X$ such that

$$1 - \varepsilon \leq \|f|_{X_f^0}\|^2.$$

We now enlarge X_f^0 suitably. First we add finitely many points to get a subset $X_f^1 \supset X_f^0$ so that $(X_f^1, b|_{X_f^1 \times X_f^1})$ is connected. (This is possible as (X, b) is connected and so there is a finite path connecting each of the finitely many pairs $(x, y) \in X_f^0 \times X_f^0$.)

By assumption on Ω we know that

$$X \subset \bigcup_{p \in D} U_R(p).$$

Therefore, we find a finite subset $D^0 \subset D$ so that

$$X_f^1 \subset \bigcup_{p \in D^0} U_R(p).$$

By adding the points of D^0 as well as the points of finitely many suitably chosen paths, we get a finite subset $X_f^1 \cup D^0 \subset Y$ such that

$$Y \subset \bigcup_{p \in D^0} U_R^Y(p),$$

where the superscript Y indicates that we are concerned with balls in the induced subgraph $(Y, b|_{Y \times Y}, m|_Y)$. This latter graph is finite, so in particular it satisfies the assumptions of Proposition 8.4.2, and we get a Voronoi decomposition with centres from Y. Now on each V_p, $p \in Y$, of the Voronoi decomposition we can appeal to Theorem 8.3.2 to get a lower bound for the form on V_p, $p \in Y$. Given this we can now proceed as discussed following the definition of Voronoi decomposition (with Y instead of D) to obtain

$$1 - \varepsilon \leq \|f|_Y\|^2 \leq R \cdot \sup\{m(U_R^Y) \mid y \in Y\} \cdot \mathcal{E}^Y(f).$$

This implies $(*)$ which finishes the proof. □

8.5 Metric Poincaré Inequalities and Optimality of Lower Bounds

In this section we put our results in the context of Poincaré inequalities on graphs; see [12, 17] for related recent results as well. This will be used to discuss optimality.

We say the a pseudo-metric p on (X, b) *satisfies a metric Poincaré inequality*, provided

$$(f(x) - f(y))^2 \leq p(x, y)\mathcal{E}(f) \qquad \text{(MPI)}$$

holds for all $f \in \mathbb{R}^X$, $x, y \in X$. Clearly, our basic Proposition 8.3.1 says that d satisfies a metric Poincaré inequality.

The best constant $r(x, y)$ in (MPI) can easily be seen to satisfy $r(x, y) = \rho(x, y)^2$, where

$$\rho(x, y) = \sup\{f(x) - f(y) \mid \mathcal{E}(f) \leq 1\}.$$

This latter metric also goes back to [4], where it was called *distance extrémale* and has later been rediscovered in different contexts, e.g., in [6]. While the validity of the triangle inequality is evident for ρ, r also itself defines a pseudo-metric, a fact that is not so obvious. For details (and further references) we refer to [18], in particular Prop. 2.2.

We define the following semi-norm on $\ell^\infty(X)$:

$$\|f\|_{\mathcal{V}} := \sup f - \inf f \text{ for } f \in \ell^\infty(X).$$

Moreover, we define

$$\mathcal{D} := \left\{f \in \mathbb{R}^X \mid \mathcal{E}(f) < \infty\right\}$$

and note the following equivalence.

Proposition 8.5.1 *Let (X, b) be a graph. Then the following statements are equivalent:*

(i) *A global variation norm Poincaré inequality holds, i.e., there is $c \geq 0$ such that for all $f \in \mathcal{D}$:*

$$\|f\|_{\mathcal{V}}^2 \leq c\mathcal{E}(f) \qquad \text{(GVPI)}$$

(ii) *The diameter of (X, b) w.r.t. r is finite, i.e.,*

$$\operatorname{diam}_r(X) = \sup\left\{r(x, y) \mid x, y \in X\right\} < \infty.$$

(iii) *The graph (X, b) satisfies*

$$\mathcal{D} \subset \ell^\infty(X).$$

Moreover, the best constant C_P in (GVPI) *equals* $\operatorname{diam}_r(X)$ *and the square norm $\|J\|^2$ of the inclusion map*

$$J : \mathcal{D}/\mathbb{R}\cdot 1 \to \ell^\infty/\mathbb{R}\cdot 1$$

of quotients modulo the constant functions $\mathbb{R}\cdot 1$.

The equivalence of (i) and (ii) is rather obvious; the equivalence of (i) and (iii) follows from a closed graph theorem together with the observation:

$$\begin{aligned}\|f\|_{\mathcal{V}} &= \sup_{s\in[\inf f, \sup f]} \|f - s\cdot 1\|_\infty \\ &= 2\inf_{t\in\mathbb{R}} \|f - t\cdot 1\|_\infty\end{aligned}$$

See, again, [19] for details. Now, we can easily deduce the following optimal Poincaré inequality, where we write $\mathcal{P}(X)$ for the set of all probability measures on X.

Theorem 8.5.2 *Let* (X, b) *satisfy* (GVPI). *Then the best possible* C_P *in* (GVPI) *satisfies:*

$$\frac{4}{C_P} = \inf\{\lambda_1^N(H(X,b,m)) \mid m \in \mathcal{P}(X) \text{ s.t } \operatorname{supp}(m) = X\}.$$

Hence we have:

Corollary 8.5.3 *Let* (X, b) *be a graph and* $m : X \to (0,\infty)$ *such that* $\operatorname{vol}(X) < \infty$. *Then*

$$\lambda_1^N(H(X,b,m)) \geq \frac{4}{\operatorname{diam}_r(X)\cdot\operatorname{vol}(X)}$$

and the estimate is optimal.

It is not too hard to see that $d = r$ in case that (X, b) is a locally finite tree (see [10]), which means that we cannot do better than in Theorem 8.3.4.

8.6 Appendix: Forms and Associated Operators, Domains and Spectra

In this section we present the operator theoretic background for our discussion.
We first mention that

$$\mathcal{E}(f) := \frac{1}{2}\sum_{x,y\in X} b(x,y)(f(x) - f(y))^2$$

defines a closed form on its maximal domain

$$\operatorname{dom}(\mathcal{E}) = \mathcal{D}\cap\ell^2(X,m) = \{f \in \ell^2(X,m) \mid \mathcal{E}(f) < \infty\}.$$

In fact, $\mathcal{E}$ can be regarded as the sum of the bounded forms $\mathcal{E}_{x,y}(f) = \frac{1}{2}b(x,y)(f(x) - f(y))^2$ and so is lower semi-continuous; an appeal to [22], Theorem S.18 gives the closedness.

Of course, in the general situation considered here, the domain $\mathrm{dom}(\mathcal{E})$ is not necessarily dense in $\ell^2(X, m)$. However, for

$$\mathcal{H}_{\mathcal{E}} := \overline{\mathrm{dom}(\mathcal{E})},$$

the closure in $\ell^2(X, m)$ is a Hilbert space and $(\mathcal{E}, \mathrm{dom}(\mathcal{E}))$ defines a closed, densely defined form on $\mathcal{H}_{\mathcal{E}}$. By the form representation theorem, Thm VIII.15 in [22], there is a unique self-adjoint operator $H = H^N(X, b, m)$ in $\mathcal{H}_{\mathcal{E}}$ that is associated with this form. In analogy with the continuum situation we call this operator Neumann Laplacian as it is associated with the maximal form.

Proposition 8.6.1 *Let* (X, b, m) *be as above and, additionally,* $m(X) < \infty$*. Then*

(a) *The function* 1 *belongs to* $\mathrm{dom}(H)$ *with* $H1 = 0$ *and* 0 *is an eigenvalue of multiplicity one,*

$$\inf \sigma(H) = 0.$$

(b)

$$\lambda_1^N = \inf \sigma(H) \setminus \{0\}.$$

If, furthermore $\mathrm{diam}(X) < \infty$*, the latter is an eigenvalue.*

Proof. Ad (a): Since $m(X)$ is finite, $1 \in \mathrm{dom}(\mathcal{E})$ and $\mathcal{E}(1) = 0$. Therefore, $H1 = 0$. As, by our standing assumption, (X, b, m) is connected, any element in the kernel of H has to be constant, so the multiplicity of the eigenvalue 0 is one. It is the bottom of the spectrum, since $H \geq 0$.

Ad (b): This follows from the min–max principle (see, e.g., Theorem XIII.2 in [23]). In case that the diameter is finite, the graph is canonically compactifiable according to [10] (see also [18]), and hence H has compact resolvent, so λ_1^N is the first non-zero eigenvalue in this case. □

Many of our results deal with the *Dirichlet Laplacian* H_Ω, where $\Omega \subset X$ is a subset and we imagine the Dirichlet boundary condition on $D := X \setminus \Omega$ as given by an infinite potential barrier. Therefore we get the form

$$\mathcal{E}_\Omega(\cdot, \cdot) = \mathcal{E}(\cdot, \cdot) \text{ on } \mathrm{dom}(\mathcal{E}_\Omega) = \{f \in \mathrm{dom}(\mathcal{E}) \mid f = 0 \text{ on } D\}.$$

We identify $\ell^2(\Omega, m)$ with $\{f \in \ell^2(X, m) \mid f = 0 \text{ on } D\}$ and get an associated self-adjoint operator H_Ω living in a subspace of $\ell^2(\Omega, m)$ in analogy to what we saw for the Neumann Laplacian above. Note that $\mathcal{E}_\Omega$ and H_Ω are always to be understood relative to the bigger ambient graph X.

Proposition 8.6.2 *Let (X, b, m) be as above and $\Omega \subset X$. Then*

$$\lambda_0^D(\Omega) = \inf \sigma(H_\Omega).$$

This, again, is a consequence of the min–max principle.

Remark 8.6.3 *Another natural choice for the relevant forms would be to consider what might be thought of as Dirichlet boundary conditions at infinity, given by the form domain*

$$\operatorname{dom}\left(\mathcal{E}^{D,\infty}\right) := \overline{\{f \in \mathcal{D} \mid \operatorname{supp}(f) \text{ is a finite set}\}}^{\mathcal{E}},$$

where the support of f is given by $\operatorname{supp}(f) := \{x \in X \mid f(x) \neq 0\}$ *and the closure is meant with respect to the form norm given by the energy. Then, the restriction*

$$\mathcal{E}^{D,\infty} := \mathcal{E}|_{\operatorname{dom}(\mathcal{E}^{D,\infty})}$$

is a closed form. We do not study the associated operator $H^{D,\infty}$ here but only mention that it is bounded below by the Neumann Laplacian. So, our estimates hold for this operator as well. Similar consideration apply to the restriction to Ω.

References

[1] N. Alon, V. D. Milman, *λ_1 isoperimetric inequalities for graphs, and superconcentrators*, J. Combin. Theory Ser. B **38** (1985), 73–88.

[2] M. Barlow, T. Coulhon and A. Grigor'yan, *Manifolds and graphs with slow heat kernel decay*, Invent. Math. **144** (2001), 609–649

[3] F. Bauer, M. Keller, R. Wojciechowski, *Cheeger inequalities for unbounded graph Laplacians*, J. Eur. Math. Soc. (JEMS) **17** (2015), 259–271.

[4] A. Beurling, J. Deny, *Espaces de Dirichlet. I. Le cas élémentaire*, Acta Math. **99** (1958), 203–224.

[5] F. R. K. Chung, *Spectral graph theory*, CBMS Regional Conference Series in Mathematics, **92**, American Mathematical Society, Providence, RI, 1997.

[6] E. B. Davies, *Analysis on graphs and noncommutative geometry*, J. Funct. Anal. **111** (1993), 398–430.

[7] J. Dodziuk, *Difference equations, isoperimetric inequality and transience of certain random walks*, Trans. Amer. Math. Soc. **284** (1984), 787–794.

[8] J. Dodziuk, W. S. Kendall, *Combinatorial laplacians and isoperimetric inequality*, in: From local times to global geometry, control and physics, Pitman Res. Notes Math. Ser., vol. **150**, Longman Sci. Tech., Harlow, (1986), 68–74.

[9] A. Georgakopoulos, *Graph topologies induced by edge lengths*, in: Infinite Graphs: Introductions, Connections, Surveys. Special issue of Discrete Math., **311** (2011), 1523–1542.

[10] A. Georgakopoulos, S. Haeseler, M. Keller. D. Lenz and R. K. Wojciechowski, *Graphs of finite measure*, J. Math. Pures Appl. (9) 103, No 5, (2015), 1093 – 1131

[11] S. Haeseler, M. Keller. D. Lenz and R. K. Wojciechowski, *Laplacians on infinite graphs: Dirichlet and Neumann boundary conditions*, J. Spectr. Theory, **2** (2012), 397– 432.

[12] B. Hua, M. Keller, M. Schwarz, M. Wirth, *Sobolev-type inequalities and eigenvalue growth on graphs with finite measure*, preprint 2018, arxiv1804.08353.

[13] M. Keller, *Intrinsic metrics on graphs: a survey*, Mathematical technology of networks, 81–119, Springer Proc. Math. Stat. **128**, Springer, Cham, 2015

[14] M. Keller and D. Lenz, *Dirichlet forms and stochastic completeness of graphs and subgraphs*, J. Reine Angew. Math, **666** (2012), 189–223.

[15] M. Keller and D. Lenz, *Unbounded laplacians on graphs: Basic spectral properties and the heat equation* , Math. Model. Nat. Phenom. **5** (2010), 198–224.

[16] M. Keller, D. Lenz, and R. Wojciechowski, *Graphs and discrete Dirichlet spaces*, monograph, in preparation.

[17] M. Keller, M. Schwarz, *The Kazdan-Warner equation on canonically compactifiable graphs*, Calc. Var. Partial Differential Equations, **57** (2018), 18 pages.

[18] D. Lenz, M. Schmidt and P. Stollmann, *Topological Poincaré type inequalities and lower bounds on the infimum of the spectrum for graphs*, Preprint 2018, arxiv:1801.09279.

[19] D. Lenz, P. Stollmann and G. Stolz: An uncertainty principle and lower bounds for the Dirichlet Laplacian on graphs. arXiv: 1606.07476 Journal of Spectral Theory, Published online first: 2019-09-16 DOI: 10.4171/JST/287

[20] B. Mohar, *Eigenvalues, diameter, and mean distance in graphs.* Graphs Combin. **7** (1991), no. 1, 53–64.

[21] L.E. Payne and H. F. Weinberger, *An optimal Poincarẽ inequality for convex domains.* Arch. Rational Mech. Anal. **5** (1960), 286–292.

[22] M. Reed and B. Simon, *Methods of modern mathematical physics I*, Functional analysis. New York: Academic, 1980.

[23] M. Reed and B. Simon, *Methods of modern mathematical physics IV*: Analysis of Operators, Elsevier, 1978.

[24] R. Samavat, P. Stollmann, I. Veselić, *Lifshitz asymptotics for percolation Hamiltonians*, Bull. Lond. Math. Soc. **46** (2014), 1113–1125.

9

Critical Hardy Inequalities on Manifolds and Graphs

Matthias Keller, Yehuda Pinchover, and Felix Pogorzelski

Abstract

In this expository chapter we give an overview of recent developments in the study of optimal Hardy-type inequalities in the continuum and in the discrete setting. In particular, we present the technique of the *super-solution construction* that yield 'as large as possible' Hardy weights which is made precise in terms of the notion of criticality. Instead of presenting the most general setting possible, we restrict ourselves to the case of the Laplacian on smooth manifolds and bounded combinatorial graphs. Although the results hold in far greater generality, the fundamental phenomena as well as the core ideas of the proofs become especially clear in these basic settings.

9.1 Introduction

9.1.1 History and Background

Hardy's inequality fascinated generations of mathematicians up till today. It started all with the original inequality conjectured by Hardy

$$\sum_{n=0}^{\infty} |\varphi(n) - \varphi(n+1)|^2 \geq \frac{1}{4} \sum_{n=1}^{\infty} \frac{1}{n^2} |\varphi(n)|^2$$

for all $\varphi : \mathbb{N}_0 \to \mathbb{R}$ such that $\varphi(0) = 0$, where $\mathbb{N}_0 := \{0\} \cup \mathbb{N}$. Its counterpart in the continuum in $\mathbb{R}^d$, $d \geq 3$, reads as

$$\int_{\mathbb{R}^d} |\nabla\varphi|^2 \, \mathrm{d}x \geq \frac{(d-2)^2}{4} \int_{\mathbb{R}^d} \frac{1}{|x|^2} |\varphi(x)|^2 \, \mathrm{d}x$$

for all smooth functions $\varphi \in C_0^\infty(\mathbb{R}^d \setminus \{0\})$.

The fascination for this inequality has many sources. First of all it is an analytic inequality where the sharp constant (i.e., $1/4$ in the case of $\mathbb{N}_0$, and $(d-2)^2/4$ in $\mathbb{R}^d$) can be computed explicitly. Another significance comes from the fact that it can be seen as a strengthened version of Heisenberg's uncertainty principle. In particular, if for a state φ the L^2-integral of the momentum is bounded, i.e., $\int_{\mathbb{R}^d} |\nabla\varphi|^2 dx \le C$, then the particle in the state φ cannot be localized at $x = 0$.

The literature on Hardy inequalities is extensive and we refer here to a marvellous survey article [KMP06] elaborating the 'prehistory' of Hardy's inequality as well as to the textbooks [BEL15, OK90]. In general, a Hardy inequality provides an estimate from below for a quadratic form (which usually models an energy) by a weighted L^2 norm. The weight function w of this L^2 norm which is $w(n) = 1/n^2$ and $w(x) = 1/|x|^2$ in the examples above is called a *Hardy weight*. When Hardy's inequality is studied in different contexts, a particular focus lies in finding the sharp constant, while the Hardy weight is typically given as the inverse square of the distance function to a closed set.

In [DFP14] Devyver, Fraas, and Pinchover took a different approach, by proving *optimal* Hardy inequalities for general (not necessarily symmetric) second-order linear non-negative elliptic operators P on noncompact Riemannian manifolds. In the simplest case, this operator P is the Laplace–Beltrami operator. Here optimality means not only that the Hardy constant is optimal, but also that the Hardy weight w (which includes the constant) cannot be improved. This is referred to as the Hardy weight (or more precisely the operator $P - w$) being *critical*. The key idea is to define the Hardy weight w intrinsically in terms of the operator P. This was achieved by the so-called super-solution construction and the ground-state transform and these Hardy weights are given explicitly in terms of two (super-)harmonic functions of the operator P.

In addition, the notion of optimality also includes *optimality at infinity*, i.e., that the Hardy constant cannot be improved in a neighbourhood of infinity, and *null-criticality*, which means in the symmetric case that the Agmon ground state of $P - w$ is not square integrable with respect to w. Later this work was generalized to the p-Laplacian, $p \in (1, \infty)$, in [DP16] by using coarea techniques. Furthermore, various results were achieved in [DFP14] concerning the spectrum of the corresponding weighted operator, the completeness of the associated Agmon's metric, and related optimal Rellich-type inequalities.

Recently, this approach of the super-solution construction with the square root was introduced to the non-local realm of graphs for $p = 2$. As a first step towards optimality, a criticality theory for Schrödinger operators on graphs had to be developed ([KPP17], confer also [Tak14]). Then the major challenge in non-local situations is the absence of the chain rule which was heavily used in [DFP14]. However, as it turns out for the square root function, a version of the chain rule also holds true in the case of graphs [BHL$^+$15]. This chain rule together with the a coarea formula (as it was used in [DP16]) served as a remedy to prove a corresponding result for Schrödinger operators on weighted graphs [KPP18b]. Surprisingly, while [DFP14] recovers the original Hardy weight $w(x) = (d-2)^2/4|x|^2$ in $\mathbb{R}^d$, the results in [KPP17] for the standard discrete Laplacian on $\mathbb{N}_0$ provide a series expansion w as

$$w(n) = \frac{1}{4n^2} + \frac{5}{64n^4} + \frac{21}{512n^6} + \cdots$$

for $n \geq 2$ and $w(1) = 2 - \sqrt{2}$. Clearly, this optimal weight is strictly larger than the weight in the classical Hardy inequality $w_{\text{class}}(n) := 1/(4n^2)$ (see Example 9.3.9).

In this chapter we aim to review these recent developments, and present them in a rather self-contained context. Specifically, to make our presentation accessible, we restrict ourselves to the Laplacian on non-compact smooth manifolds and on bounded combinatorial graphs. Moreover, our main focus is to show criticality of the Hardy weight. Furthermore, in the continuum case, we present a new and simpler proof for the spectral structure of the corresponding weighted operator based on a recent Shnol-type theorem [BP17] proved by Beckus and Pinchover (see Theorem 9.2.12).

The chapter is structured as follows. After a discussion of the notions that appear both in the manifold and the graph case, the chapter is divided into two subsequent sections. First we treat the case of smooth manifolds and secondly we treat the case of bounded graphs. In each of these sections we first introduce the notions specific to the corresponding setting, and then state the result about criticality. The proofs are given in sections we refer to as 'Toolbox and proofs', where the main tools for the proof are presented. Apart from the material that is taken from standard text books, the proofs are self-contained within the chapter. In the manifold section we then also discuss further concepts which enter, the concept of optimality of [DFP14], and some spectral theoretic considerations. Finally we apply our results to some basic examples of manifolds and graphs.

9.1.2 General Notation

Let us introduce the notion of critical Hardy weights. In this chapter X will either be a smooth non-compact Riemannian manifold or a countable discrete set, and μ is either the Riemannian volume or the counting measure, respectively. By $C(X)$ we denote the real-valued continuous functions on X.

Let q be a positive quadratic form on $L^2(X, \mu)$ whose domain contains a dense subspace D_0 which is included in $C(X)$. In this chapter, D_0 is either the space of smooth compactly supported functions on a manifold or the space of finitely supported functions on a graph.

A function $u \in C(X)$ is called *positive* if $u(x) \geq 0$ for all $x \in X$ and u does not vanish identically on X. We say that u is *strictly positive* if $u(x) > 0$ for all $x \in X$.

We say that a quadratic form q is *positive*, and write $q \geq 0$, if $q \geq 0$ on D_0. A *Hardy weight* for q is a positive function $w \in C(X)$ such that $(q - w) \geq 0$ on D_0 which is

$$q(\varphi) - w(\varphi) \geq 0\,, \qquad \varphi \in D_0,$$

where

$$w(\varphi) := \int_X w\varphi^2 \,\mathrm{d}\mu.$$

Definition 9.1.1 A positive quadratic form q is said to be *critical* in X if q does not admit a Hardy weight, and *sub-critical* otherwise. Consequently, a Hardy weight w for q is said to be *critical* if $q - w$ is critical, i.e., for all Hardy weights w' for q such that $w' \geq w$ we have $w' = w$.

In both cases, the manifold and graph case, there is a (positive) Laplace operator denoted by Δ on a vector space D which gives rise to a positive quadratic form q. In the context of our chapter, D are the smooth functions $C^\infty(X)$ in the manifold case, and the continuous (and, thus all) functions $C(X)$ in the case of graphs. We call a function $u \in D$ *harmonic* if $\Delta u = 0$ and *super-harmonic* if $\Delta u \geq 0$.

We denote by ∞ the ideal point which is added to X to obtain the one-point compactification. Then convergence in X to ∞, denoted by $x \to \infty$, means that one eventually leaves every compact set. We write $X_1 \Subset X_2$ if the set X_2 is open in X, the set $\overline{X_1}$ is compact, and $\overline{X_1} \subset X_2$.

Remark 9.1.2 As mentioned above, in the continuum case we restrict ourselves to the smooth case. Once in a while, along Section 9.2, we refer to some literature where results are stated under weaker regularity assumptions. Due to standard elliptic regularity results (see, for example, [GT01]), it follows that the results are also valid in the smooth realm.

9.2 Manifolds

In the present section we prove optimal Hardy inequalities for the Laplace–Beltrami operator on a smooth non-compact Riemannian manifold. This setting is a special case of the more general results covered in [DFP14], where such inequalities are proved for general second-order, linear, elliptic differential operators.

In the first sub-section, we introduce the basic setting and state the main result. In the toolbox sub-section (Section 9.2.2), we present the main tools needed for the proof, such as the Allegretto-Piepenbrink-type theorem, the generalized maximum principle, a criterion for criticality via ground states, and a Khas'minskiĭ-type criterion. We then prove the criticality result. In further sub-sections (Sections 9.2.3 and 9.2.4), we discuss the notion of optimality of Hardy weights and basic spectral theoretic results. Finally, in Section 9.2.5, we apply our results to the Euclidean space, bounded domains, and the three-dimensional hyperbolic space.

9.2.1 Critical Hardy Inequalities on Manifolds

Let X be a smooth noncompact Riemannian manifold of dimension d, and let μ be the Riemannian volume. Let Δ be the positive Laplace–Beltrami operator on $C^\infty(X)$. For an in-depth background on the analysis on manifolds, see [Gri09]. The Laplacian Δ gives then rise to a quadratic form on $C_c^\infty(X)$

$$q(\varphi) := \int_X |\nabla\varphi|^2 \, \mathrm{d}\mu \,, \qquad \varphi \in C_c^\infty(X),$$

via Green's formula

$$q(\varphi) = \langle \Delta\varphi, \varphi \rangle = \| |\nabla\varphi| \|^2,$$

where $\langle \cdot, \cdot \rangle$ and $\| \cdot \|$ denote the standard inner product and norm on $L^2(X, \mu)$, respectively. In this setting, a positive continuous function w is a *Hardy weight* if $(q - w)(\varphi) \geq 0$ for all $\varphi \in C_c^\infty(X)$.

A function u defined on a subset of X is said to be in $C(U)$ or respectively in $C^\infty(U)$ for $U \subseteq X$ if u is continuous on U or respectively smooth on U in the case when U is an open set.

A function u in $C^\infty(U)$ is called *harmonic* on an open set $U \subseteq X$ if $\Delta u = 0$ on U and *super-harmonic* on U if $\Delta u \geq 0$ on U.

We recall that if $w, f \in C^\infty(X)$, and a function $u \in L^2_{\mathrm{loc}}(X)$ satisfies the equation $(\Delta - w)u = f$ on U weakly, then u is already in $C^\infty(U)$ by [Gri09, Corollary 7.3].

The following theorem is the main result of the section, which shows how certain super-harmonic functions yield critical Hardy weights.

Theorem 9.2.1 ([DFP14]) *Let X be a non-compact smooth Riemannian manifold. Let $u \in C^\infty(X)$ be a strictly positive super-harmonic function on X which is harmonic outside of a compact set such that $u(x) \to 0$ as $x \to \infty$. Then the following Hardy inequality holds*

$$\int |\nabla \varphi|^2 \, \mathrm{d}\mu \geq \int w\varphi^2 \, \mathrm{d}\mu$$

for all $\varphi \in C_c^\infty(X)$, where the weight function w given by

$$w := \frac{\Delta\left(u^{1/2}\right)}{u^{1/2}}$$

is a critical Hardy weight. Furthermore,

$$w = \frac{|\nabla u^{1/2}|^2}{u} = \frac{|\nabla u|^2}{4u^2} = \frac{1}{4}|\nabla \log u|^2$$

on the set where the function u is harmonic.

In the case of non-parabolic manifolds X, or equivalently on manifolds where the corresponding Brownian motion is transient, the *(positive) minimal Green function*

$$G(x, y) := \int_0^\infty p_t(x, y) \, \mathrm{d}t$$

is finite for all $x \neq y$, where $p_t(x, y) \geq 0$ denotes the heat kernel at $x, y \in X$ and time $t \geq 0$, [Gri99, Section 4.2 and Theorem 5.1 (5)]. For a smooth function φ with compact support the function

$$G\varphi(x) = \int_X G(x, y)\varphi(y) \, \mathrm{d}y \,, \qquad x \in X,$$

yields a smooth bounded solution u of minimal growth (see Definition 9.2.4) to the equation $\Delta u = \varphi$, [Gri09, Lemma 13.1]. Furthermore, $G\varphi$ is positive whenever $\varphi \geq 0$ and it is not hard to show that $\liminf_{x\to\infty} G\varphi(x) = 0$, [Gri99, Section 4.2]. In this case $G\varphi$ is called a *Green potential associated with the density* φ. Clearly, whenever there is a u which is a positive solution near infinity and $u(x) \to 0$ as $x \to \infty$ (as in the assumption of Theorem 9.2.1), then $G\varphi(x) \to 0$ as $x \to \infty$.

9.2.2 Toolbox for Manifolds and Proofs

In the present sub-section we prove Theorem 9.2.1. To this end, we need some well-known tools. First, we recall a characterization of positivity of quadratic forms by means of super-harmonic functions and a generalized maximum principle. Secondly, we present a criterion for the criticality of Hardy weights in terms of minimal solutions.

For a function $w \in C^\infty(X)$, we say that $u \in C^\infty(X)$ is a *solution for* $\Delta - w$ if $(\Delta - w)u = 0$ in X. If $(\Delta - w)u \geq 0$ for some $u \in C^\infty(X)$, then we call u a *super-solution* for $\Delta - w$. We also say that $u \in C^\infty(X)$ is a (super-)solution *in a neighbourhood of infinity* if there exists a compact set $K \subset X$ such that $(\Delta - w)u = 0$, (respectively, $(\Delta - w)u \geq 0$) in $X \setminus K$.

9.2.2.1 The Allegretto–Piepenbrink Theorem

For the proof of the main theorem we use the so-called Allegretto–Piepenbrink theorem. It states that for energies below and up to the bottom of the spectrum there exist positive (super-)solutions. Here we refer to [Sul87, Theorem 2.1] and [Gri09, Theorem 13.16] for the case of manifolds, to [CFKS87, Theorem 2.12] for the case of Schrödinger operators, and to [LSV09] for the general case of strongly local Dirichlet forms that are perturbed by a measure.

Proposition 9.2.2 (Allegretto–Piepenbrink theorem) *Let X be a non-compact connected smooth Riemannian manifold, and $w \in C^\infty(X)$. Then the following statements are equivalent:*

(i) $(q - w) \geq 0$ *on* $C_c^\infty(X)$.
(ii) *There is a positive solution* $u \in C^\infty(X)$ *to* $(\Delta - w)u = 0$.
(iii) *There is a positive super-solution* $u \in C^\infty(X)$ *to* $(\Delta - w)u \geq 0$.

In this case the function u is strictly positive.

Note that the various settings of the literature mentioned above do not yield the smoothness of u in general. However, as discussed above, the validity of the equation $(\Delta - w)u = 0$ in a weak sense for smooth w on a smooth manifold yields the smoothness of u.

9.2.2.2 The Generalized Maximum Principle

We recall the *generalized maximum principle* which is essential for the considerations that follow. It is a classical statement that is found in various text books in various contexts. Here we refer to [Gri09, Theorem 5.13] for the Laplacian on manifolds and for elliptic operators on Euclidean domains see [PW84, Chapter 2.5] and [GT01, Chapters 3 and 8].

Proposition 9.2.3 (Generalized maximum principle) *Let $w \in C^\infty(X)$. Then $(q - w) \geq 0$ in X if and only if for any domain $U \Subset X$ and any u in $C(U \cup \partial U) \cap C^\infty(U)$ which is non-negative on ∂U and satisfies $(\Delta - w)u \geq 0$ on U, we have either $u > 0$ or $u = 0$ on U.*

Moreover, the validity of the above generalized maximum principle is also equivalent to the existence and to uniqueness of a non-negative solution to the following Dirichlet problem in any sub-domain $U \Subset X$ with smooth boundary ∂U

$$\begin{cases} (\Delta - w)v = f & \text{in } U, \\ v = g & \text{on } \partial U, \end{cases}$$

for all non-negative $f \in C^\infty(U)$ and $g \in C^\infty(\partial U)$.

Remark Proposition 9.2.3 implies that there are no non-trivial positive harmonic functions u in X such that $u(x) \to 0$ as $x \to \infty$. Indeed, let K_n be an increasing sequence of pre-compact domains with smooth boundary exhausting X. Then $c_n - u$ with $c_n = \max_{\partial K_n} u$ is harmonic in K_n and $c_n - u \geq 0$ on ∂K_n and, thus, $c_n \geq u$ on K_n by the generalized maximum principle above. However, $c_n \to 0$ by the assumption $u(x) \to 0$ as $x \to \infty$. Hence, $u = 0$. So, in view of our assumption in Theorem 9.2.1, this implies that the functions u under consideration are indeed super-harmonic and not harmonic.

9.2.2.3 Ground States and Khas'minskiĭ-Type Criterion

We introduce some notions that are used to prove Theorem 9.2.1. These notions appear first in the work of Agmon [Agm83] in the setting of second-order elliptic operators on weighted manifolds.

Definition 9.2.4 Let $w \in C^\infty(X)$ be a function. A positive function u is said to be a *positive solution of minimal growth in a neighbourhood of infinity in* X for $\Delta - w$ if there is $L \Subset X$ such that $(\Delta - w)u = 0$ on $X \setminus L$, and if for all compact $L \Subset K \Subset X$ with smooth boundary, for any positive function v in $C^\infty(X \setminus K) \cap C(\overline{X \setminus K})$ such that $(\Delta - w)v \geq 0$ on $X \setminus K$ that satisfies $u \leq v$ on ∂K, we have $u \leq v$ in $X \setminus K$.

A positive solution of minimal growth in a neighbourhood of infinity in X for $\Delta - w$ that is a global positive solution u of $(\Delta - w)u = 0$ is called an *Agmon ground state* of $\Delta - w$ in X.

The following criterion is indeed an equivalence (see, for example, [PT06]). However, here we only prove the direction we need for the proof of Theorem 9.2.1.

Lemma 9.2.5 (Criticality criterion) *If $\Delta - w$ admits a ground state in X, then w is a critical Hardy weight.*

Proof. Let $u \in C^{\infty}(X)$ be a ground state for $\Delta - w$, and assume w is not critical. Then there is $w' \geq w$ such that $w' \neq w$ and $q - w' \geq 0$. By the Allegretto–Piepenbrink theorem, Proposition 9.2.2 (iii), there is a strictly positive $v \in C^{\infty}(X)$ such that $(\Delta - w')v \geq 0$. Hence,

$$(\Delta - w)v \geq (w' - w)v \geq 0\,.$$

Let $K \subseteq X$ be compact with smooth boundary. By the strict positivity of u and v which follows from the generalized maximum principle, there is $C_K > 0$ such that $C_K u \leq v$ on ∂K. By the minimality of u, we have $C_K u \leq v$ on $X \setminus K$ and by the generalized maximum principle in K, we have $C_K u \leq v$ on K. Hence, $C_K u \leq v$ on X. Let $C > 0$ be the supremum over all constants such that

$$Cu \leq v$$

on X. Then, $v' = v - Cu \geq 0$ satisfies

$$(\Delta - w)v' = (\Delta - w)v - C(\Delta - w)u = (\Delta - w)v \geq (w' - w)v \gneqq 0.$$

Hence, v' is a non-trivial positive super-solution for $\Delta - w$ which is strictly positive by the generalized maximum principle. By the same argument as above, there exists $C' > 0$ such that $C'u \leq v'$. So,

$$(C + C')u \leq Cu + v' = v$$

which is a contradiction to the maximality of C. □

The previous lemma allows us to prove criticality by presenting a ground state. The following criterion for the minimality at infinity is indeed an equivalence in the symmetric case. Once again, we only prove the direction we need for the proof of Theorem 9.2.1 (for the proof of this direction in different contexts, see [DFP14, Has80, Pin95]). The other direction follows from [Anc02, Theorem 1].

Lemma 9.2.6 (Khas'minskiĭ-type criterion) *Let $w \in C^{\infty}(X)$ and suppose that u_0 (resp. u_1) is a smooth positive solution (resp. super-solution) for $\Delta - w$ in a neighbourhood of infinity in X such that*

$$\lim_{x \to \infty} \frac{u_0(x)}{u_1(x)} = 0.$$

Then u_0 is a solution for $\Delta - w$ of minimal growth in a neighbourhood of infinity in X.

Proof. Let L be a compact set such that $(\Delta - w)u_0 = 0$ and $(\Delta - w)u_1 \geq 0$ on $X \setminus L$. Let a compact set K with $L \Subset K \subset X$ and smooth boundary be given. Let (X_k) be an *exhaustion* of X, i.e., a sequence of relatively compact domains in X with smooth boundary such that $K \subseteq X_1$, $X_k \Subset X_{k+1}$, and $\cup_{k=1}^{\infty} X_k = X$. By the generalized maximum principle (Proposition 9.2.3), the Dirichlet problem

$$\begin{cases} (\Delta - w)\eta_k = 0 & \text{in } X_k \setminus K, \\ \eta_k = u_0 & \text{on } \partial K, \\ \eta_k = 0 & \text{on } \partial X_k. \end{cases}$$

has a unique positive solution which we denote by η_k. Then by the generalized maximum principle (Proposition 9.2.3), we have $0 < \eta_k \leq u_0$ and $(\eta_k)_{k \in \mathbb{N}}$ is an increasing sequence of positive functions. Therefore, (η_k) converges pointwise to a solution η for $\Delta - w$ on $X \setminus K$ which satisfies $0 < \eta \leq u_0$.

We claim that η has minimal growth in X. Indeed, take any relatively compact set K_1 with smooth boundary and $K \Subset K_1$. Furthermore, let $v \in C^{\infty}(X \setminus K_1) \cap C(X \setminus K_1 \cup \partial K_1)$ be positive and $(\Delta - w)$-super-harmonic such that $\eta \leq v$ on ∂K_1. Since $\eta_k \leq \eta$ on ∂K_1 and $\eta_k = 0$ on ∂X_k, we have $\eta_k \leq v$ on $\partial(X_k \setminus K_1)$ and by the generalized maximum principle (Proposition 9.2.3), we have $\eta_k \leq v$ in $X_k \setminus K_1$. Consequently, $\eta \leq v$ in $X \setminus K_1$ and η has minimal growth in X a neighbourhood of infinity.

Thus, it is enough to show that $u_0 \leq \eta$ in $X \setminus K$. Indeed, for every $\varepsilon > 0$ there is k_ε such that $u_0 \leq \varepsilon u_1$ on ∂X_k, for every $k \geq k_\varepsilon$ by the hypothesis. Again by the generalized maximum principle, this implies that $u_0 \leq \eta_k + \varepsilon u_1$ in $X_k \setminus K$ and it follows $u_0 \leq \eta + \varepsilon u_1$ in $X \setminus K$. By letting $\varepsilon \to 0$, we conclude that $u_0 \leq \eta$ in $X \setminus K$. Thus, $u_0 = \eta$ in $X \setminus K$ and u_0 has minimal growth in X. □

9.2.2.4 Proof of Critical Hardy Weights for Manifolds

With the Khas'minskiĭ-type criterion at hand we are ready to prove a lemma which, in combination with Lemma 9.2.5, immediately yields Theorem 9.2.1. The proof we present here is, in fact, one of the given proofs of Theorem 6.2 in [DFP14].

Lemma 9.2.7 *Let $u \in C^{\infty}(X)$ be strictly positive super-harmonic on X which is harmonic outside a compact set, and satisfies $u(x) \to 0$ as $x \to \infty$. Let*

$$w := \frac{\Delta\left(u^{1/2}\right)}{u^{1/2}}.$$

Then $u^{1/2}$ is a ground state of $\Delta - w$.

Proof. Let

$$u_0 = u^{1/2}, \qquad u_1 = -u^{1/2}\log u.$$

Clearly, u_0, u_1 are both positive in a neighbourhood of infinity. Furthermore, u_0 is positive $(\Delta - w)$-harmonic in X since

$$(\Delta - w)u_0 = \Delta u^{1/2} - \frac{\Delta\left(u^{1/2}\right)}{u^{1/2}} u^{1/2} = 0.$$

We now verify that $(\Delta - w)u_1 = 0$ in a neighbourhood of infinity. Keeping in mind that the Laplacian is positive (i.e., '$-\Delta = \nabla \cdot \nabla$'), we note first that

$$\Delta\big(- u^{1/2}\log u\big) = -\Delta\left(u^{1/2}\right)\log u + 2\left(\nabla u^{1/2}\right)\cdot(\nabla \log u) - u^{1/2}\Delta\log u.$$

We further compute

$$\begin{aligned}(\Delta - w)u_1 &= \Delta\big(- u^{1/2}\log u\big) - \frac{\Delta\left(u^{1/2}\right)}{u^{1/2}}(-u^{1/2})\log u \\ &= 2\left(\nabla u^{1/2}\right)\cdot\left(\nabla \log u\right) - u^{1/2}\cdot\Delta\log u \\ &= u^{-3/2}|\nabla u|^2 - u^{-3/2}|\nabla u|^2 - u^{-1/2}\cdot\Delta u \\ &= -u^{-1/2}\cdot\Delta u = 0,\end{aligned}$$

and so, u_1 is a positive solution in a neighbourhood of infinity. Note further, since $u(x) \to 0$ as $x \to \infty$, we have

$$\lim_{x\to\infty}\frac{u_0(x)}{u_1(x)} = 0.$$

By the Khas'minskiĭ-type criterion (Lemma 9.2.6), we infer that u_0 is a global positive solution for $\Delta - w$ of minimal growth in a neighbourhood of infinity. Hence, by definition, u_0 is a ground state. □

Proof of Theorem 9.2.1. By the previous lemma $u^{1/2}$ is ground state of $\Delta - w$. We conclude that w is a critical Hardy weight by Lemma 9.2.5. □

9.2.3 Optimality of the Hardy Weights

One can show that the Hardy weight w of Theorem 9.2.1 is not only critical but also optimal in the following sense. First of all one can show that if for the Hardy weight w given by Theorem 9.2.1, Hardy's inequality holds in a neighbourhood of infinity in X with Hardy weight λw, then one still has $\lambda \leq 1$. This property is called *optimality at infinity* of w. Furthermore, one can show *null-criticality* of $(\Delta - w)$ with respect to the weight w, which means that the ground state $u^{1/2}$ of $(\Delta - w)$ is not in $L^2(X, w\mu)$. The proofs of the following theorems are adaptations of the proofs given in [DFP14] to our particular situation.

Theorem 9.2.8 *Let X be a non-compact Riemannian manifold. Let $u \in C^\infty(X)$ be a strictly positive super-harmonic function that is harmonic outside of a compact set and $u(x) \to 0$ as $x \to \infty$. Let*

$$w := \frac{\Delta\left(u^{1/2}\right)}{u^{1/2}}$$

be the critical Hardy weight from Theorem 9.2.1. Let $\lambda \geq 0$ and $W \subseteq X$ be compact. If the following Hardy inequality holds true

$$\int |\nabla\varphi|^2 \,\mathrm{d}\mu \geq \lambda \int w\varphi^2 \,\mathrm{d}\mu$$

for all $\varphi \in C_c^\infty(X \setminus W)$, then $\lambda \leq 1$.

Proof. For a smooth and complex-valued function g on X, we let $\Delta g = \Delta \mathrm{Re}\, g + \mathrm{i}\,\Delta \mathrm{Im}\, g$. So, for $\alpha \in \mathbb{C}$, $\alpha \neq 0, 1$, and u smooth and real valued, we calculate

$$\Delta u^\alpha = \alpha u^{\alpha-1}\Delta u - \alpha(\alpha-1)u^{\alpha-2}|\nabla u|^2 = \alpha u^{\alpha-1}\Delta u + 4\alpha(\alpha-1)u^\alpha w$$

where we used the equality $|\nabla u|^2/(4u^2) = \Delta u^{1/2}/u^{1/2} = w$ on the set where u is harmonic. Hence, for given $\lambda \geq 0$ we obtain on the set where u is harmonic

$$(\Delta - \lambda w)u^\alpha = \alpha u^{\alpha-1}\Delta u + (4\alpha(1-\alpha) - \lambda)u^\alpha w = 0$$

if and only if $\lambda = 4\alpha(1-\alpha)$, i.e.,

$$\alpha = \frac{1}{2}(1 \pm \sqrt{1-\lambda}),$$

where for a negative number a, the term $\sqrt{a}$ denotes the complex number $\mathrm{i}\sqrt{|a|}$. Since the real and the imaginary part of a complex harmonic function are harmonic by linearity, we infer that the function

$$v := \mathrm{Re}\, u^{\frac{1}{2}(1+\sqrt{1-\lambda})} = \begin{cases} u^{\frac{1}{2}(1+\sqrt{1-\lambda})} & : \lambda \leq 1 \\ u^{\frac{1}{2}} \cdot \cos\left(\frac{1}{2}\sqrt{\lambda-1}\log u\right) & : \lambda > 1, \end{cases}$$

solves $(\Delta - \lambda w)v = 0$ outside of a compact set.

For $\lambda > 1$ the function v changes its sign since $u(x) \to 0$ for $x \to \infty$. Let U be a connected component of $\{x \in X \mid v(x) < 0\}$. Since $u(x) \to 0$ as $x \to \infty$, the set U is bounded. Moreover, $v \geq 0$ on ∂U, and $(\Delta - \lambda w)v = 0$, but $v < 0$ on U. Thus, by the generalized maximum principle (Proposition 9.2.3), we infer that the quadratic form $q - \lambda w$ associated with $\Delta - \lambda w$ is not positive. This proves the claim. □

Theorem 9.2.9 *Let X be a non-compact Riemannian manifold. Let $u \in C^\infty(X)$ be a strictly positive super-harmonic function that is harmonic outside of a compact set and satisfies $u(x) \to 0$ as $x \to \infty$. Let*

$$w := \frac{\Delta\left(u^{1/2}\right)}{u^{1/2}}$$

be the critical Hardy weight from Theorem 9.2.1. Then $u^{1/2} \notin L^2(X, w\mu)$.

Proof. Let $U \Subset X$ be an open pre-compact set such that u is harmonic outside of U. Let (t_k) be a null-sequence of positive numbers and let

$$X_k = \{x \in X \setminus U \mid u(x) \geq t_k\}, \qquad k \geq 0.$$

Since $u(x) \to 0$ as $x \to \infty$, the sets X_k are compact. Furthermore, let (u_k) be a sequence of positive functions in $C_c^\infty(X)$ that are equal to u on X_k. We estimate $\|u^{\frac{1}{2}}\|_{L^2(X,w\mu)}$ using u, $w \geq 0$ and $w = \frac{\Delta u^{1/2}}{u^{1/2}} = \frac{|\nabla u|^2}{4u^2}$ on X_k

$$\|u^{\frac{1}{2}}\|^2_{L^2(X,w\mu)} \geq \int_{X_k} uw \,\mathrm{d}\mu \geq \int_{X_k} \frac{1}{4u}|\nabla u|^2 \,\mathrm{d}\mu = \int_X \frac{1_{X_k}}{4u}|\nabla u_k|^2 \,\mathrm{d}\mu.$$

Consider the sub-manifold (which we may assume to be smooth by Sard's lemma) given by the level set $\{u_k = t\}$ with surface measure σ. Then the outer normal vector n_t of the level set is parallel to ∇u_k and we obtain $|\nabla u_k| = \nabla u_k \cdot n_t$. By the coarea formula, [Cha84, Chapter IV, Theorem 1], we infer

$$\ldots = \int_0^\infty \int_{\{u_k=t\}} \frac{1_{X_k}}{4u}(\nabla u_k \cdot n_t) \,\mathrm{d}\sigma \,\mathrm{d}t \geq \int_{t_k}^\infty \frac{1}{4t} \int_{\{u_k=t\}} \nabla u_k \cdot n_t \,\mathrm{d}\sigma \,\mathrm{d}t.$$

By Green's formula we have

$$\ldots = \int_{t_k}^\infty \frac{1}{4t} \int_{\{u_k \geq t\}} \Delta u \,\mathrm{d}\mu \,\mathrm{d}t \geq 0.$$

We assumed that $u(x) \to 0$ and u is positive and super-harmonic. In light of Remark 9.2.2.2, we conclude that u is not harmonic. Thus, there is a t' and a constant C such $\int_{\{u_k \geq t\}} \Delta u \,\mathrm{d}\mu \geq C$ for $t \leq t'$ and we obtain

$$\ldots \geq C \int_{t_k}^{t'} \frac{1}{4t} \,\mathrm{d}t \to \infty$$

as $t_k \to 0$ for $k \to \infty$. Hence, $\|u^{\frac{1}{2}}\|^2_{L^2(X,w\mu)} = \infty$ which finishes the proof. □

Remark 9.2.10 It follows from [KP18, Theorem 3.1 or Theorem 3.4] that if the operator $\Delta - w$ is null-critical in X with respect to w, then w is also optimal at infinity. We note that this phenomenon holds in more general frameworks, such as general second-order linear elliptic operators, p-Laplacian-type operators (see [KP18]), and Schrödinger operators on weighted graphs (see [KPP18b, the proof of Theorem 5.1]).

9.2.4 Spectral Theory

Let $u \in C^\infty(X)$ be a positive super-harmonic function such that

$$w := \frac{\Delta\left(u^{1/2}\right)}{u^{1/2}}$$

is strictly positive in X. Then $w\mu$ is a measure of full support on X. Hence, $q - w$ on $C_c^\infty(X)$ is positive and densely defined on $L^2(X, w\mu)$. Thus, $q - w$ is closable on $L^2(X, w\mu)$ and there is a positive self-adjoint operator acting as $w^{-1}\Delta - 1$ on $L^2(X, w\mu)$. In this sub-section we study the spectral theory of the self-adjoint operator $w^{-1}\Delta$ (since the -1 is only a spectral shift) on the *complex* Hilbert space $L^2(X, w\mu)$.

Specifically, we prove the following theorem about the essential spectrum $\sigma_{\text{ess}}(w^{-1}\Delta)$ of $w^{-1}\Delta$. For the corresponding result in a more general situation, see [DFP14, Section 9].

Theorem 9.2.11 *Let X be a non-compact Riemannian manifold. Let u be a strictly positive super-harmonic function that is harmonic outside of a compact set and $u(x) \to 0$, $x \to \infty$. Assume that*

$$w := \frac{\Delta\left(u^{1/2}\right)}{u^{1/2}} > 0.$$

Then,

$$\sigma(w^{-1}\Delta) = \sigma_{\text{ess}}(w^{-1}\Delta) = [1, \infty).$$

The proof of $\sigma(w^{-1}\Delta) \subseteq [1, \infty)$ is an application of the Allegretto–Piepenbrink theorem (Theorem 9.2.2). To prove the reverse inclusion we need a recent Shnol-type theorem of Beckus and Pinchover [BP17]. For an alternative proof, see [DFP14].

Proposition 9.2.12 (Shnol-type theorem) *Let X be a non-compact Riemannian manifold. Let $w > 0$ be a critical Hardy weight, and denote by ψ the ground state of $\Delta - w$. Let $\lambda \in \mathbb{R}$ and $v \in C^\infty(X)$ be such that*

$$|v| \le \psi \text{ on } X$$

and

$$(\Delta - \lambda w)v = \varphi v$$

for some $\varphi \in C_c^\infty(X)$. *Then,* $\lambda \in \sigma(w^{-1}(\Delta - \varphi))$.

Proof of Theorem 9.2.11. Let $\lambda \geq 0$ and $\alpha = (1+\sqrt{1-\lambda})/2$. As in the proof of Theorem 9.2.8, we observe that the function

$$v := u^{(1+\sqrt{1-\lambda})/2} = u^\alpha$$

satisfies

$$(\Delta - \lambda w)v = \alpha u^{\alpha-1}\Delta u.$$

Thus, v is $(\Delta - \lambda w)$-harmonic outside of a compact set in X.

In particular, for $0 \leq \lambda \leq 1$ the function v is a positive super-solution of $\Delta - \lambda w$ in X. Therefore, by the Allegretto–Piepenbrink theorem (Proposition 9.2.2), we conclude $\lambda \notin \sigma(w^{-1}\Delta)$ whenever $0 \leq \lambda < 1$. Hence,

$$\sigma(w^{-1}\Delta) \subseteq [1, \infty).$$

For $\lambda \geq 1$, the complex valued function $v = u^{(1+\mathrm{i}\sqrt{\lambda-1})/2}$ satisfies

$$|v| = u^{1/2}$$

in X, so, by the null-criticality with respect to w (Theorem 9.2.9), we have $v \notin L^2(X, w\mu)$. Moreover, by a similar computation as in the proof of Theorem 9.2.8, the function v satisfies the equation

$$(\Delta - \lambda w)v = \varphi v$$

on X, where

$$\varphi = \alpha\frac{u^{\alpha-1}\Delta u}{v} = \alpha\frac{\Delta u}{u} \in C_c^\infty(X).$$

By the Shnol-type theorem (Proposition 9.2.12), we obtain that

$$\lambda \in \sigma(w^{-1}(\Delta - \varphi)).$$

In light of Weyl theorem concerning the invariance of the spectrum of a self-adjoint operator under a compact perturbation, we have $\lambda \in \sigma_{\mathrm{ess}}(w^{-1}\Delta)$, and the theorem is proven. □

9.2.5 Examples of Optimal Hardy Weights on Manifolds

In this sub-section we discuss simple examples of Riemannian manifolds such as $\mathbb{R}^d \setminus \{0\}$, $d \geq 3$, bounded domains in $\mathbb{R}^d$, $d \geq 2$ and the hyperbolic space $\mathbb{H}^3$.

Example 9.2.13 For $X := \mathbb{R}^d \setminus \{0\}$, $d \geq 3$, the classical Hardy-type inequality reads

$$\int_X |\nabla \varphi|^2 \, \mathrm{d}x \geq \left(\frac{d-2}{2}\right)^2 \int_X \frac{|\varphi(x)|^2}{|x|^2} \, \mathrm{d}x \,, \qquad \varphi \in C_c^\infty(X).$$

Indeed, take $u := G(0,x) = C_d |x|^{2-d}$, the positive minimal Green function of the Laplace operator Δ with a singularity at zero. Then $\Delta u = 0$ on X and $u^{1/2}$ is positive and super-harmonic. We obtain the Hardy weight

$$w(x) := \frac{\Delta u^{1/2}(x)}{u^{1/2}(x)} = \left(\frac{d-2}{2}\right)^2 \frac{1}{|x|^2}.$$

By the Allegretto–Piepenbrink theorem (Proposition 9.2.2), we get the classical Hardy inequality above. It can be shown that $\left(\frac{d-2}{2}\right)^2 |x|^{-2}$ is indeed an optimal Hardy weight in X (see, [DFP14, Example 3.1]). Note that $G(0,x) \to \infty$ as $x \to 0$, so $u(x) \not\to 0$ as x leaves every compact set of X. Thus, the theorems proven here are not formally applicable. However, the arguments presented here can be directly used to show the statement.

Example 9.2.14 Let $X \subset \mathbb{R}^d$ be a bounded domain with smooth boundary, where $d \geq 2$. Let G be the positive minimal (Dirichlet) Green function of the Laplace operator Δ in X. For a smooth positive function φ with compact support in X, the Green potential

$$u(x) = G\varphi(x) := \int_X G(x,y)\varphi(y) \, \mathrm{d}y \,, \qquad x \in X,$$

is a positive super-harmonic function in X, harmonic outside the support of φ, and satisfies $u(x) \to 0$ as $x \to \partial X$, i.e., $x \to \infty$ with respect to the one-point compactification of X. Hence, the following optimal Hardy inequality holds true

$$\int_X |\nabla \varphi|^2 \, \mathrm{d}x \geq \int_X w |\varphi|^2 \, \mathrm{d}x \,, \qquad \varphi \in C_c^\infty(X),$$

where

$$w = \frac{\Delta \left(u^{1/2}\right)}{u^{1/2}}.$$

Moreover, Theorem 9.2.11 implies that

$$\sigma(w^{-1}\Delta) = \sigma_{\mathrm{ess}}(w^{-1}\Delta) = [1, \infty).$$

On the other hand, Hopf's lemma implies (see, for example, [GT01, PW84, LaPi16])

$$w(x) \sim \frac{1}{4\delta(x)^2} \qquad \text{as } x \to \partial X,$$

where δ is the distance function to ∂X. It follows that $\sigma_{\text{ess}}(\delta^2\Delta) = [1/4, \infty)$ (see [Dev14] and references therein).

Example 9.2.15 Consider $\mathbb{H}^3$, the three-dimensional hyperbolic space, with the canonical hyperbolic metric embedded in $\mathbb{R}^3_+$. Fix a point $p = (0, 0, 1)$, and let $d(x)$ be the Riemannian distance of a point x from p. Recall that the bottom of the spectrum of the Laplace–Beltrami operator on $\mathbb{H}^3$ is 1, and the corresponding Green function with singularity at p is given explicitly by

$$G(x) = C\frac{\mathrm{e}^{-d(x)}}{\sinh d(x)} \qquad x \in X := \mathbb{H}^3 \setminus \{p\},$$

where C is a positive constant. Then it follows that

$$w(x) = \frac{\Delta G^{1/2}(x)}{G^{1/2}(x)} = \frac{|\nabla G(x)|^2}{4G(x)^2} = \frac{(1 + \coth d(x))^2}{4}$$

is an optimal Hardy weight in X (see [DFP14]).

The computations in $\mathbb{H}^d$, for $d = 2$ and $d \geq 4$, are more intricate since it involves special functions; see [Mat01] for an explicit formula for the positive minimal Green function in any dimension, and [BGGP] for another type of an optimal Hardy inequality in X.

9.3 Graphs

In the present section, we discuss critical Hardy inequalities on bounded combinatorial graphs with standard weights. This setting is a special case of the one covered in [KPP18b] and the proofs simplify here to some degree.

In the first sub-section, we introduce the relevant notions of graphs with their canonical Laplace operator and the associated quadratic form. Then we state the main results for graphs, which is the existence of *critical* Hardy weights via positive super-harmonic functions. In the toolbox sub-section (Section 9.3.2), we discuss Harnack's inequality, a criterion for criticality via null-sequences and a coarea formula. Finally we prove our main result in this section using these tools. In Section 9.3.3 we show how our results apply to the half line, the d-dimensional Euclidean lattice, $d \geq 3$, as well as to regular trees.

9.3.1 Critical Hardy Inequalities on Graphs

A combinatorial graph is a countable (infinite) set X called the *vertex set* with a symmetric relation $\sim$. We say two vertices $x, y \in X$ are *connected by an edge* (x, y) if $x \sim y$. We assume that there are no loops, i.e., $x \not\sim x$ for all $x \in X$, and that the graph is connected, i.e., for any two distinct vertices $x, y \in X$ there are vertices $x_0, \dots, x_n \in X$ such that $x = x_0 \sim \dots \sim x_n = y$. Furthermore, we assume that the vertex degree $\deg : X \to \mathbb{N}_0$

$$\deg(x) := \#\{y \in X \mid x \sim y\}$$

is bounded, i.e.,

$$D := \sup_{x \in X} \deg(x) < \infty.$$

We let $C(X)$ be the real-valued functions and $C_c(X)$ be the subspace of functions with finite support. Similarly, for $U \subseteq X$, we denote the corresponding subspaces, by $C(U)$ and $C_c(U)$.

There is a natural (positive) *graph Laplacian* $\Delta : C(X) \to C(X)$ given by

$$\Delta f(x) := \sum_{y \sim x} \big(f(x) - f(y)\big).$$

The Laplacian is connected to the quadratic form

$$q(\varphi) := \frac{1}{2} \sum_{x \sim y} \big(\varphi(x) - \varphi(y)\big)^2 \qquad \varphi \in C_c(X)$$

defined on $C_c(X)$ via Green's formula

$$q(\varphi) = \langle \Delta\varphi, \varphi \rangle,$$

where $\langle \cdot, \cdot \rangle$ denotes the standard inner product on $\ell^2(X)$. A positive function $w : X \to [0, \infty)$ is a *Hardy weight* on $U \subseteq X$ if $(q - w)(\varphi) \geq 0$ for all $\varphi \in C_c(U)$.

Let $U \subseteq X$. A function $u \in C(X)$ is called *(super-)harmonic* in U if $\Delta u = 0$ (respectively, $\Delta u \geq 0$) in U. Note that due to the local Harnack inequality (see Lemma 9.3.3), every positive super-harmonic function on a connected set $U \subset X$ is strictly positive. Finally, recall that a function $u : X \to \mathbb{R}$ is called *proper* on $W \subseteq X$ if $u^{-1}(I)$ is compact (i.e. finite) for all compact $I \subseteq (\inf u, \sup u)$.

The next theorem is the main result of this section. It shows that certain positive super-harmonic functions yield critical Hardy weights.

Theorem 9.3.1 *Let X be a connected combinatorial graph with bounded vertex degree, and let $K \subseteq X$ be a finite set. Let $u : X \to [0, \infty)$ be strictly* *positive proper and super-harmonic on $X \setminus K$, vanishing on K, and harmonic outside of a further finite set. Then the following Hardy-type inequality holds*

$$\frac{1}{2}\sum_{\substack{x,y\in X\\ x\sim y}} (\varphi(x)-\varphi(y))^2 \geq \sum_{x\in X\setminus K} w(x)\varphi(x)^2$$

for all finitely supported functions φ with support in $X \setminus K$, where the weight function w is given by

$$w := \frac{\Delta u^{1/2}}{u^{1/2}}.$$

For all $x \in X \setminus K$, where u is harmonic, one has the formula

$$w(x) := \frac{1}{2u(x)} \sum_{y\sim x} \left(u(x)^{1/2} - u(y)^{1/2}\right)^2.$$

Moreover, w is a critical Hardy weight in $X \setminus K$.

In various cases it is hard to find explicit non-trivial harmonic functions. However, for transient graphs there are plenty of positive super-harmonic functions in terms of the positive minimal Green function:

$$G(x,y) := \sum_{n=0}^{\infty} p_n(x,y).$$

Here $p_n(x,y)$ are the matrix elements of the nth power of the transition matrix $p_1(x,y)$, where $p_1(x,y) := 1/\deg(x)$ if $x \sim y$ and 0 otherwise. The case when the sum converges for all $x, y \in X$ is called *transient*. In this case, $u = G(o,\cdot)$ is known to be a strictly positive super-harmonic and harmonic outside of $o \in X$ (see, for example, [Woe00]). By the minimality of $G(o,\cdot)$ it follows that $\inf G(o,\cdot) = 0$. As a special case of Theorem 9.3.1, we obtain the following critical Hardy inequality.

Corollary 9.3.2 *Let a transient connected graph X with bounded vertex degree be given, and let $o \in X$ be fixed. Let $G(o,\cdot) : X \to (0,\infty)$ be the positive minimal Green function for o, and assume that $G(o,\cdot)$ is proper. Then the following Hardy-type inequality holds true*

$$\frac{1}{2}\sum_{x,y\in X,\, x\sim y} (\varphi(x)-\varphi(y))^2 \geq \sum_{x\in X} w(x)\varphi(x)^2$$

for all finitely supported functions φ on X, where

$$w(x) := \frac{\Delta\left(G(o,x)^{1/2}\right)}{G(o,x)^{1/2}}$$

is a critical Hardy weight in X, and for all $x \neq o$

$$w(x) = \frac{1}{2G(o,x)} \sum_{y \sim x} \left(G(o,x)^{1/2} - G(o,y)^{1/2}\right)^2.$$

9.3.2 Toolbox for Graphs and Proofs

In this section, we collect the tools for proving the aforementioned Hardy inequalities on graphs. In particular, we discuss a local Harnack inequality, null-sequences, a coarea formula, as well as Hardy weights arising from the super-solution construction.

9.3.2.1 Local Harnack Inequality

For bounded graphs with standard weights there is a rather simple proof of a local Harnack inequality. It is found in [Dod84, Lemma 1.7]. For versions in a more general setting of weighted graphs or Schrödinger operators, see [HK11, KPP17].

Lemma 9.3.3 (Local Harnack inequality) *Let u be a positive super-harmonic function on U. Then, for all $x \sim y$ in U,*

$$\frac{1}{D} u(x) \leq u(y) \leq D u(x),$$

where $D := \sup_{x \in X} \deg(x) < \infty$.

Proof. The inequality $0 \leq \Delta u(x)$ yields for all $y \sim x$

$$u(y) \leq \sum_{z \sim x} u(z) \leq \sum_{z \sim x} u(x) = \deg(x) u(x) \leq D u(x)$$

which gives the upper bound. Interchanging the roles of x and y yields the lower bound. □

9.3.2.2 Criticality and Null-Sequences

Recall that a Hardy weight w is called critical if all positive functions w' satisfying $w' \geq w$ and $q - w' \geq 0$ coincide with w. This amounts to criticality of the form $q - w$, i.e., for a weight $w' \gneq w$, the inequality $q - w' \geq 0$ fails to hold. The following lemma gives a sufficient criterion for criticality in terms of null-sequences, i.e., pointwise convergent sequences of finitely supported functions annihilating the form in the limit. In fact, the existence of null-sequences is also necessary (cf., e.g., [KPP17, Theorem 5.3]) but we do not need the reverse direction here.

Lemma 9.3.4 (Criticality and null-sequence) *Let w be a Hardy weight for q on $U \subseteq X$, and let (e_n) be a sequence in $C_c(U)$ such that e_n converges pointwise to a strictly positive function e on U. If*

$$(q-w)(e_n) \to 0,$$

then $q-w$ is critical on U.

Proof. Suppose that there is w' such that $(q-w') \geq 0$ on U with $w' \geq w$. Then for arbitrary $x \in U$

$$(q-w)(e_n) = (q-w')(e_n) + (w'-w)(e_n) \geq (w'-w)(x)e_n^2(x).$$

Hence,

$$0 = \lim_{n\to\infty}(q-w)(e_n) \geq (w'-w)(x)e^2(x).$$

Since $e > 0$, $w' \geq w$ and $x \in U$ was chosen arbitrarily, we conclude that $w' = w$ on U. □

9.3.2.3 A Coarea Formula

In order to verify criticality of quadratic forms, the following coarea formula will be an essential tool. In [KPP18b] a corresponding formula is proven for general weighted graphs.

Lemma 9.3.5 (Coarea formula) *Let $K \subseteq X$ be finite, and let $u \geq 0$ be proper, super-harmonic on $X \setminus K$, $u = 0$ on K such that u is harmonic outside of a finite subset of $X \setminus K$. Then there exists a monotone decreasing function $g : (0,\infty) \to (0,\infty)$ such that for all Riemann-integrable functions f*

$$\frac{1}{2}\sum_{x\sim y}(u(x)-u(y))\int_{u(y)}^{u(x)} f(t)\,\mathrm{d}t = \int_{\inf u}^{\sup u} f(t)g(t)\,\mathrm{d}t.$$

If u is harmonic on X, then g is constant. If u is harmonic outside of a finite set, then g is piecewise constant with finitely many jumps.

Proof. For $t \in (\inf u, \sup u)$ let

$$g(t) := \sum_{\substack{x\sim y\\ u(y)<t\leq u(x)}} (u(x)-u(y)),$$

and for $0 < s \leq t$, set

$$A(s,t) := \{x \in X \mid s < u(x) \leq t\}.$$

By the properness of u, the set $A(s,t)$ is finite for all $s \leq t$. Moreover, $A(s,t) \subseteq X \setminus K$ for $s > 0$ and, therefore, u is super-harmonic on $A(s,t)$.

We first show that $0 < g(t) < \infty$ for all $t \in (\inf u, \sup u)$. Let $x, y \in X$ be such that $x \sim y$ and $u(y) < t \leq u(x)$, i.e., $x \in X \setminus K$. Then the local Harnack inequality (Lemma 9.3.3), yields

$$u(x) \geq t > u(y) \geq \frac{1}{D} u(x) \geq \frac{t}{D},$$

for $y \in X \setminus K$ since u is super-harmonic on $X \setminus K$. Hence, for $s \leq t/D$, we have that all $x, y \in X \setminus K$, $x \sim y$ and $u(y) < t \leq u(x)$ are contained in $A(s,t)$ which is a finite set. Furthermore, there are only finitely many $x \in X$, $y \in K$, $x \sim y$ with $0 = u(y) < t \leq u(x)$ since K is finite. Thus, the value $g(t)$ is given by a finite sum and, therefore, it is finite. On the other hand, g does not vanish since due to the connectedness of the graph, the set of edges (x, y), i.e., $x \sim y$, such that $u(y) < t \leq u(x)$ is never empty for $t \in (\inf u, \sup u)$.

Next, we show that g is a monotone decreasing simple function. Let $s \leq t$ be given. Then the edges (x, y) for which terms that occur in both of the sums of $g(s)$ and $g(t)$ are included in the set

$$\{(x,y) \in X \times X \mid u(y) < s \leq t \leq u(x)\}.$$

Thus,

$$\begin{aligned} g(t) - g(s) &= \sum_{\substack{x \sim y \\ s \leq u(y) \leq t \leq u(x)}} (u(x) - u(y)) - \sum_{\substack{x \sim y \\ u(y) < s \leq u(x) < t}} (u(x) - u(y)) \\ &= - \sum_{x \in A(s,t)} \sum_{y \sim x} (u(x) - u(y)) = - \sum_{x \in A(s,t)} \Delta u(x) \leq 0 \end{aligned}$$

where the last inequality follows from super-harmonicity of u. Thus, the monotonicity of g is proven. Moreover, the equalities show that the value of g changes if and only if a vertex where u is not harmonic enters $A(s,t)$ by increasing t (or decreasing s). Since this happens only finitely many times by the assumptions on u, the function g is simple.

With the function g at our disposal, we conclude the proof. We denote the characteristic function of the interval $I_{x,y} := \big(u(x) \wedge u(y), u(x) \vee u(y)\big]$ by $1_{x,y}$. For an edge (x, y) we have $u(y) < t \leq u(x)$ or $u(x) < t \leq u(y)$ if and only if $t \in I_{x,y}$. Consequently, we get

$$\sum_{x\sim y}(u(x)-u(y))\int_{u(y)}^{u(x)} f(t)\,\mathrm{d}t = \sum_{x\sim y}|u(x)-u(y)|\int_{\inf u}^{\sup u} f(t)1_{x,y}(t)\,\mathrm{d}t.$$

We apply Tonelli's theorem in order to arrive at

$$\begin{aligned}
\ldots &= \int_{\inf u}^{\sup u} f(t)\sum_{x\sim y}|u(x)-u(y)|1_{x,y}(t)\,\mathrm{d}t \\
&= 2\int_{\inf u}^{\sup u} f(t)\sum_{\substack{x\sim y\\ u(y)<t\le u(x)}}|u(x)-u(y)|\,\mathrm{d}t \\
&= 2\int_{\inf u}^{\sup u} f(t)\sum_{\substack{x\sim y\\ u(y)<t\le u(x)}}(u(x)-u(y))\,\mathrm{d}t \\
&= 2\int_{\inf u}^{\sup u} f(t)g(t)\,\mathrm{d}t,
\end{aligned}$$

and the coarea formula is proven. □

9.3.2.4 Super-solution Construction and Ground-State Transform

The formulas for the critical Hardy weight stated in Theorem 9.3.1 hinge on the following chain-type rule for the square root function. This rule was first used in [BHL$^+$15] in the context of Li–Yau inequalities as well as in [KPP18b] where it was utilized for Hardy inequalities.

Lemma 9.3.6 (Super-solution construction) *If u is a positive super-harmonic function on a connected subset U of X, then $u^{1/2}$ is super-harmonic on U, and we have*

$$\frac{\Delta u^{1/2}}{u^{1/2}}(x) = \frac{1}{2u(x)}\sum_{y\sim x}\left(u(x)^{1/2}-u(y)^{1/2}\right)^2$$

at all points x where $\Delta u(x) = 0$.

Proof. Note that it is a straightforward application of Harnack's inequality (Lemma 9.3.3) that a positive super-harmonic function must in fact be strictly positive, that is, $u(x) > 0$ for all $x \in U$. Hence, we do not divide by zero in the formula above. To show the validity of the lemma, we compute

$$\begin{aligned}
2(u^{1/2}\Delta u^{1/2})(x) &= \sum_{y\sim x}\Big(\big(u(x)-u(y)\big)+\big(u(x)-2(u(x)u(y))^{1/2}+u(y)\big)\Big) \\
&= \Delta u(x) + \sum_{y\sim x}\left(u(x)^{1/2}-u(y)^{1/2}\right)^2.
\end{aligned}$$

□

We use positive super-harmonic functions in order to construct Hardy weights. For this purpose, we employ the ground-state transform as stated in the following lemma. Various versions of the ground-state transform for which the following is a special case are found in the literature [FLW14, FS08, HK11, KPP17].

Lemma 9.3.7 (Ground-state transform) *Let $v \in C(X)$ be a function such that $(\Delta - w)v = 0$ on the set where v does not vanish. Then,*

$$(q - w)(v\varphi) = \frac{1}{2}\sum_{x \sim y} v(x)v(y)\Big(\varphi(x) - \varphi(y)\Big)^2, \qquad \varphi \in C_c(X).$$

Proof. Since $(\Delta - w)v = 0$ on the set where v does not vanish, we get by a straightforward computation that for all x in this set

$$(\Delta - w)(v\varphi)(x) = \sum_{y \sim x} v(y)(\varphi(x) - \varphi(y)). \tag{9.3.1}$$

Let $\varphi \in C_c(X)$. Green's formula (that can be easily verified using algebraic manipulations) applied twice and (9.3.1) yield

$$\begin{aligned}(q - w)(v\varphi) &= \sum_{x \in X} (v\varphi)(x)(\Delta - w)(v\varphi)(x) \\ &= \sum_{x \in X} \varphi(x) \sum_{y \sim x} v(x)v(y)(\varphi(x) - \varphi(y)) \\ &= \frac{1}{2}\sum_{x \sim y} v(x)v(y)\big(\varphi(x) - \varphi(y)\big)^2.\end{aligned}$$

This finishes the proof. □

With the ground-state transform at our disposal, we can construct Hardy weights as follows.

Lemma 9.3.8 *Let $U \subseteq X$ be a connected set. If u is a positive super-harmonic function on U, then $u^{1/2}$ is a strictly positive $(\Delta - w)$-harmonic function on U, where*

$$w = \frac{\Delta u^{1/2}}{u^{1/2}}.$$

Moreover, w is a Hardy weight on U, that is, $(q - w) \geq 0$ on $C_c(U)$.

Proof. Note that by the local Harnack inequality (Lemma 9.3.3) the function u is strictly positive on U and so is $u^{1/2}$. It follows from the super-solution construction (Lemma 9.3.6) that $w \geq 0$. Furthermore, we note that

$$(\Delta - w)u^{1/2} = \Delta u^{1/2} - \frac{\Delta u^{1/2}}{u^{1/2}} u^{1/2} = 0,$$

so, $u^{1/2}$ is $(\Delta - w)$-harmonic. We use the ground-state transform (Lemma 9.3.7) to arrive at

$$(q - w)(\varphi) = \frac{1}{2} \sum_{x \sim y} u^{1/2}(x) u^{1/2}(y) \left(\frac{\varphi(x)}{u^{1/2}(x)} - \frac{\varphi(y)}{u^{1/2}(y)} \right)^2 \geq 0$$

for all $\varphi \in C_c(U)$. In other words, w is a Hardy weight on U. □

9.3.2.5 Criticality of Hardy Weights for Graphs

With the tools developed above at hand, we are in the position to prove the main result of this section.

Proof of Theorem 9.3.1. Let $K \subseteq X$ be a finite set such that $X \setminus K$ is connected. Let $u : X \to [0, \infty)$ be a proper function which is super-harmonic on $X \setminus K$ and $u = 0$ on K as well as harmonic outside of a finite subset of $X \setminus K$. By the super-solution construction (Lemma 9.3.6), the function $v := u^{1/2}$ is strictly positive and super-harmonic on $X \setminus K$. Moreover, by Lemma 9.3.8, the function $w := (\Delta u^{1/2})/u^{1/2}$ is a non-negative Hardy weight on $C_c(X \setminus K)$ and $(\Delta - w)v = 0$ on $X \setminus K$. In light of Lemma 9.3.4, to establish criticality, it is enough to construct a null-sequence with respect to the form $q - w$. In view of the ground-state transform, Lemma 9.3.7, and the strict positivity of v, it is enough to show that there is a sequence (e_n), $e_n \in C_c(X \setminus K)$ such that $e_n \to 1$ pointwise on $X \setminus K$ and $(q - w)_v(e_n) \to 0$, where

$$(q - w)_v(\varphi) := (q - w)(v\varphi) = \frac{1}{2} \sum_{x \sim y} (u(x)u(y))^{1/2} \big(\varphi(x) - \varphi(y)\big)^2$$

for $\varphi \in C_c(X)$ since $v = u^{1/2}$.

Let $\psi_n : \mathbb{R} \to \mathbb{R}$,

$$\psi_n(t) := \left(2 + \frac{\log t}{\log n}\right) 1_{[\frac{1}{n^2}, \frac{1}{n}]}(t) + 1_{[\frac{1}{n}, n]}(t) + \left(2 - \frac{\log t}{\log n}\right) 1_{[n, n^2]}(t),$$

and define

$$e_n := \psi_n \circ u.$$

Clearly, each ψ_n has compact support in $(0, \infty)$. By properness of u, we have $\sup_{X \setminus K} u = \infty$ or $\inf_{X \setminus K} u = 0$ which in turn gives $e_n \in C_c(X \setminus K)$ for each $n \in \mathbb{N}$. By construction, we have $e_n \to 1$ pointwise on $X \setminus K$ as $n \to \infty$. It remains to show that $(q - w)_v(e_n) \to 0$ as $n \to \infty$. In order to do so, we will apply the coarea formula (Lemma 9.3.5). For this purpose, we set

$$c(x, y) := \frac{(u(x)u(y))^{1/2}\Big(\psi_n(u(x)) - \psi_n(u(y))\Big)^2}{(u(x) - u(y))\int_{u(y)}^{u(x)} t\psi_n'(t)^2\,\mathrm{d}t}$$

whenever the denominator is non-zero and $c(x, y) = 0$, otherwise. It follows that

$$(q - w)_v(e_n) = \frac{1}{2}\sum_{x\sim y} c(x, y)\,(u(x) - u(y))\left(\int_{u(y)}^{u(x)} t\psi_n'(t)^2\,\mathrm{d}t\right).$$

Our next aim is to prove the estimate $0 \le c(x, y) \le \sqrt{D}$. As a first step in this direction we prove the following claim.

Claim. For $x \sim y$

$$\frac{\big(\psi_n(u(y)) - \psi_n(u(x))\big)^2}{\int_{u(x)}^{u(y)} t\psi_n'(t)^2\,\mathrm{d}t} \le \log u(x) - \log u(y).$$

Proof of the claim. Notice that by the local Harnack inequality (Lemma 9.3.3), we have $\sup_{x\sim y} u(x)/u(y) \le D$, where D is the uniform vertex degree bound for the graph X. Hence, for $n > \sqrt{D}$, we have $u(x), u(y) \in (0, n]$ or $u(x), u(y) \in [1/n, \infty)$ for $x \sim y$.

Let us assume first that $x \sim y$, $u(x), u(y) \in (0, n]$, and $u(y) < u(x)$. The proof follows from two basic calculations. Consider first the contraction $C_{[0,1]}(t) = 0 \vee t \wedge 1$, $t \in \mathbb{R}$, which satisfies $C_{[0,1]}(t) - C_{[0,1]}(s) \le t - s$ for $t \ge s$, and observe that

$$\psi_n = C_{[0,1]} \circ \left(2 + \frac{\log}{\log n}\right) \quad \text{on } [0, n].$$

Hence, under the assumption $u(x), u(y) \in (0, n]$ and $u(y) < u(x)$ we get

$$\begin{aligned}
&\psi_n(u(x)) - \psi_n(u(y)) \\
&= \left(C_{[0,1]} \circ \left(2 + \frac{\log}{\log n}\right) \circ u\right)(x) - \left(C_{[0,1]} \circ \left(2 + \frac{\log}{\log n}\right) \circ u\right)(y) \\
&\le \frac{1}{\log n}\left(\log(u(x)) - \log(u(y))\right).
\end{aligned}$$

Secondly, we get by the virtue of the fundamental theorem of calculus and since ψ_n is constant on $[0, 1/n^2]$ and $[1/n, n]$

$$\frac{(\psi_n(u(x)) - \psi_n(u(y)))}{\int_{u(y)}^{u(x)} t\psi_n'(t)^2 \,\mathrm{d}t} = \frac{\int_{u(y)}^{u(x)} \psi_n'(t)\,\mathrm{d}t}{\int_{u(y)}^{u(x)} t\psi_n'(t)^2 \,\mathrm{d}t}$$
$$= \log n \cdot \frac{\int_{u(y)\vee 1/n^2}^{u(x)\wedge 1/n} \log'(t)\,\mathrm{d}t}{\int_{u(y)\vee 1/n^2}^{u(x)\wedge 1/n} t\log'(t)^2\,\mathrm{d}t} = \log n.$$

Hence, the statement follows. The considerations for the case $u(x), u(y) \in [1/n, \infty)$ follow along the same lines. This proves the claim. □

Combining the claim with the elementary inequality

$$\frac{\log b - \log a}{b - a} \leq \log'(a) = \frac{1}{a}, \quad 0 < a \leq b < \infty,$$

we obtain

$$c(x, y) \leq \frac{(u(x)u(y))^{1/2}(\log u(x) - \log u(y))}{(u(x) - u(y))} \leq \sup_{x\sim y} \left(\frac{u(x)}{u(y)}\right)^{1/2} \leq \sqrt{D}$$

for all $n > \sqrt{D}$ and $x, y \in X$ with $x \sim y$.

We will now use the coarea formula (Lemma 9.3.5), applied to $f(t) = t\psi_n'(t)^2$. Hence, there exists a constant C_0 which only depends on the graph (in particular on the vertex degree bound D) and the function u such that

$$\begin{aligned}
&(q - w)(ve_n)\\
&\quad = (q - w)_v(e_n) \leq \sqrt{D} \sum_{x\sim y} (u(x) - u(y))\Big(\int_{u(y)}^{u(x)} t\psi_n'(t)^2\,\mathrm{d}t\Big)\\
&\quad \leq C_0 \int_{\inf u}^{\sup u} t\psi_n'(t)^2\,\mathrm{d}t = \frac{C_0}{(\log n)^2}\Big(\int_{1/n^2}^{1/n} \frac{\mathrm{d}t}{t} + \int_n^{n^2} \frac{\mathrm{d}t}{t}\Big) = \frac{2C_0}{\log n}
\end{aligned}$$

for all $n \in \mathbb{N}$. The term on the right-hand side tends to 0 as $n \to \infty$. Thus, (ve_n) is indeed a null-sequence for $q - w$ which implies the criticality of the form $q - w$ on $X \setminus K$. □

9.3.3 Examples of Graphs

The first example is found in all detail in [KPP18b, KPP18a]. It shows that with the super-solution construction, Hardy's original inequality can be improved via higher-order terms of the Hardy weight.

Example 9.3.9 (The integer half-line) Let $X = \mathbb{N}_0$ and consider the graph given by the edge relation $n \sim m$ if and only if $|n - m| = 1$. Then the function

u given by $u(n) = n$ is harmonic on $\mathbb{N}$ and 0 on the set $K = \{0\}$. Hence, by a direct calculation and the Taylor expansion we obtain

$$\begin{aligned} w(n) &= \frac{\Delta n^{1/2}}{n^{1/2}} = 2 - \left(1 + \frac{1}{n}\right)^{1/2} - \left(1 - \frac{1}{n}\right)^{1/2} \\ &= \sum_{k=1}^{\infty} \binom{4k}{2k} \frac{1}{(4k-1)2^{(4k-1)}} \frac{1}{n^{2k}} \end{aligned}$$

for $n \geq 2$ and $w(1) = 2 - \sqrt{2}$. By Theorem 9.3.1 the function w is a critical Hardy weight. Note that the classical Hardy weight is given by $w_{\text{class}}(n) = 1/(4n^2)$, $n \geq 1$ (see [KMP06]). Our estimate improves this inequality by the positive higher-order terms.

The next example is also discussed in [KPP18b].

Example 9.3.10 (The integer lattice) Let $X = \mathbb{Z}^d$, $d \geq 3$, and consider the graph such that $x \sim y$ if and only if $|x - y| = 1$. Then the Green function $G = G(\cdot, 0)$ has the following asymptotics

$$G(x) = \frac{C_1(d)}{|x|^{d-2}} + C_2(d)\left(\sum_{i=1}^{d} \left(\frac{x_i}{|x|}\right)^4 - \frac{3}{d+2}\right)\frac{1}{|x|^d} + \mathcal{O}\left(\frac{1}{|x|^{d+2}}\right),$$

as $|x| \to \infty$ (see [Uch98]). Then G satisfies $\Delta G(0) = 1$, $\Delta G = 0$ otherwise, and $G(x) \to 0$ as $|x| \to \infty$. Thus, Theorem 9.3.1 yields that

$$w(x) = \frac{\Delta G^{1/2}(x)}{G^{1/2}(x)}, \qquad x \in \mathbb{Z}^d,$$

is a critical Hardy weight. Furthermore, a straightforward but tedious calculation yields

$$w(x) = \frac{(d-2)^2}{4}\frac{1}{|x|^2} + \mathcal{O}\left(\frac{1}{|x|^3}\right).$$

as $|x| \to \infty$ (see [KPP18b, Section 7.1]).

Example 9.3.11 (Rooted regular trees) For $k \geq 2$, we denote by (T_k, o) the $(k+1)$-regular rooted tree, where every vertex has exactly k forward neighbours, each vertex distinct from the root vertex o has one backward neighbour, and the root does not have a backward neighbour. Denote the combinatorial graph distance by d. Then the function u given by $u(x) = k^{-d(o,x)}$ is Δ-harmonic on $T_k \setminus \{o\}$ and $\Delta u(o) = k - 1 \geq 0$. Thus, the function w given by

$$w(x) = \frac{\Delta u^{1/2}(x)}{u^{1/2}(x)} = \begin{cases} k+1-2\sqrt{k}, & x \neq o, \\ k-\sqrt{k}, & x = o, \end{cases}$$

is a critical Hardy weight. It is well known that the number $\lambda_0 = k+1-2\sqrt{k}$ is the bottom of the spectrum of the Laplacian on the rooted regular tree. Hence, the function

$$w' = (\sqrt{k}-1)1_o$$

is a critical Hardy weight for the Schrödinger operator $\Delta - \lambda_0$.

Acknowledgments

This chapter is based on talks of Y. P. and F. P. given in the conference 'Analysis and Geometry on Graphs and Manifolds' held at the University of Potsdam in August 2017. Y. P. and F. P. wish to thank the conference's organizers for the invitation to give a talk in the conference, and the German Science Foundation for the financial support. The authors thank Idan Versano for pointing out Remark 9.2.10 to us. M. K. acknowledges the support of the German Science Foundation. Y. P. and F. P. acknowledge the support of the Israel Science Foundation (grant no. 970/15) founded by the Israel Academy of Sciences and Humanities.

References

[Agm83] Shmuel Agmon. On positivity and decay of solutions of second order elliptic equations on Riemannian manifolds. In *Methods of functional analysis and theory of elliptic equations (Naples, 1982)*, pages 19–52. Liguori, Naples, 1983.

[Anc02] Alano Ancona. Some results and examples about the behavior of harmonic functions and Green's functions with respect to second order elliptic operators. *Nagoya Math. J.*, 165:123–158, 2002.

[BEL15] Alexander A. Balinsky, W. Desmond Evans, and Roger T. Lewis. *The analysis and geometry of Hardy's inequality*. Universitext. Springer, Cham, 2015.

[BGGP] Elvise Berchio, Debdip Ganguly, Gabriele Grillo, and Yehuda Pinchover. An optimal improvement for the hardy inequality on the hyperbolic space and related manifolds. *To appear in Proc. Roy. Soc. Edinburgh, Sect. A, arXiv1711.08423*.

[BHL+15] Frank Bauer, Paul Horn, Yong Lin, Gabor Lippner, Dan Mangoubi, and Shing-Tung Yau. Li-Yau inequality on graphs. *J. Differential Geom.*, 99(3):359–405, 2015.

[BP17] Siegfried Beckus and Yehuda Pinchover. Shnol-type theorem for the Agmon ground state. J. Spectr. Theory (online first) DOI 10.4171/JST/296.

[CFKS87] H. L. Cycon, R. G. Froese, W. Kirsch, and B. Simon. *Schrödinger operators with application to quantum mechanics and global geometry*. Texts and Monographs in Physics. Springer-Verlag, Berlin, study edition, 1987.

[Cha84] Isaac Chavel. *Eigenvalues in Riemannian geometry*, volume 115 of *Pure and Applied Mathematics*. Academic Press, Inc., Orlando, FL, 1984. Including a chapter by Burton Randol, With an appendix by Jozef Dodziuk.

[Dev14] Baptiste Devyver. A spectral result for Hardy inequalities. *J. Math. Pures Appl. (9)*, 102(5):813–853, 2014.

[DFP14] Baptiste Devyver, Martin Fraas, and Yehuda Pinchover. Optimal Hardy weight for second-order elliptic operator: an answer to a problem of Agmon. *J. Funct. Anal.*, 266(7):4422–4489, 2014.

[Dod84] Jozef Dodziuk. Difference equations, isoperimetric inequality and transience of certain random walks. *Trans. Amer. Math. Soc.*, 284(2):787–794, 1984.

[DP16] Baptiste Devyver and Yehuda Pinchover. Optimal L^p Hardy-type inequalities. *Ann. Inst. H. Poincaré Anal. Non Linéaire*, 33(1):93–118, 2016.

[FLW14] Rupert L. Frank, Daniel Lenz, and Daniel Wingert. Intrinsic metrics for non-local symmetric Dirichlet forms and applications to spectral theory. *J. Funct. Anal.*, 266(8):4765–4808, 2014.

[FS08] Rupert L. Frank and Robert Seiringer. Non-linear ground state representations and sharp Hardy inequalities. *J. Funct. Anal.*, 255(12):3407–3430, 2008.

[Gri99] Alexander Grigor'yan. Analytic and geometric background of recurrence and non-explosion of the Brownian motion on Riemannian manifolds. *Bull. Amer. Math. Soc. (N.S.)*, 36(2):135–249, 1999.

[Gri09] Alexander Grigor'yan. *Heat kernel and analysis on manifolds*, volume 47 of *AMS/IP Studies in Advanced Mathematics*. American Mathematical Society, Providence, RI; International Press, Boston, MA, 2009.

[GT01] David Gilbarg and Neil S. Trudinger. *Elliptic partial differential equations of second order*. Classics in Mathematics. Springer-Verlag, Berlin, 2001. Reprint of the 1998 edition.

[Has80] R. Z. Has'minskii. *Stochastic stability of differential equations*, volume 7 of *Monographs and Textbooks on Mechanics of Solids and Fluids: Mechanics and Analysis*. Sijthoff & Noordhoff, Alphen aan den Rijn – Germantown, Md., 1980. Translated from the Russian by D. Louvish.

[HK11] Sebastian Haeseler and Matthias Keller. Generalized solutions and spectrum for Dirichlet forms on graphs. In *Random walks, boundaries and spectra*, volume 64 of *Progr. Probab.*, pages 181–199. Birkhäuser/Springer Basel AG, Basel, 2011.

[KMP06] Alois Kufner, Lech Maligranda, and Lars-Erik Persson. The prehistory of the Hardy inequality. *Amer. Math. Monthly*, 113(8):715–732, 2006.

[KP18] Hynek Kovařík, and Yehuda Pinchover, On minimal decay at infinity of Hardy-weights to appear in Commun. Contemp. Math. (2019). Preprint arXiv:1812.01849.

[KPP17] Matthias Keller, Yehuda Pinchover, and Felix Pogorzelski. Criticality theory for Schrödinger operators on graphs. *J. Spectr. Theory* 10 (2020), 73–114.

[KPP18a] Matthias Keller, Yehuda Pinchover, and Felix Pogorzelski. An improved discrete Hardy inequality. *Amer. Math. Monthly*, 125(4):347–350, 2018.

[KPP18b] Matthias Keller, Yehuda Pinchover, and Felix Pogorzelski. Optimal Hardy inequalities for Schrödinger operators on graphs. *Comm. Math. Phys.*, 358(2):767–790, 2018.

[LaPi16] Pier Domenico Lamberti, and Yehuda Pinchover. L^p Hardy inequality on $C^{1,\gamma}$ domains. *Ann. Sc. Norm. Super. Pisa Cl. Sci.*, (5) 19 (2019), 1135–1159.

[LSV09] Daniel Lenz, Peter Stollmann, and Ivan Veselić. The Allegretto-Piepenbrink theorem for strongly local Dirichlet forms. *Doc. Math.*, 14:167–189, 2009.

[Mat01] Hiroyuki Matsumoto. Closed form formulae for the heat kernels and the Green functions for the Laplacians on the symmetric spaces of rank one. *Bull. Sci. Math.*, 125(6-7):553–581, 2001. Rencontre Franco-Japonaise de Probabilités (Paris, 2000).

[OK90] B. Opic and A. Kufner. *Hardy-type inequalities*, volume 219 of *Pitman Research Notes in Mathematics Series*. Longman Scientific & Technical, Harlow, 1990.

[Pin95] Ross G. Pinsky. *Positive harmonic functions and diffusion*, volume 45 of *Cambridge Studies in Advanced Mathematics*. Cambridge University Press, Cambridge, 1995.

[PT06] Yehuda Pinchover and Kyril Tintarev. A ground state alternative for singular Schrödinger operators. *J. Funct. Anal.*, 230(1):65–77, 2006.

[PW84] Murray H. Protter and Hans F. Weinberger. *Maximum principles in differential equations*. Springer-Verlag, New York, 1984. Corrected reprint of the 1967 original.

[Sul87] Dennis Sullivan. Related aspects of positivity in Riemannian geometry. *J. Differential Geom.*, 25(3):327–351, 1987.

[Tak14] Masayoshi Takeda. Criticality and subcriticality of generalized Schrödinger forms. *Illinois J. Math.*, 58(1):251–277, 2014.

[Uch98] Kôhei Uchiyama. Green's functions for random walks on $\mathbb{Z}^N$. *Proc. London Math. Soc. (3)*, 77(1):215–240, 1998.

[Woe00] Wolfgang Woess. *Random walks on infinite graphs and groups*, volume 138 of *Cambridge Tracts in Mathematics*. Cambridge University Press, Cambridge, 2000.

10

Neumann Domains on Graphs and Manifolds

Lior Alon, Ram Band, Michael Bersudsky, and Sebastian Egger

Abstract

A Laplacian eigenfunction on a manifold or a metric graph imposes a natural partition of the manifold or the graph. This partition is determined by the gradient vector field of the eigenfunction (on a manifold) or by the extremal points of the eigenfunction (on a graph). The sub-manifolds (or subgraphs) of this partition are called Neumann domains. Their counterparts are the well-known nodal domains. This chapter reviews the subject of Neumann domains, as appears in [3, 9, 10, 58, 76] and points out some open questions and conjectures. The chapter concerns both manifolds and metric graphs and the exposition allows for a comparison between the results obtained for each of them.

10.1 Introduction

Given a Laplacian eigenfunction on a manifold or a metric graph, there is a natural partition of the manifold or the graph. The partition is dictated by the gradient vector field of the eigenfunction (on a manifold) or by the extremal points of the eigenfunction (on a graph). The sub-manifolds (or subgraphs) of such a partition are called *Neumann domains* and the separating lines (or points in the case of a graph) are called *Neumann lines* (or points). The counterpart of this partition is the nodal partition (with the same terminology of nodal domains, nodal lines, and nodal points). This latter partition is extensively studied in the last two decades or so (though interesting results on nodal domains appeared throughout all of the twentieth century and even earlier). When restricting an eigenfunction to a single nodal domain, one gets an eigenfunction of that domain with Dirichlet boundary conditions. Similarly, when restricting an eigenfunction to a Neumann domain, one gets a Neumann eigenfunction of that domain (Lemmata 10.3.1, 10.8.1), which explains the name *Neumann* domain and shows the most basic linkage between nodal domains and Neumann domains.

Neumann domains form a very new topic of study in spectral geometry. They were first mentioned in a paragraph of a manuscript by Zelditch [76]. Shortly afterwards (and independently) a paper by McDonald and Fulling was dedicated to Neumann domains [58]. Since then two additional papers contributed to this topic; by one of the authors with Fajman [10] and by two of the authors with Taylor [9]. The first part of this chapter serves as an exposition of the known results for Neumann domains on two-dimensional manifolds, adding a few supplementary new results and proofs. The second part focuses on Neumann domains on metric graphs and reviews the results which will appear in [3].[1] We aim to point out similarities and differences between Neumann domains on manifolds and those on graphs. For this purpose, each of the two parts of the chapter is divided to exactly the same sub-topics: definitions, topology, geometry, spectral position, and count. We also include an appendix which contains a short review of relevant results in basic Morse theory, useful for the manifold part of the chapter. The summary of the chapter provides guidelines for comparison between the manifold results and the graph results. Such a comparison had taught us a great deal in what concerns to the field of nodal domains and yielded a wealth of new results both on manifolds and graphs. As an example we only mention the topic of nodal partitions and refer the interested readers to [7, 17, 19, 21, 22, 25, 30, 43, 45] in order to learn on the evolution of this research direction. In addition to that, we believe that it is beneficial to compare problems between the fields of nodal domains and Neumann domains. We point out such similarities and differences throughout the chapter.

Although new in spectral theory, Neumann domains were used in computational geometry, where they are known as Morse–Smale complexes (see the book [78] or [23] for an extensive review). They are used as a tool to analyse sets of measurements on certain spaces and for getting a good qualitative and quantitative acquaintance with the measured functions [29, 33, 34]. Another field of relevance is computer graphics, where Morse–Smale complexes of Laplacian eigenfunctions are applied for surface segmentation [32, 42, 65]. Interestingly, recently the interaction between the fields of topological data

[1] While writing this chapter, we became aware that there is an ongoing research on the related topic of Neumann partitions on graphs. These works in progress are done by Gregory Berkolaiko, James Kennedy, Pavel Kurasov, Corentin Léna, and Delio Mugnolo.

analysis and spectral geometry went the other way around; in [64] the notion of persistence barcodes was used to study topological properties of the sub-level sets of Laplacian eigenfunctions.

PART 1 NEUMANN DOMAINS ON TWO-DIMENSIONAL MANIFOLDS

10.2 Definitions

Let (M, g) be a two-dimensional, connected, orientable, and closed Riemannian manifold. We denote by $-\Delta$ the (negative) self-adjoint Laplace–Beltrami operator. Its spectrum is purely discrete since M is compact. We order the eigenvalues $\{\lambda_n\}_{n=0}^{\infty}$ increasingly, $0 = \lambda_0 < \lambda_1 \leq \lambda_2 \leq \ldots$, and denote a corresponding complete system of orthonormal eigenfunctions by $\{f_n\}_{n=0}^{\infty}$, so that we have

$$-\Delta f_n = \lambda_n f_n. \tag{10.2.1}$$

We assume in the following that the eigenfunctions f are Morse functions, i.e., have no degenerate critical points.[2] We call such an f a *Morse eigenfunction*. Eigenfunctions are generically Morse, as shown in [1, 72]. At this point, we refer the interested readers to the appendix, where some basic Morse theory which is relevant to the chapter is presented.

In order to define Neumann domains and Neumann lines, we introduce the following construction based on the gradient vector field, ∇f. This vector field defines the flow:

$$\begin{aligned} &\varphi : \mathbb{R} \times M \to M, \\ &\partial_t \varphi(t, \boldsymbol{x}) = -\nabla f\big|_{\varphi(t, \boldsymbol{x})}, \\ &\varphi(0, \boldsymbol{x}) = \boldsymbol{x}. \end{aligned} \tag{10.2.2}$$

The following notations are used throughout the chapter. The set of critical points of f is denoted by $\mathscr{C}(f)$; the sets of saddle points and extrema of f are denoted by $\mathscr{S}(f)$ and $\mathscr{X}(f)$; the sets of minima and maxima of f are denoted by $\mathscr{M}_-(f)$ and $\mathscr{M}_+(f)$, respectively.

For a critical point $\boldsymbol{x} \in \mathscr{C}(f)$, we define its stable and unstable manifolds by

$$\begin{aligned} W^s(\boldsymbol{x}) &= \{\boldsymbol{y} \in M \big| \lim_{t\to\infty} \varphi(t, \boldsymbol{y}) = \boldsymbol{x}\} \text{ and} \\ W^u(\boldsymbol{x}) &= \{\boldsymbol{y} \in M \big| \lim_{t\to-\infty} \varphi(t, \boldsymbol{y}) = \boldsymbol{x}\}, \end{aligned} \tag{10.2.3}$$

[2] These are critical points where the determinant of the Hessian vanishes.

respectively. Intuitively, these notions may be visualized in terms of surface topography; the stable manifold, $W^s(\boldsymbol{x})$, may be thought of as a dale (where falling rain droplets would flow and reach $\boldsymbol{x}$) and the unstable manifold, $W^u(\boldsymbol{x})$, as a hill (with opposite meaning in terms of water flow). An interesting scientific account on those appeared by Maxwell already in 1870 [56].

Definition 10.2.1 [10] Let f be a Morse function.

(1) Let $\boldsymbol{p} \in \mathcal{M}_-(f)$, $\boldsymbol{q} \in \mathcal{M}_+(f)$, such that $W^s(\boldsymbol{p}) \cap W^u(\boldsymbol{q}) \neq \emptyset$. Each of the connected components of $W^s(\boldsymbol{p}) \cap W^u(\boldsymbol{q})$ is called a *Neumann domain* of f.
(2) The *Neumann line set* of f is

$$\mathcal{N}(f) := \overline{\bigcup_{\boldsymbol{r} \in \mathcal{S}(f)} W^s(\boldsymbol{r}) \cup W^u(\boldsymbol{r})}. \tag{10.2.4}$$

Note that the definition above may be applied to any Morse function and not necessarily to eigenfunctions. Indeed, some of the results to follow do not depend on f being an eigenfunction. Yet, the spectral theoretic point of view is the one which motivates us to consider the particular case of Laplacian eigenfunctions.

It is not hard to conclude from basic Morse theory that Neumann domains are two-dimensional subsets of M, whereas the Neumann line set is a union of one-dimensional curves on M (see appendix). As an example, see Figure 10.1 which shows an eigenfunction of the flat torus with its partition to Neumann domains. Further properties of Neumann domains and Neumann lines are described in the next section.

Throughout the chapter, we treat only manifolds without boundary, in order to avoid technicalities and ease the reading. It is possible to define Neumann domains for manifolds with boundary and to prove analogous results for those. The interested reader is referred to [10] for such a treatment.

10.3 Topology of Ω and Topography of $f|_\Omega$

Let f be an eigenfunction corresponding to an eigenvalue λ and let Ω be a Neumann domain. The boundary, $\partial\Omega$, consists of Neumann lines, which are particular gradient flow lines (see appendix). As the gradient ∇f is tangential to the Neumann lines, we get that $\hat{n} \cdot \nabla f|_{\partial\Omega} = 0$, where $\hat{n}$ is normal to $\partial\Omega$. As a consequence we have the following lemma.

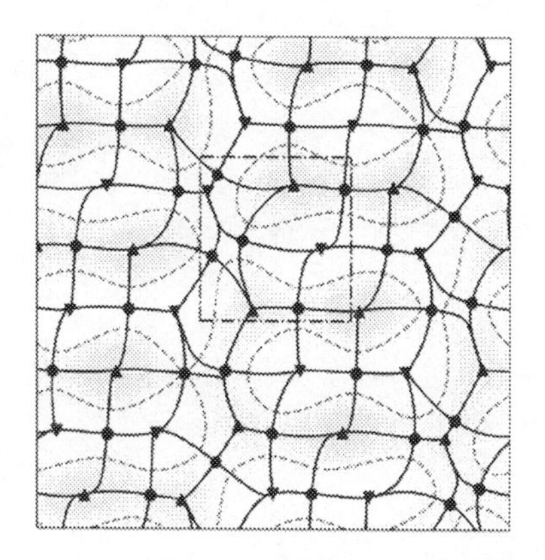

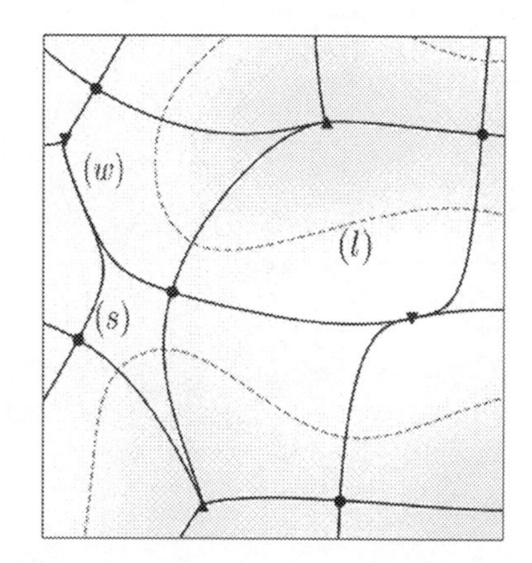

Figure 10.1 Left: An eigenfunction corresponding to eigenvalue $\lambda = 17$ of the flat torus whose fundamental domain is $[0, 2\pi] \times [0, 2\pi]$. Circles mark saddle points and triangles mark extremal points (maxima by triangles pointing upwards and vice versa for minima). The nodal set is drawn as dashed lines and the Neumann line set is marked by solid lines. The Neumann domains are the domains bounded by the Neumann line set.
Right: A magnification of the marked square from the left figure. Three Neumann domains are marked by (s), (l), and (w) according to the three distinguished Neumann domain types described in Section 10.4.1.

Lemma 10.3.1 *[8] $f|_{\Omega}$ is a Neumann eigenfunction of Ω and corresponds to the eigenvalue λ.*

This lemma is the reason for the name *Neumann* domains.

Next, we describe the topological properties of a Neumann domain Ω, as well as the topography of $f|_{\Omega}$. By topography of a function, we mean the information on its level sets and critical points.

Theorem 10.3.2 *[10, Theorem 1.4]*

Let f be a Morse function with a non-empty set of saddle points, $\mathscr{S}(f) \neq \emptyset$.

Let $\boldsymbol{p} \in \mathscr{M}_-(f)$, $\boldsymbol{q} \in \mathscr{M}_+(f)$ with $W^s(\boldsymbol{p}) \cap W^u(\boldsymbol{q}) \neq \emptyset$.

Let Ω be a connected component of $W^s(\boldsymbol{p}) \cap W^u(\boldsymbol{q})$, i.e., Ω is a Neumann domain.

The following properties hold.

(1) The Neumann domain Ω is a simply connected open set.
(2) All critical points of f belong to the Neumann line set, i.e., $\mathscr{C}(f) \subset \mathcal{N}(f)$.
(3) The extremal points which belong to $\overline{\Omega}$ are exactly $\boldsymbol{p}, \boldsymbol{q}$, i.e., $\mathscr{X}(f) \cap \partial\Omega = \{\boldsymbol{p}, \boldsymbol{q}\}$.

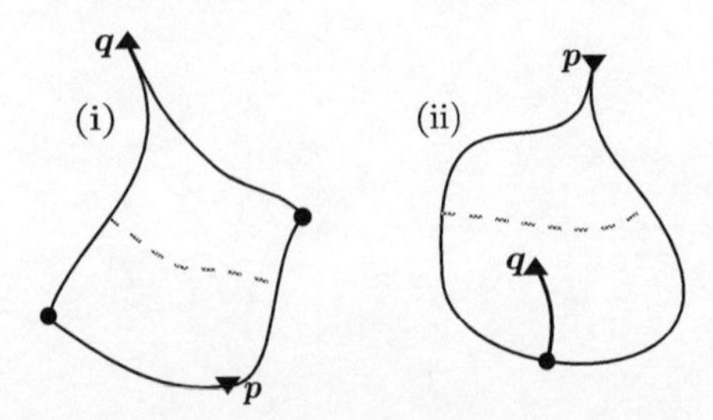

Figure 10.2 Two possible types of Neumann domains for a Morse–Smale eigenfunction. Circles mark saddle points and triangles mark extremal points (maxima by triangles pointing upwards and vice versa for minima). The nodal set is drawn as a dashed line

(4) *If f is a Morse–Smale function*[3] *then $\partial\Omega$ consists of Neumann lines connecting saddle points with $\boldsymbol{p}$ or $\boldsymbol{q}$. In particular, $\partial\Omega$ contains either one or two saddle points (see also Proposition A.7).*
(5) *Let $c \in \mathbb{R}$. such that $f(\boldsymbol{p}) < c < f(\boldsymbol{q})$. $\overline{\Omega} \cap f^{-1}(c)$ is a smooth, non-self-intersecting one-dimensional curve in $\overline{\Omega}$, with its two boundary points lying on $\partial\Omega$.*

This last theorem contains different properties of Neumann domains: claim (1) concerns the topology, claims (2),(3),(4) the critical points, and claim (5) the level sets. A special emphasize should be made for the case when f is a Morse function, which is also an eigenfunction. For Laplacian eigenfunctions we have that maxima are positive and minima are negative, i.e., $f(\boldsymbol{p}) < 0,\ \ f(\boldsymbol{q}) > 0$, in the notation of the theorem. Hence we may choose $c = 0$ in claim (5) and obtain a characterization of the nodal set which is contained within a Neumann domain.

Figure 10.2 shows the two possible schematic shapes of Neumann domains of a Morse–Smale eigenfunction, as implied from the properties above. We complement the figure by noting that there exist Morse functions with Neumann domains of type (ii) but numerical explorations have not revealed any eigenfunction with a Neumann domain of this type.

Let us compare the results above with similar properties of nodal domains. Nodal domains are not necessarily simply connected. On the contrary, it was recently found that random eigenfunctions may have nodal domains of arbitrarily high genus [66]. In addition, there in no upper bound on the number of critical points in a nodal domain. A particular nodal domain may have either

[3] See appendix for the definition of a Morse–Smale function.

minima or maxima (but not both) in its interior and saddle points both in its interior or at its boundary.

10.4 Geometry of Ω

10.4.1 Angles

The angles between Neumann lines meeting at critical points are discussed in [58]. The first two parts of the next proposition summarize the content of Theorems 3.1 and 3.2 in [58] and further generalize their result from the Euclidean case to an arbitrary smooth metric. The third part of the proposition is new and concern the angles between Neumann lines and nodal lines. The proof of the first two parts is almost the same as the one in [58] and we bring it here for completeness.

Proposition 10.4.1 *Let f be a Morse function on a two-dimensional manifold with a smooth Riemannian metric g.*

(1) Let $\mathbf{c}$ be a saddle point of f. Then there are exactly four Neumann lines meeting at $\mathbf{c}$ with angles $\pi/2$.
(2) Let $\mathbf{c}$ be an extremal point of f whose Hessian is not proportional to g. Then any two Neumann lines meet at $\mathbf{c}$ with either angle 0, π, or $\pi/2$.
(3) Further assume that f is a Morse eigenfunction.
Let $\mathbf{c}$ be an intersection point of a nodal line and a Neumann line of f. If $\mathbf{c}$ is a saddle point then the angle between those lines is $\pi/4$. Otherwise, this angle is $\pi/2$.

Proof. We start by some preliminaries that are relevant to proving all parts of the proposition. Let $\boldsymbol{c}$ be an arbitrary critical point of f. We may find a local coordinate system (x, y) around $\boldsymbol{c}$, such that $\boldsymbol{c} = (0, 0)$ and ∂_x, ∂_y is an orthonormal basis for the tangent space $T_{\boldsymbol{c}}M$ with respect to the metric g at $\boldsymbol{c}$. This means, in particular, that in those coordinates, g at $\boldsymbol{c}$ is the identity. Thus, we get that the cosine of the angle between any two vectors, $u, v \in T_{\boldsymbol{c}}M$, where $u = u_x \frac{\partial}{\partial x} + u_y \frac{\partial}{\partial y}$, $v = v_x \frac{\partial}{\partial x} + v_y \frac{\partial}{\partial y}$ is given by the usual Euclidean inner product, $u_x v_x + u_y v_y$, which we abbreviate and denote by $\langle u, v \rangle_{\mathbb{R}^2}$ allowing an abuse of notation.

Next, we analyse the Neumann lines which start or end at $\boldsymbol{c}$. To do that, we keep in mind that Neumann lines are gradient flow lines which start or end at a saddle point (see appendix), so we investigate such gradient flow lines. Using [14, Lemma 4.4] we deduce that the first (matrix-valued) coefficient in the Taylor series expansion of ∇f around $\boldsymbol{c}$ is $\text{Hess} f|_{\boldsymbol{c}}$, so that

$$\nabla f|_{(x,y)} = \operatorname{Hess} f|_c \cdot \begin{pmatrix} x \\ y \end{pmatrix} + O\left(\|(x,y)\|_{\mathbb{R}^2}^2\right).$$

Hence, if we parameterize in this local coordinate system a gradient flow line which starts or ends at $\boldsymbol{c}$ by $\begin{pmatrix} x(t) \\ y(t) \end{pmatrix}$ (so that $\begin{pmatrix} x(t) \\ y(t) \end{pmatrix} \underset{t\to\pm\infty}{\longrightarrow} \boldsymbol{c}$), the gradient flow equations (10.2.2) may be written in the vicinity of $\boldsymbol{c}$ as

$$\begin{pmatrix} x'(t) \\ y'(t) \end{pmatrix} = -\operatorname{Hess} f|_c \cdot \begin{pmatrix} x(t) \\ y(t) \end{pmatrix} + O\left(\|(x(t), y(t))\|_{\mathbb{R}^2}^2\right). \qquad (10.4.1)$$

Since the Hessian is symmetric, we may diagonalize it by an orthonormal change of the coordinates and get

$$\operatorname{Hess} f|_c = \begin{pmatrix} \alpha_x & 0 \\ 0 & \alpha_y \end{pmatrix},$$

where α_x, α_y are both non-zero since f is a Morse function. In those new coordinates, g at $\boldsymbol{c}$ is still the identity. Hence, the assumption in the second part of the proposition, that the Hessian is not proportional to g, is equivalent to $\alpha_x \neq \alpha_y$. In the vicinity of $\boldsymbol{c}$ the gradient flow equations (10.4.1) may now be approximated by

$$\begin{pmatrix} x'(t) \\ y'(t) \end{pmatrix} = \begin{pmatrix} -\alpha_x\, x(t) \\ -\alpha_y\, y(t) \end{pmatrix},$$

where we abuse notation by using (x, y) again to denote the new coordinates which diagonalize the Hessian. The solutions of the above are

$$\begin{pmatrix} x(t) \\ y(t) \end{pmatrix} = \begin{pmatrix} A_x e^{-\alpha_x t} \\ A_y e^{-\alpha_y t} \end{pmatrix}, \quad \text{with } A_x, A_y, t \in \mathbb{R}. \qquad (10.4.2)$$

Consider first the case of $\alpha_x \neq \alpha_y$ both positive, i.e., $\boldsymbol{c}$ is a minimum point. In this case, all the flow lines (10.4.2) asymptotically converge to $\boldsymbol{c}$ as $t \to \infty$. Recall that $\alpha_x \neq \alpha_y$ by assumption. This allows to assume without loss of generality that $\alpha_y > \alpha_x > 0$. If $A_x \neq 0$, we get that asymptotically as $t \to \infty$

$$\begin{pmatrix} x(t) \\ y(t) \end{pmatrix} = e^{-\alpha_x t} \begin{pmatrix} A_x \\ A_y e^{-(\alpha_y - \alpha_x)t} \end{pmatrix} \sim e^{-\alpha_x t} \begin{pmatrix} A_x \\ 0 \end{pmatrix}.$$

Any such flow line is tangential to the $\pm\hat{x}$ direction at $\boldsymbol{c}$. This gives a continuous family of gradient flow lines, some of which are actually also Neumann lines (this depends on whether or not there is a saddle point at their other end, $t \to -\infty$). Hence, the possible angles between any of those Neumann lines at $\boldsymbol{c}$ are either 0 or π. In addition, if $A_x = 0$, we get a gradient flow line which is tangential to the $\pm\hat{y}$ direction at $\boldsymbol{c}$. This gradient flow line (which is not necessarily a Neumann line) makes an angle of $\pi/2$ with all others. This proves the second part of the proposition if c is a minimum point. The case of a maximum is proven in exactly the same manner.

Next we prove the first part of the proposition. If $\boldsymbol{c}$ is a saddle point, then α_x, α_y are of different signs. The only gradient flow lines (10.4.2) which start or end at $\boldsymbol{c}$ are those for which either $A_x = 0$ or $A_y = 0$. At $\boldsymbol{c}$, these lines are either tangential to $\hat{x}$ (if $A_y = 0$) or tangential to $\hat{y}$ (if $A_x = 0$). These are indeed Neumann lines, as they are connected to a saddle point ($\boldsymbol{c}$). There are four such Neumann lines, corresponding to all possible sign choices ($A_x = 0$ and A_y is positive/negative or $A_y = 0$ and A_x is positive/negative). The angles between any neighbouring two lines out of the four is therefore $\pi/2$.

Finally, we prove the third part of the proposition. If $\boldsymbol{c}$ is a critical point, with $\nabla f|_{\boldsymbol{c}} = 0$, and $f(\boldsymbol{c}) = 0$ then it must be a saddle point, since maxima of a Laplacian eigenfunction are positive and minima are negative. As f is a Laplace–Beltrami eigenfunction, we get

$$0 = -\lambda f(\boldsymbol{c}) = \Delta f(\boldsymbol{c}) = \mathrm{traceHess} f|_{\boldsymbol{c}}. \tag{10.4.3}$$

The sum of Hessian eigenvalues is therefore zero and we may denote those by $\pm\alpha$. Choosing a coordinate system which diagonalizes the Hessian at $\boldsymbol{c} = (0, 0)$, we get

$$f(x, y) = \frac{1}{2}\left(\alpha x^2 - \alpha y^2\right) + O\left(\|(x(t), y(t))\|^3_{\mathbb{R}^2}\right).$$

This shows that the nodal lines of f at $\boldsymbol{c}$ may be approximated by $y = \pm x$. We have already seen in the previous part of the proof that the Neumann lines which are connected to a saddle point, $\boldsymbol{c}$, are tangential to either the $\hat{x}$ or the $\hat{y}$ axis and this gives an angle of $\pi/4$ between neighbouring Neumann and nodal lines.

If $\boldsymbol{c}$ is not a critical point, then $\nabla f|_{\boldsymbol{c}} \neq 0$ and we may write $\mathrm{d}f(v) = \langle \nabla f|_{\boldsymbol{c}}, v\rangle_{\mathbb{R}^2}$ for every $v \in T_{\boldsymbol{c}}M$. By taking v in the direction of the nodal line, we get that the angle between the Neumann line and the nodal line at $\boldsymbol{c}$ is given in terms of $\langle \nabla f|_{\boldsymbol{c}}, v\rangle_{\mathbb{R}^2}$, as g is the identity at $\boldsymbol{c}$. Now, since f is constant along the nodal line, we have $\mathrm{d}f(v) = 0$, and get that the angle between the nodal line and the Neumann line is $\pi/2$. □

Remark 10.4.2 It is also stated in [58, Theorem 3.1] that an angle of $\pi/2$ between Neumann lines at an extremal point is non-generic (or 'unstable special case', citing [58]). The proof of claim (10.4.1) of the proposition clarifies why it is so.

The angles between Neumann lines may be observed in Figures 10.1 and 10.2. The exact angles in Figure 10.1 are better seen when zooming in (see right part of the figure).

Proposition 10.4.1 allows to classify Neumann domains to three distinguished types, as was suggested in [9]. Each Neumann domain has one maxima and one minima on its boundary. Assume that the Neumann domain is of type (i) as depicted in Figure 10.2, i.e., it does not have an extremal point which is connected only to a single Neumann line. We call a Neumann domain

- star-like if both angles at its extremal points are 0,
- lens-like if both angles at its extremal points are π,
- wedge-like if one of those angles is 0 and the other is π.

Those three types of domains are indicated in Figure 10.1(Right) by (s), (l), (w), correspondingly.

Note that this classification requires a couple of genericity assumptions: that the Hessian at the extremal points is not proportional to the metric and that Neumann lines do not meet perpendicularly at an extremal point (see Remark 10.4.2). Indeed, our numeric explorations reveal that Neumann domains are categorized into those three types [9].

10.4.2 Area to Perimeter Ratio

Definition 10.4.3 [35] Let f be a Morse eigenfunction corresponding to the eigenvalue λ and let Ω be a Neumann domain of f. We define the normalized area to perimeter ratio of Ω by

$$\rho(\Omega) := \frac{|\Omega|}{|\partial\Omega|}\sqrt{\lambda},$$

with $|\Omega|$ being the area of Ω and $|\partial\Omega|$ the total length of its perimeter.

This parameter was introduced in [35] in order to study the geometry of nodal domains. A related quantity, $\frac{\sqrt{|\Omega|}}{|\partial\Omega|}$, is a classical one, and it is known to be bounded from above by $\frac{1}{2\sqrt{\pi}}$ (isoperimetric inequality [36]). The value $\frac{|\Omega|}{|\partial\Omega|}$ has also an interesting geometric meaning [57] – it is $\frac{1}{\pi}$ times the mean chord length of the two-dimensional shape Ω. The mean chord length is defined as follows: consider all the parallel chords in a chosen direction and take their average length. The mean chord length is then the uniform average over all directions of that average length.[4]

[4] We thank John Hannay for pointing out this interesting geometrical meaning to us.

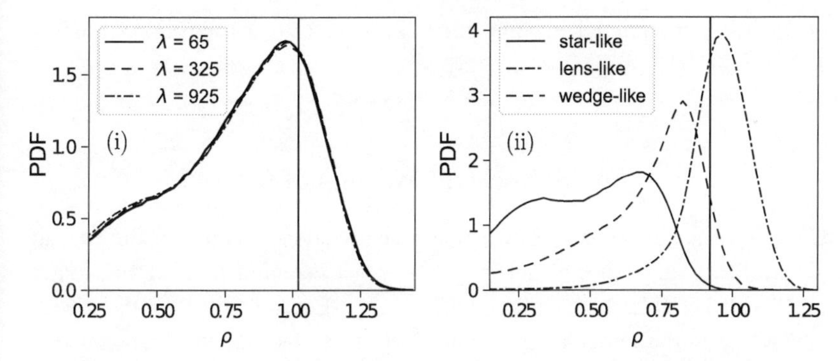

Figure 10.3 (i) A probability distribution function of ρ-values of Neumann domains for three different eigenvalues; (ii) a probability distribution function of ρ-values of Neumann domains for $\lambda = 925$ for lens-like, wedge-like, and star-like domains. The vertical black line marks the value $\rho \approx 0.9206$ (see Proposition 10.5.2,(2)). The numerical data was calculated for approximately 9000 eigenfunctions for each eigenvalue. The right plot is based on data of approximately $8.5 \cdot 10^6$ Neumann domains

There are some numerical explorations, performed to study the values of ρ for Neumann domains. In [9] the numeric was done for random eigenfunctions on the flat torus, where the eigenvalues are highly degenerate. More specifically, for a particular eigenvalue, many random eigenfunctions were chosen out of the corresponding eigenspace and the ρ value was numerically computed for all their Neumann domains. The obtained probability distribution of ρ for three different eigenvalues is shown in Figure 10.3(i). A few interesting observations can be made from those plots. First, it seems that the probability distribution does not depend on the eigenvalue, which raises the question of universality of the ρ parameter. Furthermore, in Figure 10.3(ii) the distribution was drawn separately for each of the three types of Neumann domains mentioned in the previous sub-section (star, lens, and wedge). The lens-like domains tend to get higher ρ values, star-like domains get lower values, and the wedge-like are intermediate. Another conclusion which may be drawn from these plots is related to the spectral position of the Neumann domains, which is described in detail in the next section.

We may compare those results with the ones obtained for the distribution of ρ for nodal domains [35]. It is shown in [35] that for nodal domains of separable eigenfunctions $\frac{\pi}{4} < \rho < \frac{\pi}{2}$. Furthermore, it is numerically observed there that these bounds are satisfied with probability 1 for random eigenfunctions. Also, the calculated probability distribution of ρ for nodal domains looks qualitatively different when compared to Figure 10.3 (see for example Figures 1, 2, 6 in [35]).

10.5 Spectral Position of Ω

Consider a nodal domain Ξ of some eigenfunction f corresponding to an eigenvalue λ. It is known that $f|_{\Xi}$ is the first eigenfunction (ground-state) of Ξ with Dirichlet boundary conditions [31]. Equivalently, λ is the lowest eigenvalue in the Dirichlet spectrum of Ξ. This observation is fundamental in many results concerning nodal domains and their counting. In this section we consider the analogous statement for Neumann domains. Our starting point is Lemma 10.3.1, according to which an eigenvalue λ appears in the Neumann spectrum of each of its Neumann domains. This allows the following definition.

Definition 10.5.1 Let f be a Morse eigenfunction of an eigenvalue λ and let Ω be a Neumann domain of f. We define the spectral position of Ω as the position of λ in the Neumann spectrum of Ω. It is explicitly given by

$$N_{\Omega}(\lambda) := |\{\lambda_n \in \mathrm{Spec}(\Omega) \,:\, \lambda_n < \lambda\}|\,, \tag{10.5.1}$$

where $\mathrm{Spec}(\Omega) := \{\lambda_n\}_{n=0}^{\infty}$ is the Neumann spectrum of Ω, containing multiple appearances of degenerate eigenvalues and including $\lambda_0 = 0$.

Remark.

(1) It can be shown (see [8]) that if Ω is a Neumann domain, then its Neumann spectrum is purely discrete and $f|_{\Omega}$ is a Neumann eigenfunction of Ω. This makes the above well defined.
(2) If λ is a degenerate eigenvalue of Ω, then by this definition the spectral position is the lowest position of λ in the spectrum.
(3) For any Neumann domain, $N_{\Omega}(\lambda) > 0$. Indeed, $N_{\Omega}(\lambda) = 0$ is possible only for $\lambda = 0$, but the zero eigenvalue corresponds to the constant eigenfunction and this does not have Neumann domains at all.

A qualitative feeling on the value of $N_{\Omega}(\lambda)$ might be given by Theorem 10.3.2. This theorem implies that the topography of $f|_{\Omega}$ cannot be too complex; its domain, Ω, is a simply connected domain; $f|_{\Omega}$ has no critical points in the interior of Ω; and its zero set is merely a single simple non-intersecting curve. These observations suggest that $f|_{\Omega}$ might not lie too high in the spectrum of Ω. Such a belief is also apparent in [76], where it is written that possibly, the spectral position of Neumann domains 'often' equals one, just as in the case of nodal domains. Our task is to study the possible values of $N_\Omega(\lambda)$ for various eigenfunctions and their Neumann domains and to investigate to what extent λ is indeed the first non-trivial eigenvalue of Ω ($N_\Omega(\lambda) = 1$). We proceed by relating the spectral position and the area to perimeter ratio (Definition 10.4.3).

10.5.1 Connecting Spectral Position and Area to Perimeter Ratio

The spectral position may be used to bound from above the area to perimeter ratio. This holds as the area to perimeter ratio may be written as

$$\rho(\Omega) = \frac{\sqrt{|\Omega|}}{|\partial\Omega|}\sqrt{|\Omega|\,\lambda},$$

where the first factor is bounded from above by the classical geometric isoperimetric inequality $\frac{\sqrt{|\Omega|}}{|\partial\Omega|} \leq \frac{1}{2\sqrt{\pi}}$ [36], and the second factor is bounded from above by the spectral isoperimetric inequality, once the spectral position is known. We state below the exact result, whose proof is given in [9].

Proposition 10.5.2 *[9] Let f be a Morse eigenfunction corresponding to eigenvalue λ. Let Ω be a Neumann domain of f. We have*

(1) $\rho(\Omega) \leq \sqrt{2N_\Omega(\lambda)}$.
(2) if $N_\Omega(\lambda) = 1$ *then* $\rho(\Omega) \leq \frac{j'}{2} \approx 0.9206$
(3) if $N_\Omega(\lambda) = 2$ *then* $\rho(\Omega) \leq \frac{j'}{\sqrt{2}} \approx 1.3019$,

where $j' \approx 1.8412$ is the first zero of the derivative of J_1, the first Bessel function.

The bounds above may be used to gather information on the spectral position. The calculation of $\rho(\Omega)$ is easier (either numerically or sometimes even analytically) than this of $N_\Omega(\lambda)$. As an example, consider the probability distribution of ρ given in Figure 10.3(i). The distribution was calculated numerically for random eigenfunctions on the torus. It is easy to observe that a substantial proportion of the Neumann domains have a ρ value which is larger than 0.9206, the upper bound given in Proposition 10.5.2(ii). Hence, all those Neumann domains have spectral position which is larger than one, $N_\Omega(\lambda) > 1$. We note that those results seem to be independent of the particular eigenvalue, as the ρ distribution itself seems not to depend on the eigenvalue. Those results are somewhat counter-intuitive, due to what is written above (see discussion after Definition 10.5.1). Furthermore, when calculating the ρ distribution separately for each of the three different types of Neumann domains (Figure 10.3(ii)), the higher ρ values of lens-like domains suggest that the spectral position of those domains is higher. These results call for some further investigation of the spectral position dependence on the shape of the Neumann domains.

10.5.2 Separable Eigenfunctions on the Torus

The general problem of analytically determining the spectral position is quite involved. Yet, there are some interesting results obtained for separable eigenfunctions on the torus, which we review next. We consider the flat torus with fundamental domain $\mathbb{R}^2/\mathbb{Z}^2$ equipped with the Laplace operator. The eigenvalues are

$$\lambda_{a,b} := \frac{\pi^2}{4}\left(\frac{1}{a^2}+\frac{1}{b^2}\right), \tag{10.5.2}$$

where

$$a := \frac{1}{4m_x}, \quad b := \frac{1}{4m_y}, \qquad \text{for } m_x, m_y \in \mathbb{N}. \tag{10.5.3}$$

We consider in the following only the separable eigenfunctions, which may be written as

$$f_{a,b}(x,y) = \cos\left(\frac{\pi}{2a}x\right)\cos\left(\frac{\pi}{2b}y\right). \tag{10.5.4}$$

Half of the Neumann domains of this eigenfunction are star-like and congruent to each other and the other half are lens-like and also congruent (Figure 10.4). We denote those domains by $\Omega_{a,b}^{\text{star}}$ (Figure 10.4(ii)) and $\Omega_{a,b}^{\text{lens}}$ (Figure 10.4(iii)), respectively, and in the following we investigate their spectral position. First, we may consider only the case $b \le a$, thanks to the symmetry of the problem. Second, the spectral position of either $\Omega_{a,b}^{\text{star}}$ or $\Omega_{a,b}^{\text{lens}}$ depends only on the ratio $\frac{b}{a}$, as rescaling both a and b by the same factor amounts to an appropriate rescaling of the Neumann domain together with the restriction of the eigenfunction to it. The next theorem summarizes results on the spectral positions of $\Omega_{a,b}^{\text{star}}$ and $\Omega_{a,b}^{\text{lens}}$ from [9].

Theorem 10.5.3 *[9]*

(1) The set of spectral positions of the *lens*-like domains $\left\{N_{\Omega_{a,b}^{\text{lens}}}\left(\lambda_{a,b}\right)\right\}_{a,b}$ is unbounded. In particular, $N_{\Omega_{a,b}^{\text{lens}}}\left(\lambda_{a,b}\right) \to \infty$ for $\frac{a}{b} \to \infty$.

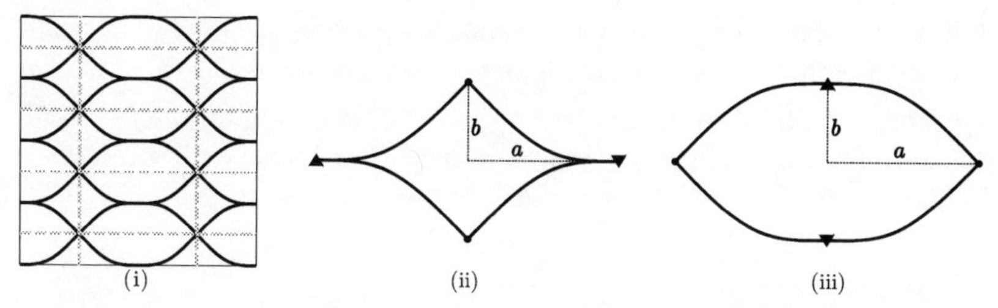

Figure 10.4 (i) Dashed grey lines indicate the nodal set and solid lines indicate the Neumann set of a torus eigenfunction $f(x, y) = \cos(2\pi x)\cos(4\pi y)$. (ii) and (iii) The star-like and lens-like Neumann domains of a separable eigenfunction (10.5.4), with the typical lengths a, b marked as dashed lines. Saddle points are marked by circles and extrema by triangles (maxima by triangles pointing upwards and vice versa for minima)

(2) There exists $c > 0$ such that if $\frac{a}{b} > c$ then the spectral position of the *star-like* domains is one, i.e., $N_{\Omega_{a,b}^{\text{star}}}\left(\lambda_{a,b}\right) = 1$. In addition, $\lambda_{a,b}$ is a simple eigenvalue of $\Omega_{a,b}^{\text{star}}$.

Remark. The condition $\frac{a}{b} > c$ in the second part of the theorem is equivalent to the condition $\frac{m_y}{m_x} > c$ (see (10.5.3)). As $m_x, m_y \in \mathbb{N}$, this means that the claim in the second part of the theorem is valid for a particular proportion of the separable eigenfunctions on the torus. In particular, combining both parts of the theorem, there is a range of a, b values for which $N_{\Omega_{a,b}^{\text{star}}}\left(\lambda_{a,b}\right) = 1$, but $N_{\Omega_{a,b}^{\text{lens}}}\left(\lambda_{a,b}\right)$ is as large as we wish.

The proofs of the two parts of this theorem are of different nature. To prove (1) one assumes by contradiction that the spectral positions $\{N_{\Omega_{a,b}^{\text{lens}}}\left(\lambda_{a,b}\right)\}_{a,b}$ are bounded. This implies an upper bound on the product $\lambda_{a,b}\left|\Omega_{a,b}^{\text{lens}}\right|$. But an asymptotic estimate of this product shows that it diverges for $\frac{a}{b} \to \infty$. Hence the contradiction.

The proof of (2) is based on two main ingredients. The first is a spectral decomposition of the eigenvalue problem on $\Omega_{a,b}^{\text{star}}$, using symmetries [6, 12, 60]. The second ingredient is the comparison of eigenvalue problems resulting from the symmetry reduction mentioned above.

The motivation which stands behind Theorem 10.5.3 is the following. As already mentioned above, it was very natural to believe that generically the spectral position equals one, just as in the case of nodal domains. The first part of the theorem shows that this belief is extremely violated in a particular case. The second part somewhat revives this belief, by showing that this violation which occurs for half of the Neumann domains is compensated by the other half. We wonder whether this compensation holds for all manifolds. For example, can it be that for any manifold, there exists a constant $0 < p \leq 1$, such that each eigenfunction would have at least a p proportion of its Neumann domains with spectral position equals to one? (See Lemma 10.6.3, where a similar assumption is employed.)

10.6 Neumann Domain Count

10.6.1 Bounds

A wealth of results exists on the number of nodal domains. We start this section by bounding the number of Neumann domains from below in terms of the number of nodal domains. Denote the number of Neumann domains of some eigenfunction f by $\mu(f)$ and the number of its nodal domains by $\nu(f)$. Observe that Theorem 10.3.2(5) implies that each Neumann domain intersects with exactly two nodal domains (see discussion following Theorem 10.3.2). This allows to conclude.

Corollary 10.6.1 *[10]*

$$\mu(f) \geq \frac{1}{2}\nu(f). \tag{10.6.1}$$

Next, we equip the Neumann lines with a graph structure which we call the Neumann set graph. This allows to provide further estimates on the number of the Neumann domains. Let f be a Morse function on a closed two-dimensional manifold and consider its Neumann set graph obtained by taking the vertices (V) to be all critical points, the edges (E) are the Neumann lines connecting critical points, and the faces (F) are the Neumann domains. Define the *valency of a critical point*, val $(\boldsymbol{x})$, as the number of Neumann lines which are connected to x.

Proposition 10.6.2 *[10] We have*

$$|E| \leq 4\,|\mathscr{S}(f)|, \tag{10.6.2}$$

$$\mu(f) \leq 2\,|\mathscr{S}(f)|, \tag{10.6.3}$$

where $\mathscr{S}(f)$ is the set of saddle points of f. If we further assume that f is a Morse–Smale function we get equalities in both (10.6.2) and (10.6.3). In addition, we have

$$\mu(f) = \frac{1}{2}\sum_{\boldsymbol{x}\in\mathscr{X}(f)} val\,(\boldsymbol{x}) \geq \frac{1}{2}\,|\mathscr{X}(f)| = \frac{1}{2}\left(\chi(M) + |\mathscr{S}(f)|\right), \tag{10.6.4}$$

where $\mathscr{X}(f)$ is the set of extremal points of f and $\chi(M)$ is the Euler characteristic of the manifold.

The proof of this proposition is done by combining Euler's formula and Morse inequalities for the Neumann set graph.

10.6.2 The Ratio $\frac{\mu_n}{n}$ – Asymptotics and Statistics

The most fundamental result for the nodal domain count is Courant's bound $\frac{\nu_n}{n} \leq 1$, where ν_n is the nodal count of the nth eigenfunction [31]. Following this, Pleijel had shown that $\limsup_{n\to\infty} \frac{\nu_n}{n} \leq (2/j_{0,1})^2$, where $j_{0,1} \approx 2.4048$ is the first zero of J_0, the zeroth Bessel function [61]. Many modern works concern the generalizations or improvements of Pleijel's result, as well as the distribution of the ratio $\frac{\nu_n}{n}$ [15, 24, 26, 28, 40, 44, 54, 63, 68]. The study of the distribution of $\frac{\nu_n}{n}$ was initiated in [24]. This distribution was presented there for separable eigenfunctions on the rectangle and the disc. Later, in [40], a more general calculation of the distribution of $\frac{\nu_n}{n}$ was performed, for the Schrödinger operator on separable systems of any dimension.

In the following, we consider the analogous quantity, $\frac{\mu_n}{n}$, the number of Neumann domains of the nth eigenfunction divided by n. We start by pointing out the connection between $\frac{\mu_n}{n}$, and the spectral position.

Lemma 10.6.3 *Let (M, g) be a two-dimensional, connected, orientable, and closed Riemannian manifold. Assume that there exist $\mathcal{N}$ and ϵ such that*

$$\sum_{\substack{\Omega \text{ s.t.} \\ N_\Omega(\lambda_n) \leq \mathcal{N}}} |\Omega| > \epsilon, \tag{10.6.5}$$

for all λ_n in the spectrum of M, where the sum above is over all Neumann domains (of an eigenfunction) of λ_n whose spectral position is at most $\mathcal{N}$. Then

$$\liminf_{n\to\infty} \frac{\mu_n}{n} \geq \frac{\epsilon}{2\mathcal{N}}. \tag{10.6.6}$$

In addition, for the special case of $\mathcal{N} = 1$ we get

$$\liminf_{n\to\infty} \frac{\mu_n}{n} \geq (2/j')^2\, \epsilon, \tag{10.6.7}$$

where $j' \approx 1.8412$ is the first zero of the derivative of J_1, the first Bessel function.

Proof. We start by proving the special case (10.6.7). The Szegö–Weinberger inequality [69, 74] is $\lambda_1(\Omega)|\Omega| \le \pi(j')^2$. Consider an eigenfunction f_n of M corresponding to an eigenvalue λ_n. For each Neumann domain Ω of f_n, for which $N_\Omega(\lambda_n) = 1$, we have $\lambda_n = \lambda_1(\Omega)$. Combining the Szegö–Weinberger inequality with the assumption in the lemma gives

$$\mu_n \pi(j')^2 \ge \sum_{\substack{\Omega \text{ s.t.} \\ N_\Omega(\lambda_n)=1}} \pi(j')^2 \ge \sum_{\substack{\Omega \text{ s.t.} \\ N_\Omega(\lambda_n)=1}} \lambda_n |\Omega| > \epsilon\lambda_n.$$

Applying Weyl asymptotics, $\lim_{n\to\infty} \frac{\lambda_n}{n} = 4\pi$ [75] we get (10.6.7).

For the case of a general value for $\mathcal{N}$, instead of the Szegö–Weinberger inequality, we employ the bound $\lambda_{\mathcal{N}}(\Omega)|\Omega| \le 8\pi\mathcal{N}$, [53], to get (10.6.6). □

The implication of this lemma is interesting since it shows that the Neumann count tends to infinity. Similar problems are investigated for the nodal count. It was asked a few years ago by Hoffmann–Ostenhof whether $\limsup_{n\to\infty} \nu_n = \infty$ holds for any manifold [73]. Following this, it was shown that for various two-dimensional surfaces, the number of nodal domains tends to infinity with the eigenvalue along almost the entire sequence of eigenvalues [38, 50, 51, 77, 39, 48, 49, 55]. On the other extreme, Jung and Zelditch recently demonstrated the possibility of a bounded number of nodal domains on some three-dimensional manifolds [52].

The validity of inequality (10.6.6) (and hence the validity of assumption (10.6.5)) may be examined by studying the distribution of $\frac{\mu_n}{n}$, which is our next task. We consider the separable eigenfunctions of the flat torus $\mathbb{T}$ with fundamental domain $\mathbb{R}^2/\mathbb{Z}^2$. For those eigenfunctions we calculate the limiting probability distribution of $\frac{\mu_n}{n}$.

Given a couple of natural numbers $m_x, m_y \in \mathbb{N}$, we have that

$$f_{m_x,m_y}(x,y) = \cos(2\pi m_x x)\cos\left(2\pi m_y y\right), \tag{10.6.8}$$

is a separable eigenfunction of the following eigenvalue

$$\lambda_{m_x,m_y} := 4\pi^2\left(m_x^2 + m_y^2\right), \tag{10.6.9}$$

(as in (10.5.2),(10.5.4)). Note that the functions $\sin(2\pi m_x x)\cos\left(2\pi m_y y\right)$, $\cos(2\pi m_x x)\sin\left(2\pi m_y y\right)$, $\sin(2\pi m_x x)\sin\left(2\pi m_y y\right)$ together with (10.6.8) are linearly independent eigenfunctions which belong to the eigenvalue (10.6.9). The set of all those separable eigenfunctions for all possible values of $m_x, m_y \in \mathbb{N}$ form an orthogonal complete set of eigenfunctions on $\mathbb{T}$. We further note that the four eigenfunctions above which correspond to a particular eigenvalue λ_{m_x,m_y} are equal on the torus up to a translation. Hence, all four have the same number of Neumann domains as f_{m_x,m_y} and we denote this number by μ_{m_x,m_y}. With this we may define the following cumulative distribution function

$$F_\lambda(c) := \frac{4}{N_{\mathbb{T}}(\lambda)} \left| \left\{ (m_x, m_y) \in \mathbb{N}^2 \ : \ \lambda_{m_x,m_y} < \lambda \ , \ \frac{\mu_{m_x,m_y}}{N_{\mathbb{T}}(\lambda_{m_x,m_y})} < c \right\} \right| , \tag{10.6.10}$$

where $N_{\mathbb{T}}(\lambda)$ is the spectral position of λ in the torus $\mathbb{T}$, as in (10.5.1), and the factor 4 stands for the four eigenfunctions which correspond to λ_{m_x,m_y}. In words, $F_\lambda(c)$ is the proportion of the separable eigenfunctions with eigenvalue less than λ, whose normalized Neumann count is smaller than c. Its limiting distribution is given by the following.

Proposition 10.6.4

$$\lim_{\lambda\to\infty} F_\lambda(c) = \begin{cases} \frac{1}{2}\int_0^c \frac{1}{\sqrt{1-(\frac{\pi}{4}x)^2}} dx, & 0 \le c < \frac{4}{\pi}, \\ 1, & \frac{4}{\pi} \le c. \end{cases} \tag{10.6.11}$$

Proof. The proof consists of a reduction to a lattice counting problem, which allows to derive the limiting distribution. First, observe that the number of Neumann domains of f_{m_x,m_y} is $\mu_{m_x,m_y} = 8m_x m_y$. This holds since f_{m_x,m_y} is Morse–Smale, so that there is an equality in (10.6.3), and the number of saddle points of f_{m_x,m_y} is the number nodal crossings which is easily shown to be $4m_x m_y$. The symmetry between m_x and m_y in the expression for μ_{m_x,m_y} motivate us to define the set

$$W := \left\{ (m_x, m_y) \in \mathbb{N}^2 \ : \ m_x < m_y \right\},$$

and observe

$$\begin{aligned} \forall\lambda \ \left| \left\{ (m_x, m_y) \in \mathbb{N}^2 : \lambda_{m_x,m_y} < \lambda \right\} \right| &= 2 \left| \left\{ (m_x, m_y) \in W \ : \ \lambda_{m_x,m_y} < \lambda \right\} \right| \\ &+ \left| \left\{ (m_x, m_y) \in \mathbb{N}^2 \ : \ m_x = m_y \text{ and } \lambda_{m_x,m_y} < \lambda \right\} \right| \end{aligned} \tag{10.6.12}$$

Plugging (10.6.12) in (10.6.10) and taking the limit $\lambda \to \infty$ give

$$\lim_{\lambda\to\infty} F_\lambda(c) = \lim_{\lambda\to\infty} \frac{8}{N_{\mathbb{T}}(\lambda)} \left| \left\{ (m_x, m_y) \in W \ : \ \lambda_{m_x,m_y} < \lambda \ , \ \frac{\mu_{m_x,m_y}}{N_{\mathbb{T}}(\lambda_{m_x,m_y})} < c \right\} \right| , \tag{10.6.13}$$

where we use the Weyl asymptotics, $\lim_{\lambda\to\infty} N_{\mathbb{T}}(\lambda) = \frac{\lambda}{4\pi}$, [75], and that the second term in the right-hand side of (10.6.12) grows like $\sqrt{\lambda}$ and hence drops when taking the limit.

We analyse (10.6.13) geometrically. First, $N_{\mathbb{T}}(\lambda)$ counts the number of $\mathbb{Z}^2$ points with non-zero coordinates that lie inside a disc of radius $\sqrt{\lambda}$ around the origin. Hence, it may be written as

$$N_{\mathbb{T}}(\lambda_{m_x,m_y}) = \pi(m_x^2 + m_y^2) + Err(m_x^2 + m_y^2), \tag{10.6.14}$$

where $Err(m_x^2 + m_y^2) = o(m_x^2 + m_y^2)$ [46]. In addition, the point $(m_x, m_y) \in W$ may be characterized by the angle it makes with the x-axis, i.e., $\frac{m_y}{m_x} = \tan\theta_{m_x,,m_y}$, so that

$$\frac{2m_x m_y}{m_x^2 + m_y^2} = 2\cos\theta_{m_x,,m_y} \cdot \sin\theta_{m_x,,m_y} = \sin 2\theta_{m_x,,m_y}. \tag{10.6.15}$$

With (10.6.14) and (10.6.15) we may write

$$\begin{aligned} \frac{\mu_{m_x,m_y}}{N_{\mathbb{T}}(\lambda_{m_x,m_y})} &= \frac{8m_x m_y}{\pi(m_x^2 + m_y^2)\left(1 + Err(m_x^2 + m_y^2)/\pi(m_x^2 + m_y^2)\right)} \\ &= \frac{1}{\left(1 + Err(m_x^2 + m_y^2)/\pi(m_x^2 + m_y^2)\right)} \frac{4}{\pi} \cdot \sin 2\theta_{m_x,,m_y}. \end{aligned}$$

Let $\varepsilon > 0$. Since $Err(m_x^2 + m_y^2) = o(m_x^2 + m_y^2)$, there exists $\Lambda > 0$ such that for all $(m_x, m_y) \in W$ satisfying $4\pi^2\left(m_x^2 + m_y^2\right) > \Lambda$, the following holds

$$\frac{1}{1+\varepsilon}\frac{4}{\pi}\sin 2\theta_{m_x,,m_y} < \frac{\mu_{m_x,m_y}}{N_{\mathbb{T}}(\lambda_{m_x,m_y})} < \frac{1}{1-\varepsilon}\frac{4}{\pi}\sin 2\theta_{m_x,,m_y}. \tag{10.6.16}$$

The limiting cumulative distribution (10.6.13) may be slightly rewritten as

$$\begin{aligned} &\lim_{\lambda\to\infty} F_\lambda(c) \\ &= \lim_{\lambda\to\infty} \frac{8}{N_{\mathbb{T}}(\lambda)} \left| \left\{ (m_x, m_y) \in W \;:\; \Lambda < \lambda_{m_x,m_y} < \lambda\,,\; \frac{\mu_{m_x,m_y}}{N_{\mathbb{T}}(\lambda_{m_x,m_y})} < c \right\} \right|, \end{aligned}$$

where the additional condition $\Lambda < \lambda_{m_x,m_y}$ removes only a finite number of points from the set and does not affect the limit. We may now use (10.6.16) to get the following inequalities by set inclusion:

$$\lim_{\lambda\to\infty} F_\lambda(c) \le \lim_{\lambda\to\infty} \frac{8}{N_{\mathbb{T}}(\lambda)} \left| \left\{ (m_x, m_y) \in W \, : \, \Lambda < \lambda_{m_x, m_y} < \lambda \text{ and} \right.\right.$$

$$\left.\left. \theta_{m_x, m_y} < \frac{1}{2} \arcsin\left(\frac{\pi c(1+\varepsilon)}{4}\right) \right\} \right|, \quad (10.6.17)$$

and

$$\lim_{\lambda\to\infty} F_\lambda(c) \ge \lim_{\lambda\to\infty} \frac{8}{N_{\mathbb{T}}(\lambda)} \left| \left\{ (m_x, m_y) \in W \, : \, \Lambda < \lambda_{m_x, m_y} < \lambda \text{ and} \right.\right.$$

$$\left.\left. \theta_{m_x, m_y} < \frac{1}{2} \arcsin\left(\frac{\pi c(1-\varepsilon)}{4}\right) \right\} \right|, \quad (10.6.18)$$

where in the above we assume that $0 \le c < \frac{4}{\pi}$ and ε is small enough so that $\pi c(1+\varepsilon)/4 \le 1$, and in particular $\arcsin(\pi c(1+\varepsilon)/4)$ is well defined.

We notice that the right-hand sides of (10.6.17) and (10.6.18) correspond to counting integer lattice points which are contained within a certain sector. This number of points grows like the area of the corresponding sector [46], i.e.,

$$\left| \left\{ (m_x, m_y) \in W \, : \, \Lambda < \lambda_{m_x, m_y} < \lambda \, , \, \theta_{m_x, n_y} < \frac{1}{2} \arcsin\left(\frac{\pi c(1\pm\varepsilon)}{4}\right) \right\} \right| =$$

$$= \underbrace{\frac{1}{4} \arcsin\left(\frac{\pi c(1\pm\varepsilon)}{4}\right) \frac{\lambda - \Lambda}{4\pi^2}}_{\text{area of a sector}} + o(\lambda). \quad (10.6.19)$$

Plugging (10.6.19) in the bounds (10.6.17),(10.6.18) and using (10.6.14) give

$$\frac{2}{\pi} \arcsin\left(\frac{\pi c(1-\varepsilon)}{4}\right) \le \lim_{\lambda\to\infty} F_\lambda(c) \le \frac{2}{\pi} \arcsin\left(\frac{\pi c(1+\varepsilon)}{4}\right).$$

As $\varepsilon > 0$ is arbitrary we get

$$\forall c < \frac{4}{\pi} \qquad \lim_{\lambda\to\infty} F_\lambda(c) = \frac{2}{\pi} \arcsin\left(\frac{\pi c}{4}\right),$$

$$= \frac{1}{2} \int_0^c \frac{1}{\sqrt{1 - (\frac{\pi}{4}x)^2}} dx,$$

which proves (10.6.11). Finally, note that we have $\lim_{c\to\frac{4}{\pi}} \lim_{\lambda\to\infty} F_\lambda(c) = 1$, and since for every value of λ, the function $F_\lambda(c)$ is a cumulative distribution function we get $\lim_{\lambda\to\infty} F_\lambda(c) = 1$ for $c \ge \frac{4}{\pi}$. □

Remark. The calculation in the proof above may be considered as a particular case of those done in [40]. The proof here is explicitly tailored for the purpose of this chapter.

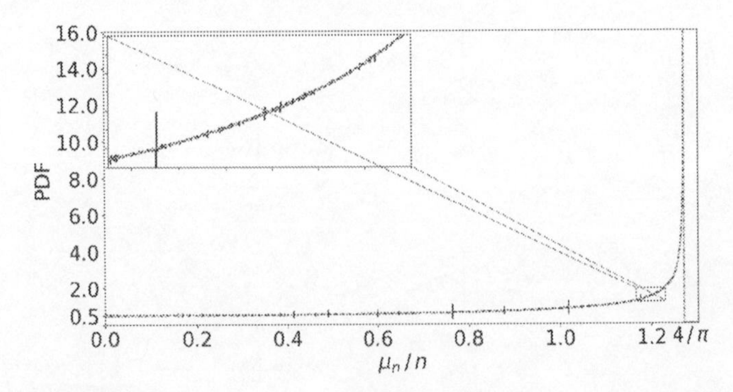

Figure 10.5 *Dashed grey curve*: the probability distribution of $\frac{\mu_n}{n}$ as given in (10.6.11). *Solid black curve*: a numerical calculation of this distribution as calculated for the first $3 \cdot 10^8$ torus eigenfunctions

The next figure shows the probability distribution given in (10.6.11) and compares it to a numerical examination of the probability distribution of $\frac{\mu_n}{n}$ for the separable eigenfunctions on the torus.

Examining the $\frac{\mu_n}{n}$ distribution leads to the following. First, we note that $\frac{\mu_n}{n}$ may get arbitrarily low values for a positive proportion of the eigenfunctions. This is in contradiction with (10.6.6) and therefore we conclude that the separable eigenfunctions on the flat torus do not satisfy assumption (10.6.5) in Lemma 10.6.3. Indeed, this can be verified directly. The separable eigenfunctions have two types of Neumann domains, star-like and lens-like. Only the spectral position of the star-like domains is bounded (Theorem 10.5.3), but it can be checked that their total area (of all star-like domains of the eigenfunction) goes to zero as the eigenvalue $\lambda_{a,b}$ tends to infinity. Therefore, assumption (10.6.5) is not satisfied.

In this context it is interesting to note that in [47] Jakobson and Nadirashvili show the existence of a sequence of eigenfunctions with a bounded number of critical points. A bounded number of critical points implies a bounded number of Neumann domains (see Proposition 10.6.2). Their result holds for particularly constructed metrics on the two-dimensional torus. Relating this to Lemma 10.6.3, we obtain that (10.6.6) does not hold for those metrics and hence (10.6.5) is not satisfied there. As an implication we get that for those manifolds the spectral positions of the Neumann domains are unbounded. Furthermore, the total area of the Neumann domains with bounded spectral positions converges to zero (at least in the lim inf sense).

Returning to Proposition 10.6.4 we observe that $\frac{\mu_n}{n} > 1$ for a positive proportion of the eigenfunctions. This means that an analogue of the strict Courant bound does not apply to the Neumann domain count. An even more extreme result is found in [27]. Buhovsky, Logunov, and Sodin show that the number of critical points may grow arbitrarily fast with the eigenvalue.[5] This implies an arbitrary growth in the Neumann domain count and hence, there is no hope to get any general form of an upper bound for the Neumann count. Yet, the metric constructed in [27] is not real analytic. It is therefore still interesting to examine the real analytic case or to restrict to particular manifolds and to determine the exact growth rate of μ_n with n and its dependence on the manifold and the metric.

PART 2 NEUMANN DOMAINS ON METRIC GRAPHS

10.7 Definitions

10.7.1 Discrete Graphs and Graph Topologies

We denote by $\Gamma = (\mathcal{V}, \mathcal{E})$ a connected undirected graph with finite sets of vertices $\mathcal{V}$ and edges $\mathcal{E}$. We allow the graph edges to connect either two distinct vertices or a vertex to itself. In the latter case, such an edge is called a *loop*.

For a vertex $v \in \mathcal{V}$, its *degree*, d_v, equals the number of edges connected to it (a loop is counted twice, if exits). The set of graph vertices of degree one turns out to be useful and we denote it by

$$\partial\Gamma := \{v \in \mathcal{V} : d_v = 1\}.$$

We call the vertices in $\partial\Gamma$, *boundary vertices*, and the rest of the vertices, $\mathcal{V}\backslash\partial\Gamma$, are called *interior vertices*.

An important topological quantity of graphs is the first Betti number (dimension of the first homology group) given, for a connected graph, by

$$\beta := |\mathcal{E}| - |\mathcal{V}| + 1. \tag{10.7.1}$$

The value of β is the number cycles needed to span the space of cycles on the graph. By definition, a graph is simply connected when $\beta = 0$, and such a graph is called a *tree graph*. Two particular examples of trees are star graphs and path graphs. A star graph is a graph with one interior vertex which is connected by edges to the other $|\mathcal{V}| - 1$ boundary vertices. A path graph is a connected graph with two boundary vertices and $|\mathcal{V}| - 2$ interior vertices which

[5] They actually show that there might even be infinitely many isolated critical points, but in that case the eigenfunctions are not Morse.

are all of degree two. The path graph which shows up later in this chapter is the simplest graph of only two vertices connected by a single edge.

10.7.2 Spectral Theory of Metric Graphs

A *metric graph* is a discrete graph for which each edge, $e \in \mathcal{E}$, is identified with a one-dimensional interval $[0, L_e]$ of a positive finite length L_e. We assign to each edge $e \in \mathcal{E}$ a coordinate, x_e, which measures the distance along the edge from one of the two boundary vertices of e.

A function on the graph is described by its restrictions to the edges, $\left\{ f|_e \right\}_{e\in\mathcal{E}}$, where $f|_e : [0, L_e] \to \mathbb{C}$. We equip the metric graphs with the differential operator,

$$- \Delta \ : \ f|_e \left(x_e\right) \mapsto -\frac{\mathrm{d}^2}{\mathrm{d}x_e^2} f|_e \left(x_e\right), \tag{10.7.2}$$

which is the Laplacian.[6] It is most common to call this setting of a metric graph and an operator by the name *quantum graph*.

To complete the definition of the operator we need to specify its domain. We consider functions which belong to the following direct sum of Sobolev spaces

$$H^2(\Gamma) := \bigoplus_{e\in\mathcal{E}} H^2([0, L_e]). \tag{10.7.3}$$

In addition, we require some matching conditions on the graph vertices. A function $f \in H^2(\Gamma)$ is said to satisfy the Neumann vertex conditions at a vertex v if

(1) f is continuous at $v \in \mathcal{V}$, i.e.,

$$\forall e_1, e_2 \in \mathcal{E}_v \quad f|_{e_1}(0) = f|_{e_2}(0), \tag{10.7.4}$$

where $\mathcal{E}_v$ is the set of edges connected to v, and for all $e \in \mathcal{E}_v$, $x_e = 0$ at v.

(2) The outgoing derivatives of f at v satisfy

$$\sum_{e\in\mathcal{E}_v} \left.\frac{\mathrm{d}f}{\mathrm{d}x_e}\right|_e (0) = 0. \tag{10.7.5}$$

Requiring these conditions at each vertex leads to the operator (10.7.2) being self-adjoint and its spectrum being real and bounded from below [18]. In addition, since we only consider compact graphs, the spectrum is discrete. We number the eigenvalues in the ascending order and denote them by $\{\lambda_n\}_{n=0}^{\infty}$ and their corresponding eigenfunctions by $\{f_n\}_{n=0}^{\infty}$. As the operator is both

[6] More general operators appear in the literature. (See, for example, [18, 41].

real and self-adjoint, we may choose the eigenfunctions to be real, which we will always do.

In this chapter, we only consider graphs whose vertex conditions are Neumann at all vertices, and call those *standard graphs.* A special attention should be given to vertices of degree two. Introducing such a vertex at the interior of an existing edge (thus splitting this edge into two) and requiring Neumann conditions at this vertex does not change the eigenvalues and eigenfunctions of the graph. The same holds when removing a degree two vertex and uniting two existing edges into one (see, e.g., [18, Remark 1.4.2]). This spectral invariance allows us to assume in the following that standard graphs do not have any vertices of degree two. Furthermore, the only graph all of whose vertices are of degree two (or equivalently has no vertices at all) is the single loop graph. We assume throughout the chapter that our graphs are different than the single loop graph and call those *non-trivial graphs*.

The spectrum of a standard graph is non-negative, which means that we may represent the spectrum by the non-negative square roots of the eigenvalues, $k_n = \sqrt{\lambda_n}$. For convenience, we abuse terminology and call also $\{k_n\}_{n=0}^{\infty}$ the eigenvalues of the graph. Most of the results and proofs in this part are expressed in terms of those eigenvalues. A Neumann graph has $k_0 = 0$ with multiplicity which equals the number of graph components. The common convention is that if an eigenvalue is degenerate (i.e., non-simple), it appears more than once in the sequence $\{k_n\}_{n=0}^{\infty}$. In addition, we choose a corresponding set of eigenfunctions, denoted by $\{f_n\}_{n=0}^{\infty}$. The choice of eigenfunctions is unique if all eigenvalues are simple. Otherwise, for any degenerate eigenvalue, we pick a basis for its eigenspace and all members of this basis appear in the sequence $\{f_n\}_{n=0}^{\infty}$. Obviously, this makes the choice of the sequence $\{f_n\}_{n=0}^{\infty}$ non-unique. Nevertheless, it is important to note that all the statements to follow hold for any choice of $\{f_n\}_{n=0}^{\infty}$.

10.7.3 Neumann Points and Neumann Domains

For metric graphs, the nodal point set of a function is the set of points at which the function vanishes. Removing the nodal point set from the graph, splits it into connected components and those are called *nodal domains*. The Neumann set and Neumann domains are similarly defined, but before doing so we need to restrict to particular classes of functions.

Definition 10.7.1 Let Γ be a non-trivial standard graph and f be an eigenfunction of Γ.

(1) We call f a *Morse eigenfunction* if for each edge e, $f|_e$ is a Morse function. Namely, at no point in the interior of e both the first and the second derivatives of f vanish.
(2) We call an eigenfunction f *generic* if it is a Morse eigenfunction and in addition it satisfies all of the following:
   (a) f corresponds to a simple eigenvalue.
   (b) f does not vanish at any vertex.
   (c) For any interior vertex $v \in \mathcal{V} \setminus \partial\Gamma$, none of the outgoing derivatives of f at v vanish.

An equivalent characterization of a Morse eigenfunction is the following.

Lemma 10.7.2 *Let f be a non-constant eigenfunction. f is Morse if and only if there exists no edge e such that $f|_e \equiv 0$.*

Proof. First, observe that a non-constant eigenfunction of the Laplacian vanishes at an interior point of an edge if and only if the second derivative vanishes at that point. Therefore, if f is a Morse eigenfunction then there is no interior point at which both the function and its derivative vanish. This means that a Morse eigenfunction cannot vanish entirely at a graph edge. As for the converse, if f is a non-Morse eigenfunction then there exists $\boldsymbol{x}$, an interior point of an edge e, such that $f|_e'(\boldsymbol{x}) = f|_e''(\boldsymbol{x}) = 0$. By the same argument as above, this means that either $f|_e(\boldsymbol{x}) = 0$ or $f|_e$ is the constant eigenfunction. The vanishing of $f|_e$ and its first derivative at the same point, together with $f|_e$ being a solution of an ordinary differential equation of second-order implies $f|_e \equiv 0$. □

We complement this lemma and note that the constant eigenfunction, corresponding to $k_0 = 0$, is not a Morse function. This, together with the lemma, implies that a Morse eigenfunction may vanish only at isolated points of the graph; the same holds for its derivative. This quality allows the following.

Definition 10.7.3 Let f be a Morse eigenfunction of Γ.

(1) A Neumann point of f is an extremal point (maximum or minimum) not located at a boundary vertex of Γ. We denote the set of Neumann points by $\mathcal{N}(f)$ (reusing the notation (10.2.4) for the Neumann lines of manifold eigenfunctions).
(2) A Neumann domain of f is a closure of a connected component of $\Gamma \setminus \mathcal{N}(f)$. The closure is done by adding vertices of degree one at the open endpoints of the connected component.

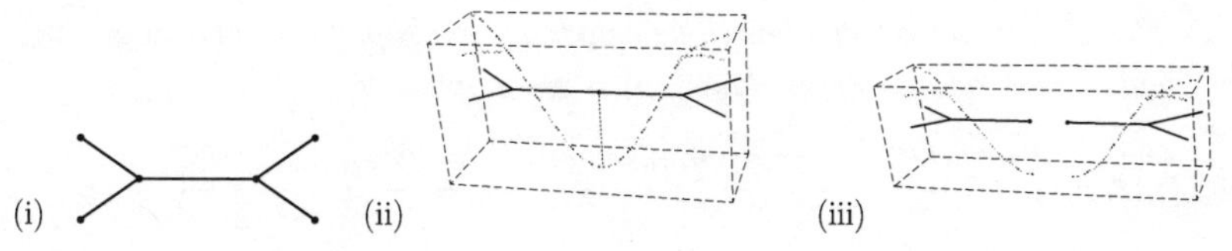

Figure 10.6 (i) A graph Γ; (ii) an eigenfunction f of Γ, with its single Neumann point marked; and (iii) a decomposition of Γ into the Neumann domains of f

Figure 10.6 shows the Neumann point and Neumann domains of a particular eigenfunction.

Remark. From the proof of Lemma 10.7.2 we learn that no point can be both a nodal point and a Neumann point.

The definition above implies that a Neumann point is either a point $\boldsymbol{x} \in \Gamma\backslash\mathcal{V}$ at some interior of an edge such that $f'(\boldsymbol{x}) = 0$, or it is a vertex $v \in \mathcal{V}$ such that all outgoing derivatives of f at that vertex vanish. The latter possibility does not occur if f is generic. Hence, for a generic f we have

$$\mathcal{N}(f) = \left\{\boldsymbol{x} \in \Gamma\backslash\mathcal{V} \, : \, f'(\boldsymbol{x}) = 0\right\}. \tag{10.7.6}$$

In other words, restricting to generic eigenfunctions allows to describe the Neumann points of an eigenfunction as the nodal points of its derivative. These points are isolated, as mentioned just before Definition 10.7.3.

All the results to follow concern generic eigenfunctions. The name generic is justified since almost every Morse eigenfunction is generic as is implied by the next proposition.[7] Furthermore, this proposition gives a quantitative estimate to the proportion of generic eigenfunctions out of a complete set of eigenfunctions. In order to do so, we need to assume that the set of edge lengths is linearly independent over the field $\mathbb{Q}$ of rational numbers. We call such lengths *rationally independent* and we will employ this assumption in some of the propositions to follow.

Proposition 10.7.4 *[2] Let Γ be a non-trivial standard graph, with rationally independent edge lengths of total length $|\Gamma|$, and denote the total length of all loops in Γ by L_{loops} (if there are no loops then $L_{loops} = 0$). Let $\{f_n\}_{n=0}^{\infty}$ be a complete set of eigenfunctions of Γ.*

[7] Actually, by the proposition, almost every eigenfunction which is not supported on a single loop is generic. In particular, if the graph has no loops then almost every eigenfunction is generic. See also discussion in [2, 3].

(1) *The density of the eigenfunctions which are not supported on a single loop among the set of all eigenfunctions is given by*

$$\lim_{N\to\infty} \frac{|\{n \le N \ : \ f_n \text{ is not supported on a loop}\}|}{N} = 1 - \frac{1}{2}\frac{L_{loops}}{|\Gamma|} \ge \frac{1}{2}. \tag{10.7.7}$$

(2) *The density of the generic eigenfunctions among the eigenfunctions which are not supported on a single loop is*

$$\lim_{N\to\infty} \frac{|\{n \le N \ : \ f_n \text{ is generic}\}|}{|\{n \le N \ : \ f_n \text{ is not supported on a loop}\}|} = 1. \tag{10.7.8}$$

Namely, almost every eigenfunction which is not supported on a loop is generic. Furthermore, since Morse eigenfunctions are a subset of eigenfunctions which are not supported on a loop and generic eigenfunctions are a subset of Morse eigenfunctions we have that almost every Morse eigenfunction is generic.

Remark.

(1) The limits in (10.7.7) and (10.7.8) exist even without assuming that the edge lengths are rationally independent. This assumption is needed to obtain the exact values of those limits.
(2) The proposition above extends Proposition A.1 of [4]. Both propositions are based on Theorem 3.6 of [20].

10.8 Topology of Ω and Topography of $f|_\Omega$

Let Γ be a non-trivial standard graph and f an eigenfunction of Γ corresponding to the eigenvalue k. Formally, every Neumann domain Ω of f may be considered as a subgraph of Γ, if we add degree two vertices to Γ at all the Neumann points of f (see discussion on those vertices in Section 10.7.2). In particular, a Neumann domain is a closed set (by Definition 10.7.3). This difference from the manifold case (where Neumann domains are open sets) is technical and serves our need to consider Ω as a metric graph on its own. Being a metric graph, we take the usual Laplacian on Ω and impose Neumann vertex conditions at all of its vertices, so that Ω is considered as a standard graph. Note that the restriction of $f|_\Omega$ to the edges of Ω trivially satisfies $f'' = -k^2 f$. It also obeys Neumann vertex conditions at all vertices of Ω, as each vertex is either a vertex of Γ or a point $\boldsymbol{x} \in \Gamma$ in an interior of an edge for which $f'(\boldsymbol{x}) = 0$. This gives the following, which is analogous to Lemma 10.3.1.

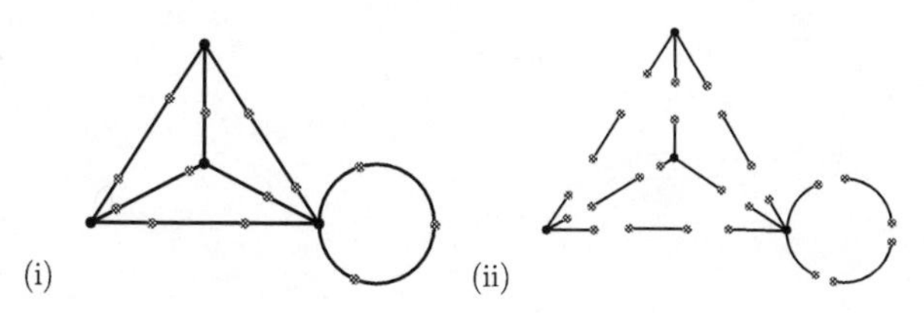

Figure 10.7 (i) A graph Γ with Neumann points of a given eigenfunction. (ii) The decomposition of the graph to the corresponding Neumann domains

Lemma 10.8.1 *$f|_{\Omega}$ is an eigenfunction of the standard graph Ω and corresponds to the eigenvalue k.*

Remark. Furthermore, it can be proved that if f is a generic eigenfunction and Ω is a tree graph then $f|_{\Omega}$ is also generic [3]. Indeed, if f is generic and k is a simple eigenvalue of Ω, then we get that $f|_{\Omega}$ is also generic. But, if Ω is a tree graph and $f|_{\Omega}$ does not vanish at vertices, then it must belong to a simple eigenvalue [18, Corollary 3.1.9].

10.8.1 Possible Topologies for Neumann Domains

In this sub-section we discuss which graphs may be obtained as a Neumann domain. The next lemma shows that if we consider an eigenfunction, f, whose eigenvalue is high enough, each of its Neumann domains is either a path graph or a star graph. A star Neumann domain contains an interior vertex of the graph, and a path Neumann domain is contained in a single edge of the graph (see Figure 10.7).

Lemma 10.8.2 *Let Γ be a non-trivial standard graph. Let f be an eigenfunction corresponding to an eigenvalue $k > \frac{\pi}{L_{min}}$, where L_{min} is the minimal edge length of Γ. Let Ω be a Neumann domain of f.*

(1) If Ω contains a vertex $v \in \mathcal{V}$ of degree $d_v > 2$, then Ω is a *star graph* with $deg\,(v)$ edges.
(2) If Ω does not contain a vertex $v \in \mathcal{V}$ of degree $d_v > 2$, then Ω is a *path graph, of length $\frac{\pi}{k}$.*

Proof. For any edge $e \in \mathcal{E}$ we have that $f|_e\,(x) = B_e \cos(kx + \varphi_e)$, where B_e, φ_e are some edge-dependent real parameters. This together with $k > \frac{\pi}{L_{min}}$ implies that the derivative of f vanishes at least once at the interior of each edge. Hence, the set of Neumann points, $\mathcal{N}\,(f)$ contains at least one point on each edge. It follows that each Neumann domain contains at most one vertex of Γ. Thus, there are two types of Neumann domains: if a Neumann domain,

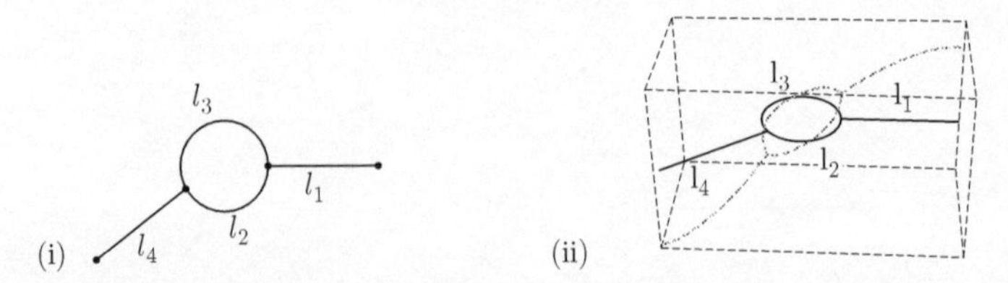

Figure 10.8 (i) A graph Γ with (ii) an eigenfunction whose single Neumann domain is not simply connected

Ω, contains a vertex with $\deg(v) > 2$ then Ω is a *star graph*, whose number of edges is d_v; otherwise Ω is a *path graph*. A Neumann domain which is a path graph can be parameterized as $\Omega = [0, l]$. Since $f'(0) = 0$ we get that $f|_\Omega(x) = \cos(kx)$ up to a multiplicative constant. Using $f'(l) = 0$ and that f' does not vanish in the interior of Ω we conclude $l = \frac{\pi}{k}$. □

Remark. Only finitely many eigenvalues do not satisfy the condition $k > \frac{\pi}{L_{min}}$ in the lemma. The number of those eigenvalues is bounded by

$$\left|\left\{n \in \mathbb{N} : 0 \le k_n \le \frac{\pi}{L_{min}}\right\}\right| \le 2\frac{|\Gamma|}{L_{min}},$$

where $|\Gamma| = \sum_{e \in \mathcal{E}} L_e$ is the total sum of all edge lengths of Γ. This can be shown using

$$\forall n \in \mathbb{N}, \quad k_n \ge \frac{\pi}{2|\Gamma|}(n+1), \tag{10.8.1}$$

which is the statement of Theorem 1 in [37].

To complement the lemma above, we note that there are also Neumann domains which are not simply connected. Indeed, consider the graph Γ depicted in Figure 10.8(i). It has an eigenfunction with no Neumann points, so that the eigenfunction has a single Neumann domain which is the whole of Γ and in particular, it is not simply connected (Figure 10.8(ii)).

10.8.2 Critical Points and Nodal Points – Number and Position

In the following we consider the critical points and nodal points of $f|_\Omega$. Note that, by definition, a Morse function on a one-dimensional interval cannot have a saddle point. Hence, all critical points of a Morse eigenfunction of a graph are extremal points. We re-use the notations from the manifold part: $\mathcal{X}(f)$ for extremal points of f and $\mathcal{M}_+(f)$ ($\mathcal{M}_-(f)$) for maxima (minima). Denote by $\phi(f|_\Omega)$ the number of nodal points of $f|_\Omega$, by E_Ω the number of edges of

Ω, by V_Ω the number of its vertices, and by $\partial\Omega$ the vertices of Ω which are of degree one.

Proposition 10.8.3 *Let f be a generic eigenfunction and Ω a Neumann domain of f. Then*

(1) The extremal points of f, which are located on Ω are exactly the boundary of Ω, i.e., $\mathscr{X}(f) \cap \Omega = \partial\Omega$.
(2) $1 \leq |\mathscr{M}_+(f) \cap \partial\Omega| \leq |\partial\Omega| - 1$ (and the same bounds for $|\mathscr{M}_-(f) \cap \partial\Omega|$).
(3) $1 \leq \phi(f|_\Omega) \leq E_\Omega - V_\Omega + |\partial\Omega|$.

Proof. The third part of the proposition is proven in [3]. The first two parts of the proposition are proven below and they actually require only the assumption that f is Morse.

Part (1) of the proposition follows from the following two observations:

(a) The extremal points of a Morse eigenfunction f are $\mathscr{X}(f) = \partial\Gamma \cup \mathscr{N}(f)$.
(b) The definition of a Neumann domain (Definition 10.7.3) implies $\Omega \cap (\partial\Gamma \cup \mathscr{N}(f)) = \partial\Omega$.

From part (1) of the proposition we get that minima and maxima of $f|_\Omega$ are attained exactly at boundary points of Ω. As Ω is compact and $f|_\Omega$ is continuous and non-constant it must attain at least one maximum and at least one minimum, which proves part (2) of the proposition. □

Remark. Note that when Ω is a path graph the proposition implies that it has exactly one maximum, one minimum, and one nodal point. Also, when Ω is a tree graph, the last part of the proposition gives $1 \leq \phi(f|_\Omega) \leq |\partial\Omega| - 1$.

10.9 Geometry of Ω

Similar to the manifold case, we use the normalized area to perimeter ratio to quantify the geometry of a Neumann domain. The following is to be compared with Definition 10.4.3.

Definition 10.9.1 Let f be a Morse eigenfunction corresponding to the eigenvalue k. Let Ω be a Neumann domain of f, whose edge lengths are $\{l_j\}_{j=1}^{E_\Omega}$. We define the normalized area to perimeter ratio of Ω to be

$$\rho(\Omega) := \frac{|\Omega|}{|\partial\Omega|}k,$$

where $|\Omega| = \sum_{j=1}^{E_\Omega} l_j$ and $|\partial\Omega|$ is the number of boundary vertices of Ω.

For graphs we are able to obtain global bounds on $\rho(\Omega)$.

Proposition 10.9.2 *[3] Let Ω be a Neumann domain. We have*

$$\frac{1}{|\partial\Omega|} \leq \frac{\rho(\Omega)}{\pi} \leq \frac{E_\Omega}{|\partial\Omega|}. \tag{10.9.1}$$

If Ω is a star graph then we have a better upper bound $\frac{\rho(\Omega)}{\pi} \leq 1 - \frac{1}{|\partial\Omega|}$. If Ω is a path graph then $\rho(\Omega) = \frac{\pi}{2}$.

Next, we study the probability distribution of ρ. We find that for this purpose, it is useful to separately consider only the Neumann domains containing a particular vertex. Let Γ be a non-trivial standard graph and let f_n be its nth eigenfunction. Assume that f_n is generic. Then, for any vertex $v \in \mathcal{V}$ there is a unique Neumann domain of f_n which contains v and we denote it by $\Omega_n^{(v)}$.

Proposition 10.9.3 *[3] Let Γ be a non-trivial standard graph, with rationally independent edge lengths and let $v \in \mathcal{V}$ of degree $d_v > 2$. The value of $\frac{1}{\pi}\rho$ on $\{\Omega_n^{(v)}\}_{n=1}^{\infty}$ is distributed according to*

$$\lim_{N\to\infty} \frac{\left|\left\{n \leq N \,:\, f_n \text{ is generic and } \frac{1}{\pi}\rho\left(\Omega_n^{(v)}\right) \in (a,b)\right\}\right|}{|\{n \leq N \,:\, f_n \text{ is generic}\}|} = \int_a^b \zeta^{(v)}(x)\,\mathrm{d}x, \tag{10.9.2}$$

where $\zeta^{(v)}$ is a probability distribution supported on $[\frac{1}{d_v}, 1 - \frac{1}{d_v}]$. Furthermore, it is symmetric around $\frac{1}{2}$, i.e., $\zeta^{(v)}(x) = \zeta^{(v)}(1-x)$.

Remark. If $d_v = 1$ then $\Omega_n^{(v)}$ is a path graph for all n, so that by Proposition 10.9.2 we get that $\zeta^{(v)}$ is a Dirac measure $\zeta^{(v)}(x) = \delta\left(x - \frac{1}{2}\right)$.

As is implied by choice of notation, the distribution $\zeta^{(v)}$ indeed depends on the particular vertex $v \in \mathcal{V}$. We demonstrate this in Figure 10.9(iii) where we compare between the probability distributions of two vertices of different degrees from the same graph. In addition, Figure 10.9(iv) shows a comparison between the probability distributions of two vertices of the same degree from different graphs. The numerics suggest that the distributions are different, which implies that $\zeta^{(v)}$ may depend on the graph connectivity and not only on the degree of the vertex. It is of interest to further investigate this distribution, $\zeta^{(v)}$, and in particular its dependence on the graph's properties.

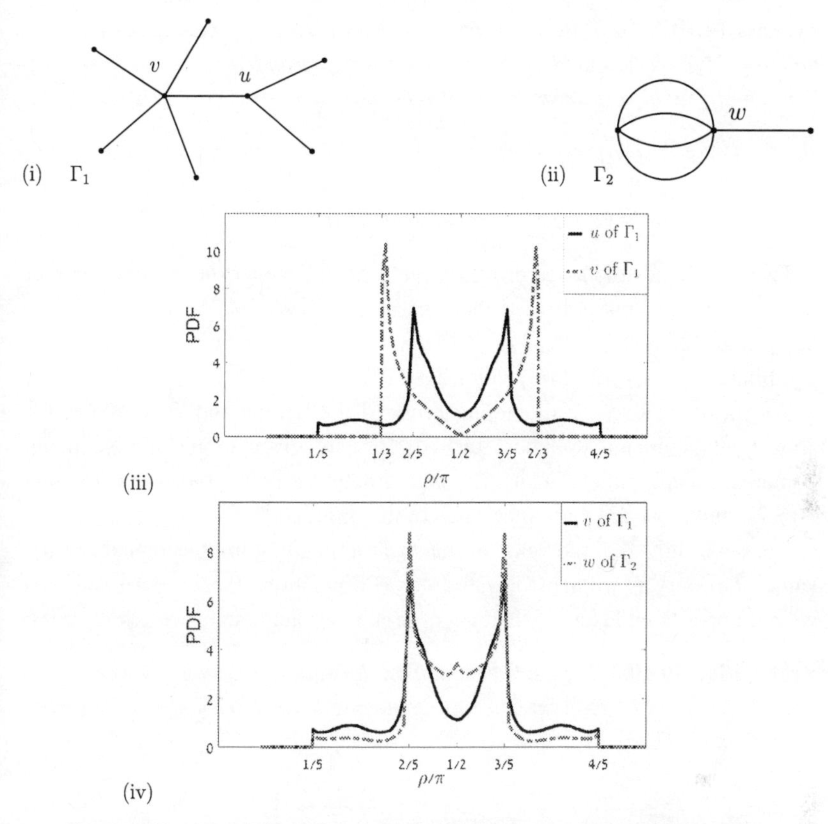

Figure 10.9 (i) Γ_1, with vertices v, u of degrees 5, 3, correspondingly. (ii) Γ_2, with vertex w of degree 5. (iii) A probability distribution function of $\frac{\rho}{\pi}$-values for the Γ_1 Neumann domains which contain v (i.e., $\zeta^{(v)}$ in (10.9.2)) compared with $\zeta^{(u)}$. (iv) Similarly, $\zeta^{(v)}$ compared with $\zeta^{(w)}$.
All the numerical data was calculated for the first 10^6 eigenfunctions and for a choice of rationally independent lengths.

10.10 Spectral Position of Ω

By Lemma 10.8.1, a graph eigenvalue k appears in the spectrum of each of its Neumann domains. Exactly as in Definition 10.5.1 for manifolds, we define the spectral position of a Neumann domain Ω, as the position of k in the spectrum of Ω and denote it by $N_\Omega(k)$. Also, for exactly the same reason as in the manifold case, we have that $N_\Omega(k) \geq 1$ for graphs (see discussion after Definition 10.5.1).

A useful tool in estimating the spectral position is the following lemma, connecting the spectral position of Ω to the nodal count of $f|_\Omega$.

Lemma 10.10.1 *Let Γ be a non-trivial standard graph, f be a generic eigenfunction of Γ corresponding to an eigenvalue k, and let Ω be a Neumann domain of f, which is a tree graph. Then*

(1) $N_\Omega(k) = \phi(f|_\Omega)$.
(2) $N_\Omega(k) \leq |\partial\Omega| - 1$.
In particular, if Ω is a path graph then $N_\Omega(k) = 1$.

The statement in (1) was proven in [16, 62, 67] under the assumption that $f|_\Omega$ is generic. This is indeed the case since f itself is generic and Ω is a tree graph (see remark after Lemma 10.8.1). The statement in (2) follows as a combination of (1) with Proposition 10.8.3 (3).

We further remark on the applicability of the lemma above; it applies for almost all Neumann domains. Indeed, for any given graph, all Neumann domains except finitely many are star graphs or path graphs (by Lemma 10.8.2), and those are particular cases of tree graphs.

Next, we show that the value of the spectral position implies bounds on the value of ρ, just as we had for manifolds (Proposition 10.5.2). For manifolds we got upper bounds on ρ, whereas for graphs we get bounds from both sides.

Proposition 10.10.2 *[3] Let Γ be a non-trivial standard graph, f be an eigenfunction of Γ corresponding to an eigenvalue k, and let Ω be a Neumann domain of f. Then*

$$\frac{\rho(\Omega)}{\pi} \geq \frac{1}{|\partial\Omega|}\left(\frac{N_\Omega(k)+1}{2}\right). \tag{10.10.1}$$

If Ω is a star graph then we further have the upper bound

$$\frac{\rho(\Omega)}{\pi} \leq \frac{1}{2} + \frac{1}{|\partial\Omega|}\left(\frac{N_\Omega(k)-1}{2}\right). \tag{10.10.2}$$

Remark. Note that if $N_\Omega(\lambda) > 1$ then the bound in (10.10.1) improves the lower bound given in Proposition 10.9.2. Similarly, if Ω is a star graph and $N_\Omega(\lambda) < |\partial\Omega| - 1$, then the bound (10.10.2) improves the upper bound given in Proposition 10.9.2 for star graphs.

Next, we show that the spectral position has a well-defined probability distribution. As in the previous section (Proposition 10.9.2), we find that this distribution is best described when one focuses on Neumann domains containing a particular graph vertex.

Proposition 10.10.3 *[3] Let Γ be a non-trivial standard graph, with rationally independent edge lengths and let $v \in \mathcal{V}$ of degree d_v. Then the following limit exists,*

$$P\left(N_{\Omega^{(v)}} = j\right) := \lim_{N\to\infty} \frac{\left|\left\{n \le N \,:\, f_n \text{ is generic and } N_{\Omega_n^{(v)}}(k_n) = j\right\}\right|}{\left|\left\{n \le N \,:\, f_n \text{ is generic}\right\}\right|}, \tag{10.10.3}$$

and defines a probability distribution for $N_{\Omega^{(v)}}$.
In addition, for $d_v > 2$

(1) $P\left(N_{\Omega^{(v)}} = j\right)$ *is supported in the set* $j \in \{1, \ldots, d_v - 1\}$.
(2) $P\left(N_{\Omega^{(v)}} = j\right)$ *is symmetric around* $\frac{d_v}{2}$, *i.e.,* $P\left(N_{\Omega^{(v)}} = j\right) = P\left(N_{\Omega^{(v)}} = d_v - j\right)$.

If $d_v = 1$ *then* $P\left(N_{\Omega^{(v)}} = j\right) = \delta_{j,1}$.

By the proposition, the support of the spectral position probability depends on the degree of the vertex. Yet, vertices of the same degree, but from different graphs, may have different probability distributions as is demonstrated in Figure 10.10(iii). In addition, we show in Figure 10.10(iv) how the conditional probability distribution of $\rho(\Omega)$ depends on the value of the spectral position N_Ω (compare with the bounds (10.10.1),(10.10.2)).

10.11 Neumann Count

In this section we present bounds on the number of Neumann points and provide some properties of the probability distribution of this number.

Definition 10.11.1 Let Γ be a non-trivial standard graph and $\{f_n\}_{n=0}^{\infty}$ a complete set of its eigenfunctions. Denote by $\mu_n := \mu(f_n)$ and $\phi_n := \phi(f_n)$ the numbers of Neumann points and nodal points, respectively. We call the sequences $\{\mu_n\}$, $\{\phi_n\}$ the *Neumann count* and *nodal count*, and the normalized quantities $\omega_n := \mu_n - n$, $\sigma_n := \phi_n - n$ are called the *Neumann surplus* and *nodal surplus*.

Proposition 10.11.2 *[3] Let* Γ *be a non-trivial standard graph. Let* f_n *be the* nth *eigenfunction of* Γ *and assume it is generic. We have the following bounds:*

$$1 - \beta \le \sigma_n - \omega_n \le \beta - 1 + |\partial\Gamma|, \tag{10.11.1}$$

and

$$1 - \beta - |\partial\Gamma| \le \omega_n \le 2\beta - 1, \tag{10.11.2}$$

where $\beta = |\mathcal{E}| - |\mathcal{V}| + 1$ *is the first Betti number of* Γ.

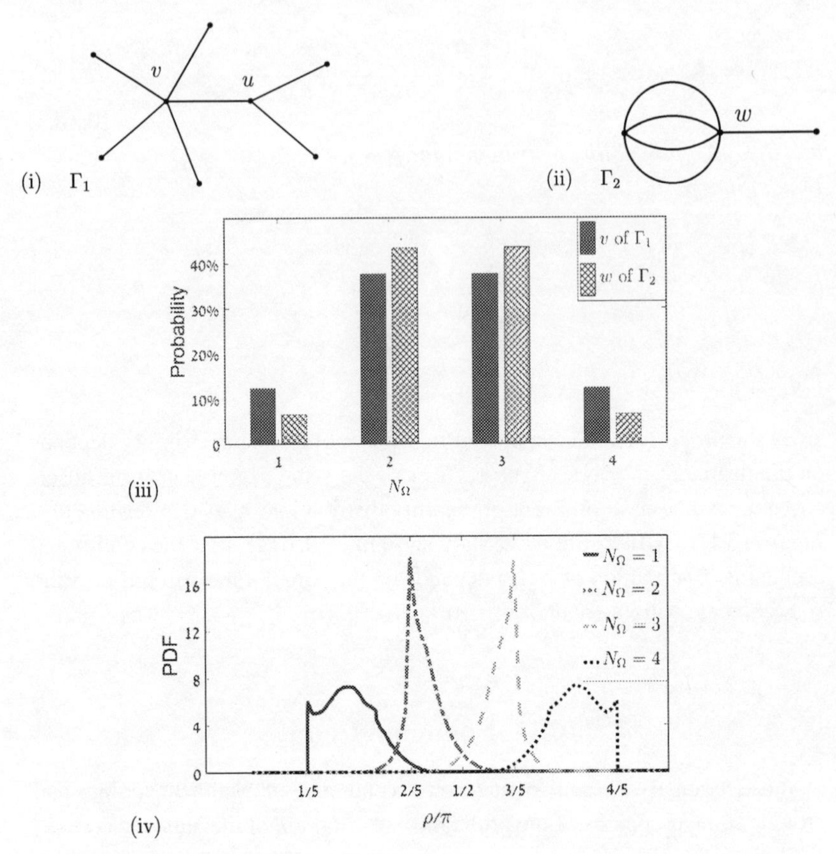

Figure 10.10 (i) Γ_1, with vertex v of degree 5. (ii) Γ_2, with vertex w of degree 5. (iii) The spectral position probability $P\left(N_{\Omega^{(v)}} = j\right)$ for v of Γ_1 compared with $P\left(N_{\Omega^{(w)}} = j\right)$ for w of Γ_2. (iv) A probability distribution function of $\frac{\rho}{\pi}$-values for the Γ_1 Neumann domains which contain v, conditioned on the value of the spectral position $N_{\Omega_n^{(v)}}$.
All the numerical data was calculated for the first 10^6 eigenfunctions for a choice of rationally independent lengths.

Moreover, both quantities $\sigma_n - \omega_n$ and ω_n have well-defined probability distributions, as stated in what follows.

Proposition 10.11.3 *[3] Let Γ be a non-trivial standard graph, with rationally independent edge lengths. Then*

(1) The following limit exists

$$P(\sigma - \omega = j) := \lim_{N \to \infty} \frac{|\{n \leq N \; : \; f_n \text{ is generic and } \sigma_n - \omega_n = j\}|}{|\{n \leq N \; : \; f_n \text{ is generic}\}|}. \tag{10.11.3}$$

and defines a probability distribution for the difference between the Neumann and nodal surplus.
Furthermore, this probability distribution is symmetric around $\frac{1}{2}|\partial\Gamma|$, *i.e.,* $P(\sigma - \omega = j) = P(\sigma - \omega = |\partial\Gamma| - j)$.

(2) *Similarly, the Neumann surplus has a well-defined probability distribution which is symmetric around* $\frac{1}{2}(\beta - |\partial\Gamma|)$.

This proposition is in the spirit of the recently obtained result for the distribution of the nodal surplus [4]. It was shown in [4] that the nodal surplus, σ, has a well-defined probability distribution which is symmetric around $\frac{1}{2}\beta$. The proof of Proposition 10.11.3 uses similar techniques to the proof of this latter result and appears in [3].

The proposition above also has an interesting meaning in terms of inverse problems. It is common to ask what one can deduce on a graph out of its nodal count sequence, $\{\phi_n\}$ [11, 13, 59]. It was found in [5] that the nodal count distinguishes tree graphs from others. More progress was made in [4] where it was shown that the nodal surplus distribution reveals the graph's first Betti number, as twice the expected value of the nodal surplus. However, it should be noted that all tree graphs have the same nodal count, so that one cannot distinguish between different trees in terms of the nodal count. Proposition 10.11.3 shows that the Neumann count, $\{\mu_n\}$, contains information on the size of the graph's boundary, $|\partial\Gamma|$. In particular, this enables the distinction between some tree graphs, which was not possible before. We summarize the discussion above in the following.

Corollary 10.11.4 *[3] Let* Γ *is a non-trivial standard graph with rationally independent edge lengths, and let* $\mathcal{G}_N := \{n \le N : \ f_n \text{ is generic}\}$. *Then*

$$\beta = 2 \lim_{N\to\infty} \frac{\sum_{n\in\mathcal{G}_N} \sigma_n}{|\mathcal{G}_N|}$$

$$|\partial\Gamma| = 2 \lim_{N\to\infty} \frac{\sum_{n\in\mathcal{G}_N} (\sigma_n - \omega_n)}{|\mathcal{G}_N|} = 2 \lim_{N\to\infty} \frac{\sum_{n\in\mathcal{G}_N} (\phi_n - \mu_n)}{|\mathcal{G}_N|}.$$

We emphasize that different tree graphs with the same boundary size, $|\partial\Gamma|$, have the same expected value for both their nodal surplus and their Neumann surplus and are not distinguishable in this sense. Furthermore, we may wonder whether the boundary size of a tree graph fully determines the probability distribution of its Neumann surplus. We do not have an answer to this question yet and carry on this exploration.

We end this section by noting that numerics lead us to believe that the bounds obtained in (10.11.2) on the Neumann surplus ω_n are not strict for graphs with high β values. Furthermore, we conjecture the following bounds on ω_n (which, for $\beta > 2$, are sharper than the bounds in (10.11.2)).

Conjecture 10.11.5 *The Neumann surplus is bounded by*

$$-1 - |\partial\Gamma| \leq \omega_n \leq \beta + 1.$$

Proving the bounds (10.11.2) on ω_n is done by combining the bounds on $\sigma_n - \omega_n$ (10.11.1) with the bounds $0 \leq \sigma_n \leq \beta$, [22]. The bounds on both $\sigma_n - \omega_n$ and σ_n are known to be strict. Hence, if indeed the bounds on ω_n are not strict, it implies that the nodal surplus, σ_n, and the Neumann surplus, ω_n, are correlated when considered as random variables, which is an interesting result on its own.

PART 3 SUMMARY

In this part we summarize the chapter's main results and focus on the comparison between analogous statements on graphs and manifolds. This is emphasized by using common terminology and notations for both graphs and manifolds.

Let f be an eigenfunction corresponding to the eigenvalue λ and Ω be a Neumann domain of f. On manifolds, we have that Ω and $f|_\Omega$ are of a rather simple form; Ω is simply connected; $f|_\Omega$ has only two nodal domains; and its critical points are all located on $\partial\Omega$ (Theorem 10.3.2). On graphs, the situation is similar, as almost all Neumann domains are either star graphs or path graphs. It is possible to have other Neumann domains, and even non-simply connected ones, only if λ is small enough (Lemma 10.8.2). For graphs, $f|_\Omega$ has two nodal domains if Ω is a path graph, but otherwise may have more, with a global bound on this number (Proposition 10.8.3,(3)).

The most basic property of Neumann domains is that $f|_\Omega$ is a Neumann eigenfunction of Ω (manifolds – Lemma 10.3.1; graphs – Lemma 10.8.1). The eigenvalue of $f|_\Omega$ is also λ and the interesting question is to find out what is the position of λ in the spectrum of Ω – a quantity which we denote by $N_\Omega(\lambda)$ (Definition 10.5.1). The intuitive feeling at the beginning of the Neumann domain study was that generically, $N_\Omega(\lambda) = 1$ or that at least the spectral position gets low values.

The general problem of determining the spectral position is quite hard for manifolds. The most general result we are able to provide for manifolds (Proposition 10.5.2) is a lower bound given in terms of the geometric quantity ρ, which is a normalized area to perimeter ratio (Definition 10.4.3). Interestingly, this result allows to estimate the spectral position numerically; a numerical calculation of ρ is rather easy compared to the involved calculation of the spectrum of an arbitrary domain, which is needed to determine a spectral position. This numerical method allows to refute the belief that for manifolds, generically, $N_\Omega(\lambda) = 1$. For graphs, the quantity ρ (Definition 10.9.1)) allows to bound the spectral position from both sides, for almost all Neumann domains (Proposition 10.10.2). Two additional results we have for the spectral position on graphs (but not for manifolds) are as follows. First, the spectral position of Ω is given explicitly by the nodal count of $f|_\Omega$, and this yields an upper bound on the spectral position (Lemma 10.10.1). Second, the spectral position has a limiting distribution which is symmetric (Proposition 10.10.3). Another point of comparison is that an upper bound on the spectral position, which we have for graphs, does not exist for manifolds. We show by means of an example that the spectral position is unbounded in the manifold case. This example is given in terms of separable eigenfunctions on the torus. For this example, we show that although the spectral position of half of the Neumann domains is unbounded, it equals one for the other half (Theorem 10.5.3). This finding might imply that even though $N_\Omega(\lambda) = 1$ does not hold generically, there might be a substantial proportion of Neumann domains for which it does hold (see e.g., (10.6.5) for such an assumption). This is indeed the case for graphs where the spectral position equals one for each path graph Neumann domain, and all of those form a substantial proportion of all Neumann domains (their number as well as their total length increase with the eigenvalue).

Finally, we discuss the Neumann domain count. On manifolds we count the number of Neumann domains, while on graphs we count the number of Neumann points. There is also a connection between the Neumann count and the nodal count. On manifolds, we have that the difference between the Neumann count and half the nodal count is non-negative (Corollary 10.6.1). On graphs, the difference between the Neumann count and the nodal count is bounded from both sides (Proposition 10.11.2). As for the Neumann count itself, it makes sense to consider it with a normalization: $\frac{\mu(f_n)}{n}$ on manifolds and $\mu(f_n) - n$ on graphs. For graphs we provide general bounds on $\omega_n = \mu(f_n) - n$ (Proposition 10.11.2), but believe that those are not sharp and conjecture sharper bounds (Conjecture 10.11.5). The validity of the conjecture would also imply a correlation between the nodal and the Neumann counts. In addition, ω_n possesses a limiting probability distribution which is symmetric

(Proposition 10.11.3). The expected value of this distribution stores information on the size of the graph's boundary, $|\partial\Gamma|$; an information that is absent from the nodal count. Which other graph properties may be revealed by this distribution is still to be found. Turning back to manifolds, we treat separable eigenfunctions on the torus and for those derive the probability distribution of $\frac{\mu(f_n)}{n}$ (Proposition 10.6.4). This is to be viewed as the beginning of the analysis of Neumann count on manifolds. Some further progress can be made in studying the asymptotic growth of the Neumann count and for example showing that $\limsup_{n\to\infty} \mu(f_n) = \infty$. This direction is related to the recent series of works on asymptotic growth of the nodal count [38, 50, 51, 77, 39, 48, 49, 55, 52] (see full description in Section 10.6.2) as well as to works on the number of critical points [47, 27].

Acknowledgements

We thank Alexander Taylor for providing the python code [70] we used to generate the figures of Neumann domains on manifolds and calculate the distribution of ρ. The authors were supported by ISF (grant no. 494/14).

Appendix A Basic Morse Theory

This section brings some basic statements in Morse theory which are useful for understanding the first part of the chapter. For a more thorough exposition, we refer the reader to [14]. Throughout the appendix we consider (M, g) to be a compact smooth Riemannian manifold of a finite dimension. At some points of the appendix we specialize for the two-dimensional case and mention explicitly when we do so.

Definition A.1 Let $f : M \to \mathbb{R}$ be a smooth function.

(1) f is a Morse function if at every critical point, $\boldsymbol{p} \in \mathscr{C}(f)$, the Hessian matrix, $\text{Hess} f|_{\boldsymbol{p}}$, is non-degenerate, i.e., it does not have any zero eigenvalues.
(2) The Morse index $\lambda_{\boldsymbol{p}}$ of a critical point $\boldsymbol{p} \in \mathscr{C}(f)$ is the number of negative eigenvalues of the Hessian matrix, $\text{Hess} f|_{\boldsymbol{p}}$.

The following three propositions may be found in [14].

Proposition A.2 [14, Lemma 3.2 and Corollary 3.3]
If f is a Morse function then the critical points of f are isolated and f has only finitely many critical points.

Next, we consider the gradient flow $\varphi : \mathbb{R} \times M \to M$ defined by (10.2.2). For a particular $\boldsymbol{x} \in M$ we call the image of $\varphi : \mathbb{R} \times \boldsymbol{x} \to M$, a gradient flow line. Note that a gradient flow line, $\{\varphi(t; \boldsymbol{x})\}_{t=-\infty}^{\infty}$, has a natural direction dictated by the order of the t values.

Proposition A.3 [14, Propositions 3.18, 3.19]

(1) Any smooth real-valued function f decreases along its gradient flow lines. The decrease is strict at non-critical points.
(2) Every gradient flow line of a Morse function f begins and ends at a critical point. Namely, for all $\boldsymbol{x} \in M$ both limits $\lim_{t \to \pm\infty} \varphi(t, \boldsymbol{x})$ exist and they are both critical points of f.

Proposition A.4 (Stable/unstable manifold theorem for a Morse function) [14, Theorem 4.2]
Let f be a Morse function and $\boldsymbol{p} \in \mathscr{C}(f)$. Then the tangent space at $\boldsymbol{p}$ splits as

$$T_{\boldsymbol{p}}M = T_{\boldsymbol{p}}^{s}M \oplus T_{\boldsymbol{p}}^{u}M,$$

where the Hessian is positive definite on $T_{\boldsymbol{p}}^{s}M$ and negative definite on $T_{\boldsymbol{p}}^{u}M$.

Moreover, the stable and unstable manifolds (10.2.3) are surjective images of smooth embeddings

$$T_p^s M \to W^s(\boldsymbol{p}) \subseteq M,$$
$$T_p^u M \to W^u(\boldsymbol{p}) \subseteq M.$$

Therefore, $W^u(\boldsymbol{p})$ is a smoothly embedded open disc of dimension $\lambda_{\boldsymbol{p}}$ and $W^s(\boldsymbol{p})$ is a smoothly embedded open disc of dimension $m - \lambda_{\boldsymbol{p}}$, where m is the dimension of M.

Let us examine the implications of the results above in the particular case of Morse functions on a two-dimensional manifold.

- If $\boldsymbol{q}$ is a maximum then $\lambda_{\boldsymbol{q}} = 2$ and so $W^u(\boldsymbol{q})$ is a two-dimensional open and simply connected set and $W^s(\boldsymbol{q}) = \{\boldsymbol{q}\}$.
- If $\boldsymbol{p}$ is a minimum then $\lambda_{\boldsymbol{p}} = 0$ and so $W^s(\boldsymbol{p})$ is a two-dimensional open and simply connected set and $W^u(\boldsymbol{p}) = \{\boldsymbol{p}\}$.
- If $\boldsymbol{r}$ is a saddle point then $\lambda_{\boldsymbol{r}} = 1$ and so both $W^s(\boldsymbol{r})$ and $W^u(\boldsymbol{r})$ are one-dimensional curves. Note that $W^s(\boldsymbol{r}) \cap W^u(\boldsymbol{r}) = \{\boldsymbol{r}\}$ and so we get that $W^s(\boldsymbol{r})$ is a union of two gradient flow lines (actually even Neumann lines) which end at $\boldsymbol{r}$. Similarly, $W^u(\boldsymbol{r})$ is a union of two gradient flow lines (Neumann lines) which start at $\boldsymbol{r}$.

By Definition 10.2.1 we get that Neumann domains are open two-dimensional sets and that the Neumann line set is a union of one-dimensional curves. Moreover, those sets are complementary. Namely, the union of all Neumann domains together with the Neumann line set gives the whole manifold [10, Proposition 1.3].

Next, we focus on a subset of the Morse functions, known as *Morse–Smale functions*, described by the following two definitions.

Definition A.5 We say that two sub-manifolds $M_1, M_2 \subset M$ intersect transversally and write $M_1 \pitchfork M_2$ if for every $\boldsymbol{x} \in M_1 \cap M_2$ the tangent space of M at $\boldsymbol{x}$ equals the sum of tangent spaces of M_1 and M_2 at $\boldsymbol{x}$, i.e.,

$$T_{\boldsymbol{x}} M = T_{\boldsymbol{x}} M_1 + T_{\boldsymbol{x}} M_2. \tag{A.1}$$

This is also called the *transversality condition*.

Definition A.6 A Morse function such that for all of its critical points $\boldsymbol{p}, \boldsymbol{q} \in \mathscr{C}(f)$ the stable and unstable sub-manifolds intersect transversely, i.e., $W^s(\boldsymbol{q}) \pitchfork W^u(\boldsymbol{p})$ is called a *Morse–Smale* function.

Let us assume now that M is a two-dimensional manifold and provide a necessary and sufficient condition for a Morse function to be a Morse–Smale function. First, for two critical points $\boldsymbol{p}, \boldsymbol{q} \in \mathscr{C}(f)$, the intersection $W^s(\boldsymbol{p}) \cap W^u(\boldsymbol{q})$ may be non-empty only for the following cases:

(1) if $\boldsymbol{p} = \boldsymbol{q}$,
(2) if $\boldsymbol{p}$ is a minimum and $\boldsymbol{q}$ is a maximum,
(3) if $\boldsymbol{p}$ is a minimum and $\boldsymbol{q}$ is a saddle point,
(4) if $\boldsymbol{p}$ is a saddle point and $\boldsymbol{q}$ is a maximum, or
(5) if both $\boldsymbol{p}$ and $\boldsymbol{q}$ are saddle points.

In the first four cases, it is straightforward to check that the transversality condition is satisfied. In the last case we have that if $W^s(\boldsymbol{p}) \cap W^u(\boldsymbol{q}) \neq \emptyset$ then $W^s(\boldsymbol{p}) \cap W^u(\boldsymbol{q})$ equals to the gradient flow line (also Neumann line in this case) which asymptotically starts at $\boldsymbol{q}$ and ends at $\boldsymbol{p}$. In such a case we get that for all $\boldsymbol{x} \in W^s(\boldsymbol{p}) \cap W^u(\boldsymbol{q})$, the tangent spaces obey $T_{\boldsymbol{x}} W^s(\boldsymbol{p}) = T_{\boldsymbol{x}} W^u(\boldsymbol{q})$ and those are one-dimensional, so their sum cannot be equal to the two-dimensional $T_{\boldsymbol{x}} M$. Therefore, in this case the transversality condition, (A.1) is not satisfied and as a conclusion we get the following.

Proposition A.7 *On a two-dimensional manifold, a Morse function is Morse–Smale if and only if there is no Neumann line connecting two saddle points.*

By the Kupka–Smale theorem (see [14]) Morse–Smale gradient vector fields are generic among the set of all vector fields. Currently, there is no similar genericity result regarding eigenfunctions of elliptic operators which are Morse–Smale (in the spirit of [71, 72]). Our preliminary numerics suggest that Morse–Smale eigenfunctions are indeed generic.

References

[1] J. H. ALBERT, *Topology of the nodal and critical point sets for eigenfunctions of elliptic operators*, ProQuest LLC, Ann Arbor, MI, 1972. Thesis (Ph.D.), Massachusetts Institute of Technology.

[2] L. ALON, *Generic eigenfunctions of quantum graphs*. In preparation.

[3] L. ALON AND R. BAND, *Neumann domains on quantum graphs*, arXiv:1911.12435.

[4] L. ALON, R. BAND, AND G. BERKOLAIKO, *Nodal statistics on quantum graphs*, Comm. Math. Phys., 362 (2018), pp. 909–948.

[5] R. BAND, *The nodal count* $\{0, 1, 2, 3, \dots\}$ *implies the graph is a tree*, Philos. Trans. R. Soc. Lond. A, 372 (2014), pp. 20120504, 24.

[6] R. Band, G. Berkolaiko, C. H. Joyner, and W. Liu, *Quotients of finite-dimensional operators by symmetry representations*, arXiv preprint arXiv:1711.00918, (2017).

[7] R. Band, G. Berkolaiko, H. Raz, and U. Smilansky, *The number of nodal domains on quantum graphs as a stability index of graph partitions*, Commun. Math. Phys., 311 (2012), pp. 815–838.

[8] R. Band, G. Cox, and S. Egger, *Spectral properties of Neumann domains via the Dirichlet-to-Neumann operator*. In preparation.

[9] R. Band, S. Egger, and A. Taylor, *The spectral position of Neumann domains on the torus*, arXiv:1707.03488.

[10] R. Band and D. Fajman, *Topological properties of Neumann domains*, Ann. Henri Poincaré, 17 (2016), pp. 2379–2407.

[11] R. Band, I. Oren, and U. Smilansky, *Nodal domains on graphs – how to count them and why?*, in Analysis on graphs and its applications, vol. 77 of Proc. Sympos. Pure Math., Amer. Math. Soc., Providence, RI, 2008, pp. 5–27.

[12] R. Band, O. Parzanchevski, and G. Ben-Shach, *The isospectral fruits of representation theory: quantum graphs and drums*, J. Phys. A, 42 (2009), pp. 175202, 42.

[13] R. Band, T. Shapira, and U. Smilansky, *Nodal domains on isospectral quantum graphs: the resolution of isospectrality?*, J. Phys. A, 39 (2006), pp. 13999–14014.

[14] A. Banyaga and D. Hurtubise, *Lectures on Morse Homology*, Kluwer Acad. Pub., 2004.

[15] P. Bérard and B. Helffer, *The weak Pleijel theorem with geometric control*, J. Spectr. Theory, 6 (2016), pp. 717–733.

[16] G. Berkolaiko, *A lower bound for nodal count on discrete and metric graphs*, Commun. Math. Phys., 278 (2007), pp. 803–819.

[17] G. Berkolaiko, *Nodal count of graph eigenfunctions via magnetic perturbation*, Anal. PDE, 6 (2013), pp. 1213–1233.

[18] G. Berkolaiko and P. Kuchment, *Introduction to quantum graphs*, vol. 186 of Mathematical Surveys and Monographs, AMS, 2013.

[19] G. Berkolaiko, P. Kuchment, and U. Smilansky, *Critical partitions and nodal deficiency of billiard eigenfunctions*, Geom. Funct. Anal., 22 (2012), pp. 1517–1540. preprint arXiv:1107.3489.

[20] G. Berkolaiko and W. Liu, *Simplicity of eigenvalues and non-vanishing of eigenfunctions of a quantum graph*, J. Math. Anal. Appl., 445 (2017), pp. 803–818. preprint arXiv:1601.06225.

[21] G. Berkolaiko, H. Raz, and U. Smilansky, *Stability of nodal structures in graph eigenfunctions and its relation to the nodal domain count*, J. Phys. A, 45 (2012), p. 165203.

[22] G. Berkolaiko and T. Weyand, *Stability of eigenvalues of quantum graphs with respect to magnetic perturbation and the nodal count of the eigenfunctions*, Philos. Trans. R. Soc. Lond. Ser. A Math. Phys. Eng. Sci., 372 (2014), pp. 20120522, 17.

[23] S. Biasotti, L. De Floriani, B. Falcidieno, P. Frosini, D. Giorgi, C. Landi, L. Papaleo, and M. Spagnuolo, *Describing shapes by*

geometrical-topological properties of real functions, ACM Comput. Surv., 40 (2008), pp. 12:1–12:87.

[24] G. Blum, S. Gnutzmann, and U. Smilansky, *Nodal domains statistics: A criterion for quantum chaos*, Phys. Rev. Lett., 88 (2002), p. 114101.

[25] V. Bonnaillie-Noël and B. Helffer, *Nodal and spectral minimal partitions—the state of the art in 2016*, in Shape optimization and spectral theory, De Gruyter Open, Warsaw, 2017, pp. 353–397.

[26] J. Bourgain, *On Pleijel's nodal domain theorem*, Int. Math. Res. Not. IMRN, (2015), pp. 1601–1612.

[27] L. Buhovsky, A. Logunov, and M. Sodin, *Eigenfunctions with infinitely many isolated critical points*, International Mathematics Research Notices, (2019). rnz181.

[28] P. Charron, B. Helffer, and T. Hoffmann-Ostenhof, *Pleijel's theorem for Schrödinger operators with radial potentials*, ArXiv e-prints, (2016).

[29] A. Chattopadhyay, G. Vegter, and C. K. Yap, *Certified computation of planar morse-smale complexes*, Journal of Symbolic Computation, 78 (2017), pp. 3 – 40. Algorithms and Software for Computational Topology.

[30] Y. Colin de Verdière, *Magnetic interpretation of the nodal defect on graphs*, Anal. PDE, 6 (2013), pp. 1235–1242.

[31] R. Courant, *Ein allgemeiner Satz zur Theorie der Eigenfuktionen selbstadjungierter Differentialausdrücke*, Nachr. Ges. Wiss. Göttingen Math Phys, (1923), pp. 81–84.

[32] S. Dong, P. Bremer, M. Garland, V. Pascucci, and J. C. Hart, *Spectral surface quadrangulation*, in ACM SIGGRAPH 2006 Papers, SIGGRAPH '06, New York, NY, USA, 2006, ACM, pp. 1057–1066.

[33] H. Edelsbrunner, J. Harer, and Z. A., *Hierarchical morse-smale complexes for piecewise linear 2-manifolds*, Discrete & Computational Geometry, 30 (2003), pp. 87–107.

[34] H. Edelsbrunner, J. Harer, V. Natarajan, and V. Pascucci, *Morse-smale complexes for piecewise linear 3-manifolds*, in Proceedings of the nineteenth annual symposium on Computational geometry, SCG '03, New York, NY, USA, 2003, ACM, pp. 361–370.

[35] Y. Elon, S. Gnutzmann, C. Joas, and U. Smilansky, *Geometric characterization of nodal domains: the area-to-perimeter ratio*, J. Phys. A: Math. Theor., 40 (2007), p. 2689.

[36] H. Federer, *Geometric measure theory*, Die Grundlehren der mathematischen Wissenschaften, Band 153, Springer-Verlag New York Inc., New York, 1969.

[37] L. Friedlander, *Extremal properties of eigenvalues for a metric graph*, Ann. Inst. Fourier (Grenoble), 55 (2005), pp. 199–211.

[38] A. Ghosh, A. Reznikov, and P. Sarnak, *Nodal domains of Maass forms i*, Geometric and Functional Analysis, 23 (2013), pp. 1515–1568.

[39] A. Ghosh, A. Reznikov, and P. Sarnak, *Nodal domains of Maass forms, II*, Amer. J. Math., 139 (2017), pp. 1395–1447.

[40] S. Gnutzmann and S. Lois, *On the nodal count statistics for separable systems in any dimension*, J. Phys. A, 46 (2013), pp. 045201, 18.

[41] S. Gnutzmann and U. Smilansky, *Quantum graphs: Applications to quantum chaos and universal spectral statistics*, Adv. Phys., 55 (2006), pp. 527–625.

[42] P. V. Gyulassy A., Bremer P.-T., *Computing morse-smale complexes with accurate geometry*, IEEE Trans-actions on Visualization and Computer Graphics, 18 (2012).

[43] B. Helffer and T. Hoffmann-Ostenhof, *On a magnetic characterization of spectral minimal partitions*, J. Eur. Math. Soc. (JEMS), 15 (2013), pp. 2081–2092.

[44] B. Helffer and T. Hoffmann-Ostenhof, *A review on large k minimal spectral k-partitions and Pleijel's theorem*, in Spectral theory and partial differential equations, vol. 640 of Contemp. Math., Amer. Math. Soc., Providence, RI, 2015, pp. 39–57.

[45] B. Helffer, T. Hoffmann-Ostenhof, and S. Terracini, *Nodal domains and spectral minimal partitions*, Ann. Inst. H. Poincaré Anal. Non Linéaire, 26 (2009), pp. 101–138.

[46] M. N. Huxley, *Area, lattice points, and exponential sums*, Clarendon Press, 1996.

[47] D. Jakobson and N. Nadirashvili, *Eigenfunctions with few critical points*, J. Differential Geom., 53 (1999), pp. 177–182.

[48] S. u. Jang and J. Jung, *Quantum unique ergodicity and the number of nodal domains of eigenfunctions*, J. Amer. Math. Soc., 31 (2018), pp. 303–318.

[49] J. Jung and M. P. Young, *Sign changes of the eisenstein series on the critical line*, International Mathematics Research Notices, (2017), p. rnx139.

[50] J. Jung and S. Zelditch, *Number of nodal domains and singular points of eigenfunctions of negatively curved surfaces with an isometric involution*, J. Differential Geom., 102 (2016), pp. 37–66.

[51] ——, *Number of nodal domains of eigenfunctions on non-positively curved surfaces with concave boundary*, Math. Ann., 364 (2016), pp. 813–840.

[52] J. Jung and S. Zelditch, *Boundedness of the number of nodal domains for eigenfunctions of generic Kaluza-Klein 3-folds*, ArXiv e-prints, (2018).

[53] P. Kröger, *Upper bounds for the Neumann eigenvalues on a bounded domain in Euclidean space*, J. Funct. Anal., 106 (1992), pp. 353–357.

[54] C. Léna, *Pleijel's nodal domain theorem for Neumann and Robin eigenfunctions*, ArXiv e-prints, (2016).

[55] M. Magee, *Arithmetic, zeros, and nodal domains on the sphere*, Comm. Math. Phys., 338 (2015), pp. 919–951.

[56] J. C. Maxwell, *On hills and dales*, The London, Edinburgh, and Dublin Philosophical Magazine and Journal of Science, 40 (1870), pp. 421–427.

[57] A. Mazzolo, C. de Mulatier, and A. Zoia, *Cauchy's formulas for random walks in bounded domains*, Journal of Mathematical Physics, 55 (2014), p. 083308.

[58] R. B. McDonald and S. A. Fulling, *Neumann nodal domains*, Philos. Trans. R. Soc. Lond. Ser. A Math. Phys. Eng. Sci., 372 (2014), pp. 20120505, 6.

[59] I. Oren and R. Band, *Isospectral graphs with identical nodal counts*, J. Phys. A, 45 (2012), p. 135203. preprint `arXiv:1110.0158`.

[60] O. Parzanchevski and R. Band, *Linear representations and isospectrality with boundary conditions*, J. Geom. Anal., 20 (2010), pp. 439–471.

[61] A. Pleijel, *Remarks on Courant's nodal line theorem*, Communications on Pure and Applied Mathematics, 9 (1956), pp. 543–550.

[62] Y. V. POKORNYĬ, V. L. PRYADIEV, AND A. AL′-OBEĬD, *On the oscillation of the spectrum of a boundary value problem on a graph*, Mat. Zametki, 60 (1996), pp. 468–470.

[63] I. POLTEROVICH, *Pleijel's nodal domain theorem for free membranes*, vol. 137 of Proc. Amer. Math. Soc., 2009.

[64] I. POLTEROVICH, L. POLTEROVICH, AND V. STOJISAVLJEVIĆ, *Persistence barcodes and Laplace eigenfunctions on surfaces*, Geometriae Dedicata, (2018).

[65] M. REUTER, *Hierarchical shape segmentation and registration via topological features of Laplace-Beltrami eigenfunctions*, International Journal of Computer Vision, 89 (2010), pp. 287–308.

[66] P. SARNAK AND I. WIGMAN, *Topologies of nodal sets of random band limited functions*, in Advances in the theory of automorphic forms and their L-functions, vol. 664 of Contemp. Math., AMS, 2016, pp. 351–365.

[67] P. SCHAPOTSCHNIKOW, *Eigenvalue and nodal properties on quantum graph trees*, Waves in Random and Complex Media, 16 (2006), pp. 167–78.

[68] S. STEINERBERGER, *A geometric uncertainty principle with an application to Pleijel's estimate*, Ann. Henri Poincaré, 15 (2014), pp. 2299–2319.

[69] G. SZEGÖ, *Inequalities for certain eigenvalues of a membrane of given area*, J. Rational Mech. Anal., 3 (1954), pp. 343–356.

[70] A. J. TAYLOR, *pyneumann toolkit*. `https://github.com/inclement/neumann`, 2017.

[71] K. UHLENBECK, *Eigenfunctions of Laplace operators*, Bull. Amer. Math. Soc., 78 (1972), pp. 1073–1076.

[72] ———, *Generic properties of eigenfunctions*, Amer. J. Math., 98 (1976), pp. 1059–1078.

[73] M. VAN DEN BERG, D. GRIESER, T. HOFFMANN-OSTENHOF, AND I. POLTEROVICH, *Geometric aspects of spectral theory*. Oberwolfach Report 33/2012, July 2012.

[74] H. F. WEINBERGER, *An isoperimetric inequality for the N-dimensional free membrane problem*, J. Rational Mech. Anal., 5 (1956), pp. 633–636.

[75] H. WEYL, *Über die asymptotische Verteilung der Eigenwerte*, Nachrichten von der Gesellschaft der Wissenschaften zu Göttingen, Mathematisch-Physikalische Klasse, 1911 (1911), pp. 110–117.

[76] S. ZELDITCH, *Eigenfunctions and nodal sets*, Surveys in Differential Geometry, 18 (2013), pp. 237–308.

[77] S. ZELDITCH, *Logarithmic lower bound on the number of nodal domains*, J. Spectr. Theory, 6 (2016), pp. 1047–1086.

[78] A. ZOMORODIAN, *Topology for Computing*, Cambridge Monographs on Applied and Computational Mathematics, Cambridge University Press, 2005.

11

On the Existence and Uniqueness of Self-Adjoint Realizations of Discrete (Magnetic) Schrödinger Operators

Marcel Schmidt

Abstract

In this expository chapter we answer two fundamental questions concerning discrete magnetic Schrödinger operator associated with weighted graphs. We discuss when formal expressions of such operators give rise to self-adjoint operators, i.e., when they have self-adjoint realizations. If such self-adjoint realizations exist, we explore when they are unique.

11.1 Introduction

Discrete Laplacians and discrete magnetic Schrödinger operators feature in many different areas of mathematics. They are used in combinatorics and computer science, appear as discretizations of (pseudo-)differential operators on Riemannian manifolds, serve as toy models for Hamiltonians in mathematical physics, and play an important role in the study of random walks – just to name a few. Even though discrete operators are used for very different means, their basic structure is always the same. Given a graph $G = (X, E)$ consisting of vertices X and edges E one defines the discrete Laplacian Δ acting on functions on X (up to a sign) by

$$\Delta f(x) = \sum_{(x,y)\in E} (f(x) - f(y)).$$

Since this operator itself is often not flexible enough, one then introduces various weights to decorate the discrete Laplacian to obtain a discrete magnetic Schrödinger operator $\mathcal{M}$ of the form

$$\mathcal{M}f(x) = \frac{1}{\mu(x)} \sum_{(x,y)\in E} b(x,y)(f(x) - e^{i\theta(x,y)} f(y)) + V(x)f(x).$$

Here, μ is a weight on the vertices, b is an edge weight, $e^{i\theta(x,y)}$ is a magnetic field, and V is a potential. The operator with vanishing V and θ is called *weighted discrete Laplacian*. For other appropriate choices of weights these operators include the adjacency operator and the transition operator of Markov chains.

From the viewpoint of applications a very important case is when the formal expression for $\mathcal{M}$ gives rise to a self-adjoint operator on $\ell^2(X, \mu)$. Of course, this puts some symmetry restrictions on the weights; the edge weight b has to be symmetric, θ has to be antisymmetric, and V needs to be real-valued. In fact, if X is finite, any self-adjoint matrix is a discrete magnetic Schrödinger operator of this form and if X is infinite, the same holds true for self-adjoint finite range operators on $\ell^2(X, \mu)$.

This text discusses the two most basic questions about discrete magnetic Schrödinger operators.

1. Given μ, b, θ, V does there exist a self-adjoint realization of $\mathcal{M}$ on $\ell^2(X, \mu)$?
2. If there exists a self-adjoint realization of $\mathcal{M}$ on $\ell^2(X, \mu)$, is it unique?

It is our goal to provide expository answers to both questions, which are as comprehensive as possible. We present the best known answer to the first question with two restrictions on the considered operators. We only treat the existence of lower semi-bounded realizations of $\mathcal{M}$ whose domain of the associated quadratic form contains the finitely supported functions; we call such operators *standard realizations* of $\mathcal{M}$. While the first assumption is technical, the second is natural since on discrete spaces functions with finite support can be seen as test functions. For the second question there is vast amount of results and we had to make some choices. Here we focus on current results that involve the geometry of the weighted graph and omit some more abstract criteria.

For the sake of being expository we repeat several known arguments and summarize some basic definitions and properties of quadratic forms in Appendix A. However, we would like to stress that everything that we call a theorem in this text goes in one form or another beyond what can be found in the literature.

The weighted graphs that we treat are not assumed to be locally finite. This has several reasons which are based on applications; here we name three. First of all, the weighted discrete Laplacian of any given graph can be approximated (in the strong resolvent sense) by bounded weighted discrete Laplacians. If the given graph is connected, the graphs of the approximating operators are not locally finite. Secondly, approximations of long-range non-local pseudo-differential operators by weighted discrete Laplacians often make use of graphs that are not locally finite. For concrete examples we refer to [2, 4], where convergence to the generator of stable-like processes is studied. Thirdly, fractional powers of discrete Laplacians, the discrete analogue of generators of stable-like processes, tend to be weighted discrete Laplacians of not locally finite graphs.

Since there has recently been some interest in discrete magnetic Schrödinger operators on functions taking values in vector bundles (see [43, 15] and references therein), everything is formulated in this slightly more general setting (see Section 11.2).

It is folklore that the existence (and uniqueness) of bounded realizations of the weighted discrete Laplacian is equivalent to the weighted vertex degree being bounded (see, e.g., [28]). In Section 11.4 we extend this result to magnetic Schrödinger operators. Here, the difficulty lays in treating potentials without a fixed sign. If the weighted vertex degree is unbounded, the weighted graph Laplacian is an unbounded operator and so answers to our questions naturally become more difficult.

In the case when the underlying graph is locally finite, it is possible to obtain the existence of standard realizations of a magnetic Schrödinger operator by studying self-adjoint extensions of the corresponding minimal magnetic Schrödinger operator acting on finitely supported functions (see Section 11.3.3). Since finitely supported functions are the test functions on discrete spaces, in principle this theory works exactly as for Schrödinger operators on open subsets of Euclidean space.

The real challenge for answering the first question lies with graphs that are not locally finite. In this case discrete magnetic Schrödinger operators need not map finitely supported functions to $\ell^2(X, \mu)$. Therefore, the strategy of the locally finite case, namely, seeking for self-adjoint extensions of the minimal operator, does not work. A first systematic study of discrete Schrödinger operators with non-negative potentials on not necessarily locally finite graphs via Dirichlet forms is given in [28]. Discrete magnetic Schrödinger operators on such graphs with potentials without a fixed sign but with small negative part were then treated in [14, 15] by means of perturbation theory of quadratic forms. The results on the existence of standard realizations of [28, 14, 15] are presented in Section 11.3.1. For discrete (magnetic) Schrödinger operators whose negative part of the potential cannot be treated through perturbation theory, the existence of standard realizations was still open. We settle this issue and prove existence for a very general class of so-called admissible potentials in Section 11.3.2. In the scalar case without magnetic field our results are optimal, see Section 11.3.4. As a by-product we obtain an entirely analytic proof for the closability of associated quadratic forms on finitely supported functions (see Theorems 11.3.8 and 11.3.13), which was only previously obtained through probabilistic arguments (see [14]).

Our second question on uniqueness of self-adjoint realizations of discrete Laplacians and discrete magnetic Schrödinger operators has seen quite some attention in recent years, see [5, 6, 7, 11, 13, 14, 18, 21, 23, 33, 38, 40, 41, 42, 43, 53]. Some of these references also discuss situations where multiple realizations are present. Moreover, uniqueness of realizations plays a role in the study of the discrete heat equation (see [27, 28, 54, 56, 57]), and spectral theory of discrete operators (see [12, 24]).

There are different classes of operators in which uniqueness can be studied. If the graph is locally finite, we study essential self-adjointness of the minimal operator – the magnetic Schrödinger operator on finitely supported functions. This corresponds to uniqueness in the class of all self-adjoint operators. As discussed above, for general not locally finite graphs the minimal operator need not exist and so the concept of essential self-adjointness of the minimal operator makes no sense. In this case one can still ask for uniqueness of standard realizations. By definition (see above) this corresponds to uniqueness in the class of semi-bounded self-adjoint operators. For both cases we present two criteria (and corollaries) involving the geometry of the weighted graph and properties of the underlying measure.

In Section 11.5.1 we show that if the measure is large enough along infinite paths, the discussed uniqueness properties hold (see Theorem 11.5.2 and Corollary 11.5.3). Our presentation is based on results for Schrödinger operators with non-negative potentials from [28] and for scalar magnetic Schrödinger operators from [12]. Section 11.5.2 is devoted to uniqueness criteria that involve the metric geometry with respect to an intrinsic metric. We prove uniqueness if there is no metric boundary or if the potential is growing fast enough towards the boundary (see Theorem 11.5.4 and Corollary 11.5.5). The presented proofs are based on [43], which only treats path metrics. We allow arbitrary intrinsic metrics. As one possible application for this more general situation we discuss metrics coming from embeddings of the graph into Euclidean space or into complete Riemannian manifolds.

In the scalar case with vanishing magnetic field and non-negative potential an important class of standard realizations are non-negative ones whose associated quadratic form is a Dirichlet form. They correspond to Markov processes on the underlying space through the Feynman–Kac formula and are therefore called *Markovian realizations*. Section 11.6 focuses on on them. In Section 11.6.1 we show that the quadratic form of a Markovian realization lies between a minimal Dirichlet form, the one with 'Dirichlet boundary conditions at infinity', and a maximal Dirichlet form, the one with 'Neumann boundary conditions at infinity'. In this generality our result is new, for locally finite graphs it is given in [18]. Our discussion is based on methods developed in [49].

In Section 11.6.2 we study Markov uniqueness, i.e., uniqueness of Markovian realizations. By the previous discussion this is equivalent to the maximal and the minimal Dirichlet form being equal. Of course, the results of Section 11.5 can always be applied, but in general their assumptions are stronger than what is actually needed for Markov uniqueness. We show that smallness of 'the boundary', in the sense that its capacity vanishes, is equivalent to Markov uniqueness (see Theorem 11.6.9). Here, 'the boundary' can have two forms; it can be either a boundary with respect to a compactification or, in the locally finite case, a metric boundary with respect to a path metric. For the abstract boundary coming from a compactification the characterization of Markov uniqueness is taken from [49], while the path metric case is treated in [21]. In Theorem 11.6.15 we characterize Markov uniqueness in the case when the underlying measure is finite. In particular, we relate Markov uniqueness for all finite measures to uniqueness of arbitrary standard realizations and the existence of intrinsic metrics with finite distance balls with respect to particular finite measures. This part is taken from [45].

Even though we try to give comprehensive answers to our two questions, there are several problems that remain unresolved. We collect the – in our view – most important ones in Section 11.7 and comment on their relevance and expected outcome.

For the convenience of the reader (and to fix notation) we recall some basic properties of quadratic forms in Appendix A.

11.2 Discrete (Magnetic) Schrödinger Operators: The Setup and Basic Properties

In this section we introduce weighted graphs, discrete magnetic Schrödinger operators, and Schrödinger forms. We discuss their connections via Green's formulae and Kato's inequality. The contents of this section are all known in one form or another.

11.2.1 Graphs and Discrete Schrödinger Operators

In this section we introduce graphs and the associated (formal) discrete Schrödinger operators.

A *(symmetric) weighted graph* (X, b) consists of a countable set of *vertices* $X \neq \emptyset$ and an *edge weight function* $b : X \times X \to [0, \infty)$ with the following properties:

(b0) $b(x, x) = 0$ for all $x \in X$,
(b1) $b(x, y) = b(y, x)$ for all $x, y \in X$,
(b2) $\sum_{z \in X} b(x, z) < \infty$ for all $x \in X$.

The *vertex degree* of a vertex $x \in X$ is

$$\deg(x) := \sum_{z \in X} b(x, z).$$

We say that two vertices $x, y \in X$ are *connected by an edge* if $b(x, y) > 0$. In this case, we write $x \sim y$. A *path* is a finite or infinite sequence of vertices $(x_1, x_2, \ldots)$ such that $x_j \sim x_{j+1}$ for all $j = 1, 2, \ldots$

For $U \subseteq X$ we define the *combinatorial neighbourhood of* U by

$$N(U) := U \cup \{x \in X \mid \text{there ex. } y \in U \text{ with } y \sim x\}.$$

A weighted graph (X, b) is called *locally finite*, if for any $x \in X$ the set $N(\{x\})$ is finite, i.e., if all vertices have only finitely many neighbours. A vertex $x \in X$ is *isolated* if $N(\{x\}) = \emptyset$.

We equip the vertex set X with the discrete topology and write $C(X)$ for the space of all complex-valued functions on X (the space of continuous complex-valued functions with respect to the discrete topology). Its subspace of functions with finite support (the functions of compact support for the discrete topology) is denoted by $C_c(X)$. Moreover, we write $\ell^\infty(X)$ for the space of bounded complex-valued functions on X.

Given two real-valued functions $f, g \in C(X)$ we denote by $f \wedge g := \min\{f, g\}$ their minimum and by $f \vee g := \max\{f, g\}$ their maximum. Moreover, we let $f_+ = f \vee 0$ and $f_- = (-f) \vee 0$ be the *positive part* and the *negative part* of f, respectively.

For a set $K \subseteq X$ we let $1_K : X \to \{0, 1\}$ be the *indicator function of* K, i.e., $1_K(x) = 1$, if $x \in K$, and $1_K(x) = 0$, else. The indicator function of the singleton set $\{x\}$ is denoted by δ_x.

A strictly positive function $\mu : X \to (0, \infty)$ is called *weight*. Any weight μ induces a Radon measure of full support on all subsets of X by letting

$$\mu(A) := \sum_{x \in A} \mu(x), \quad A \subseteq X.$$

In what follows we shall not distinguish between the weight and the measure it induces.

If we equip $C(X)$ with the topology of pointwise convergence, its continuous dual space $C(X)'$ is isomorphic to $C_c(X)$. Given a weight μ, the dual pairing $(\cdot, \cdot) : C_c(X) \times C(X) \to \mathbb{C}$,

$$(\varphi, f) := \sum_{x \in X} \overline{\varphi(x)} f(x) \mu(x)$$

induces an anti-linear isomorphism via $C_c(X) \to C(X)', \varphi \mapsto (\varphi, \cdot)$. Note that $(\cdot, \cdot)$ depends on the choice of μ.

The space of square integrable functions with respect to a weight μ is denoted by

$$\ell^2(X, \mu) := \{f \in C(X) \mid \sum_{x \in X} |f(x)|^2 \mu(x) < \infty\}.$$

It is a Hilbert space when equipped with the inner product

$$\langle f, g \rangle_2 := \sum_{x \in X} \overline{f(x)} g(x) \mu(x).$$

The norm induced by $\langle \cdot, \cdot \rangle_2$ is denoted by $\| \cdot \|_2$. Note that for $\varphi \in C_c(X)$ and $f \in \ell^2(X, \mu)$ we have $\langle \varphi, f \rangle_2 = (\varphi, f)$. As for the inner product $\langle \cdot, \cdot \rangle_2$, throughout the text all inner products are assumed to be linear in the second argument and anti-linear in the first.

To a graph (X, b), a weight μ and a real-valued $V \in C(X)$ we associate the *formal discrete Schrödinger operator* $\mathcal{H} = \mathcal{H}_{\mu, V} : \mathcal{F} \to C(X)$, where

$$\mathcal{F} := \{f \in C(X) \mid \sum_{y \in X} b(x, y)|f(y)| < \infty \text{ for all } x \in X\},$$

on which $\mathcal{H}$ acts by

$$\mathcal{H} f(x) := \frac{1}{\mu(x)} \sum_{y \in X} b(x, y)(f(x) - f(y)) + V(x) f(x), \quad x \in X.$$

The definition of $\mathcal{F}$ ensures that this sum converges absolutely. Bounded functions are always contained in $\mathcal{F}$ and we have $\mathcal{F} = C(X)$ if and only if the graph (X, b) is locally finite. Sometimes, a real-valued function V as above is simply called *potential*.

11.2.2 Discrete Magnetic Schrödinger Operators

In this section we introduce discrete magnetic Schrödinger operators. They are vector-valued versions of discrete Schrödinger operators with an additional magnetic interaction term.

A *Hermitian vector bundle* over a countable discrete base space X is a collection $E = (E_x)_{x\in X}$ of isomorphic complex Hilbert spaces of dimension greater or equal to 1. We denote the inner product on E_x by $\langle\cdot,\cdot\rangle_x$ and the induced norm by $|\cdot|_x$. Often it is clear from the context in which of the E_x we consider the inner product and the norm. If this is the case, we drop the subscript x. We write

$$\Gamma(X; E) := \{f : X \to \bigsqcup_{x\in X} E_x \mid f(x) \in E_x\}$$

for the space of all *sections* of the bundle E. The subspace of all *sections with finite support* is denoted by $\Gamma_c(X; E)$. For a given weight μ on X we define $(\cdot,\cdot)_E : \Gamma_c(X; E) \times \Gamma(X; E) \to \mathbb{C}$ by

$$(\varphi, f)_E = \sum_{x\in X} \langle\varphi(x), f(x)\rangle\mu(x).$$

Scalar-valued functions naturally act on the space of all sections by pointwise scalar multiplication so that $\Gamma(X; E)$ is a module over $C(X)$. More precisely, for $\varphi \in C(X)$ and $f \in \Gamma(X; E)$ we define $\varphi f \in \Gamma(X; E)$ by $(\varphi f)(x) := \varphi(x)f(x)$.

For $f \in \Gamma(X; E)$ we let $|f| \in C(X)$ be given by $|f|(x) = |f(x)|$ and $\operatorname{sgn} f \in \Gamma(X; E)$ by

$$\operatorname{sgn} f(x) = \begin{cases} \frac{1}{|f(x)|} f(x) & \text{if } f(x) \neq 0 \\ 0 & \text{if } f(x) = 0 \end{cases}.$$

With this notation any $f \in \Gamma(X; E)$ can be written as $f = |f|\operatorname{sgn} f$.

Given a weight μ on X the corresponding space of *ℓ^2-sections* is

$$\ell^2(X, \mu; E) := \{f \in \Gamma(X; E) \mid \sum_{x\in X} |f(x)|^2\mu(x) < \infty\}.$$

It is a Hilbert space when equipped with the inner product

$$\langle f, g\rangle_{2;\,E} := \sum_{x\in X} \langle f(x), g(x)\rangle\mu(x).$$

The norm induced by $\langle\cdot,\cdot\rangle_{2;\,E}$ is denoted by $\|\cdot\|_{2;\,E}$.

Remark The identity $\langle\varphi, f\rangle_{2;\,E} = (\varphi, f)_E$ holds for $\varphi \in \Gamma_c(X; E)$ and $f \in \ell^2(X, \mu; E)$ and a similar statement is valid for $\langle\cdot,\cdot\rangle_2$ and $(\cdot,\cdot)$ in the scalar case (cf. Section 11.2.1). Because of this it is tempting to abuse notation and denote both pairings by the same symbol. In this text we introduced different notation for a reason. Often we can easily perform pointwise computations that lead to an identity or inequality on the level of $(\varphi, f)_E$ for pairs $\varphi \in \Gamma_c(X; E)$, $f \in S$, where S is some subspace of $\Gamma(X; E)$. However, in applications, we are truly interested in these identities on the level of $\langle g, f\rangle_{2;\,E}$ for pairs $g \in G$ and $f \in F$, where F and G are subspaces of $\ell^2(X, \mu; E)$ with $\Gamma_c(X; E) \subseteq F, G$. In order to lift results from $(\cdot, \cdot)_E$ to $\langle \cdot, \cdot\rangle_{2;\,E}$ usually additional arguments are needed. In order to make such arguments more transparent, we distinguish the pairings.

A *unitary connection* on a Hermitian bundle $E = (E_x)_{x\in X}$ over X is a family $\Phi = (\Phi_{x,y})_{x,y\in X}$ of unitary maps $\Phi_{x,y} : E_y \to E_x$ such that $\Phi_{x,y} = \Phi_{y,x}^{-1}$. A *bundle endomorphism* on a Hermitian vector bundle $E = (E_x)_{x\in X}$ over X is a collection $W = (W_x)_{x\in X}$ of bounded linear maps $W_x : E_x \to E_x$. It is called *self-adjoint*, if for all $x \in X$ the operator W_x is self-adjoint.

Let (X, b) be a graph and E be a Hermitian vector bundle over X. To a weight μ on X, a self-adjoint bundle endomorphism W and a unitary connection Φ on E we associate the *formal magnetic Schrödinger operator* $\mathcal{M} = \mathcal{M}_{\mu,\Phi,W} : \mathcal{F}_E \to \Gamma(X; E)$, where

$$\mathcal{F}_E := \{f \in \Gamma(X; E) \mid \sum_{y\in X} b(x, y)|f(y)| < \infty \text{ for all } x \in X\},$$

on which $\mathcal{M}$ acts by

$$\mathcal{M}f(x) := \frac{1}{\mu(x)} \sum_{y\in X} b(x, y)(f(x) - \Phi_{x,y} f(y)) + W_x f(x), \quad x \in X.$$

Note that since $|\Phi_{x,y} f(y)|_x = |f(y)|_y$, for each $f \in \mathcal{F}_E$ the sum in the definition of $\mathcal{M}f(x)$ converges absolutely in the Hilbert space E_x.

Remark (Scalar magnetic Schrödinger operators) If $E_x = \mathbb{C}$ for all $x \in X$, then $\Gamma(X; E)$ can be identified with $C(X)$. In this case, a self-adjoint bundle endomorphism W of E acts on $\Gamma(X; E)$ as multiplication by a potential, i.e., there exists a real-valued $V \in C(X)$ such that $W_x f(x) = V(x) f(x)$ for $x \in X$. Moreover, any connection Φ of E is parametrized by a function

$$\theta : X \times X \to \mathbb{R}/2\pi\mathbb{Z},$$

with $\theta(x, y) = -\theta(x, y)$ for all $x, y \in X$, via the identity

$$\Phi_{x,y} z = e^{i\theta(x,y)} z, \quad z \in \mathbb{C}.$$

The corresponding magnetic Schrödinger operator is called *scalar magnetic Schrödinger operator* and is denoted by $\mathcal{M}_{\mu,\theta,V}$. If $\theta = 0$, then $\Phi_{x,y} = \mathrm{Id}$ and $\mathcal{M}_{\mu,0,V} = \mathcal{H}_{\mu,V}$. Therefore, the discrete Schrödinger operators discussed in Section 11.2.1 are a special case of discrete magnetic Schrödinger operators.

11.2.3 Standing Assumptions and Notation

Unless otherwise specified we always assume the following.

- (X, b) is a weighted graph.
- μ is a weight on X and $V \in C(X)$ is real-valued.
- $\mathcal{H} = \mathcal{H}_{\mu,V}$ is the associated formal discrete Schrödinger operator.
- $E = (E_x)_{x\in X}$ is a Hermitian vector bundle over X, $\Phi = (\Phi_{x,y})_{x,y\in X}$ is a unitary connection on E and W is a self-adjoint bundle endomorphism on E.
- $\mathcal{M} = \mathcal{M}_{\mu,\Phi,W}$ is the associated formal discrete Schrödinger operator.

Whenever no confusion can arise we suppress the dependence of objects from these data in our notation. Often we deal with complex-valued functions and sections (E-valued functions) at the same time. Therefore, a subscript E indicates spaces of E-valued functions or functionals on E-valued functions.

11.2.4 Green's Formula and Domination

In this section we discuss Green's formula for the magnetic Schrödinger operator $\mathcal{M}$, which is in principle an integration-by-parts formula. Moreover, we give a version of Kato's inequality on the level of the pairing $(\cdot, \cdot)_E$.

The following integration-by-parts formula is an extension of [16, Lemma 4.7] to magnetic Schrödinger operators.

Lemma 11.2.1 (Green's formula) *For all $f \in \mathcal{F}_E$ and $\varphi \in \Gamma_c(X; E)$ the following sums converge absolutely and satisfy the stated identities.*

$$\begin{aligned}(\varphi, \mathcal{M}f)_E &= \sum_{x\in X}\langle\varphi(x), \mathcal{M}f(x)\rangle\mu(x) = \sum_{x\in X}\langle\mathcal{M}\varphi(x), f(x)\rangle\mu(x)\\ &= \frac{1}{2}\sum_{x,y\in X} b(x,y)\langle\varphi(x) - \Phi_{x,y}\varphi(y), f(x) - \Phi_{x,y}f(y)\rangle\\ &\quad + \sum_{x\in X}\langle W_x\varphi(x), f(x)\rangle\mu(x).\end{aligned}$$

Proof. The first identity is just writing out the definition of the pairing $(\cdot, \cdot)_E$. Since the support of φ is finite, we have

$$\sum_{x,y \in X} b(x,y)\,|\langle \varphi(x), f(x)\rangle| = \sum_{x \in X} |\langle \varphi(x), f(x)\rangle| \deg(x) < \infty.$$

Moreover, the finiteness of the support of φ, $f \in \mathcal{F}_E$ and the Cauchy–Schwarz inequality yield

$$\sum_{x,y \in X} b(x,y)|\langle \Phi_{x,y}\varphi(y), f(x)\rangle| \le \sum_{y \in X} |\varphi(y)| \sum_{x \in X} b(x,y)|f(x)| < \infty.$$

With these integrability properties at hand, the convergence of the sums and the other identities follow from re-arranging the sum

$$\frac{1}{2}\sum_{x,y \in X} b(x,y)\langle \varphi(x) - \Phi_{x,y}\varphi(y), f(x) - \Phi_{x,y} f(x)\rangle + \sum_{x \in X} \langle W_x \varphi(x), f(x)\rangle \mu(x)$$

with the help of Fubini's theorem. □

For a bundle endomorphism W and a potential V we write $W \ge V$ if for all sections $f \in \Gamma(X; E)$ and $x \in X$ the inequality $\langle W_x f(x), f(x)\rangle \ge V(x)|f(x)|^2$ holds. The following lemma provides some form of Kato's inequality for the magnetic operator $\mathcal{M} = \mathcal{M}_{\mu,\Phi,W}$ and the discrete Schrödinger operator $\mathcal{H} = \mathcal{H}_{\mu,V}$ when $W \ge V$. The presented proof is taken from [34].

Lemma 11.2.2 (Kato's inequality) *Let $W \ge V$. If $f \in \mathcal{F}_E$ and $\varphi \in \Gamma_c(X; E)$ with $\langle f(x), \varphi(x)\rangle = |f(x)||\varphi(x)|$ for all $x \in X$, then*

$$\operatorname{Re}(\varphi, \mathcal{M} f)_E \ge (|\varphi|, \mathcal{H}|f|).$$

Proof. It follows from the definitions that $f \in \mathcal{F}_E$ implies $|f| \in \mathcal{F}$. We have $\langle f(x), \varphi(x)\rangle = |f(x)||\varphi(x)|$ by assumption. Moreover, the Cauchy–Schwarz inequality and $|\Phi_{x,y} f(y)|_x = |f(y)|_y$ yields

$$-\operatorname{Re}\langle \Phi_{x,y} f(y), \varphi(x)\rangle \ge -|f(y)||\varphi(x)|.$$

Summing up these inequalities, multiplying them by $b(x,y)$ and summing over $x, y \in X$ shows the desired inequality for the operators $\mathcal{M}_{\mu,\Phi,0}$ and $\mathcal{H}_{\mu,0}$.

It remains to prove the inequality $\operatorname{Re}\langle W_x f(x), \varphi(x)\rangle \ge V(x)|f(x)||\varphi(x)|$. Without loss of generality we can assume $f(x) \ne 0 \ne \varphi(x)$. Since

$$|f(x)||\varphi(x)| = \langle f(x), \varphi(x)\rangle = |f(x)||\varphi(x)|\langle \operatorname{sgn} f(x), \operatorname{sgn}\varphi(x)\rangle,$$

we have $\operatorname{sgn} f(x) = \operatorname{sgn}\varphi(x)$. Using $W \geq V$ and $|\operatorname{sgn} f(x)| = 1$, we obtain

$$\langle W_x f(x), \varphi(x)\rangle = |f(x)||\varphi(x)|\langle W_x \operatorname{sgn} f(x), \operatorname{sgn} f(x)\rangle \geq |f(x)||\varphi(x)|V(x).$$

This finishes the proof. □

As a corollary we obtain that solutions to the eigenvalue equation with respect to $\mathcal{M}$ yield non-negative sub-solutions to the eigenvalue equation with respect to $\mathcal{H}$.

Corollary 11.2.3 *Let $f \in \mathcal{F}_E$ and $\lambda \in \mathbb{R}$ with $\mathcal{M}f = \lambda f$. Then $\mathcal{H}|f| \leq \lambda|f|$.*

Proof. For $z \in X$ let $\xi \in E_z$ with $|\xi| = 1$ and consider the section $\varphi \in \Gamma_c(X; E)$ that is given by

$$\varphi(x) = \begin{cases} \frac{1}{\mu(z)|f(z)|} f(z) & \text{if } x = z \text{ and } f(z) \neq 0 \\ \frac{1}{\mu(z)}\xi & \text{if } x = z \text{ and } f(z) = 0\,. \\ 0 & \text{else} \end{cases}$$

It satisfies $|\varphi(z)| = \mu(z)^{-1}$ and $\varphi(x) = 0$ for $x \neq z$. Therefore, $\langle f(x), \varphi(x)\rangle = 0 = |\varphi(x)||f(x)|$ if $x \neq z$ and $\langle \varphi(z), f(z)\rangle = \mu(z)^{-1}|f(z)| = |\varphi(z)||f(z)|$. From Kato's inequality we infer

$$\lambda|f(z)| = (\varphi, \lambda f)_E = (\varphi, \mathcal{M}f)_E \geq (|\varphi|, \mathcal{H}|f|) = \mathcal{H}|f|(z).$$

This finishes the proof. □

11.2.5 Schrödinger Forms and a Ground-State Transform

In this section we introduce Schrödinger forms, which are related to magnetic Schrödinger operators by Green's formula. One possible domain for such forms are the functions of finite support another choice are functions of finite energy.

If A is bounded and self-adjoint, the operator $|A|$ is defined by means of the functional calculus. Moreover, we let $A_+ = \frac{1}{2}(|A|+A)$ and $A_- = \frac{1}{2}(|A|-A)$. Then $|A|$, A_+, A_- are non-negative self-adjoint operators and $A = A_+ - A_-$.

The space of *sections of finite magnetic energy* $\mathcal{D}_E = \mathcal{D}_{\Phi,W;\,E}$ is defined by

$$\mathcal{D}_E := \{f \in \Gamma(X; E) \mid \sum_{x,y\in X} b(x,y)|f(x) - \Phi_{x,y} f(y)|^2 + \sum_{x\in X}\langle |W_x| f(x), f(x)\rangle \mu(x) < \infty\}.$$

On it we introduce the sesquilinear *magnetic Schrödinger form* $\mathcal{Q}_E = \mathcal{Q}_{\Phi,W;\,E} : \mathcal{D}_E \times \mathcal{D}_E \to \mathbb{C}$, which acts by

$$\mathcal{Q}_E(f,g) := \frac{1}{2}\sum_{x,y\in X} b(x,y)\langle f(x) - \Phi_{x,y}f(y), g(x) - \Phi_{x,y}g(y)\rangle + \sum_{x\in X}\langle W_x f(x), g(x)\rangle \mu(x).$$

It follows from the summability condition (b2) of the graph (X,b) that $\Gamma_c(X;E) \subseteq \mathcal{D}_E$. We denote the restriction of $\mathcal{Q}_E$ to $\Gamma_c(X;E)$ by $\mathcal{Q}_E^c = \mathcal{Q}^c_{\Phi,W;\,E}$. The next lemma is a consequence of Green's formula. It relates $\mathcal{Q}_E$ and $\mathcal{Q}_E^c$ to the operator $\mathcal{M}$.

Lemma 11.2.4 *The inclusion $\mathcal{D}_E \subseteq \mathcal{F}_E$ holds and for all $\varphi \in \Gamma_c(X;E)$ and $f \in \mathcal{D}_E$ we have $\mathcal{Q}_E(\varphi, f) = (\varphi, \mathcal{M}f)_E$. In particular, if $f \in \Gamma_c(X;E)$, then $\mathcal{Q}_E^c(\varphi, f) = (\varphi, \mathcal{M}f)_E$.*

Proof. Because of Lemma 11.2.1 it suffices to prove the inclusion $\mathcal{D}_E \subseteq \mathcal{F}_E$. The identity $|\Phi_{x,y}f(y)| = |f(y)|$ and the Cauchy–Schwarz inequality yield

$$\begin{aligned}\sum_{y\in X} b(x,y)|f(y)| &\leq \sum_{y\in X} b(x,y)|f(x) - \Phi_{x,y}f(y)| + |f(x)|\deg(x)\\ &\leq \deg(x)^{1/2}\left(\sum_{y\in X} b(x,y)|f(x) - \Phi_{x,y}f(y)|^2\right)^{1/2}\\ &\quad + |f(x)|\deg(x).\end{aligned}$$

This finishes the proof. □

As remarked in Section 11.2.2, if $E = (\mathbb{C})_{x\in X}$, then $\Gamma(X;E) = C(X)$ and $\Gamma_c(X;E) = C_c(X)$. Moreover, the unitary connection $\Phi = (\Phi_{x,y})_{x,y}$ is parametrized by an anti-symmetric function $\theta : X \times X \to \mathbb{R}/2\pi\mathbb{Z}$ via $\Phi_{x,y} = e^{i\theta(x,y)}$ and the bundle endomorphism W is induced by a real-valued function V. In this case, we drop the subscript E and write $\mathcal{Q}_{\theta,V}$ respectively $\mathcal{Q}^c_{\theta,V}$ for the associated *scalar magnetic Schrödinger forms*.

If, moreover, $\Phi_{x,y} = \mathrm{Id}$, which is the same as $\theta = 0$, we also drop the subscript θ and write $\mathcal{Q} = \mathcal{Q}_V := \mathcal{Q}_{0,V}$ respectively $\mathcal{Q}^c = \mathcal{Q}^c_V := \mathcal{Q}^c_{0,V}$ for the associated *Schrödinger forms*. The corresponding space of *functions of finite energy* $\mathcal{D} = \mathcal{D}_V$ is

$$\mathcal{D} = \{f \in C(X) \mid \sum_{x,y\in X} b(x,y)|f(x) - f(y)|^2 + \sum_{x\in X}|V(x)||f(x)|^2\mu(x) < \infty\},$$

and $\mathcal{Q}$ acts upon $\mathcal{D}$ by

$$\mathcal{Q}(f,g) = \frac{1}{2} \sum_{x,y \in X} b(x,y)\overline{(f(x) - f(y))}(g(x) - g(y))$$

$$+ \sum_{x \in X} V(x)\overline{f(x)}g(x)\mu(x).$$

In this situation the formula discussed in Lemma 11.2.4 reads

$$\mathcal{Q}^c(f,g) = (f, \mathcal{H}g), \quad f,g \in C_c(X).$$

We finish this section with the discussion of a ground-state transform for the operator $\mathcal{H}$ and the form $\mathcal{Q}^c$. For a non-negative function f we set

$$b^{(f)} : X \times X \to [0,\infty),\ b^{(f)}(x,y) = f(x)f(y)b(x,y).$$

Clearly, $b^{(f)}(x,x) = 0$ and $b^{(f)}(x,y) = b^{(f)}(y,x)$ for all $x, y \in X$. Therefore, $b^{(f)}$ satisfies assumptions (b0) and (b1). It satisfies (b2) if and only if $f \in \mathcal{F}$. In this case, $(X, b^{(f)})$ is a weighted graph in the sense of Section 11.2.1. The Schrödinger form on $C_c(X)$ that is associated with the graph $(X, b^{(f)})$ and the potential $V = 0$ is denoted by $\mathcal{Q}^{c,f}$. If f is a sub-solution to an eigenvalue equation, then $\mathcal{Q}^c$ and $\mathcal{Q}^{c,f}$ are related through the following lemma. The presented proof is taken from [16], which contains a version for non-negative potentials.

Lemma 11.2.5 (Ground-state transform) *Let $f \in \mathcal{F}$ be non-negative and $\lambda \in \mathbb{R}$ such that $\mathcal{H}f \le \lambda f$. For all $\varphi \in C_c(X)$ the inequality*

$$\mathcal{Q}^c(f\varphi) \le \mathcal{Q}^{c,f}(\varphi) + \lambda \|f\varphi\|_2^2.$$

holds.

Proof. Since f is non-negative and $f|\varphi|^2$ has compact support, the inequality $\mathcal{H}f \le \lambda f$ yields

$$\lambda \|f\varphi\|_2^2 = (f|\varphi|^2, \lambda f) \ge (f|\varphi|^2, \mathcal{H}f).$$

From Green's formula (Lemma 11.2.1) we infer

$$(f|\varphi|^2, \mathcal{H}f) = \frac{1}{2} \sum_{x,y \in X} b(x,y)(f(x) - f(y))(f(x)|\varphi(x)|^2 - f(y)|\varphi(y)|^2)$$

$$+ \sum_{x \in X} V(x)|f\varphi|(x)^2.$$

With this at hand, the desired inequality follows from the identity

$$(f(x)-f(y))(f(x)|\varphi(x)|^2 - f(y)|\varphi(y)|^2) = |f(x)\varphi(x) - f(y)\varphi(y)|^2 - f(x)f(y)|\varphi(x)-\varphi(y)|^2$$

and the fact that all occurring sums converge due to $f \in \mathcal{F}$ and the finite support of $f\varphi$. □

11.3 Existence of Realizations of $\mathcal{H}$ and $\mathcal{M}$

In this section we discuss the existence of self-adjoint realizations of the formal operators $\mathcal{H}$ and $\mathcal{M}$ with two additional properties. The operators that we seek for are lower semi-bounded and they reflect the discrete nature of the underlying space, i.e., we assume that test functions are cores in the domains of the associated quadratic forms. This is made precise in Definition 11.3.1. In Section 11.3.1 we prove that such realizations exist for non-negative bundle endomorphisms and potentials (see Proposition 11.3.4), and we use perturbation theory of quadratic forms to lift this result to bundle endomorphisms and potentials with small negative part (see Proposition 11.3.7). In Section 11.3.2 we prove the existence of realizations for scalar Schrödinger operators with admissible potentials (see Theorem 11.3.13), and lift this result to magnetic Schrödinger operators with the help of domination of the associated quadratic forms (see Theorem 11.3.8). Domination can be avoided when the underlying graph is locally finite or satisfies the weaker finiteness condition (FC); this is the content of Section 11.3.3. Instead of discussing literature and examples in the main text, for the most part we postpone them to Section 11.3.4.

Definition 11.3.1 (Discrete magnetic Schrödinger operator) A self-adjoint operator M on $\ell^2(X,\mu;E)$ is called a *realization of* $\mathcal{M}$ if $D(M) \subseteq \mathcal{F}_E$ and $Mf = \mathcal{M}f$ for all $f \in D(M)$. A realization M of $\mathcal{M}$ is called a *standard realization* if, additionally, the following conditions are satisfied.

(R1) M is semi-bounded from below.
(R2) The domain of the quadratic form of M contains $\Gamma_c(X;E)$.

A self-adjoint operator is called *discrete magnetic Schrödinger operator* if it is a standard realization of a formal magnetic Schrödinger operator.

Remark 1. Condition (R1) is technical and restricts the class of operators to the ones that can be treated with quadratic form methods. In most applications the considered discrete magnetic Schrödinger operators satisfy this assumption. However, there is one instance where this is not the case. The adjacency operator that acts formally on functions in $\mathcal{F}$ by

$$Af(x) = \frac{1}{\mu(x)} \sum_{y \in X} b(x, y) f(y)$$

may have self-adjoint realizations without being lower semi-bounded (see [11]). Note that A equals the scalar magnetic Schrödinger operator $\mathcal{M}_{\mu,-\pi,-\mathrm{Deg}}$, where $\mathrm{Deg} = \mu^{-1} \deg$.

2. Condition (R2) pays tribute to the fact that we deal with discrete operators. The space $\Gamma_c(X; E)$ can be seen as the space of test functions (sections). It is tempting to replace (R2) by the stronger $\Gamma_c(X; E) \subseteq D(M)$. However, we shall see below that for graphs which are not locally finite there exist standard realizations M of $\mathcal{M}$ that satisfy (R2) but not $\Gamma_c(X; E) \subseteq D(M)$. For the existence part see Theorem 11.3.8 and for the statement about the domain see Corollary 11.3.18.
3. Since the operator $\mathcal{H}$ is a special instance of the operator $\mathcal{M}$ (see Section 11.2.2), this also defines standard realizations of $\mathcal{H}$. They are called *discrete Schrödinger operators*.

If there exist realizations of $\mathcal{M}$, then they are necessarily restrictions of the following operator.

Definition 11.3.2 (Maximal restriction of $\mathcal{M}$) The domain of the *maximal restriction* $M^{\max}$ *of* $\mathcal{M}$ is $D(M^{\max}) = \{f \in \mathcal{F}_E \cap \ell^2(X, \mu; E) \mid \mathcal{M}f \in \ell^2(X, \mu; E)\}$, on which it acts by $M^{\max} f = \mathcal{M}f$. In the scalar case the maximal restriction of $\mathcal{H}$ is denoted by $H^{\max}$.

Remark In general $\Gamma_c(X; E)$ does not belong to $D(M^{\max})$ (see Section 11.3.3). Nevertheless, it can be checked that it is always densely defined. It is unclear whether $M^{\max}$ is always a closed (or even closable) operator on $\ell^2(X, \mu; E)$.

Typically standard realizations of $\mathcal{M}$ and $\mathcal{H}$ come from closed lower semi-bounded quadratic forms on $\ell^2(X, \mu; E)$ that extend $\mathcal{Q}_E^c$, respectively, forms on $\ell^2(X, \mu)$ that extend $\mathcal{Q}^c$. If this is the case, $\mathcal{Q}_E^c$ and $\mathcal{Q}^c$ need to be semi-bounded and closable. Their closures and associated operators deserve a name.

Definition 11.3.3 (Closure of $\mathcal{Q}_E^c$ and $\mathcal{Q}^c$) If $\mathcal{Q}_E^c = \mathcal{Q}_{\Phi,W;E}^c$ is lower semi-bounded and closable on $\ell^2(X, \mu; E)$, we denote its closure by $Q_E^0 = Q_{\mu,\Phi,W;E}^0$ and write $M^0 = M_{\mu,\Phi,W}^0$ for the associated self-adjoint operator. Likewise, if $\mathcal{Q}^c = \mathcal{Q}_V^c$ is lower semi-bounded and closable on $\ell^2(X, \mu)$, we denote its closure by $Q^0 = Q_{\mu,V}^0$ and write $H^0 = H_{\mu,V}^0$ for the associated self-adjoint operator.

Remark In all the cases where we can prove the closability of $\mathcal{Q}_E^0$ it will turn out that M^0 is indeed a standard realization of $\mathcal{M}$. We can think of M^0 as the standard realization of $\mathcal{M}$ with Dirichlet boundary conditions at infinity.

The remaining course of this section is as follows. The simplest situation for proving the existence of standard realizations of $\mathcal{M}$ is when the bundle endomorphism W is non-negative. In this case, the closability of $\mathcal{Q}_E^c$ follows from Fatou's lemma and that M^0 is a standard realization of $\mathcal{M}$ is a consequence of Green's formula. For endomorphisms with form small negative part (see Definition 11.3.5), the corresponding statements follow from standard perturbation theory.

If the negative part of the endomorphism is not form small but only such that $\mathcal{Q}_E^c$ is lower semi-bounded on $\ell^2(X, \mu; E)$, then the situation is more delicate. In this case, unconditionally we can only treat scalar Schrödinger operators; if V is such that $\mathcal{Q}_V^c$ is lower semi-bounded on $\ell^2(X, \mu)$, we prove its closability and that H^0 is a standard realization of $\mathcal{H}$. Given such a potential V and a bundle endomorphism $W \geq V$, we then use domination of the associated resolvents to prove the closability of $\mathcal{Q}_E^c$ and that M^0 is a standard realization of $\mathcal{M}$.

If the underlying weighted graph satisfies the finiteness condition of Definition 11.3.16, which depends on b and μ, no condition on W except that $\mathcal{Q}_E^c$ is lower semi-bounded is necessary. In this case, the form $\mathcal{Q}_E^c$ is induced by a symmetric lower semi-bounded operator and we employ Friedrichs' extension theorem to obtain a standard realization of $\mathcal{M}$. Locally finite graphs satisfy the finiteness condition for all choices of μ.

11.3.1 Non-negative Endomorphisms and Small Perturbations

Proposition 11.3.4 (Standard realization for non-negative endomorphisms) *If $W \geq 0$, then $\mathcal{Q}_E^c$ is non-negative and closable. Moreover, Q_E^0 is a restriction of $\mathcal{Q}_E$ and M^0 is a standard realization of $\mathcal{M}$.*

Proof. Since $W \geq 0$, the form $\mathcal{Q}_E^c$ is non-negative and, in particular, lower semi-bounded. We prove that $\mathcal{Q}_E^c$ is closable by showing that it is lower semi-continuous on its domain $\Gamma_c(X; E)$ with respect to $\ell^2(X, \mu; E)$-convergence. Let $\varphi, \varphi_n \in \Gamma_c(X; E)$ with $\varphi_n \to \varphi$ in $\ell^2(X, \mu; E)$. Since $W_x : E_x \to E_x$ and $\Phi_{x,y} : E_y \to E_x$ are continuous, for $x, y \in X$ we have $|\varphi_n(x) - \Phi_{x,y}\varphi_n(y)| \to |\varphi(x) - \Phi_{x,y}\varphi(y)|$ and $\langle W_x\varphi_n(x), \varphi_n(x)\rangle \to \langle W_x\varphi(x), \varphi(x)\rangle$, as $n \to \infty$. With this and $W \geq 0$ at hand, we infer from Fatou's lemma

$$\mathcal{Q}_E^c(\varphi) \leq \liminf_{n\to\infty} \mathcal{Q}_E^c(\varphi_n).$$

This shows closability of $\mathcal{Q}_E^c$.

Next we prove $D(Q_E^0) \subseteq \mathcal{D}_E$ and $Q_E^0 = \mathcal{Q}_E$ on $D(Q_E^0)$. To this end, let $f \in D(Q_E^0)$ and let $\varphi_n \in \Gamma_c(X; E)$ with $\varphi_n \to f$ with respect to the form norm. In particular, $\mathcal{Q}_E^c(\varphi_m - \varphi_n)$ is arbitrarily small for large enough n, m and $\mathcal{Q}_E^c(\varphi_n) \to Q_E^0(f)$, as $n \to \infty$. Let $f_n := f - \varphi_n$. Another application of Fatou's lemma as above shows

$$\frac{1}{2} \sum_{x,y \in X} b(x, y)|f_n(x) - \Phi_{x,y} f_n(y)|^2 + \sum_{x \in X} \langle W_x f_n(x), f_n(x)\rangle \mu(x) \leq \liminf_{m \to \infty} \mathcal{Q}_E^c(\varphi_m - \varphi_n).$$

Therefore, $f \in \mathcal{D}_E$ and the left-hand side of the previous inequality equals $\mathcal{Q}_E(f - \varphi_n)$. Since the square roots of $\mathcal{Q}_E^c$ and $\mathcal{Q}_E$ are semi-norms that agree on $\Gamma_c(X; E)$, the above also implies

$$|\mathcal{Q}_E(f)^{1/2} - \mathcal{Q}_E^c(\varphi_n)^{1/2}| \leq \mathcal{Q}_E(f - \varphi_n)^{1/2} \leq \liminf_{m \to \infty} \mathcal{Q}_E^c(\varphi_m - \varphi_n)^{1/2}.$$

This shows

$$Q_E^0(f) = \lim_{n \to \infty} \mathcal{Q}_E^c(\varphi_n) = \mathcal{Q}_E(f).$$

Since the off-diagonal values of $\mathcal{Q}_E$ and Q_E^0 can be recovered from on-diagonal values via polarization, we obtain that Q_E^0 is a restriction of $\mathcal{Q}_E$.

It remains to prove the claim about the operator M^0. Since Q_E^0 is non-negative, M^0 is lower semi-bounded and therefore satisfies (R1). By definition we have $\Gamma_c(X; E) \subseteq D(Q_E^0)$ such that (R2) also holds. Moreover, the general inclusion $D(M^0) \subseteq D(Q_E^0)$ and the already proven $D(Q_E^0) \subseteq \mathcal{D}_E$ combined with Lemma 11.2.4 yield $D(M^0) \subseteq \mathcal{F}_E$. With all this at hand, another application of Lemma 11.2.4 shows that for $f \in D(M^0)$ and $\varphi \in \Gamma_c(X; E) \subseteq D(Q_E^0)$ we have

$$\langle \varphi, M^0 f\rangle_{2;\,E} = Q_E^0(\varphi, f) = \mathcal{Q}_E(\varphi, f) = (\varphi, \mathcal{M} f)_E.$$

This implies $M^0 f = \mathcal{M} f$ and finishes the proof. □

As already mentioned at the beginning of this section, we can only treat special classes of bundle endomorphism that are not non-negative. This is discussed next.

To a self-adjoint bundle endomorphism W we associate the quadratic form q_W on $\ell^2(X, \mu; E)$ with domain

$$D(q_W) = \{f \in \ell^2(X, \mu; E) \mid \sum_{x \in X} \langle |W_x| f(x), f(x)\rangle \mu(x) < \infty\},$$

on which it acts by

$$q_W(f) = \sum_{x \in X} \langle W_x f(x), f(x) \rangle \mu(x).$$

Recall that q_{W_-} is *relatively form bounded* with respect to $\mathcal{Q}^c_{\Phi, W_+; E}$ on $\ell^2(X, \mu; E)$ with relative bound $\alpha \geq 0$, if there exists $C \in \mathbb{R}$ such that

$$q_{W_-}(f) \leq \alpha \mathcal{Q}^c_{\Phi, W_+; E}(f) + C \|f\|^2_{2; E} \text{ for all } f \in \Gamma_c(X; E).$$

We shall deal with the following classes of endomorphisms.

Definition 11.3.5 (Admissible endomorphisms) A self-adjoint bundle endomorphism W is called *admissible* with respect to μ and Φ, if $\mathcal{Q}^c_{\Phi, W; E}$ is lower semi-bounded on $\ell^2(X, \mu; E)$. The class of all admissible bundle endomorphisms with respect to μ and Φ is denoted by $\mathcal{A}_{\mu, \Phi; E}$. We say that an admissible endomorphism W has a *form small negative part* if q_{W_-} is relatively form bounded with respect to $\mathcal{Q}^c_{\Phi, W_+; E}$ on $\ell^2(X, \mu; E)$ with relative bound $\alpha < 1$. The class of endomorphisms with form small negative part is denoted by $\mathcal{S}_{\mu, \Phi; E}$.

In the scalar case when $E = (\mathbb{C})_{x \in X}$ and $\Phi = (\mathrm{Id})_{x, y \in X}$ we drop the subscripts for the connection and the bundle and write $\mathcal{A}_\mu$ for the class of *admissible potentials* $\mathcal{A}_{\mu, \mathrm{Id}; (\mathbb{C})}$ and we write $\mathcal{S}_\mu$ for the class of potentials with form small negative part $\mathcal{S}_{\mu, \mathrm{Id}; (\mathbb{C})}$.

Remark The name admissible potentials is borrowed from [30].

To a self-adjoint bundle endomorphism W on E, we associate the function $W_{\min} : X \to \mathbb{R}$ that is given by

$$W_{\min}(x) := \inf\{\langle W_x \xi, \xi \rangle \mid \xi \in E_x \text{ with } |\xi| = 1\}.$$

Since W_x is bounded and self-adjoint, $W_{\min}(x)$ is the minimum of the spectrum of W_x. If W acts on sections by multiplication with a real-valued $V \in C(X)$, then $W_{\min}(x) = V(x)$.

In general it is non-trivial to determine whether a given endomorphism belongs to one of the discussed classes. However, if there is some information in the scaler case, Kato's inequality provides some information in the magnetic case. More precisely, the following proposition characterizes relative boundedness (uniform in the connection) of magnetic Schrödinger forms with endomorphism W in terms of the scalar Schrödinger form with potential $W_{\min}$.

Proposition 11.3.6 *The following assertions are equivalent.*

(i) $W_{\min} \in \mathcal{A}_\mu$, *i.e.,* $\mathcal{Q}^c_{W_{\min}}$ *is lower semi-bounded on* $\ell^2(X, \mu)$.
(ii) There exists $C \in \mathbb{R}$ *such that for all unitary connections* Φ *on* E *and all* $\varphi \in \Gamma_c(X; E)$ *we have*

$$C\|\varphi\|^2_{2;\,E} \leq \mathcal{Q}^c_{\Phi,W;\,E}(\varphi).$$

Proof. (i) $\Rightarrow$ (ii): Let $\varphi \in \Gamma_c(X; E)$. We use Green's formula (Lemma 11.2.4) and Kato's inequality together with the bound $W \geq W_{\min}$ to obtain

$$\mathcal{Q}^c_{\Phi,W;E}(\varphi) = (\varphi, \mathcal{M}_{\mu,\Phi,W}\varphi)_E \geq (|\varphi|, \mathcal{H}_{\mu,W_{\min}}|\varphi|)_E = \mathcal{Q}^c_{W_{\min}}(|\varphi|).$$

Since $\||\varphi|\|_2 = \|\varphi\|_{2;\,E}$, assertion (ii) follows from this inequality and $W_{\min} \in \mathcal{A}_\mu$.

(ii) $\Rightarrow$ (i): Let $\varepsilon > 0$. Choose $\eta \in \Gamma(X; E)$ with $|\eta| = 1$ such that for all $x \in X$ it satisfies $W_{\min}(x) \geq \langle W_x\eta(x), \eta(x)\rangle - \varepsilon$. Moreover, choose a unitary connection Φ such that for all $x, y \in X$ we have $\Phi_{x,y}\eta(y) = \eta(x)$. By assumption there exists $C \in \mathbb{R}$ such that for all $\varphi \in C_c(X)$ we have

$$C\|\varphi\|^2_2 = C\|\varphi\eta\|^2_{2;\,E} \leq \mathcal{Q}^c_{\Phi,W;E}(\varphi\eta) = \mathcal{Q}^c_0(\varphi) + q_W(\varphi\eta),$$

where in the last step we used

$$|\varphi(x)\eta(x) - \varphi(y)\Phi_{x,y}\eta(y)|^2 = |\varphi(x)\eta(x) - \varphi(y)\eta(x)|^2 = |\varphi(x) - \varphi(y)|^2.$$

Since

$$q_W(\varphi\eta) = \sum_{x\in X} |\varphi(x)|^2\langle W_x\eta(x), \eta(x)\rangle\mu(x) \leq \sum_{x\in X} |\varphi(x)|^2(W_{\min} + \varepsilon)\mu(x)$$

and $\varepsilon > 0$ was arbitrary, this shows that $\mathcal{Q}^c_{W_{\min}}$ is lower semi-bounded on $\ell^2(X, \mu)$. □

Remark 1. The previous proposition shows that $W_{\min} \in \mathcal{A}_\mu$ implies $W \in \mathcal{A}_{\mu,\Phi;E}$ for all unitary connections Φ on E. For scalar magnetic Schrödinger operators also the converse implication is true, but for general magnetic Schrödinger operators the situation is unclear.
2. Even if $W_{\min} \notin \mathcal{A}_\mu$, there can still exist connections Φ for which $\mathcal{Q}^c_{\Phi,W:E}$ is lower semi-bounded on $\ell^2(X, \mu; E)$ (see Example 11.3.19).

The following proposition shows that closability of forms with non-negative endomorphisms is preserved under perturbations by small negative endomorphisms.

Proposition 11.3.7 (Standard realization for form small negative parts) *Let $W \in \mathcal{S}_{\Phi;E}$. Then $\mathcal{Q}_E^c = \mathcal{Q}_{\Phi,W;E}^c$ is lower semi-bounded and closable on $\ell^2(X,\mu;E)$. Its closure Q_E^0 is the restriction of $\mathcal{Q}_E$ to $D(Q_{\mu,\Phi,W_+;E}^0)$ and the associated operator M^0 is a standard realization of $\mathcal{M}$.*

Proof. Since $W \in \mathcal{S}_{\mu,\Phi;E}$, there exists an $\varepsilon > 0$ and constants $K' > K > 0$ such that for all $\varphi \in \Gamma_c(X;E)$ we have

$$0 \leq \varepsilon \mathcal{Q}_{\Phi,W_+;E}^c(\varphi) + K\|\varphi\|_{2;E}^2 \leq \mathcal{Q}_E^c(\varphi) + K'\|\varphi\|_{2;E}^2 \leq \mathcal{Q}_{\Phi,W_+;E}^c(\varphi) + K'\|\varphi\|_{2;E}^2.$$

Therefore, the form norms of $\mathcal{Q}_E^c$ and $\mathcal{Q}_{\Phi,W_+;E}^c$ are equivalent. Since $\mathcal{Q}_{\Phi,W_+;E}^c$ is closable on $\ell^2(X,\mu;E)$ with closure $Q_{\mu,\Phi,W_+;E}^0$, it follows that $\mathcal{Q}_E^c$ is closable and that its closure Q_E^0 satisfies $D(Q_E^0) = D(Q_{\mu,\Phi,W_+;E}^0)$.

Next we prove that Q_E^0 is a restriction of $\mathcal{Q}_E$. Let $f \in D(Q_E^0) = D(Q_{\mu,\Phi,W_+;E}^0)$ and choose a sequence (φ_n) in $\Gamma_c(X;E)$ that converges to f with respect to the form norm of Q_E^0. We have $\varphi_n \to f$ in $\ell^2(X,\mu;E)$, $\mathcal{Q}_E^c(\varphi_n) \to Q_E^0(f)$ and, by the equivalence of the form norms, $\mathcal{Q}_{\Phi,W_+;E}^c(\varphi_n) \to Q_{\mu,\Phi,W_+;E}^0(f)$, as $n \to \infty$. Fatou's lemma and the properties of (φ_n) imply

$$\begin{aligned}\sum_{x\in X}\langle W_-(x)f(x), f(x)\rangle &\leq \liminf_{n\to\infty} q_{W_-}(\varphi_n)\\ &\leq \liminf_{n\to\infty}\left(\mathcal{Q}_{\Phi,W_+;E}^c(\varphi_n) + K'\|\varphi_n\|_{2;E}^2\right)\\ &= Q_{\mu,\Phi,W_+;E}^0(f) + K'\|f\|_{2;E}^2.\end{aligned}$$

This shows $f \in D(q_{W_-})$ and, since $f \in D(Q_E^0)$ was arbitrary, also

$$q_{W_-}(f-\varphi_n) \leq Q_{\mu,\Phi,W_+;E}^0(f-\varphi_n) + C\|f-\varphi_n\|_{2;E}^2. \qquad (\Delta)$$

In Proposition 11.3.4 we proved $D(Q_{\mu,\Phi,W_+;E}^0) \subseteq \mathcal{D}_{\Phi,W_+;E}$. Since $\mathcal{D}_{\Phi,W_+;E} \cap D(q_{W_-}) \subseteq \mathcal{D}_{\Phi,W;E}$, we obtain $f \in \mathcal{D}_{\Phi,W;E}$. The properties of (φ_n) and that q_{W_-} is a quadratic form yield

$$\begin{aligned}|Q_E^0(f)^{1/2} - \mathcal{Q}_E(f)^{1/2}| &= \lim_{n\to\infty}|\mathcal{Q}_E^c(\varphi_n)^{1/2} - \mathcal{Q}_E(f)^{1/2}|\\ &\leq \limsup_{n\to\infty}|\mathcal{Q}_{\Phi,W_+;E}^c(\varphi_n)^{1/2} - \mathcal{Q}_{\Phi,W_+;E}(f)^{1/2}|\\ &\quad + \limsup_{n\to\infty}|q_{W_-}(\varphi_n)^{1/2} - q_{W_-}(f)^{1/2}|.\\ &\leq |Q_{\mu,\Phi,W_+;E}^0(f)^{1/2} - \mathcal{Q}_{\Phi,W_+;E}(f)^{1/2}|\\ &\quad + \limsup_{n\to\infty} q_{W_-}(\varphi_n - f)^{1/2} = 0.\end{aligned}$$

The last equality follows from the fact that $Q^0_{\mu,\Phi,W_+;E}$ is a restriction of $\mathcal{Q}_{\Phi,W_+;E}$ (see Proposition 11.3.4, and Inequality (Δ)). Therefore, Q^0_E is a restriction of $\mathcal{Q}_E$. With this at hand, the statement on the operator M^0 follows from Green's formula (Lemma 11.2.4; cf. the end of the proof of Proposition 11.3.4). □

If the endomorphism only belongs to $\mathcal{A}_{\mu,\Phi;E}$ and not to $\mathcal{S}_{\mu,\Phi;E}$, the closability of $\mathcal{Q}^c_E$ and the existence of standard realizations of $\mathcal{M}$ is more delicate. Even if $\mathcal{Q}^c_E$ is closable, the existence of standard realizations of $\mathcal{M}$ is non-trivial. This is due to the following observation. Suppose that $\mathcal{Q}^c_E$ is closable. In this case it can happen that there exist $\varphi_n \in \Gamma_c(X; E)$ and $f \in D(Q^0_E)$ such that $\varphi_n \to f$ with respect to the form norm, $q_{W_-}(\varphi_n) \to \infty$ and $\mathcal{Q}^c_{\Phi,W_+;E}(\varphi_n) \to \infty$, as $n \to \infty$ (see Example 11.3.20). For such a sequence we still have

$$Q^0_E(f) = \lim_{n\to\infty} \mathcal{Q}^c_E(\varphi_n),$$

but f does not belong to $\mathcal{D}_E$. In particular, Q^0_E is not a restriction if $\mathcal{Q}_E$ and we cannot infer from Green's formula that M^0 is a realization of $\mathcal{M}$. There are essentially two ways for dealing with this situation, which are discussed in the subsequent sub-sections.

11.3.2 Admissible Endomorphisms and Domination

In this section we use domination of resolvents of magnetic operators by resolvents of scalar operators to prove closability of $\mathcal{Q}^c_E$ and to show that the operator associated with its closure is a standard realization of $\mathcal{M}$. The main result of this section is the following.

Theorem 11.3.8 (Realization of discrete magnetic Schrödinger operators) *Let $W_{\min} \in \mathcal{A}_\mu$. Then $\mathcal{Q}^c_E$ is lower semi-bounded and closable on $\ell^2(X,\mu; E)$ and M^0 is a standard realization of $\mathcal{M}$.*

In order to employ domination for proving closability, we need several technical lemmas, which may be of interest on their own right. For a background on the Beurling–Deny criteria, we refer to Appendix A.3.

Lemma 11.3.9 *Let $V \in \mathcal{A}_\mu$. If $\mathcal{Q}^c$ is closable on $\ell^2(X,\mu)$, its closure Q^0 satisfies the first Beurling–Deny criterion. In particular, for $\alpha > -\lambda_0(H^0)$ the resolvent $(H^0+\alpha)^{-1}$ is positivity preserving, i.e., $f \geq 0$ implies $(H^0 + \alpha)^{-1} f \geq 0$.*

Proof. Let $f \in D(Q^0)$ and let (φ_n) be a sequence in $C_c(X)$ that converges to f with respect to the form norm. It follows from the definition of $\mathcal{Q}^c$ that

$$\mathcal{Q}^c(|\varphi_n|) \leq \mathcal{Q}^c(\varphi_n), \quad n \in \mathbb{N}.$$

The ℓ^2-lower semi-continuity of Q^0 implies

$$\begin{aligned} Q^0(|f|) &\leq \liminf_{n\to\infty} Q^0(|\varphi_n|) = \liminf_{n\to\infty} \mathcal{Q}^c(|\varphi_n|) \\ &\leq \liminf_{n\to\infty} \mathcal{Q}^c(\varphi_n) = Q^0(f) < \infty. \end{aligned}$$

This shows $|f| \in D(Q^0)$ and the desired inequality. The statement on the resolvent follows from Proposition A.4. □

The following lemma shows that self-adjoint realizations whose forms satisfy the first Beurling–Deny criterion are indeed standard realizations in the sense of Definition 11.3.1.

Lemma 11.3.10 *Let $V \in C(X)$ be real valued. Let H be a lower semi-bounded realization of $\mathcal{H}$ such that the associated quadratic form Q satisfies the first Beurling–Deny criterion. Then Q is an extension of $\mathcal{Q}^c$, $V \in \mathcal{A}_\mu$, and H is a standard realization of $\mathcal{H}$.*

Proof. For $\alpha > -\lambda_0(H)$ let $G_\alpha := (H + \alpha)^{-1}$ be the associated resolvent. Since H is a realization of $\mathcal{H}$, $D(H) \subseteq \mathcal{F}$ so that G_α maps $\ell^2(X, \mu)$ to $\mathcal{F}$.

We show that for $\varphi \in C_c(X)$ the convergence $\mathcal{H}\alpha G_\alpha\varphi \to \mathcal{H}\varphi$ holds pointwise, as $\alpha \to \infty$. Once this is proven, Green's formula (Lemma 11.2.4) and the characterization of Q via approximating forms (cf. Appendix A.1) yield

$$\begin{aligned} \mathcal{Q}^c(\varphi) &= (\varphi, \mathcal{H}\varphi) = \lim_{\alpha\to\infty} (\varphi, \mathcal{H}\alpha G_\alpha\varphi) = \lim_{\alpha\to\infty} \langle \varphi, H\alpha G_\alpha\varphi\rangle_2 \\ &= \lim_{\alpha\to\infty} \alpha\langle \varphi - \alpha G_\alpha\varphi, \varphi\rangle_2 = Q(\varphi). \end{aligned}$$

This shows $\varphi \in D(Q)$ and that Q is an extension of $\mathcal{Q}^c$. Since Q is lower semi-bounded, so is $\mathcal{Q}^c$ and we obtain $V \in \mathcal{A}_\mu$. Therefore, H is a standard realization of $\mathcal{H}$ in the sense of Definition 11.3.1.

Let now $\varphi \in C_c(X)$. The strong continuity of the resolvent implies $\alpha G_\alpha\varphi \to \varphi$ pointwise, as $\alpha \to \infty$. We construct a function $f \in \mathcal{F}$ that dominates $\alpha G_\alpha\varphi$ for all α large enough. The pointwise convergence $\mathcal{H}\alpha G_\alpha\varphi \to \mathcal{H}\varphi$, as $\alpha \to \infty$, then follows from Lebesgue's theorem.

We choose a non-negative $\psi \in \ell^2(X, \mu)$ and $\beta > -\lambda_0(H)$ such that $G_\beta\psi \geq |\varphi|$. Such ψ and β exist because (G_α) is positivity preserving

(Proposition A.4) and strongly continuous and φ has finite support. Using the resolvent identity and that G_α is positivity preserving, for $\alpha > 2\beta$ we obtain

$$\alpha/2|G_\alpha\varphi| \leq (\alpha-\beta)G_\alpha|\varphi| \leq (\alpha-\beta)G_\alpha G_\beta\psi = G_\beta\psi - G_\alpha\psi \leq G_\beta\psi.$$

This shows that for $\alpha > 2\beta$ the function $\alpha G_\alpha\varphi$ is dominated by $f = 2G_\beta\psi \in \mathcal{F}$ and finishes the proof. □

Remark It is important to note that in the lemma we do not assume that H is a standard realization of $\mathcal{H}$. We only assume that H is a lower semi-bounded realization. The lemma says that this implies $C_c(X) \subseteq D(Q)$ if Q satisfies the first Beurling–Deny criterion.

Lemma 11.3.11 *Let $V \in \mathcal{A}_\mu$. If $\mathcal{Q}^c$ is closable on $\ell^2(X,\mu)$, the self-adjoint operator H^0 associated with the closure Q^0 is a standard realization of $\mathcal{H}$.*

Proof. By definition H^0 is lower semi-bounded and $C_c(X) \subseteq D(Q^0)$. Hence, it suffices to prove $D(H^0) \subseteq \mathcal{F}$ and $H^0 f = \mathcal{H}f$ for $f \in D(H^0)$.

Let $f \in D(H^0)$ and let $\alpha > -\lambda_0(H^0)$. Then $f = (H^0+\alpha)^{-1}g$ for some $g \in \ell^2(X,\mu)$. Since $(H^0+\alpha)^{-1}$ is positivity preserving, this shows that there exist non-negative $f_i \in D(H^0)$, $i = 1,\ldots,4$, with $f = f_1 - f_2 + i(f_3 - f_4)$. We can therefore assume $f \geq 0$. Let now (φ_n) be a sequence in $C_c(X)$ that converges to f with respect to the form norm. Then $\psi_n := (\varphi_n \wedge f) \vee 0$ belongs to $C_c(X)$ and converges to f in $\ell^2(X,\mu)$. Since by Lemma 11.3.9 the form Q^0 satisfies the first Beurling–Deny criterion, we obtain

$$\|\psi_n\|_{Q^0} \leq \|\varphi_n\|_{Q^0} + \|f\|_{Q^0}$$

from Lemma A.5. Thus (after taking a suitable subsequence) we can assume $\psi_n \to f$ weakly with respect to the form inner product. Using Green's formula (Lemma 11.2.1), for $x \in X$ we obtain

$$\begin{aligned}\mu(x)H^0 f(x) &= Q^0(\delta_x, f) = \lim_{n\to\infty} Q^0(\delta_x,\psi_n)\\ &= \lim_{n\to\infty} \mathcal{Q}^c(\delta_x,\psi_n) = \lim_{n\to\infty} \mu(x)\mathcal{H}\psi_n(x). \qquad (\diamond)\end{aligned}$$

In particular, this shows that $\lim_n \mu(x)\mathcal{H}\psi_n(x)$ exists. Moreover,

$$\mu(x)\mathcal{H}\psi_n(x) = \deg(x)\psi_n(x) - \sum_{y\in X} b(x,y)\psi_n(y) + \mu(x)V(x)\psi_n(x).$$

Since the ψ_n also converge pointwise towards f, we obtain that $\sum_{y\in X} b(x,y)\psi_n(y)$ converges, as $n \to \infty$. Fatou's lemma yields

$$0 \leq \sum_{y\in X} b(x,y)f(y) \leq \liminf_{n\to\infty} \sum_{y\in X} b(x,y)\psi_n(y) < \infty.$$

Since $x \in X$ was arbitrary, we infer $f \in \mathcal{F}$. By construction we have $|\psi_n| \leq f$. Therefore, Lebesgue's dominated convergence theorem yields $\mathcal{H}\psi_n(x) \to \mathcal{H}f(x)$, as $n \to \infty$. With equation ($\diamondsuit$) we arrive at $H^0 f(x) = \mathcal{H}f(x)$ and the lemma is proven. $\square$

Lemma 11.3.12 (Domination) *Let $V \in \mathcal{A}_\mu$ and let $W \geq V$. Suppose that $\mathcal{Q}^c$ and $\mathcal{Q}^c_E$ are closable. For any $\alpha > -\lambda_0(H^0)$ the value $-\alpha$ belongs to the resolvent set of M^0 and the resolvent $(H^0+\alpha)^{-1}$ dominates $(M^0+\alpha)^{-1}$, i.e., for any $f \in \ell^2(X, \mu; E)$ the following inequality holds*

$$|(M^0+\alpha)^{-1} f| \leq (H^0+\alpha)^{-1}|f|.$$

Proof. We employ the theory of domination developed in [37]. In contrast to our situation this chapter deals with (not necessarily densely defined) forms on functions taking values in a fixed Hilbert space. If we let $\widetilde{E} = \bigoplus_{x \in X} E_x$ the direct sum of Hilbert spaces, then $\ell^2(X, \mu; E)$ isometrically embeds into the Hilbert space of square summable $\widetilde{E}$-valued functions $\ell^2(X, \mu; \widetilde{E})$. Therefore, $\mathcal{Q}^0_E$ can be viewed as a not densely defined closed form on the Hilbert space $\ell^2(X, \mu; \widetilde{E})$ and the theory of [37] can be applied.

According to [37, Theorem 4.1] we need to prove that there are form norm dense subspaces $U \subseteq D(Q^0_E)$ and $V \subseteq D(Q^0)$ such that the following holds:

(a) U is a generalized ideal in V, i.e.,
- $f \in U$ implies $|f| \in V$,
- $f \in U, \varphi \in V$ and $|\varphi| \leq |f|$ implies $\varphi \operatorname{sgn} f \in U$.

(b) For all $f \in U$ and $\varphi \in V$ with $0 \leq \varphi \leq |f|$ we have

$$\operatorname{Re} Q^0_E(\varphi \operatorname{sgn} f, f) \geq Q^0(\varphi, |f|).$$

We choose $U = \Gamma_c(X; E)$ and $V = C_c(X)$. By definition they are form dense subspaces of the forms Q^0_E respectively Q^0. Moreover, $\Gamma_c(X; E)$ is a generalized ideal in $C_c(X)$. The required inequality follows from Kato's inequality and Green's formula. More precisely, for $f \in \Gamma_c(X; E)$ and $\varphi \in C_c(X)$ with $0 \leq \varphi \leq |f|$ we have $\langle f(x), \varphi(x) \operatorname{sgn} f(x) \rangle = |\varphi(x)||f(x)|$ and $|\varphi(x) \operatorname{sgn} f(x)| = |\varphi(x)| = \varphi(x)$. Since also $W \geq V$, Kato's inequality (Lemma 11.2.2) implies

$$\operatorname{Re}(\varphi \operatorname{sgn} f, \mathcal{M}f)_E \geq (\varphi, \mathcal{H}|f|).$$

With this at hand Green's formula (Lemma 11.2.4) yields the desired statement. $\square$

Theorem 11.3.13 (Realization of discrete Schrödinger operator) *Let $V \in \mathcal{A}_\mu$. Then $\mathcal{Q}^c$ is lower semi-bounded and closable on $\ell^2(X,\mu)$ and H^0 is a standard realization of $\mathcal{H}$. Moreover, the associated quadratic form Q^0 satisfies the first Beurling–Deny criterion.*

Proof. Let $\lambda_0 := \lambda_0(\mathcal{Q}^c)$ be the largest lower bound of $\mathcal{Q}^c$ on $\ell^2(X,\mu)$. For $n \in \mathbb{N}_0$ we let $V_n := \max\{V, -n\}$. Since $(V_n)_-$ is bounded, it is readily verified that V_n has a form small negative part so that $V_n \in \mathcal{S}_\mu$. By Proposition 11.3.7 the form $\mathcal{Q}^c_{V_n}$ is closable; we denote its closure by Q^0_n and the associated operator by H^0_n. Since $V \le V_n$, we have $\lambda_0 \le \lambda_0(Q^0_n)$, so that for $\alpha > -\lambda_0$ the resolvent $G^n_\alpha := (H^0_n + \alpha)^{-1}$ exists. As a first step we prove that for each $\alpha > -\lambda_0$ the sequence $(G^n_\alpha)_n$ converges strongly to some operator G_α and that $(G_\alpha)_{\alpha > -\lambda_0}$ is a strongly continuous resolvent family.

If $m \ge n$, we have $V_+ = V_0 \ge V_n \ge V_m$. For $f \ge 0$ and $\alpha > -\lambda_0$ Lemma 11.3.12 applied to the scalar situation yield

$$0 \le G^0_\alpha f \le G^n_\alpha f \le G^m_\alpha f.$$

The bound $\lambda_0 \le \lambda_0(H^0_n)$ implies the bound for the operator norm $\|G^n_\alpha\| \le (\lambda_0 + \alpha)^{-1}$. Hence, it follows from the monotone convergence theorem that for each $f \in \ell^2(X,\mu)$ and $\alpha > -\lambda_0$ the limit

$$G_\alpha f := \lim_{n\to\infty} G^n_\alpha f$$

exists. It is readily verified that (G_α) is a family of self-adjoint bounded operators with $\|G_\alpha\| \le (\lambda_0 + \alpha)^{-1}$ and that it satisfies the resolvent identity. Next we prove that it is strongly continuous. To this end, we let $f \ge 0$ and use domination to estimate

$$\begin{aligned} \|\alpha G_\alpha f - \alpha G^0_\alpha f\|^2 &= \|\alpha G_\alpha f\|^2 - 2\langle \alpha G_\alpha f, \alpha G^0_\alpha f\rangle + \|\alpha G^0_\alpha f\|^2 \\ &\le \frac{\alpha^2}{(\lambda_0 + \alpha)^2}\|f\|_2^2 - \|\alpha G^0_\alpha f\|^2. \end{aligned}$$

Since (G^0_α) is strongly continuous, this inequality shows $\alpha G_\alpha f \to f$, as $\alpha \to \infty$, i.e., the strong continuity of (G_α).

Let now Q be the lower semi-bounded closed quadratic form that is associated with (G_α) and denote by H the associated self-adjoint operator. We prove that H is a realization of $\mathcal{H}$.

Since G_α is surjective onto $D(H)$, it suffices to show that G_α maps $\ell^2(X,\mu)$ to $\mathcal{F}$ and to verify the equality $\mathcal{H} G_\alpha f = f - \alpha G_\alpha f$. To this end, let $f \in \ell^2(X,\mu)$ be non-negative and set $\mathrm{Deg} = \mu^{-1}\deg$. We already know from Proposition 11.3.7 that H^0_n is a realization of $\mathcal{H}_{\mu,V_n}$. For $x \in X$ the monotone convergence theorem yields

$$\begin{aligned}
\frac{1}{\mu(x)} & \sum_{y \in X} b(x,y) G_\alpha f(y) \\
&= \lim_{n\to\infty} \frac{1}{\mu(x)} \sum_{y \in X} b(x,y) G_\alpha^n f(y) \\
&= \lim_{n\to\infty} \left(-\mathcal{H}_{\mu,V_n} G_\alpha^n f(x) + \mathrm{Deg}(x) G_\alpha^n f(x) + V_n(x) G_\alpha^n f(x) \right) \\
&= \lim_{n\to\infty} \left(-f(x) + \alpha G_\alpha^n f(x) + \mathrm{Deg}(x) G_\alpha^n f(x) + V_n(x) G_\alpha^n f(x) \right) \\
&= -f + \alpha G_\alpha f(x) + \mathrm{Deg}(x) G_\alpha f(x) + V(x) G_\alpha f(x).
\end{aligned}$$

This computation implies $G_\alpha f \in \mathcal{F}$ and $\mathcal{H} G_\alpha f = f - \alpha G_\alpha f$.

The operator H is a realization of $\mathcal{H}$ and as a monotone limit of positivity preserving resolvents, its resolvent is positivity preserving. Therefore, the associated form Q satisfies the first Beurling–Deny criterion (see Proposition A.4). Lemma 11.3.10 yields that Q is an extension of $\mathcal{Q}^c$ and, in particular, that $\mathcal{Q}^c$ is closable. According to Lemma 11.3.11 the corresponding operator H^0 is a standard realization of $\mathcal{H}$. This finishes the proof. □

Remark 1. The arguments for proving that (G_α^n) converges monotone to a strongly continuous resolvent are taken from [30].
2. The form Q constructed in the proof is indeed the closure of $\mathcal{Q}^c$. Proving this would have saved us from using Lemma 11.3.10 and Lemma 11.3.11. We chose this alternative presentation for two reasons. Lemma 11.3.10 is used below and Lemma 11.3.11 might be interesting on its own right. If anyone comes up with a different proof of the closability of $\mathcal{Q}^c$, Lemma 11.3.11 shows that the associated operator is a standard realization of $\mathcal{H}$. One such alternative proof, which uses the Feynman–Kac formula, can be found in [14].

Corollary 11.3.14 *The operator $\mathcal{H}$ has a standard realization whose associated quadratic form satisfies the first Beurling–Deny criterion if and only if $V \in \mathcal{A}_\mu$.*

Proof. If $V \in \mathcal{A}_\mu$, the previous theorem and Lemma 11.3.9 show that H^0 is a standard realization of $\mathcal{H}$ whose associated form satisfies the first Beurling–Deny criterion.

If H is a standard realization of $\mathcal{H}$ whose associated quadratic form Q satisfies the first Beurling–Deny criterion, then Lemma 11.3.10 implies that Q is an extension of $\mathcal{Q}^c$. Hence, $\mathcal{Q}^c$ is lower semi-bounded, i.e., $V \in \mathcal{A}_\mu$. □

Proof of Theorem 11.3.8. By assumption we have $W_{\min} \in \mathcal{A}_\mu$. Therefore, Theorem 11.3.13 shows that $H^0 := H^0_{\mu,W_{\min}}$ is a standard realization of

$\mathcal{H}_{\mu,W_{\min}}$. For $\alpha > -\lambda_0(H^0)$ we denote the associated resolvent by $R_\alpha := (H^0+\alpha)^{-1}$.

Let $K_n := \{x \in X \mid \lambda_0(W_x) \geq -n\}$ and let $W_n := W_+ - 1_{K_n} W_-$. Since $1_{K_n} W_-$ is uniformly bounded, it is readily verified that $W_n \in \mathcal{S}_{\mu,\Phi;E}$. We denote the closure of $\mathcal{Q}^c_{\Phi,W_n;E}$ on $\ell^2(X,\mu;E)$ by Q^n_E and write (G^n_α) for the corresponding resolvent; their existence follows from Proposition 11.3.7. Since $W_n \geq W_{\min}$, Proposition 11.3.6 shows that all the Q^n_E have a common lower bound. Hence, the forms (Q^n_E) fulfil the assumptions of Lemma A.3 on monotone convergence of quadratic forms. We obtain that the resolvents (G^n_α) strongly converge to a resolvent (G_α) with associated lower semi-bounded closed quadratic form Q_E and self-adjoint operator M.

We prove that M is a standard realization of $\mathcal{M}$. Condition (R1) is trivially satisfied since Q_E is lower semi-bounded. For $\varphi \in \Gamma_c(X;E)$ it follows from Lemma A.3 that

$$Q_E(\varphi) \leq \liminf_{n\to\infty} Q^n_E(\varphi) = \mathcal{Q}^c(\varphi) < \infty,$$

showing (R2). Since $W_n \geq W_{\min}$, Lemma 11.3.12 yields that the resolvent R_α dominates G^n_α and hence also G_α, i.e.,

$$|G^n_\alpha f|, |G_\alpha f| \leq R_\alpha |f|, \quad f \in \ell^2(X,\mu;E).$$

The operator H^0 is a standard realization of $\mathcal{H}$ so that $R_\alpha|f| \in D(H^0) \subseteq \mathcal{F}$ for all $f \in \ell^2(X,\mu;E)$. These two observations imply that the image of (G_α) is contained in $\mathcal{F}_E$ so that $D(M) = G_\alpha \ell^2(X,\mu;E) \subseteq \mathcal{F}_E$. Moreover, domination of the resolvents and Lebesgue's dominated convergence theorem show that for $x \in X$ and $f \in \ell^2(X,\mu;E)$ in the space E_x we have

$$\sum_{y\in X} b(x,y)\Phi_{x,y} G_\alpha f(y) = \lim_{n\to\infty} \sum_{y\in X} b(x,y)\Phi_{x,y} G^n_\alpha f(y).$$

For $x \in X$ this implies

$$\begin{aligned}
\mathcal{M} G_\alpha f(x) &= \lim_{n\to\infty} \mathcal{M} G^n_\alpha f(x)\\
&= \lim_{n\to\infty} \big(\mathcal{M}_{\mu,\Phi,W_n} G^n_\alpha f(x) + (W_x - (W_n)_x) G^n_\alpha f(x)\big)\\
&= \lim_{n\to\infty} \big(f(x) - \alpha G^n_\alpha f(x) + (1_{K_n}(x) - 1)(W_-)_x G^n_\alpha f(x)\big)\\
&= f(x) - \alpha G_\alpha f(x)\\
&= M G_\alpha f(x).
\end{aligned}$$

For the third to last equality we used the definition of W_n and that the operator associated with Q^n_E is a realization of $\mathcal{M}_{\mu,\Phi,W_n}$ (see Proposition 11.3.7). From this computation it follows that M is a realization of $\mathcal{M}$.

It remains to show $Q_E = Q_E^0$. [35, Corollary 2.4] implies that $\Gamma_c(X; E)$ is dense in $D(Q_E)$. More precisely, the resolvent (G_α) of Q_E is dominated by the resolvent (R_α) of $Q^0_{\mu, W_{\min}}$ and $C_c(X) = \ell^2_c(X, \mu)$ (the ℓ^2-functions with compact support in X) is dense in $D(Q^0_{\mu, W_{\min}})$ with respect to the form norm. It then follows from [35, Corollary 2.4] that $\Gamma_c(X; E) = \ell^2_c(X, \mu; E)$ (the ℓ^2-sections with compact support in X) is dense in Q_E with respect to the form norm.

Let now $\varphi \in \Gamma_c(X; E)$. Below we prove that for every $x \in X$ we have $\mathcal{M}\alpha G_\alpha \varphi(x) \to \mathcal{M}\varphi(x)$ in E_x, as $\alpha \to \infty$. Since M is a standard realization of $\mathcal{M}$ and φ has finite support, this implies

$$\begin{aligned} Q_E(\varphi) &= \lim_{\alpha\to\infty} Q_E(\varphi, \alpha G_\alpha \varphi) = \lim_{\alpha\to\infty} \langle \varphi, \mathcal{M}\alpha G_\alpha \varphi\rangle_{2;E} \\ &= (\varphi, \mathcal{M}\varphi)_E = \mathcal{Q}^c_E(\varphi), \end{aligned}$$

where for the last equality we used Green's formula (Lemma 11.2.4). This shows that Q_E is an extension of $\mathcal{Q}^c$. Together with $\Gamma_c(X; E)$ being dense in $D(Q_E)$, we arrive at $Q_E = Q_E^0$.

To finish the proof, we show $\mathcal{M}\alpha G_\alpha \varphi(x) \to \mathcal{M}\varphi(x)$, as $\alpha \to \infty$, with almost the same arguments as in the proof of Lemma 11.3.10. Let $\psi \in \ell^2(X, \mu)$ be non-negative and $\beta > -\lambda_0(H^0)$ such that $R_\beta \psi \geq |\varphi|$. It follows from the domination of G_α by R_α and the resolvent identity for R_α, that for $\alpha > \max\{2\beta, -\lambda_0(M)\}$ we have

$$\alpha/2|G_\alpha \varphi| \leq \alpha/2 R_\alpha |\varphi| \leq R_\beta \psi,$$

cf. the proof of Lemma 11.3.10. Since $R_\beta \psi \in \mathcal{F}$ and (G_α) is strongly continuous, an application of Lebesgue's dominated convergence theorem yields

$$\sum_{y\in X} \Phi_{x,y} \alpha G_\alpha \varphi(y) \to \sum_{y\in X} \Phi_{x,y} \varphi(y), \text{ as } \alpha \to \infty.$$

This implies the desired convergence $\mathcal{M}\alpha G_\alpha \varphi(x) \to \mathcal{M}\varphi(x)$ in E_x, as $\alpha \to \infty$, and finishes the proof. □

Remark In the previous proof we showed that the form Q_E constructed there equals Q_E^0. As a short cut we employed the theory developed in [35] to show that $\Gamma_c(X; E)$ is dense in $D(Q_E)$. However, for proving Theorem 11.3.8 this is not necessary. It is also possible to argue the same way as at the end of the proof of Theorem 11.3.13: First show that the operator associated with Q_E is a restriction of $\mathcal{M}$ and then use domination to obtain that Q_E is an extension of $\mathcal{Q}^c_E$ (the latter statement is a version of Lemma 11.3.10 for magnetic operators, which is one step in the presented proof of Theorem 11.3.8). After that prove a version of Lemma 11.3.11 for magnetic forms with the help of domination instead of using that resolvents are positivity preserving.

The known proofs for Theorem 11.3.8 (the discussed one and a probabilistic one for scalar magnetic Schrödinger operators in [14], cf. Section 11.3.4) have in common that they use domination of forms. Such a perturbative approach has the drawback that it always has two steps: One needs to first prove the existence of standard realizations for scalar Schrödinger operators before one can treat the magnetic case.

11.3.3 Admissible Endomorphisms and Graphs with a Finiteness Condition

The last existence result of standard realizations deals with the situation when $\mathcal{Q}_E^c$ is induced by a symmetric operator, i.e., when $\mathcal{M}$ maps $\Gamma_c(X;E)$ to $\ell^2(X,\mu;E)$. We first put this condition into perspective.

Lemma 11.3.15 *The following assertions are equivalent.*

(i) $\mathcal{M}\Gamma_c(X;E) \subseteq \ell^2(X,\mu;E)$.
(ii) For all $x \in X$ *the function* $X \to \mathbb{R}$, $y \mapsto b(x,y)/\mu(y)$ *belongs to* $\ell^2(X,\mu)$.

In this case, $\ell^2(X,\mu;E) \subseteq \mathcal{F}_E$. *In particular, both assertions are satisfied if the graph* (X,b) *is locally finite.*

Proof. For $x \in X$ and $\xi \in E_x$ with $|\xi| = 1$ we let $\delta_{x,\xi}$, be the compactly supported section with $\delta_{x,\xi}(x) = \xi$ and $\delta_{x,\xi}(y) = 0$ if $y \neq x$. It follows from the definitions that $\mathcal{M}\delta_{x,\xi} \in \ell^2(X,\mu)$ if and only if $X \to \mathbb{R}$, $y \mapsto b(x,y)/\mu(y)$ belongs to $\ell^2(X,\mu)$. This shows the equivalence of (i) and (ii).

Moreover, if (ii) holds, the Cauchy–Schwarz inequality implies

$$\sum_{y\in X} b(x,y)|f(y)| \le \|b(x,\cdot)/\mu\|_2 \left(\sum_{y\in X} |f(y)|^2\mu(y)\right)^{1/2}.$$

This proves the inclusion $\ell^2(X,\mu;E) \subseteq \mathcal{F}_E$. □

The previous lemma shows that the inclusion $\mathcal{M}\Gamma_c(X;E) \subseteq \ell^2(X,\mu;E)$ only depends on (X,b) and the weight μ and not on the connection nor on the endomorphism. If it is satisfied, the restriction of $\mathcal{M}$ to $\Gamma_c(X;E)$ is a densely defined operator on $\ell^2(X,\mu;E)$.

Definition 11.3.16 (Finiteness condition and the minimal restriction of $\mathcal{M}$) The triplet (X,b,μ) satisfies the *finiteness condition* (FC) if for all $x \in X$ the function $X \to \mathbb{R}$, $y \mapsto b(x, y)/\mu(y)$ belongs to $\ell^2(X, \mu)$. In this case, the operator $M^{\min} : D(M^{\min}) \to \ell^2(X, \mu; E)$ with $D(M^{\min}) = \Gamma_c(X; E)$ and $M^{\min} f = \mathcal{M} f$ for $f \in D(M^{\min})$ is called the *minimal restriction of* $\mathcal{M}$.

The following proposition is the main result of this sub-section.

Proposition 11.3.17 (Realization under finiteness condition) *Suppose that (FC) holds.*

(a) $(M^{\min})^* = M^{\max}$. *In particular,* $(M^{\min})^*$ *is the restriction of* $\mathcal{M}$ *to*

$$D((M^{\min})^*) = \{f \in \ell^2(X, \mu; E) \mid \mathcal{M} f \in \ell^2(X, \mu; E)\}.$$

(b) If $W \in \mathcal{A}_{\mu,\Phi;E}$, *the form* $\mathcal{Q}_E^c$ *is lower semi-bound and closable, and the operator* M^0 *is a standard realization of* $\mathcal{M}$.

(c) If M *is a standard realization of* $\mathcal{M}$, *then* M *is an extension of* $M^{\min}$ *and the associated quadratic form is an extension of* $\mathcal{Q}_E^c$.

Proof. (a) Let $f \in D((M^{\min})^*)$. By Lemma 11.3.15 we have $f \in \mathcal{F}_E$. For $\varphi \in \Gamma_c(X; E)$ we infer from Green's formula (Lemma 11.2.1) and the definition of $M^{\min}$ that

$$\langle \varphi, (M^{\min})^* f\rangle_{2;E} = \langle M^{\min}\varphi, f\rangle_{2;E} = \sum_{x\in X} \langle \mathcal{M}\varphi(x), f(x)\rangle \mu(x) = (\varphi, \mathcal{M} f)_E.$$

This shows $\mathcal{M} f = (M^{\min})^* f \in \ell^2(X, \mu; E)$. Therefore, $(M^{\min})^*$ is a restriction of $M^{\max}$.

Let now $f \in D(M^{\max})$. Since by definition $D(M^{\max}) \subseteq \mathcal{F}_E$, Green's formula (Lemma 11.2.1) implies that for $\varphi \in D(M^{\min}) = \Gamma_c(X; E)$ we have

$$\langle M^{\min}\varphi, f\rangle_{2;E} = \sum_{x\in X} \langle \mathcal{M}\varphi(x), f(x)\rangle \mu(x) = (\varphi, \mathcal{M} f)_E = \langle \varphi, M^{\max} f\rangle_{2;E}.$$

This shows $f \in D((M^{\min})^*)$ and that $M^{\max}$ is a restriction of $(M^{\min})^*$.

The 'In particular'-statement follows from the definition of $M^{\max}$ and Lemma 11.3.15.

(b) For $\varphi, \psi \in \Gamma_c(X; E)$ Green's formula (Lemma 11.2.1) and the definition of $M^{\min}$ yield

$$\begin{aligned}\langle M^{\min}\varphi, \psi\rangle_{2;E} &= \sum_{x\in X} \langle \mathcal{M}\varphi(x), \psi(x)\rangle \mu(x) = \mathcal{Q}_E^c(\varphi, \psi)\\ &= (\varphi, \mathcal{M}\psi)_E = \langle \varphi, M^{\min}\psi\rangle_{2;E}.\end{aligned}$$

Therefore, $\mathcal{Q}_E^c$ is the quadratic form of the symmetric operator $M^{\min}$. Since $W \in \mathcal{A}_{\mu,\Phi;E}$, it is also lower semi-bounded. It follows from Friedrichs' extension theorem that $\mathcal{Q}_E^c$ is closable and that the self-adjoint operator M^0 that is associated with the closure Q_E^0 is an extension of $M^{\min}$. With this at hand, (a) implies that M^0 is a restriction of $M^{\max}$ and therefore a realization of $\mathcal{M}$. Since $\Gamma_c(X; E) \subseteq D(Q_E^0)$, the operator M^0 is a standard realization of $\mathcal{M}$.

(c) We first prove that the associated quadratic from, which we denote by Q, is an extension of $\mathcal{Q}_E^c$. Let $\varphi \in \Gamma_c(X; E)$. By the definition of standard realizations we have $\varphi \in D(Q)$. The domain of M is dense in $D(Q)$ with respect to the form norm. Hence, there exists a sequence $f_n \in D(M)$ such that $f_n \to \varphi$ with respect to the form norm. Since $\mathcal{M}\varphi \in \ell^2(X, \mu; E)$ and $f_n \in D(M) \subseteq \mathcal{F}_E$, we obtain with the help of Green's formula (Lemma 11.2.1) that

$$\begin{aligned} Q(\varphi) &= \lim_{n\to\infty} Q(\varphi, f_n) = \lim_{n\to\infty} \langle \varphi, Mf_n\rangle_{2;E} \\ &= \lim_{n\to\infty} \langle \mathcal{M}\varphi, f_n\rangle_{2;E} = \langle \mathcal{M}\varphi, \varphi\rangle_{2;E} = \mathcal{Q}_E^c(\varphi). \end{aligned}$$

It remains to prove the statement about M. Let $\varphi \in \Gamma_c(X; E)$. Since $\ell^2(X, \mu; E) \subseteq \mathcal{F}_E$, for $f \in D(M)$ Green's formula (Lemma 11.2.1) yields

$$\langle \mathcal{M}\varphi, f\rangle_{2;E} = (\varphi, \mathcal{M}f)_E = \langle \varphi, Mf\rangle_{2;E}.$$

holds. Since $\mathcal{M}\varphi \in \ell^2(X, \mu; E)$, this implies $\varphi \in D(M^*)$ and $M^*\varphi = \mathcal{M}\varphi$. Now the claim follows because M is self-adjoint. □

Remark 1. This proposition allows more general endomorphisms than the corresponding results for graphs without (FC) in Section 11.3.2 and its proof is much simpler. The reason for this is that under (FC) Green's formula is valid for sections with compact support, i.e., one has

$$\langle \varphi, \mathcal{M}\psi\rangle_{2;E} = \mathcal{Q}_E^c(\varphi, \psi), \quad \varphi, \psi \in \Gamma_c(X; E).$$

2. The proposition also shows that under (FC) the operator $M^{\max}$ is closed. It would be interesting to know whether or not this is true for graphs which do not satisfy (FC).

The following lemma shows why we formulated condition (R2) in the definition of standard realizations for the domain of the associated quadratic form and not for the domain of the operator. Otherwise, we could have only dealt with graphs (X, b) and weights μ that satisfy (FC).

Corollary 11.3.18 *Let $W \in \mathcal{A}_{\mu,\Phi;E}$. The following assertions are equivalent.*

(i) (FC) holds.
(ii) For any standard realization M of $\mathcal{M}$ we have $\Gamma_c(X; E) \subseteq D(M)$.
(iii) There exists a standard realization M of $\mathcal{M}$ with $\Gamma_c(X; E) \subseteq D(M)$.

Proof. (i) $\Rightarrow$ (ii): Proposition 11.3.17 (c) shows that any standard realization of $\mathcal{M}$ is an extension of $M^{\min}$ so that $\Gamma_c(X; E) = D(M^{\min}) \subseteq D(M)$.

(ii) $\Rightarrow$ (iii): Proposition 11.3.17 (b) yields the existence of a standard realization .

(iii) $\Rightarrow$ (i): Let M be a standard realization of $\mathcal{M}$ with $\Gamma_c(X; E) \subseteq D(M)$. Then $\mathcal{M}\Gamma_c(X; E) = M\Gamma_c(X; E) \subseteq \ell^2(X, \mu; E)$, i.e., (FC) holds. □

11.3.4 Summary and Examples

In this sub-section we summarize the results of this section and put them into perspective of the existing literature. Moreover, we discuss some examples that show the optimality of our results.

The whole section was devoted to proving the existence of standard realizations of $\mathcal{M}$ (and $\mathcal{H}$) under the following conditions.

(a) $W \in \mathcal{A}_{\mu,\Phi;E}$ and (FC) – admissible endomorphisms and graphs with a finiteness condition.
(b) $W \in \mathcal{S}_{\mu,\Phi;E}$ – endomorphisms with small negative part.
(c) $W_{\min} \in \mathcal{A}_{\mu}$ – admissible endomorphisms dominated by an admissible potential.

Remark (Existing literature) (a) The condition (FC), Lemma 11.3.15, Proposition 11.3.17, and their proofs are basically taken from [28], which contains versions of these results for Schrödinger operators with non-negative potentials. With the same arguments a version of Proposition 11.3.17 is proven in [14] for scalar magnetic Schrödinger operators with admissible potentials.

(b) The stability of closability under form small perturbations is a standard result in perturbation theory of quadratic forms. Thus, the closability of $\mathcal{Q}_E^c$ in Proposition 11.3.7 is well known. The arguments for proving that in this case M^0 is a standard realization of $\mathcal{M}$ are taken from [14], which treats scalar magnetic Schrödinger operators. That they can be extended to general magnetic Schrödinger operators has also been observed in [15].

(c) For scalar magnetic Schrödinger operators the closability of $\mathcal{Q}_{\theta,V}^c$ on $\ell^2(X, \mu)$ when $V \in \mathcal{A}_\mu$ is one of the main results of [14]. There, the proof is based on domination and explicit computations involving a Feynman–Kac–Ito formula for the corresponding semigroups. Our approach to proving the closability statements in Theorems 11.3.13 and 11.3.8 also uses domination but, in contrast, is entirely analytic. The statement that

$W_{\min} \in \mathcal{A}_\mu$ implies that M^0 is a realization of $\mathcal{M}$ is new even for scalar (magnetic) Schrödinger operators. In [14, 31] the authors could only prove this result under additional conditions.

Remark (Optimality of the results) (a) For graphs with (FC) the condition $W \in \mathcal{A}_{\mu,\Phi;E}$ is optimal. In this case, the quadratic form of any standard realization of $\mathcal{M}$ is an extension of $\mathcal{Q}^c_E$ (see Proposition 11.3.17). Thus, if (FC) holds and $\mathcal{M}$ has a standard realization, then $\mathcal{Q}^c_E$ is necessarily lower semi-bounded, i.e., $W \in \mathcal{A}_{\mu,\Phi;E}$.

(c) For general graphs it is unclear whether or not $W_{\min} \in \mathcal{A}_\mu$ is optimal for the existence of standard realizations of $\mathcal{M}$. In particular, the existence of standard realizations of $\mathcal{M}$ remains unresolved when the graph does not satisfy (FC) and $W \in \mathcal{A}_{\mu,\Phi;E}$ but $W_{\min} \notin \mathcal{A}_\mu$. Even in the case of scalar Schrödinger operators, it is unclear whether $V \in \mathcal{A}_\mu$ is necessary for the existence of standard realizations of $\mathcal{H}$. We only proved that $V \in \mathcal{A}_\mu$ is equivalent to the existence of a standard realization whose associated quadratic form satisfies the first Beurling–Deny criterion (cf. Corollary 11.3.14). The problem is that in general we were not able prove that the quadratic form of an arbitrary standard realization of $\mathcal{H}$ is an extension of $\mathcal{Q}^c$; we needed to assume the first Beurling–Deny criterion (cf. Lemma 11.3.10).

The following example shows that Proposition 11.3.6 is only valid with some uniform control over all magnetic fields. It is taken from [11].

Example 11.3.19 Let $K_n = (X_n, b_n)$ be the complete graph on n-vertices X_n, i.e., $|X_n| = n$ and $b_n(x, y) = 1$ for all $x, y \in X_n$, and let μ_n be the counting measure on X_n. The corresponding adjacency operator

$$A_n f(x) := \sum_{y \in X_n} b_n(x, y) f(x) = \sum_{y \in X_n} f(y)$$

has the eigenvalues n and -1.

Let (X, b) be the direct sum of (X_n, b_n), $n \in \mathbb{N}$, i.e., $X = \bigsqcup_{n \geq 1} X_n$ and $b(x, y) = 1$, if $x, y \in X_n$ for some $n \in \mathbb{N}$, and $b(x, y) = 0$, else. Moreover, let μ be the counting measure on X and let $A : C(X) \to C(X)$ be the formal adjacency operator

$$Af(x) = \sum_{y \in X} f(y).$$

It equals the scalar formal magnetic Schrödinger operator $\mathcal{M}_{\mu,-\pi,-\deg}$. Since all the A_n are lower semi-bounded by -1 on $\ell^2(X_n, \mu_n)$, for $\varphi \in C_c(X)$ we obtain

$$\mathcal{Q}^c_{-\pi,-\deg}(\varphi) = \sum_{n=1}^{\infty} \langle A_n(\varphi 1_{X_n}), \varphi 1_{X_n} \rangle_2 \geq \sum_{n=1}^{\infty} -\|1_{X_n}\varphi\|_2^2 = -\|\varphi\|^2.$$

This shows that $\mathcal{Q}^c_{-\pi,-\deg}$ is semi-bounded from below so that $-\deg \in \mathcal{A}_{\mu,(-\mathrm{Id});(\mathbb{C})}$. However, $\mathcal{Q}^c_{-\pi,-\deg}$ is not bounded from above (test, e.g., with the sequence of normalized eigenfunctions $f_n = n^{-1/2} 1_{X_n}$). Hence, $\mathcal{Q}^c_{-\deg} = \mathcal{Q}^c_{0,-\deg} = -\mathcal{Q}^c_{-\pi,-\deg}$ is not bounded from below on $\ell^2(X,\mu)$, so that $-\deg \notin \mathcal{A}_\mu = \mathcal{A}_{\mu,(\mathrm{Id});(\mathbb{C})}$.

The graph constructed in this example is not connected. However, one can modify the graph to make it connected as follows. If for each $n \in \mathbb{N}$ one adds a single edge of weight 1 from some vertex in X_n to some vertex in X_{n+1}, then the adjacency operator A' of the resulting connected graph is a bounded perturbation of A. Therefore, the discussed boundedness properties of A are passed on to A'.

When the endomorphism W has a small negative part (i.e., $W \in \mathcal{S}_{\mu,\Phi;E}$) we do not only obtain that M^0 is a standard realization of $\mathcal{M}$ but also gain some information about the domain of the associated quadratic form; it is contained in $\mathcal{D}_{\Phi,W_+;E}$ and Q^0_E is a restriction of $\mathcal{Q}_E$. In the following example we construct a scalar potential for which this fails. Typically this is the case for optimal Hardy weights. As a consequence we also obtain that the classes $\mathcal{A}_{\mu,\Phi;E}$ and $\mathcal{S}_{\mu,\Phi;E}$ are different.

Example 11.3.20 A function $w : X \to [0,\infty)$ is called a *Hardy weight* for the graph (X,b) if

$$\mathcal{Q}^c_0(\varphi) \geq \sum_{x\in X} |\varphi(x)|^2 w(x), \quad \varphi \in C_c(X).$$

In [31] a Hardy weight w is called optimal if there exists a sequence (e_n) in $C_c(X)$ with the following properties.

- $\lim_{n\to\infty} \left(\mathcal{Q}^c_0(e_n) - \sum_{x\in X} |e_n(x)|^2 w(x) \right) = 0.$
- The sequence (e_n) converges pointwise to a non-negative function G.
- $G \notin \ell^2(X,w)$, i.e.,

$$\sum_{x\in X} |G(x)|^2 w(x) = \infty.$$

If (X,b) is connected, the function G is unique up to multiplication by a constant. It is called the *Agmon ground state* for (X,b) and the weight w. The existence of optimal Hardy weights is established in [31]. On the graph $\mathbb{Z}^d$ with weight $b : \mathbb{Z}^d \times \mathbb{Z}^d \to \{0,1\}$ given by $b(x,y) = 1$ if $|x-y| = 1$

and $b(x, y) = 0$ else, they construct an optimal Hardy weight provided that $d \geq 3$.

Let (X, b) be connected and suppose that w is an optimal Hardy weight. Let $\mu : X \to (0, \infty)$ be a weight such that the Agmon ground state G satisfies $G \in \ell^2(X, \mu)$ and let $V := -w\mu^{-1}$. Since w is a Hardy weight, the form

$$Q_V^c(\varphi) = \frac{1}{2} \sum_{x,y \in X} b(x, y)|\varphi(x) - \varphi(y)|^2 - \sum_{x \in X} |\varphi(x)|^2 w(x), \quad \varphi \in C_c(X),$$

is non-negative. According to Theorem 11.3.13 it is closable on $\ell^2(X, \mu)$. We prove that its closure $Q^0 = Q^0_{\mu,V}$ is not a restriction of $\mathcal{Q}_V$. Since $G \notin \ell^2(X, w)$, it suffices to prove $G \in D(Q^0)$ and $Q^0(G) = 0$.

Let (e_n) be a sequence in $C_c(X)$ as in the definition of optimal Hardy weights and consider $f_n := (e_n \vee 0) \wedge G$, which also has compact support. The choice of μ and Lebesgue's dominated convergence theorem imply that (f_n) converges in $\ell^2(X, \mu)$ towards G. Moreover, the ℓ^2-lower semi-continuity of Q^0 and it satisfying the first Beurling–Deny criterion imply

$$\begin{aligned} Q^0(f_n)^{1/2} &\leq \liminf_{m \to \infty} Q^0((e_n \vee 0) \wedge e_m)^{1/2} \\ &\leq \liminf_{m \to \infty} \left(Q^0(e_n)^{1/2} + Q^0(e_m)^{1/2} \right) = Q^0(e_n)^{1/2} \end{aligned}$$

(see Lemma A.5). Hence, (f_n) is also a sequence as in the definition of optimal Hardy weights, which additionally converges in $\ell^2(X, \mu)$ towards G. These properties and the inequality $Q^0(f_n - f_m)^{1/2} \leq Q^0(f_n)^{1/2} + Q^0(f_m)^{1/2}$ show that (f_n) is Cauchy with respect to the form norm. Since Q^0 is closed, it follows that $G \in D(Q^0)$ and

$$Q^0(G) = \lim_{n \to \infty} Q^0(f_n) = 0.$$

This finishes the proof.

11.4 Bounded Realizations of $\mathcal{M}$ and $\mathcal{H}$

In this section we discuss when $\mathcal{M}$ and $\mathcal{H}$ have bounded realizations. For this the function $B = B_{\mu,W} : X \to [0, \infty)$ that is given by

$$B(x) = \sup\{|\mu(x)^{-1} \deg(x) + \langle W_x \xi, \xi \rangle| \mid \xi \in E_x \text{ with } |\xi| = 1\}, \quad x \in X,$$

plays an important role. Our main theorem regarding bounded realizations reads as follows.

Theorem 11.4.1 *The following assertions are equivalent.*

(i) The function B is bounded and $W \in \mathcal{A}_{\mu,\Phi;E} \cap \mathcal{A}_{\mu,-\Phi;E}$.
(ii) The form $\mathcal{Q}^c_E = \mathcal{Q}^c_{\Phi,W;E}$ is bounded on $\ell^2(X,\mu;E)$.
(iii) $\mathcal{M}$ has a bounded realization.
(iv) $D(M^{\max}) = \ell^2(X,\mu;E)$.
(v) The operator $M^{\max}$ is a bounded realization of $\mathcal{M}$.

If the above are satisfied, then (FC) holds and $M^{\min}$ is essentially self-adjoint.

Proof. We let $\mathrm{Deg} = \mu^{-1}\deg$. With the same symbol we denote the bundle endomorphism that acts upon $\Gamma(X;E)$ by pointwise scalar multiplication with Deg. For $\varphi \in \Gamma_c(X;E)$ we note the identity

$$\mathcal{Q}^c_{\Phi,W;E}(\varphi) = 2q_{\mathrm{Deg}+W}(\varphi) - \mathcal{Q}^c_{-\Phi,W;E}(\varphi). \qquad (\heartsuit)$$

(i) $\Rightarrow$ (ii): The identity ($\heartsuit$) and the lower semi-boundedness of $\mathcal{Q}^c_{\Phi,W;E}$ and $\mathcal{Q}^c_{-\Phi,W;E}$ imply the existence of $C \geq 0$ such that

$$-C\|\varphi\|^2_{2;\,E} \leq \mathcal{Q}^c_{\Phi,W;E}(\varphi) \leq 2q_{\mathrm{Deg}+W}(\varphi) + C\|\varphi\|^2_{2;\,E}.$$

Since $q_{\mathrm{Deg}+W}(\varphi) \leq q_B(|\varphi|) \leq \sup B\|\varphi\|^2_{2;\,E}$, we arrive at (ii).

(ii) $\Rightarrow$ (i): Clearly, the boundedness of forms implies lower semi-boundedness. Hence, it suffices to prove that the function B and the form $\mathcal{Q}^c_{-\Phi,W;E}$ are bounded. For $x \in X$ and $\xi \in E_x$ we denote by $\delta_{x,\xi}$ the finitely supported section with $\delta_{x,\xi}(x) = \xi$ and $\delta_{x,\xi}(y) = 0$ for $y \neq x$. If $|\xi| = 1$, then

$$\mathcal{Q}^c_{\Phi,W;E}(\delta_{x,\xi}) = \deg(x) + \langle W_x\xi, \xi\rangle\mu(x).$$

From the boundedness of $\mathcal{Q}^c_{\Phi,W;E}$ we infer the existence of $C \geq 0$ such that

$$|\mathcal{Q}^c_{\Phi,W;E}(\delta_{x,\xi})| \leq C\|\delta_{x,\xi}\|^2_{2;E} = C\mu(x).$$

Combining both inequalities yields that B is bounded. With this at hand, the boundedness of $\mathcal{Q}^c_{-\Phi,W;E}$ follows from the boundedness of $\mathcal{Q}^c_{\Phi,W;E}$ and the identity ($\heartsuit$).

(ii) $\Rightarrow$ (iv): Since $\mathcal{Q}^c_{\Phi,W;E}$ is bounded, its closure Q^0 is a continuous quadratic form on $\ell^2(X,\mu;E)$ and the associated self-adjoint operator M^0 is bounded. For $\varphi, \psi \in \Gamma_c(X;E)$ Green's formula (Lemma 11.2.1) implies

$$\langle \varphi, M^0\psi\rangle_{2;E} = Q^0(\varphi,\psi) = \mathcal{Q}^c_{\Phi,W;E}(\varphi,\psi) = (\varphi, \mathcal{M}\psi)_E.$$

This shows (FC), i.e., $\mathcal{M}\Gamma_c(X;E) \subseteq \ell^2(X,\mu;E)$, and that M^0 is an extension of $M^{\min}$. Therefore, $M^0 = (M^0)^*$ is a restriction of $(M^{\min})^* = M^{\max}$; for the last equality we used Proposition 11.3.17. We arrive at $\ell^2(X,\mu;E) = D(M^0) \subseteq D(M^{\max})$.

(iv) $\Rightarrow$ (v): Assertion (iv) implies $\mathcal{M}\Gamma_c(X; E) \subseteq \ell^2(X, \mu; E)$ and so (FC) holds. It follows from Proposition 11.3.17 that $M^{\max}$ is closed. Since by assumption $D(M^{\max}) = \ell^2(X, \mu; E)$, the closed graph theorem implies that $M^{\max}$ is continuous. It remains to prove that $M^{\max}$ is self-adjoint. Green's formula (Lemma 11.2.1) and (FC) yield

$$\langle M^{\max} f, g\rangle_{2;E} = \langle f, M^{\max} g\rangle_{2;E}$$

for $f, g \in \Gamma_c(X; E)$. By continuity this identity extends to $f, g \in \ell^2(X, \mu; E)$.

(v) $\Rightarrow$ (iii): This is trivial.

(iii) $\Rightarrow$ (ii): Let M be a bounded realization of $\mathcal{M}$. For $\varphi \in \Gamma_c(X; E)$ Green's formula (Lemma 11.2.1) implies

$$|\mathcal{Q}^c_{\Phi,W;E}(\varphi)| = |(\varphi, \mathcal{M}\varphi)_E| = |\langle\varphi, M\varphi\rangle_{2;E}| \leq \|M\| \|\varphi\|^2_{2;E}.$$

This proves (ii).

Suppose now that one of the assertions holds. That they imply (FC) was proven along the way. Moreover, (v) shows that $M^{\max}$ is self-adjoint. According to Proposition 11.3.17 it satisfies $M^{\max} = (M^{\min})^*$, so that $M^{\min}$ is essentially self-adjoint. □

Remark 1. Bounded realizations are always standard realizations.

2. As remarked in Section 11.3.2, in general it is hard to determine whether a given endomorphism W belongs to $\mathcal{A}_{\mu,\Phi;E} \cap \mathcal{A}_{\mu,-\Phi;E}$ or not. Proposition 11.3.6 gives the sufficient condition $W_{\min} \in \mathcal{A}_\mu$, which might be easier to check. Note that this is always satisfied if $W \geq 0$.

For scalar Schrödinger operators $\mathcal{H}_{\mu,V}$ the function B can be easily computed. It is given by $B = |\mu^{-1}\deg + V|$. Moreover, the lower bound on the spectrum of the endomorphism $W_{\min}$ is given by V itself (cf. the discussion after the definition of $W_{\min}$ in Section 11.3.1. Therefore, the theorem and the previous remark yield the following.

Corollary 11.4.2 *The following assertions are equivalent.*

(i) The function $\mu^{-1}\deg + V$ is bounded and $V \in \mathcal{A}_\mu$.
(ii) $\mathcal{Q}^c = \mathcal{Q}^c_V$ is bounded on $\ell^2(X, \mu)$.
(iii) $\mathcal{H}$ has a bounded realization.
(iv) $H^{\max}$ is a bounded realization of $\mathcal{H}$.
(v) $D(H^{\max}) = \ell^2(X, \mu)$.

Remark 1. For Schrödinger operators with non-negative potentials this characterization of boundedness is given in [27].

2. Example 11.3.19 shows that the assumption $W \in \mathcal{A}_{\mu,\Phi;E} \cap \mathcal{A}_{\mu,-\Phi;E}$ in (i) cannot be weakened. In the example we constructed a graph with counting measure μ (i.e., $\mu(x) = 1, x \in X$) such that the scalar magnetic Schrödinger operator with magnetic field $\theta = -\pi$ and potential $-\deg$ (the adjacency operator) is bounded from above but not from below on $\ell^2(X,\mu)$. As discussed there, this means that $-\deg \in \mathcal{A}_{\mu,(-\mathrm{Id});(\mathbb{C})}$ but $-\deg \notin \mathcal{A}_\mu = \mathcal{A}_{\mu,(\mathrm{Id});(\mathbb{C})}$. Moreover, it satisfies $B(x) = |\mu(x)^{-1}\deg(x) - \deg(x)| = 0$.

11.5 Uniqueness of Realizations

In this section we discuss two criteria that guarantee the uniqueness of standard realizations of $\mathcal{M}$ and $\mathcal{H}$. More precisely, we prove the absence of non-negative sub-solutions for scalar operators and then extend this to magnetic operators with the help of Kato's inequality. That this in turn yields uniqueness of standard realizations is guaranteed by the following lemma.

Lemma 11.5.1 (Abstract criterion for uniqueness) *Let $V \in C(X)$ be real valued. Assume that there exists $C \in \mathbb{R}$ such that all non-negative $f \in \ell^2(X,\mu) \cap \mathcal{F}$ with $\mathcal{H}f \leq Cf$ satisfy $f = 0$.*

(a) If $W \geq V$, then $\mathcal{M}$ has at most one standard realization. If, moreover, $W_{\min} \in \mathcal{A}_\mu$, then $\mathcal{M}$ has exactly one standard realization.
(b) If $W \in \mathcal{A}_{\mu,\Phi;E}$ with $W \geq V$ and (FC) holds, then $M^{\min}$ is essentially self-adjoint.

Proof. (a): Suppose that $\mathcal{M}$ has two standard realizations M_1 and M_2. Since both are by definition lower semi-bounded, their resolvents $(M_1 - \lambda)^{-1}$ and $(M_2 - \lambda)^{-1}$ exist for λ small enough. Let such a λ with $\lambda \leq C$ be given and let $g \in \ell^2(X,\mu;E)$. Since both are standard realizations, the function

$$f := (M_1 - \lambda)^{-1}g - (M_2 - \lambda)^{-1}g$$

belongs to $\ell^2(X,\mu;E) \cap \mathcal{F}_E$ and satisfies $(\mathcal{M} - \lambda)f = 0$. Kato's inequality (Lemma 11.2.2) and $W \geq V$ imply $|f| \in \ell^2(X,\mu) \cap \mathcal{F}$ and $\mathcal{H}|f| \leq \lambda|f|$. Since $\lambda \leq C$ and $|f| \geq 0$, we obtain $\mathcal{H}|f| \leq C|f|$. By our assumption this implies $|f| = 0$, i.e., $f = 0$. Hence, the resolvents agree and we conclude $M_1 = M_2$. If $W_{\min} \in \mathcal{A}_\mu$, the existence of standard realizations is guaranteed by Theorem 11.3.8.

(b) We need to prove that $(M^{\min})^*$ is self-adjoint. According to Proposition 11.3.17 we have $(M^{\min})^* = M^{\max}$, where $M^{\max}$ acts as $\mathcal{M}$ on the domain

$D(M^{\max}) = \{f \in \ell^2(X, \mu; E) \mid \mathcal{M}f \in \ell^2(X, \mu; E)\}$. Since M^0 is a self-adjoint and $M^{\max}$ is an extension of M^0 (see Proposition 11.3.17), it suffices to prove $D(M^{\max}) \subseteq D(M^0)$ to settle the claim. Let $f \in D(M^{\max})$ and for λ small enough consider $g := (M^0 - \lambda)^{-1}(\mathcal{M} - \lambda)f$, which exists since $\mathcal{M}f \in \ell^2(X, \mu; E)$. Since M^0 is a realization of $\mathcal{M}$, we conclude that

$$(\mathcal{M} - \lambda)(f - g) = 0.$$

With the same arguments as in (a) we obtain $f = g \in D(M^0)$ and the claim is proven. □

Remark For graphs satisfying (FC) essential self-adjointness is a stronger property than uniqueness of standard realizations. This is because standard realizations are always semi-bounded. Indeed, there are graphs and magnetic Schrödinger operators where $M^{\min}$ is not semi-bounded (neither from above nor from below) but essentially self-adjoint. For example, similar to Example 11.3.19 one can consider the adjacency operator on the disjoint union $\bigsqcup_{n=2}^{\infty} K_{n,n}$ of complete bipartite graphs on n vertices $K_{n,n}$. We leave the details to the reader.

It is unclear whether there are graphs with (FC) and $W \in \mathcal{A}_{\mu,\Phi;E}$ such that the associated magnetic Schrödinger operator has a unique standard realization but $M^{\min}$ is not essentially self-adjoint.

11.5.1 A Measure Space Criterion

In this section we present a uniqueness criterion that is based on combinatorics and the discreteness of the measure space. It seems to have no counter-part for operators on smooth spaces.

Theorem 11.5.2 (Measure space criterion for uniqueness) *Suppose that (X, b) has no isolated vertices and $W_{\min} \in \mathcal{A}_{\mu}$. If there exists $\alpha \in \mathbb{R}$ such that for each infinite path (x_n) we have*

$$\sum_{n=1}^{\infty} \mu(x_n) \prod_{j=0}^{n-1} \left(1 + \frac{\mu(x_j)(W_{\min}(x_j) - \alpha)}{\deg(x_j)}\right)^2 = \infty, \qquad (\clubsuit)$$

then $\mathcal{M}$ has exactly one standard realization. If, additionally, (FC) holds, then $M^{\min}$ is essentially self-adjoint.

Proof. To simplify notation we set $\mathrm{Deg} := \mu^{-1} \deg$. We start with proving the following observation. Under the given assumptions there exists an $\alpha \in \mathbb{R}$ such that $\mathrm{Deg} + W_{\min} \geq \alpha$ and ($\clubsuit$) holds for all infinite paths.

Since $\mathcal{Q}^c_{W_{\min}}$ is lower semi-bounded on $\ell^2(X,\mu)$, the function $\mathrm{Deg} + W_{\min}$ is bounded from below. This can be easily inferred from the identity

$$\mathcal{Q}^c_{W_{\min}}(\delta_x) = \deg(x) + W_{\min}(x)\mu(x), \quad x \in X.$$

Thus, the bound $\mathrm{Deg} + W_{\min} \geq \alpha$ is satisfied for small enough $\alpha \in \mathbb{R}$. We also show that (♣) holds for all α small enough. Let $\alpha_0 \in \mathbb{R}$ for which (♣) holds for all infinite paths. It suffices to show that for all small enough α and all $x \in X$ we have

$$(\mathrm{Deg}(x) + W_{\min}(x) - \alpha)^2 \geq (\mathrm{Deg}(x) + W_{\min}(x) - \alpha_0)^2.$$

This, however, is a consequence of $\mathrm{Deg} + W_{\min}$ being bounded from below.

We now use Lemma 11.5.1 to deduce uniqueness. Let $\alpha \in \mathbb{R}$ such that (♣) holds for all infinite paths and $\mathrm{Deg} + W_{\min} \geq \alpha$, and let $f \in \ell^2(X,\mu) \cap \mathcal{F}$ be non-negative with $\mathcal{H}_{\mu,W_{\min}} f \leq \alpha f$. The definition of $\mathcal{H}_{\mu,W_{\min}}$ then shows that for each $x \in X$ we have

$$(\mathrm{Deg}(x) + W_{\min}(x) - \alpha) f(x) \leq \frac{1}{\mu(x)} \sum_{y \in X} b(x,y) f(y).$$

Assume that there exists some $x_0 \in X$ with $f(x_0) > 0$. By the previous inequality there exists some $x_1 \in X$ with $x_1 \sim x_0$ and

$$\frac{\mathrm{Deg}(x_0) + W_{\min}(x_0) - \alpha}{\mathrm{Deg}(x_0)} f(x_0) \leq f(x_1).$$

Iterating this argument and using $\mathrm{Deg} + W_{\min} - \alpha \geq 0$ yields an infinite path (x_n) such that for each $n \in \mathbb{N}$ we have

$$f(x_n) \geq f(x_0) \prod_{j=0}^{n-1} \left(1 + \frac{W_{\min}(x_j) - \alpha}{\mathrm{Deg}(x_j)}\right).$$

Since (♣) holds, this inequality and $f(x_0) > 0$ contradict $f \in \ell^2(X,\mu)$. □

The assumption on the divergence of the sum in the previous theorem is a bit technical. For non-negative endomorphisms it reduces to infinite paths having infinite measure, which is satisfied if $\inf_{x \in X} \mu(x) > 0$.

Corollary 11.5.3 *If $W \geq 0$ and every infinite path (x_n) satisfies*

$$\sum_{n=1}^{\infty} \mu(x_n) = \infty,$$

then $\mathcal{M}$ has exactly one standard realization. If, additionally, (FC) holds, then $M^{\min}$ is essentially self-adjoint.

Remark 1. The most important application of Corollary 11.5.3 is when $\inf_{x \in X} \mu(x) > 0$. In this case, it shows uniqueness of standard realizations for all magnetic Schrödinger operators with non-negative bundle endomorphism.

If one path has infinite measure, the whole space has infinite measure. Therefore, Corollary 11.5.3 cannot be used to infer uniqueness of standard realizations or essential self-adjointness when $\mu(X) < \infty$.

2. The given abstract criterion for essential self adjointness in Lemma 11.5.1 and the basic idea for the proof of Theorem 11.5.2 go back to [57], which treats graphs with weights $b \in \{0, 1\}$ and the counting measure. In the presented form Corollary 11.5.3 is taken from [28], which treats Schrödinger operators with non-negative potentials. Variants of Theorem 11.5.2 are contained in [12, 14] for scalar magnetic Schrödinger operators.

Compared to [14], our Theorem 11.5.2 is a bit stronger because we also obtain the existence of standard realizations by means of Theorem 11.3.8.

In [12] only locally finite graphs are considered, but essential self-adjointness of $M^{\min}$ is proven without assuming it to be lower semi-bounded. In this case (and more generally if (FC) holds), for establishing essential self-adjointness of $M^{\min}$ without assumptions on W it suffices to prove that for some $\alpha \in \mathbb{R}$, $\gamma > 0$ all solutions to

$$((M^{\min})^* + \alpha \pm \gamma i)f = (M^{\max} + \alpha \pm \gamma i)f = 0$$

satisfy $f = 0$ (see, e.g., [47, Theorem X.1]). If (♣) holds for all infinite paths, the vanishing of such solutions can be proven along the same lines as in [12], where scalar magnetic Schrödinger operators are treated. Since our chapter focuses on lower semi-bounded realizations, we refrain from giving details.

11.5.2 A Metric Space Criterion

In this section we prove a criterion on uniqueness of standard realizations that is based on intrinsic metrics. The philosophy, which is inspired by corresponding results on manifolds, is the following. A magnetic Schrödinger operator can have unique realizations for two reasons: 1. The space has no boundary so that it is impossible to have different realizations from imposing different boundary conditions. 2. The space has a boundary but a strong growth of the potential (or the endomorphism) forces functions in the domain of the operator to vanish at the boundary. Also in this case boundary conditions cannot lead to different realizations. It turns out that a possible boundary to make this work is the Cauchy boundary with respect to an intrinsic metric. Theorem 11.5.4 is a unified approach to both perspectives and Corollary 11.5.5 is a precise form of the first. The vanishing of the boundary (completeness) is replaced by balls with respect to an intrinsic metric being finite. For path metrics on locally finite graphs this is equivalent to completeness by a discrete version of the Hopf–Rinow theorem (see Proposition 11.5.6).

A *pseudo metric* on X is a symmetric function $\rho : X \times X \to [0, \infty)$ that vanishes on the diagonal and satisfies the triangle inequality. We let $\overline{X}^{\rho}$ be the completion of X with respect to ρ and $\partial_\rho X := \overline{X}^{\rho} \setminus X$ the corresponding *Cauchy boundary*. By $D_\rho : X \to [0, \infty]$ we denote the distance to the boundary, i.e.,

$$D_\rho(x) := \rho(x, \partial_\rho X) := \inf\{\rho(x, z) \mid z \in \partial_\rho X\}.$$

Here we use the convention $D_\rho = \infty$ if $\partial_\rho X = \emptyset$. Note that $D_\rho(x) > 0$ for all $x \in X$ if and only if $\partial_\rho X$ is closed in $\overline{X}^{\rho}$.

For a graph (X, b) and a weight μ a pseudo metric ρ on X is called *intrinsic (with respect to b and μ)* if

$$\sum_{y \in X} b(x, y)\rho(x, y)^2 \leq \mu(x), \quad \text{for all } x \in X.$$

Remark For regular Dirichlet forms intrinsic metrics were introduced and systematically studied in [9]. For graphs and other non-local operators related concepts, so-called adapted metrics, were independently introduced in [8, 20, 39]. In recent years they have been used to solve several open problems in global analysis on graphs. We refer to the survey [26] for a detailed discussion.

The possibility that D_ρ is infinite is implicit in the statement of following theorem, where by convention dividing by infinity yields zero.

Theorem 11.5.4 *Let ρ be an intrinsic pseudo-metric with the following properties.*

- *$\partial_\rho X$ is closed in $\overline{X}^{\rho}$.*
- *For all $\varepsilon > 0$ all ρ-bounded subsets of $\{x \in X \mid D_\rho(x) \geq \varepsilon\}$ are finite.*

If $W_{\min} \geq \frac{1}{2D_\rho^2} + V$ with $V \in \mathcal{A}_\mu$, then $\mathcal{M}$ has exactly one standard realization. If, additionally, (FC) holds, then $M^{\min}$ is essentially self-adjoint on $\ell^2(X, \mu; E)$.

Proof. We use Lemma 11.3.9 to prove the statement. Since constant functions belong to $\mathcal{A}_\mu$, the assumption implies that

$$W_{\min} \geq \frac{1}{2} \max\left\{1, \frac{1}{D_\rho^2}\right\} + V$$

for some $V \in \mathcal{A}_\mu$ (which is not the same as in the statement of the theorem). We consider the discrete Schrödinger operator $\mathcal{H}' := \mathcal{H}_{\mu,V'}$ with potential $V' := \frac{1}{2}\max\{1, D_\rho^{-2}\}+V$. Using $V \in \mathcal{A}_\mu$ we choose $\lambda \in \mathbb{R}$ such that $\lambda\|\varphi\|^2 \leq \mathcal{Q}_V^c(\varphi)$ for all $\varphi \in C_c(X)$ and let $C := \lambda - 1$. Furthermore, we let $f \in \mathcal{F} \cap \ell^2(X,\mu)$ non-negative with $\mathcal{H}'f \leq Cf$. By Lemma 11.3.9 it suffices to show $f = 0$.

For $\varphi \in C_c(X)$ the ground-state transform (Lemma 11.2.5) yields

$$\begin{aligned}\mathcal{Q}_{V'}^c(f\varphi) - C\|\varphi f\|^2 &\leq \frac{1}{2}\sum_{x,y\in X} b(x,y)f(x)f(y)|\varphi(x)-\varphi(y)|^2\\ &\leq \frac{1}{2}\sum_{x\in X} f(x)^2 \sum_{y\in X} b(x,y)|\varphi(x)-\varphi(y)|^2.\end{aligned}$$

Moreover, from the choice of C and the definition of V' we obtain

$$\frac{1}{2}\sum_{x\in X} |\varphi(x)f(x)|^2 \max\left\{1, D_\rho(x)^{-2}\right\}\mu(x) + \|\varphi f\|^2 \leq \mathcal{Q}_{V'}^c(f\varphi) - C\|\varphi f\|^2.$$

We choose φ to obtain the desired statement from these estimates. We fix a point $o \in X$. Let $0 < \varepsilon, R$ and set

$$X_{\varepsilon,R} := \{x \in X \mid D_\rho(x) \geq \varepsilon \text{ and } \rho(o,x) \leq R\}.$$

By assumption these sets are finite. We consider the piecewise affine functions $F : \mathbb{R}\cup\{\infty\} \to \mathbb{R}$, $F(t) = (t-\varepsilon)_+ \wedge 1$ and $G : \mathbb{R}_+ \to \mathbb{R}$, $G(t) = (2-t/R)_+ \wedge 1$ and we define $\varphi : X \to \mathbb{R}$ by $\varphi(x) := F(D_\rho(x))G(\rho(o,x))$. It is straightforward that φ is ρ-Lipschitz with Lipschitz constant $1 + 1/R$. Moreover, it is supported in $X_{\varepsilon,2R}$, which is finite. Combining these observations with the above inequalities and using that ρ is intrinsic we arrive at

$$\frac{1}{2}\sum_{x\in X} |\varphi(x)f(x)|^2 \max\left\{1, D_\rho(x)^{-2}\right\}\mu(x) + \|\varphi f\|^2 \leq \frac{1}{2}\left(1+\frac{1}{R}\right)^2 \|f\|^2.$$

For $x \in X_{\varepsilon,R}$ we have $\varphi(x) = (D_\rho(x)-\varepsilon)\wedge 1$ and hence

$$\max\left\{1, D_\rho(x)^{-2}\right\}|\varphi(x)|^2 \geq (1-\varepsilon/D_\rho(x))^2.$$

This amounts to

$$\frac{1}{2}\sum_{x\in X_{\varepsilon,R}}(1-\varepsilon/D_\rho(x))^2|f(x)|^2\mu(x)+\|\varphi f\|_2^2\leq\frac{1}{2}\left(1+\frac{1}{R}\right)^2\|f\|_2^2.$$

Now we let $\varepsilon\to 0+$ and $R\to\infty$. Since $\partial_\rho X$ is closed, we have $D_\rho(x)>0$ for all $x\in X$, such that under these limits $X_{\varepsilon,R}\nearrow X$ and $\varphi\to D_\rho\wedge 1$ pointwise. From this we obtain $\|(D_\rho\wedge 1)f\|_2=0$ and we arrive at $f=0$, which was to be proven. □

Corollary 11.5.5 *Let ρ be an intrinsic pseudo-metric. Assume that all ρ-balls are finite. If $W_{\min}\in\mathcal{A}_\mu$, then $\mathcal{M}$ has exactly one standard realization. If, moreover, (FC) holds, then $M^{\min}$ is essentially self-adjoint.*

Proof. Finiteness of ρ-balls implies completeness of (X,ρ) so that $\partial_\rho X=\emptyset$ and $D_\rho\equiv\infty$. With this at hand, the statement follows from the previous theorem. □

Remark 1. For path metrics (see below) on locally finite graphs finiteness of balls and completeness coincide (see Proposition 11.5.6). Hence, Corollary 11.5.5 is a discrete version of a classical theorem of Strichartz on manifolds [52], which says that on a complete Riemannian manifold Laplacians on functions, forms, and tensors are essentially self-adjoint. For Schrödinger operators on open subsets of Euclidean space variants of Theorem 11.5.4 are well known (with D_ρ replaced by the distance to the topological boundary of the domain; see, e.g., [44]).

2. For discrete Schrödinger operators with vanishing potential, Corollary 11.5.5 was first proven in [21] and then extended to scalar magnetic Schrödinger operators in [14]. Related but somewhat weaker results with additional assumptions on $\mu^{-1}\deg$ or particular metrics are given in [53, 40]. For magnetic Schrödinger operators on bundles over locally finite graphs Theorem 11.5.4 is proven in [43], which was inspired by results on Schrödinger operators in [5]. In [43] only path metrics (see below) are considered and the potentials need to be uniformly bounded from below. Moreover, a somewhat stronger assumption than finiteness of bounded subsets of $\{x\in X\mid D_\rho(x)\geq\varepsilon\}$, called *regularity of the graph*, is needed. The presented proof of Theorem 11.5.4 is a simplified version of the one given in [43].

3. If $\inf_{x\in X}\mu(x)>0$ and $W\geq 0$, the criteria in Section 11.5.1 always yield uniqueness of standard realizations while Theorem 11.5.2 may not be applicable. For finite measures the situation is opposite. In this case, Theorem 11.5.2 and its corollary may give uniqueness results while the criteria

from Section 11.5.1 fail (see, e.g., Example 11.5.10). As a rule of thumb one can say that Theorem 11.5.4 and its corollary are most interesting for the finite measure case.

In the remainder of this section we put the assumptions on closedness of $\partial_\rho X$, finiteness of bounded subsets of $\{x \in X \mid D_\rho(x) \geq \varepsilon\}$, and finiteness of bounded subsets of X into perspective. There are basically two kinds of (pseudo-)metrics on X for which this is possible; path metrics on locally finite graphs and metrics induced from embeddings of X into Euclidean spaces (or more generally complete Riemannian manifolds).

Let $\sigma : X \times X \to [0, \infty)$ be a symmetric function with $\sigma(x, y) > 0$ if and only if $x \sim y$. The length of a finite path $\gamma = (x_0, x_1, \ldots, x_n)$ with respect to σ is defined by

$$L_\sigma(\gamma) := \sum_{i=1}^{n} \sigma(x_{i-1}, x_i).$$

If the graph (X, b) is connected, the associated *path pseudo-metric* $\rho_\sigma : X \times X \to [0, \infty)$ is defined by

$$\rho_\sigma(x, y) := \inf\{L_\sigma(\gamma) \mid \gamma = (x_0, \ldots, x_n) \text{ is a path with } x_0 = x, x_n = y\}.$$

Pseudo-metrics that arise in this way are called *path pseudo-metrics*. One way to guarantee that ρ_σ is intrinsic (with respect to b and μ) is to demand that

$$\sum_{y \in X} b(x, y)\sigma(x, y)^2 \leq \mu(x), \qquad \text{for all } x \in X.$$

If σ satisfies this property the path pseudo-metric ρ_σ is called *strongly intrinsic*. One edge weight for which this assumption holds is $\sigma_H : X \times X \to [0, \infty)$ with $\sigma_H(x, y) := 0$ if $x \not\sim y$ and

$$\sigma_H(x, y) := \min\left\{\frac{\mu(x)}{\deg(x)}, \frac{\mu(y)}{\deg(y)}\right\}^{1/2}, \quad \text{if } x \sim y.$$

The corresponding path pseudo-metric ρ_{σ_H} was introduced in [20].

The following lemma characterizes some of the required properties to apply Theorem 11.5.4 and Corollary 11.5.5 for path metrics on locally finite graphs.

Proposition 11.5.6 *Let (X, b) be a locally finite connected graph and let $\sigma : X \times X \to [0, \infty)$ be symmetric with $\sigma(x, y) > 0$ if and only if $x \sim y$.*

(a) ρ_σ is a metric on X that induces the discrete topology and $\partial_{\rho_\sigma} X$ is closed in $\overline{X}^{\rho_\sigma}$.

(b) The following assertions are equivalent.

(i) *All* ρ_σ*-balls are finite.*
(ii) (X, ρ_σ) *is complete.*

Proof. (a) For $x \in X$ we let $\sigma_x := \inf\{\sigma(x, y) \mid y \sim x\}$. Since (X, b) is locally finite, it satisfies $\sigma_x > 0$. If $y \neq x$ we have $\rho_\sigma(x, y) \geq \sigma_x > 0$. Hence, $\rho_\sigma(x, y) = 0$ implies $x = y$ so that ρ_σ is a metric. Moreover, it follows from the definitions that any ρ_σ-ball of radius less than σ_x around x only contains x. This shows that $\{x\}$ is open in (X, ρ_σ) and in $(\overline{X}^{\rho_\sigma}, \rho_\sigma)$. Thus, ρ_σ induces the discrete topology on X and $\partial_{\rho_\sigma} X$ is closed in $\overline{X}^{\rho_\sigma}$.

(b) This is given in [21, Theorem A.1]. □

The assumption that bounded subsets of $\{x \in X \mid D_\rho(x) \geq \varepsilon\}$ are finite may or may not be satisfied for path metrics ρ on locally finite graphs. In [6, 43] it is claimed that all locally finite weighted trees (and more generally graphs of finite first Betti number) have this property. However, the following example shows that this is not true. It was communicated to us by Matthias Keller.

Example 11.5.7 On $X := \{(n, i) \mid n \in \mathbb{N}, i = 1, 2\}$ let the graph $b : X \times X \to [0, \infty)$ be given by

$$b((n, i), (m, j)) = \begin{cases} 1 & \text{if } n = m \text{ and } i \neq j \\ \min\{n, m\}^{-2} & \text{if } |n - m| = 1 \text{ and } i = j = 0\,. \\ 0 & \text{else} \end{cases}$$

We consider the path metric ρ_b that is induced by b. Then (X, b) consists of the infinite path of finite length $((n, 0))_{n \in \mathbb{N}}$ with an edge of length 1 attached to each of the vertices in the path. In particular, (X, b) is a tree.

The diameter of (X, ρ_b) is bounded by $1 + \frac{\pi^2}{6}$; hence, any of its subsets is bounded. The Cauchy boundary is one point ∂ and the distance to it satisfies

$$D_{\rho_b}((n, i)) = \rho_b((n, i), \partial) = i + \sum_{k=n}^{\infty} \frac{1}{k^2}.$$

This shows that for $0 < \varepsilon < 1$ we have $\{(n, 1) \mid n \in \mathbb{N}\} \subseteq \{(n, i) \in X \mid D_{\rho_b}((n, i)) \geq \varepsilon\}$. Since $\{(n, 1) \mid n \in \mathbb{N}\}$ is bounded, bounded subsets of $\{(n, i) \in X \mid D_{\rho_b}((n, i)) \geq \varepsilon\}$ are not necessarily finite.

We finish this section by discussing metrics that arise from embeddings into Euclidean spaces. Let $\iota : X \to \mathbb{R}^n$ be an injective function. We define the metric $d_\iota : X \times X \to [0, \infty)$ by $d_\iota(x, y) := |\iota(x) - \iota(y)|$. Then ι is an isometry from (X, d_ι) to $(\mathbb{R}^n, |\cdot|)$ that maps X to $\iota(X)$. It is readily verified that it extends uniquely to a surjective isometry $\hat{\imath} : (\overline{X}^{d_\iota}, d_\iota) \to (\overline{\iota(X)}, |\cdot|)$, where

$\overline{\iota(X)}$ is the closure of $\iota(X)$ in $\mathbb{R}^n$. Under this map the Cauchy boundary $\partial_{d_\iota} X$ is one-to-one with $\overline{\iota(X)} \setminus \iota(X)$. In particular,

$$D_{d_\iota}(x) = \inf\{|\iota(x) - a| \mid a \in \overline{\iota(X)} \setminus \iota(X)\}.$$

We say that $\lim_{|x|\to\infty} \iota(x) = \infty$ if for every $R > 0$ there exists a finite $K \subseteq X$ such that $|\iota(x)| \geq R$ for all $x \in X \setminus K$. Recall that $D \subseteq \mathbb{R}^n$ is called *discrete* if every $x \in D$ has an open neighbourhood U such that $U \cap D = \{x\}$. The following lemma summarizes properties of the metric space (X, d_ι), which are relevant for an application of Theorem 11.5.4.

Proposition 11.5.8 *Let $\iota : X \to \mathbb{R}^d$ be injective.*

(a) The following assertions are equivalent.
  (i) $\partial_{d_\iota} X$ is closed in $\overline{X}^{d_\iota}$.
  (ii) $\iota(X)$ is open in $\overline{\iota(X)}$.

(b) The following assertions are equivalent.
  (i) $\iota(X)$ is discrete.
  (ii) $\partial_{d_\iota} X$ is closed and for every $\varepsilon > 0$ all d_ι-bounded subsets of

$$\{x \in X \mid D_{d_\iota}(x) \geq \varepsilon\} = \{x \in X \mid |\iota(x) - a| \geq \varepsilon \text{ for all } a \in \overline{\iota(X)} \setminus \iota(X)\}$$

  are finite.

(c) The following assertions are equivalent.
  (i) All d_ι-balls are finite.
  (ii) $\lim_{|x|\to\infty} \iota(x) = \infty$.
  (iii) $\iota(X)$ is discrete and (X, d_ι) is complete.

Proof. (a) (i) $\Leftrightarrow$ (ii): As mentioned above, $\partial_{d_\iota} X$ and $\overline{\iota(X)} \setminus \iota(X)$ are one-to-one under the surjective isometry $\hat{\iota} : \overline{X}^{d_\iota} \to \overline{\iota(X)}$. Therefore, $\partial_{d_\iota} X$ is closed if and only if $\overline{\iota(X)} \setminus \iota(X)$ is closed in $\overline{\iota(X)}$. This in turn is equivalent to $\iota(X)$ being open in $\overline{\iota(X)}$.

(b) (i) $\Rightarrow$ (ii): The discreteness of $\iota(X)$ implies that singleton sets $\{\iota(x)\}$ are open in $\iota(X)$ in the relative topology. Hence, $\iota(X)$ is open in $\overline{\iota(X)}$ and (a) implies that $\partial_{d_\iota} X$ is closed.

Let now B be a d_ι-bounded subset of $\{x \in X \mid D_{d_\iota}(x) \geq \varepsilon\}$. By assumption $\iota(B)$ is a bounded discrete set in $\mathbb{R}^d$ that has positive distance from $\overline{\iota(X)} \setminus \iota(X)$. If $\iota(B)$ were infinite, it would contain a sequence (a_n) of pairwise different points. Since $\iota(B)$ is bounded, without loss of generality we can assume that (a_n) converges to some point $a \in \overline{\iota(B)}$. However, since $\iota(X)$ is discrete, we also have $a \in \mathbb{R}^d \setminus \iota(X)$. This contradicts the fact that $\iota(B)$ has positive distance to $\overline{\iota(X)} \setminus \iota(X)$.

(ii) $\Rightarrow$ (i): Suppose that $\iota(X)$ is not discrete. Then there exists $o \in X$ such that every Euclidean ball around $f(o)$ contains infinitely many elements of $\iota(X)$. According to (a) our assumption implies that $\iota(X)$ is open in $\overline{\iota(X)}$. Hence, there exists an $\varepsilon > 0$ such that $|\iota(o) - a| \geq 2\varepsilon$ for every $a \in \overline{\iota(X)} \setminus \iota(X)$. Now consider the d_ι-bounded set

$$B := \{x \in X \mid d_\iota(o, x) \leq \varepsilon\} = \{x \in X \mid |\iota(o) - \iota(x)| \leq \varepsilon\}.$$

By the choice of o the set B is infinite. Moreover, $x \in B$ satisfy

$$|\iota(x) - a| \geq |\iota(o) - a| - |\iota(x) - \iota(o)| \geq \varepsilon$$

for every $a \in \overline{\iota(X)} \setminus \iota(X)$ so that $B \subseteq \{x \in X \mid D_{d_\iota}(x) \geq \varepsilon\}$. This contradicts (ii).

(c) (i) $\Leftrightarrow$ (ii): This is straightforward from the definitions.

(i) $\Rightarrow$ (iii): Completeness follows from the fact that Cauchy sequences are bounded and finite sets are compact. The discreteness of $\iota(X)$ follows from (b).

(iii) $\Rightarrow$ (ii): Let $o \in X$ and for $r > 0$ let $B_r(o) := \{x \in X \mid d_\iota(o, x) \leq r\}$. Then $\iota(B_r(o))$ is obviously bounded in $\mathbb{R}^d$. Since (X, d_ι) is complete, it is also complete and hence even compact in $\mathbb{R}^d$. Thus, if $\iota(B_r(o))$ were not finite, there would be an infinite sequence (a_n) of pairwise different elements of $\iota(B_r(o))$ that converges to some $a \in \overline{\iota(B_r(o))} = \iota(B_r(o))$. This however contradicts the discreteness of $\iota(X)$. □

Remark 1. The injectivity of ι is not so important. It would have been possible to deal with functions for which the preimages of singleton sets are finite. In this case, the resulting distance functions are only pseudo-metrics and not metrics.

2. In view of Proposition 11.5.6 (b) one could ask whether finiteness of d_ι balls alone is equivalent to completeness of (X, d_ι). This is not the case. Consider, e.g., $X = \mathbb{N}_0$ and $\iota : \mathbb{N} \to \mathbb{R}$ given by $\iota(0) = 2$ and $\iota(n) = 2 - 1/n$, $n \in \mathbb{N}$. Then $(\mathbb{N}, d_\iota)$ is complete but $\mathbb{N}$ is bounded with respect to d_ι. In particular, not all d_ι-balls are finite.
3. Instead of functions with values in Euclidean spaces we could have considered functions with values in complete metric spaces whose bounded sets are pre-compact. For example, this is the case for complete Riemannian manifolds.

For a given graph (X, b) without isolated vertices and an injective function $\iota : X \to \mathbb{R}^n$ there is a smallest weight $\mu_\iota : X \to (0, \infty)$ such that d_ι is intrinsic with respect to b and μ_ι. It is given by

$$\mu_\iota(x) = \sum_{y \in X} b(x, y)|\iota(x) - \iota(y)|^2$$

and has the property that for all weights $\mu : X \to (0, \infty)$ that satisfy $\mu \geq \mu_\iota$ the metric d_ι is intrinsic with respect to b and μ. Even though this observation is elementary, it allows us to construct many interesting examples of graphs with finite measures for which Theorem 11.5.4 can be applied. We finish this section with two such examples.

The following example is taken from [45], which is based on [1].

Example 11.5.9 (Nerves of circle packings) A *circle packing* in $\mathbb{R}^2$ (or $\mathbb{C}$) is a collection of circles $\mathcal{C} = (C_j)_{j \in J}$, with $C_j = \{x \in \mathbb{R}^2 \mid |x - x_j| = r_j\}$ for some $x_j \in \mathbb{R}^2$ and $r_j > 0$, such that for $i \neq j$ the interiors of the circles C_i and C_j do not intersect. We say that $\mathcal{C}$ is *bounded* if $\cup_{j \in J} C_j$ is bounded in $\mathbb{R}^2$. The *nerve* (or contact graph) of a circle packing $\mathcal{C} = (C_j)_{j \in J}$ is the graph $(X_\mathcal{C}, b_\mathcal{C})$, where $X_\mathcal{C} = \{x_j \mid j \in J\}$ and

$$b_\mathcal{C}(x_i, x_j) = \begin{cases} 1 & \text{if } S_j \cap S_i \neq \emptyset \\ 0 & \text{else} \end{cases}.$$

The graph $(X_\mathcal{C}, b_\mathcal{C})$ satisfies the condition (b2) if and only if it is locally finite, i.e., $\deg(x_i) = \#\{j \in J \mid C_i \cap C_j \neq \emptyset\} < \infty$ for all $x_i \in X_\mathcal{C}$. We say that $\mathcal{C}$ is *connected* if $(X_\mathcal{C}, b_\mathcal{C})$ is a connected graph.

Let $\iota : X_\mathcal{C} \to \mathbb{R}^2$ be the identity, i.e., $\iota(x_j) = x_j$ for $j \in J$. Then $d_\iota(x_i, x_j) = |x_i - x_j| = r_i + r_j$ so that $\iota(X_\mathcal{C}) = X_\mathcal{C}$ is discrete. If $\mathcal{C}$ is bounded and connected and the function deg is bounded, then the measure μ_ι is finite. Indeed, if K is an upper bound for deg we obtain

$$\begin{aligned} \mu_\iota(X) &= \sum_{x \in X_\mathcal{C}} \sum_{y \in X_\mathcal{C}} b_\mathcal{C}(x, y)|\iota(x) - \iota(y)|^2 \\ &= \sum_{j \in J} \sum_{i \in J : S_j \cap S_i \neq \emptyset} (r_i + r_j)^2 \\ &\leq 4K \sum_{j \in J} r_j^2. \end{aligned}$$

Since πr_j^2 equals the area of the circle C_j and different circles have disjoint interiors, the above computation yields $\mu_\iota(X) \leq \frac{4K}{\pi} \mathrm{Area}(\mathcal{C})$. Here, $\mathrm{Area}(\mathcal{C})$ is the area covered by the interiors of the circles in $\mathcal{C}$. This shows that for the nerve of a bounded connected circle packing with bounded degree the Euclidean metric is an intrinsic metric with respect to a finite measure. Moreover, it satisfies the assumptions of Theorem 11.5.4.

The following example shows how perturbation by a large potential forces a non-essentially self-adjoint operator to become essentially self-adjoint. It also shows that for finite measures Theorem 11.5.4 is stronger than Theorem 11.5.2. The first part of the discussion without the potential is taken from [21].

Example 11.5.10 Consider the graph $(\mathbb{Z}, b)$ with $b(k,l) = 1$ if $|k-l| = 1$. For a weight $\mu : \mathbb{Z} \to (0,\infty)$ the associated formal Schrödinger operator $\mathcal{H}_{\mu,0}$ acts on $C(\mathbb{Z})$ by

$$\mathcal{H}_{\mu,0} f(k) = \frac{1}{\mu(k)} (2f(k) - f(k+1) - f(k-1)), \quad k \in \mathbb{Z}.$$

Consider the weight $\nu : \mathbb{Z} \to (0,\infty)$, $\nu(k) = 2k^{-4}$ for $k \neq 0$ and $\nu(0) = 2$. Then $H_{\nu,0}^{\min}$ is not essentially self-adjoint. This can be seen as follows. The function $h : \mathbb{Z} \to \mathbb{R}$ with $h(k) = k$ belongs to $\ell^2(X,\nu)$ and satisfies $\mathcal{H}_{\nu,0} h = 0$. Hence, $h \in D(H_{\nu,0}^{\max})$. However, h has infinite energy because

$$\sum_{k,l \in \mathbb{Z}} b(k,l)(h(k) - h(l))^2 = \sum_{k \in \mathbb{Z}} 2 = \infty.$$

Since all functions in $D(H_{\nu,0}^{0})$ have finite energy (see Proposition 11.3.4), this implies $h \notin D(H_{\nu,0}^{0})$. If $H_{\nu,0}^{\min}$ were essentially self-adjoint, all self-adjoint extensions of $H_{\nu,0}^{\min}$ would coincide with $(H_{\nu,0}^{\min})^* = H_{\nu,0}^{\max}$. The previous discussion shows that this is not the case.

Let now $\iota : \mathbb{Z} \to \mathbb{R}$ given by $\iota(k) = 2 - 1/k$ if $k \neq 0$ and $\iota(0) = 0$. Then $\iota(\mathbb{Z})$ is discrete in $\mathbb{R}$. The Cauchy boundary of $\mathbb{Z}$ with respect to d_ι consists of exactly one point ∂ and the isometric extension of ι to the completions is given by $\hat{\iota} : \mathbb{Z} \cup \{\partial\} \to \overline{\iota(\mathbb{Z})} = \iota(\mathbb{Z}) \cup \{2\}$ with $\hat{\iota}(k) = \iota(k)$ for $k \in \mathbb{Z}$ and $\hat{\iota}(\partial) = 2$. We obtain that

$$D_{d_\iota}(k) = |\iota(k) - 2| = \begin{cases} 1/|k| & \text{for } k \neq 0 \\ 2 & \text{for } k = 0 \end{cases}.$$

Moreover, the metric d_ι is intrinsic with respect to ν, since

$$\mu_\iota(k) = \sum_{l \in \mathbb{Z}} b(k,l) d_\iota(k,l)^2 = \sum_{l \in \mathbb{Z}} b(k,l)|\iota(k) - \iota(l)|^2 = \begin{cases} \frac{2}{k^2(k^2-1)} & \text{if } |k| \geq 2 \\ \frac{5}{4} & \text{if } |k| = 1 \\ 2 & \text{if } k = 0 \end{cases}.$$

Then Theorem 11.5.4 and Proposition 11.5.8 imply that for any $V : \mathbb{Z} \to \mathbb{R}$ with $V(k) \geq k^2/2$ the operator $H_{\nu,V}^{\min}$ is essentially self-adjoint $\ell^2(\mathbb{Z},\nu)$.

Moreover, for the case $V : \mathbb{Z} \to \mathbb{R}$, $V(k) = k^2/2$ the assumptions of Theorem 11.5.2 are not satisfied, since for all $\alpha \in \mathbb{R}$ the infinite path $(x_n)_{n\geq 0} = (n)_{n\geq 0}$ satisfies

$$\sum_{n=1}^{\infty} \nu(n) \prod_{k=0}^{n-1} \left(1 + \frac{\nu(k)(k^2/2 - \alpha)}{2}\right)^2 = 2\sum_{n=1}^{\infty} \frac{1}{n^4} \prod_{k=0}^{n-1} \left(1 + \frac{(k^2/2 - \alpha)}{2k^4}\right)^2 < \infty.$$

11.6 Markovian Realizations of $\mathcal{H}$ and Markov Uniqueness

In this section we study standard realizations of $\mathcal{H}$ whose associated quadratic forms are Dirichlet forms. The semigroups generated by these standard realizations are Markovian and therefore correspond to Markov processes on X through the Feynman–Kac formula. Such standard realizations on possibly non-locally finite graphs have received quite some attention in recent years (see, e.g., [27, 28, 18, 50] and references therein). Here we discuss their basic structure and give criteria for their uniqueness. It turns out that if the potential is non-negative, there always is a minimal and a maximal Markovian realization in the sense of quadratic forms (see Theorem 11.6.5). Since Markovian realizations are special standard realizations, theorems that ensure their uniqueness tend to have weaker assumptions than the ones for essential self-adjointness or uniqueness of standard realizations, which we discussed in Section 11.5. This is the content of Section 11.6.2.

11.6.1 The Structure of Markovian Realizations

In this section we construct a minimal and a maximal Markovian realization of $\mathcal{H}$ and show that all other Markovian realization lie between them in the sense of quadratic forms.

For the definition and basic properties of Dirichlet forms we refer to Appendix A.3. Since Dirichlet forms are real quadratic forms, for the purpose of this section it suffices to consider real-valued functions. We use the following convention.

Convention In this whole section all functions are real valued. In particular, we abuse notation and denote by $C(X)$ the real-valued functions on X, by $C_c(X)$ the real-valued functions on X with finite support, and by $\ell^2(X, \mu)$ the Hilbert space of real-valued square summable functions.

Definition 11.6.1 (Markovian realization) A standard realization of $\mathcal{H}$ is called *Markovian* if the associated quadratic form is a Dirichlet form.

The following lemma shows that for studying Markovian realizations it suffices to consider the case $V \geq 0$.

Lemma 11.6.2 *If $\mathcal{H}$ has a Markovian realization, then $V \geq 0$.*

Proof. Let H be a Markovian realization of $\mathcal{H}$ with associated quadratic form Q. It follows from Lemma 11.3.10 that Q is an extension of $\mathcal{Q}^c$. This implies $\mathcal{Q}^c(\varphi \wedge 1) \leq \mathcal{Q}^c(\varphi)$ for all $\varphi \in C_c(X)$. Let $K \subseteq X$ be finite and let $x \in K$. Lemma A.7 shows $\mathcal{Q}^c(1_K, \delta_x) \geq 0$ and, therefore,

$$0 \leq \mathcal{Q}^c(1_K, \delta_x) = \sum_{y \in X \setminus K} b(x, y) + V(x).$$

Letting $K \nearrow X$ yields $V \geq 0$. □

Recall the definition of $\mathcal{Q}$ and $\mathcal{D}$ from Section 11.2.5. The following well-known proposition gives the two most prominent Markovian realizations of $\mathcal{H}$ when $V \geq 0$ (see, e.g., [28, 50]).

Proposition 11.6.3 (Existence of Markovian realizations) *Let $V \geq 0$. Then Q^0 and the restriction of $\mathcal{Q}$ to $\mathcal{D} \cap \ell^2(X, \mu)$ are Dirichlet forms and the associated operators are Markovian realizations of $\mathcal{H}$.*

Proof. Let $Q^{(N)}$ denote the restriction of $\mathcal{Q}$ to $\mathcal{D} \cap \ell^2(X, \mu)$. We first prove that $Q^{(N)}$ is a Dirichlet form. Its lower semi-continuity with respect to $\ell^2(X, \mu)$-convergence follows from Fatou's lemma. Hence, it is closed by Lemma A.1. The definition of $\mathcal{Q}$ and $V \geq 0$ imply that for $f \in \mathcal{D}$ and any normal contraction $C : \mathbb{R} \to \mathbb{R}$ we have $C \circ f \in \mathcal{D}$ and $\mathcal{Q}(C \circ f) \leq \mathcal{Q}(f)$. Since $f \in \ell^2(X, \mu)$ also implies $C \circ f \in \ell^2(X, \mu)$, we obtain that $Q^{(N)}$ is a Dirichlet form. Let $H^{(N)}$ be the associated self-adjoint operator. For $f \in D(H^{(N)})$ and $\varphi \in C_c(X) \subseteq D(Q^{(N)})$ Green's formula (Lemma 11.2.4) shows

$$\langle \varphi, H^{(N)} f \rangle_2 = Q^{(N)}(\varphi, f) = \mathcal{Q}(\varphi, f) = (\varphi, \mathcal{H} f).$$

This implies that $H^{(N)}$ is a realization of $\mathcal{H}$.

We have already seen in Proposition 11.3.4 that H^0 is a realization of $\mathcal{H}$. It remains to show that Q^0 is a Dirichlet form. To this end, let $f \in D(Q^0)$ and choose a sequence (φ_n) in $C_c(X)$ that converges to f with respect to the form norm. For a normal contraction $C : \mathbb{R} \to \mathbb{R}$ the sequence $(C \circ \varphi_n)$ belongs to $C_c(X)$ and converges in $\ell^2(X, \mu)$ to f. Therefore, the lower semi-continuity of Q^0 and its definition yield

$$\begin{aligned} Q^0(C \circ f) &\le \liminf_{n\to\infty} Q^0(C \circ \varphi_n) = \liminf_{n\to\infty} \mathcal{Q}^c(C \circ \varphi_n) \\ &\le \liminf_{n\to\infty} \mathcal{Q}^c(\varphi_n) = Q^0(f). \end{aligned}$$

This shows that Q^0 satisfies the first Beurling–Deny criterion and finishes the proof. □

Definition 11.6.4 (Neumann realization) Let $V \ge 0$. The Dirichlet form on $\ell^2(X, \mu)$ that is the restriction of $\mathcal{Q}$ to the domain $\mathcal{D} \cap \ell^2(X, \mu)$ is denoted by $Q^{(N)} = Q^{(N)}_{\mu,V}$. The associated self-adjoint operator is called $H^{(N)} = H^{(N)}_{\mu,V}$.

Remark We already noted that H^0 can be thought of being the realization of $\mathcal{H}$ with 'Dirichlet boundary conditions at infinity'. As the notation suggests, $H^{(N)}$ should be thought of as the realization with 'Neumann boundary conditions at infinity'. On appropriate compactifications of X this intuition can be made precise (see [29, 19]).

Markovian realizations are ordered in terms of their quadratic forms (cf. Appendix A.1). The following theorem shows that in this sense H^0 is the minimal Markovian realization and $H^{(N)}$ is the maximal Markovian realization.

Theorem 11.6.5 (Structure of Markovian realizations) *Let H be a Markovian realization of $\mathcal{H}$ with associated Dirichlet form Q. Then $V \ge 0$ and $Q^0 \le Q \le Q^{(N)}$. In particular, Q is an extension of Q^0.*

Corollary 11.6.6 (Markov uniqueness) *The operator $\mathcal{H}$ has exactly one Markovian realization if and only if $V \ge 0$ and $Q^0 = Q^{(N)}$.*

Remark 1. We would like to stress that in our convention $Q^0 \le Q \le Q^{(N)}$ means $D(Q^0) \subseteq D(Q) \subseteq D(Q^{(N)})$ and $Q^0(f) \ge Q(f) \ge Q^{(N)}(f)$ for all $f \in \ell^2(X, \mu)$ (with the forms being equal ∞ outside their domains). It emphasizes the size of the form domain; forms with large domains and small values are large in the sense of this ordering. This convention seems to be standard in Dirichlet form theory (see, e.g., [10, 3]).

2. For locally finite graphs this theorem and its corollary are given in [18]. It is new for general graphs. The proof given below is based on results of [49].
3. Since the space is discrete, any Dirichlet form Q that extends Q^0 is a Silverstein extension, i.e., for $f \in D(Q^0) \cap \ell^\infty(X)$ and $g \in D(Q) \cap \ell^\infty(X)$ we have $fg \in D(Q^0) \cap \ell^\infty(X)$. This is clear whenever $f \in C_c(X)$, the case of general f follows by an approximation argument.

   It also follows from abstract results in Dirichlet form theory that when $V = 0$ the form $Q^{(N)}$ is the maximal (Silverstein) extension of Q^0 (see, e.g., [3, Theorem 6.6.9]). In the case $V \ne 0$ the abstract results contained in the literature are wrong; they claim that any (Silverstein) extension Q of Q^0 satisfies $Q \le Q^{(N)}$. However, in general there are extensions of Q^0 for which $Q^0 \le Q^{(N)}$ does no hold; for example, see [29, Section 5]. Note that the class of Dirichlet forms treated by Theorem 11.6.5 is a bit smaller. We only consider extensions Q of Q^0 whose associated operator is a Markovian realization of $\mathcal{H}$.

4. Dirichlet forms Q on $\ell^2(X,\mu)$ with $Q^0 \le Q \le Q^{(N)}$ are parametrized by certain Dirichlet forms on the Royden boundary of the graph. This is discussed in [29].

Let H be a Markovian realization of $\mathcal{H}$ and let Q be the associated Dirichlet form. For $f \in D(Q) \cap \ell^\infty(X)$ and $\varphi \in D(Q)$ with $0 \le \varphi \le 1$ we define the concatenated form

$$Q_\varphi(f) := Q(\varphi f) - Q(\varphi f^2, \varphi).$$

Since $D(Q) \cap \ell^\infty(X)$ is an algebra (see, e.g., [10, Theorem 1.4.2]), this is well defined. For $f \in D(Q) \cap \ell^\infty(X)$ we define the *main part of Q* by

$$Q^m(f) := \sup\{Q_\varphi(f) \mid \varphi \in D(Q) \text{ with } 0 \le \varphi \le 1\},$$

and the *killing part of Q* by

$$Q^k(f) := Q(f) - Q^m(f).$$

The following lemma shows that both are well defined. It is a special case of the theory developed in [49, Chapter 3].

Lemma 11.6.7 *Let $f, g \in D(Q) \cap \ell^\infty(X)$ and let $\varphi, \psi \in D(Q)$ with $0 \le \varphi, \psi \le 1$. Then the following holds.*

(a) $0 \le Q_\varphi(f) \le Q^m(f) \le Q(f)$ and $0 \le Q^k(f) \le Q(f)$.
(b) $\varphi \le \psi$ implies $Q_\varphi(f) \le Q_\psi(f)$.
(c) $|f| \le |g|$ implies $Q^k(f) \le Q^k(g)$.

Proof. (a) and (b) We show the inequalities $0 \le Q_\varphi(f) \le Q_\psi(f) \le Q(f)$, the rest then follows from the definitions. We denote by $G_\alpha := (H+\alpha)^{-1}$ the resolvents and we use the approximating forms (see Appendix A.1) to compute

$$\begin{aligned} Q_\varphi(f) &= \lim_{\alpha\to\infty} \alpha(\langle (I-\alpha G_\alpha)(\varphi f), \varphi f\rangle_2 - \langle (I-\alpha G_\alpha)(\varphi f^2), \varphi\rangle_2) \\ &= \lim_{\alpha\to\infty} \alpha(\langle \alpha G_\alpha(\varphi f^2), \varphi\rangle_2 - \langle \alpha G_\alpha(\varphi f), \varphi f\rangle_2) \end{aligned}$$

and

$$Q(f) = \lim_{\alpha\to\infty} \alpha\langle (I-\alpha G_\alpha) f, f\rangle_2.$$

Since the involved quantities are continuous in f, it suffices to prove

$$Q^\alpha_\varphi(f) := \langle \alpha G_\alpha(\varphi f^2), \varphi\rangle_2 - \langle \alpha G_\alpha(\varphi f), \varphi f\rangle_2 \leq \langle (I-\alpha G_\alpha) f, f\rangle_2 =: Q^\alpha(f)$$

and the monotonicity in φ for $f \in C_c(X)$. Let S be the finite support of f. An elementary computation shows

$$Q^\alpha_\varphi(f) = \sum_{x,y\in S} b^\alpha_\varphi(x,y)(f(x)-f(y))^2 + \sum_{x\in S} c^\alpha_\varphi(x) f(x)^2,$$

where for $x, y \in S$ the coefficients satisfy

$$\begin{aligned} b^\alpha_\varphi(x,y) &= \begin{cases} -Q^\alpha_\varphi(\delta_x,\delta_y) & \text{if } x \neq y \\ 0 & \text{if } x = y \end{cases} \\ &= \begin{cases} \langle \alpha G_\alpha(\varphi\delta_x), \varphi\delta_y\rangle_2 & \text{if } x \neq y \\ 0 & \text{if } x = y \end{cases}, \end{aligned}$$

and

$$c^\alpha_\varphi(x) = Q^\alpha_\varphi(1_S, \delta_x) = \langle \alpha G_\alpha(1_{X\setminus S}\varphi), \varphi\delta_x\rangle_2.$$

Similarly, we obtain

$$Q^\alpha(f) = \sum_{x,y\in S} b^\alpha(x,y)(f(x)-f(y))^2 + \sum_{x\in S} c^\alpha(x) f(x)^2,$$

where for $x, y \in S$ the coefficients satisfy

$$b^\alpha(x,y) = \begin{cases} -Q^\alpha(\delta_x,\delta_y) & \text{if } x \neq y \\ 0 & \text{if } x = y \end{cases} = \begin{cases} \langle \alpha G_\alpha \delta_x, \delta y\rangle_2 & \text{if } x \neq y \\ 0 & \text{else} \end{cases},$$

and

$$c^\alpha(x) = Q^\alpha(1_S, \delta_x) = \langle 1_S - \alpha G_\alpha 1_S, \delta_x\rangle_2.$$

Since resolvents of Dirichlet forms are positivity preserving, these computations immediately yield $0 \leq b^\alpha_\varphi \leq b^\alpha_\psi \leq b^\alpha$ and $0 \leq c^\alpha_\varphi \leq c^\alpha_\psi$. For $x \in S$ we obtain

$$\begin{aligned} c^\alpha(x) - c^\alpha_\varphi(x) &= \langle 1_S - \alpha G_\alpha 1_S, \delta_x\rangle_2 - \langle \alpha G_\alpha(1_{X\setminus S}\varphi), \varphi\delta_x\rangle_2 \\ &\geq \langle 1_S - \alpha G_\alpha 1_S, \delta_x\rangle_2 - \langle \alpha G_\alpha(1_{X\setminus S}\varphi), \delta_x\rangle_2 \\ &= \langle 1_S - \alpha G_\alpha(1_S + \varphi 1_{X\setminus S}), \delta_x\rangle_2. \end{aligned}$$

Since αG_α is Markovian and $0 \leq \varphi \leq 1$, we have $\alpha G_\alpha(1_S + \varphi 1_{X\setminus S}) \leq 1$. Therefore, the above computation shows $c^\alpha_\varphi(x) \leq c^\alpha(x)$ for $x \in S$ and the claims are proven.

(c) We prove this statement in two steps. First, we additionally assume that there exists $\psi \in D(Q)$ such that $1_{\{|g|>0\}} \leq \psi \leq 1$. Let $\varepsilon > 0$. We use the definition of Q^k and that Q_φ is monotone in φ to choose $\varphi \in D(Q)$ with $\psi \leq \varphi \leq 1$ such that

$$|Q^m(f) - Q_\varphi(f)| \leq \varepsilon \text{ and } |Q^m(g) - Q_\varphi(g)| \leq \varepsilon.$$

Note that φ equals one on the supports of f and g. We obtain

$$\begin{aligned} Q^k(g) - Q^k(f) &\geq Q(g) - Q_\varphi(g) - Q(f) + Q_\varphi(f) - 2\varepsilon \\ &= Q(\varphi g) - Q_\varphi(g) - Q(\varphi f) + Q_\varphi(f) - 2\varepsilon \\ &= Q(\varphi g^2, \varphi) - Q(\varphi f^2, \varphi) \\ &= Q(g^2 - f^2, \varphi) - 2\varepsilon. \end{aligned}$$

Since $g^2 - f^2$ is non-negative and φ equals one on the support of $g^2 - f^2$, Lemma A.7 yields $Q(g^2 - f^2, \varphi) \geq 0$ and the claim is proven for these special f and g.

Let now $f, g \in D(Q) \cap \ell^\infty(X)$ with $|f| \leq |g|$ arbitrary. For $\alpha > 0$ consider the functions $f_\alpha := f - (f \wedge \alpha) \vee (-\alpha)$ and $g_\alpha := g - (g \wedge \alpha) \vee (-\alpha)$. They satisfy $|f_\alpha| \leq |g_\alpha|$ and [10, Theorem 1.4.2] shows $f_\alpha, g_\alpha \in D(Q)$ and $g_\alpha \to g$ and $f_\alpha \to f$ with respect to the form norm, as $\alpha \to 0+$. Since Q is a Dirichlet form, the function $\psi_\alpha := (\alpha^{-1}|g|) \wedge 1$ belongs to $D(Q)$ and since $\{|g_\alpha| > 0\} = \{|g| > \alpha\}$, it equals one on the support of g_α. We can therefore apply the already proven inequality to the functions f_α and g_α. Moreover, the quadratic form Q^k is smaller than Q and therefore continuous with respect to Q-convergence. With all of these properties we conclude

$$Q^k(f) = \lim_{\alpha \to 0+} Q^k(f_\alpha) \leq \lim_{\alpha \to 0+} Q^k(g_\alpha) = Q^k(g).$$

This finishes the proof. □

For the following proof recall that $\mathcal{Q}_0$ is the Schrödinger form with respect to the potential $V = 0$.

Proof of Theorem 11.6.5. By definition we have $Q = Q^m + Q^k$ on $D(Q) \cap \ell^\infty(X)$ and $Q^{(N)} = \mathcal{Q}_0 + q_V$ on $D(Q^{(N)})$. Since bounded functions are dense in the domains of Dirichlet forms (see, e.g., [10, Theorem 1.4.2]), it suffices to prove $\mathcal{Q}_0(f) \leq Q^m(f)$ and $q_V(f) \leq Q^k(f)$ for $f \in D(Q) \cap \ell^\infty(X)$.

Let $K \subseteq X$ be finite and let $f \in D(Q) \cap \ell^\infty(X)$. By Lemma 11.3.10 the form Q is an extension of $\mathcal{Q}^c$ and 1_K, $1_K f$, and $1_K f^2$ have finite support. Therefore, Lemma 11.6.7 yields

$$Q^m(f) \geq Q_{1_K}(f) = \mathcal{Q}^c(1_K f) - \mathcal{Q}^c(1_K f^2, 1_K)$$
$$= \frac{1}{2} \sum_{x,y \in K} b(x,y)(f(x) - f(y))^2.$$

Letting $K \nearrow X$ yields $Q^m(f) \geq \mathcal{Q}_0(f)$ for all $f \in D(Q) \cap \ell^\infty(X)$.

It remains to prove the inequality $Q^k(f) \geq q_V(f)$ for $f \in D(Q) \cap \ell^\infty(X)$. To this end, let $K \subseteq X$ be finite and let $\varepsilon > 0$. According to Lemma 11.6.7, we can choose $\psi \in D(Q)$ with $1_K \leq \psi \leq 1$ such that for each $\varphi \in D(Q)$ with $\psi \leq \varphi \leq 1$ we have $Q^k(1_K f) \geq Q(1_K f) - Q_\varphi(1_K f) - \varepsilon$. The monotonicity of Q^k (Lemma 11.6.7 (c)) then implies

$$Q^k(f) \geq Q^k(1_K f) \geq Q(1_K f) - Q_\varphi(1_K f) - \varepsilon = Q(1_K f^2, \varphi) - \varepsilon.$$

Since $0 \leq \varphi \leq 1$ is bounded and the resolvent $G_\alpha := (H+\alpha)^{-1}$ is Markovian, we have $0 \leq \alpha G_\alpha \varphi \leq 1$. It follows from Lebesgue's dominated convergence theorem that $\mathcal{H}\alpha G_\alpha \varphi \to \mathcal{H}\varphi$ pointwise, as $\alpha \to \infty$. Moreover, $\mathcal{H}$ is a realization of H and $1_K f^2$ has finite support so that we obtain

$$Q(1_K f^2, \varphi) = \lim_{\alpha \to \infty} Q(1_K f^2, \alpha G_\alpha \varphi)$$
$$= \lim_{\alpha \to \infty} (1_K f^2, \mathcal{H}\alpha G_\alpha \varphi) = (1_K f^2, \mathcal{H}\varphi).$$

Combining these computations we arrive at

$$Q^k(f) \geq (1_K f^2, \mathcal{H}\varphi) - \varepsilon,$$

whenever $\varphi \in D(Q)$ with $\psi \leq \varphi \leq 1$. Letting $\varphi \nearrow 1$ pointwise and using Lebesgue's dominated convergence theorem yields $\mathcal{H}\varphi \to V$ pointwise. This shows

$$Q^k(f) \geq (1_K f^2, V) - \varepsilon = \sum_{x \in K} f(x)^2 V(x) \mu(x) - \varepsilon.$$

Since K and ε were arbitrary, this finishes the proof. □

11.6.2 Markov Uniqueness

In this section we discuss criteria for the uniqueness of Markovian realizations of $\mathcal{H}$, which by Theorem 11.6.5 is equivalent to $Q^0 = Q^{(N)}$. Of course, the results of Section 11.5 can always be applied, but in general their assumptions are stronger than what is actually needed for Markov uniqueness. There are various abstract characterization for $Q^0 = Q^{(N)}$ in the context of graphs (see, e.g., [18, 50]). In this text the focus is a bit different. We present characterizations of Markov uniqueness that are related to the geometry of the graph and therefore have the same spirit as the results in Section 11.5.

The first criterion deals with capacities of boundaries. As a preparation we discuss some of their elementary properties. Let $V \geq 0$. A function $h : X \to [0, \infty)$ is called 1-*excessive* (with respect to the form $Q^{(N)}$) if for every $\beta > 0$ we have $\beta(H^{(N)} + \beta + 1)^{-1}h \leq h$. Here the resolvent is extended to arbitrary non-negative functions by monotonicity, i.e.,

$$(H^{(N)}+\beta+1)^{-1}h := \sup\{(H^{(N)}+\beta+1)^{-1}f \mid f \in \ell^2(X, \mu) \text{ with } 0 \leq f \leq h\}.$$

Since the resolvent of $Q^{(N)}$ is Markovian, the constant function 1 is always 1-excessive. If $\mu(X) < \infty$ and $(V \cdot \mu)(X) < \infty$, it belongs to $D(Q^{(N)})$. We say that a function $f : X \to \mathbb{R}$ is *strictly positive* if $f(x) > 0$ for all $x \in X$. We need the following characterizations of excessive functions.

Lemma 11.6.8 (Characterization of excessive functions) *Let $h : X \to [0, \infty)$. The following assertions are equivalent.*

(i) h is 1*-excessive.*
(ii) For every $f \in D(Q^{(N)})$ we have $f \wedge h \in D(Q^{(N)})$ and

$$Q^{(N)}(f \wedge h) + \|f \wedge h\|_2^2 \leq Q^{(N)}(f) + \|f\|_2^2.$$

If, additionally, $h \in D(Q^{(N)})$, then these are equivalent to the following.

(iii) $Q^{(N)}(h, f) + \langle h, f\rangle_2 \geq 0$ for all non-negative $f \in D(Q^{(N)})$.

In particular, there always exists a strictly positive 1*-excessive function in $D(Q^{(N)})$.*

Proof. The equivalence of the assertions (i)–(iii) follows from [36, Proposition III.1.2] and [25, Proposition 4].

For the 'In particular'-part we let $g \in \ell^2(X, \mu)$ be strictly positive and consider $h := (H^{(N)} + 1)^{-1}g$. By (iii) it is 1-excessive. We show that it is strictly positive. Suppose that there exists an $x \in X$ with $h(x) = 0$. For $\beta > 0$ let $G_\beta := (H^{(N)} + 1)^{-1}$. The resolvent identity and that G_β is positivity preserving yield for $\beta > 1$

$$0 = h(x) = G_1 g(x) = (\beta - 1)G_1 G_\beta g(x) + G_\beta g(x) \geq G_\beta g(x) \geq 0.$$

Since (G_β) is strongly continuous, we obtain

$$0 = \lim_{\beta \to \infty} \beta G_\beta g(x) = g(x) > 0,$$

a contradiction. □

Let now h be a 1-excessive function. We define the *capacity* of a set $U \subseteq X$ (with respect to h and $Q^{(N)}$) by

$$\mathrm{cap}_h(U) := \inf\{\|f\|^2_{Q^{(N)}} \mid f \in D(Q^{(N)}) \text{ with } f \geq 1_U h\},$$

where $\|f\|^2_{Q^{(N)}} = Q^{(N)}(f) + \|f\|_2^2$ is the square of the form norm. We use the convention $\mathrm{cap}_h(U) = \infty$ if the infimum is taken over an empty set. If the convex set $\{f \in D(Q^{(N)}) \mid f \geq 1_U h\}$ is non-empty, by the Hilbert space projection theorem there exists a unique minimizer $h_U \geq 1_U h$ such that $\mathrm{cap}_h(U) = Q^{(N)}(h_U) + \|h_U\|_2^2$. It is called the *equilibrium potential of* U (with respect to h). Since by Lemma 11.6.8 we have $\||h_U| \wedge h\|^2_{Q^{(N)}} \leq \|h_U\|^2_{Q^{(N)}}$, it satisfies $0 \leq h_U \leq h$. In particular, $h_U = h$ on U.

The *capacity of the boundary of* X is defined by

$$\mathrm{cap}_h(\partial X) := \inf\{\mathrm{cap}_h(X \setminus K) \mid K \subseteq X \text{ finite}\}.$$

Let ρ be a pseudo-metric. The *capacity of the Cauchy boundary* is defined by

$$\mathrm{cap}_h(\partial_\rho X) := \inf\{\mathrm{cap}_h(X \cap O) \mid O \subseteq \overline{X}^\rho \text{ open neighbourhood of } \partial_\rho X\}.$$

Remark Let $\hat{X}$ be a compactification of X. For an open neighbourhood U of $\hat{X} \setminus X$ in $\hat{X}$ the set $\hat{X} \setminus U = X \setminus U$ is compact in X. Since X carries the discrete topology, this means that $X \setminus U$ is finite. This implies that $X \cap U = X \setminus K$ for some finite $K \subseteq X$. On the other hand, if $K \subseteq X$ is finite, the set $\hat{X} \setminus K$ is an open neighbourhood of $\hat{X} \setminus X$. These considerations show

$$\mathrm{cap}_h(\partial X) = \inf\{\mathrm{cap}_h(X \cap O) \mid O \subseteq \hat{X} \text{ open neighbourhood of } \hat{X} \setminus X\}.$$

Therefore, $\mathrm{cap}_h(\partial X)$ and $\mathrm{cap}_h(\partial_\rho X)$ have basically the same definition with the only difference that open neighbourhoods of 'the boundary' are defined by two different topologies in topological spaces containing X.

The results of Section 11.5.2 could be read that the absence of a boundary implies uniqueness of standard realizations. The following theorem is similar in spirit. It says that smallness of the boundary yields Markov uniqueness.

Theorem 11.6.9 *If $V \geq 0$, the following assertions are equivalent.*

(i) $\mathcal{H}$ has a unique Markovian realization.
(ii) For any 1-excessive function $h \in D(Q^{(N)})$ we have $\mathrm{cap}_h(\partial X) = 0$.
(iii) There exists one strictly positive 1-excessive function $h \in D(Q^{(N)})$ with $\mathrm{cap}_h(\partial X) = 0$.

If, additionally, (X, b) is connected and locally finite, then these are equivalent to the following.

(iv) *For one/any strongly intrinsic path metric* ρ *and for any* 1*-excessive function* $h \in D(Q^{(N)})$ *we have* $\mathrm{cap}_h(\partial_\rho X) = 0$.
(v) *For one/any strongly intrinsic path metric* ρ *and for one strictly positive* 1*-excessive function* $h \in D(Q^{(N)})$ *we have* $\mathrm{cap}_h(\partial_\rho X) = 0$.

We prove the theorem through several lemmas, which may be of interest on their own right. For some subset $U \subseteq X$ we denote by $\pi_U : \ell^2(X,\mu) \to \ell^2(U,\mu|_U)$ the restriction $f \mapsto f|_U$ and by $\iota_U = (\pi_U)^* : \ell^2(U,\mu|_U) \to \ell^2(X,\mu)$ the embedding that is given by $\iota_U f(x) = f(x)$ if $x \in U$ and $\iota_U f(x) = 0$ if $x \in X \setminus U$. Let $Q_U^{(N)}$ be the Dirichlet form on $\ell^2(U,\mu|_U)$ with domain $D(Q_U^{(N)}) = \{f \in \ell^2(U,\mu|_U) \mid \iota_U f \in D(Q^{(N)})\}$ on which it acts by $Q_U^{(N)}(f) := Q^{(N)}(\iota_U f)$. Moreover, let Q_U^0 be the restriction of $Q_U^{(N)}$ to the closure of $C_c(U)$ with respect to the form norm $\|\cdot\|_{Q_U^{(N)}}$. For $f \in D(Q_U^{(N)})$ we have

$$Q_U^{(N)}(f) = \frac{1}{2}\sum_{x,y\in U} b(x,y)(f(x)-f(y))^2 + \sum_{x\in U} f(x)^2(V(x)+d_U(x))\mu(x),$$

with $d_U(x) = \mu(x)^{-1}\sum_{y\in X\setminus U} b(x,y)$. Hence, $Q_U^{(N)}$ and Q_U^0 are the maximal and the minimal Dirichlet form on $\ell^2(U,\mu|_U)$ associated with the graph $(U, b|_{U\times U})$ with potential $V|_U + d_U$. The associated operators are Markovian realizations of the Schrödinger operator $\mathcal{H}_{\mu|_U, V|_U + d_U}$ on the graph $b|_{U\times U}$.

Lemma 11.6.10 *Let* (X,b) *be a connected locally finite graph and let* ρ *be a strongly intrinsic path metric. If* $U \subseteq X$ *is complete with respect to* ρ*, then* $Q_U^0 = Q_U^{(N)}$.

Proof. We first show that it suffices to consider the connected components of U. Let $(U_i)_{i\in I}$ be the connected components of U. A simple computation using the aforementioned formula for $Q_U^{(N)}$ shows

$$Q_U^{(N)}(f) = \sum_{i\in I} Q_{U_i}^{(N)}(f|_{U_i}).$$

Hence, if for each U_i we have $Q_{U_i}^{(N)} = Q_{U_i}^0$, then $Q_U^{(N)} = Q_U^0$.

Let now σ be an edge weight such that $\rho = \rho_\sigma$ and let W be a connected component of U. By ρ_σ^W we denote the path metric that $\sigma|_{W\times W}$ induces on W. Since there are less paths in W than in X, we have $\rho_\sigma^W \geq \rho_\sigma$ on $W \times W$. Moreover, since ρ_σ is strongly intrinsic, the metric ρ_σ^W is intrinsic with respect to $b|_{W\times W}$ and $\mu|_W$.

Claim: (W, ρ_σ^W) is complete.

Proof of the claim. Let (x_n) be a Cauchy sequence in (W, ρ_σ^W). Since $\rho_\sigma^W \geq \rho_\sigma$ and W is complete in with respect to ρ_σ, it converges with respect to ρ_σ to some $x \in W$. According to Proposition 11.5.6 the singleton set $\{x\}$ is open in (X, ρ_σ). Hence, (x_n) is eventually constant so that it also converges in (W, ρ_σ^W) to x. This proves the claim.

As discussed in Proposition 11.5.6 completeness of (W, ρ_σ^W) implies that bounded sets in (W, ρ_σ^W) are finite. Hence, by Corollary 11.5.5 the Schrödinger operator $\mathcal{H}_{\mu|_W, V|_W + d_W}$ (with respect to the graph $b|_{W\times W}$) has a unique standard realization on $\ell^2(W, \mu|_W)$. The discussion preceding this theorem shows that this implies $Q_W^0 = Q_W^{(N)}$. □

Lemma 11.6.11 *Let $h \in D(Q^{(N)})$ be 1-excessive. For any $U \subseteq X$ we have*

$$\operatorname{cap}_h(U) = \inf\{\|h - f\|_{Q^{(N)}}^2 \mid f \in D(Q^{(N)}) \text{ with } f 1_U = 0\}.$$

Proof. Let h_U be the equilibrium potential of U. Since $h_U = h$ on U, we have $h - h_U = 0$ on U so that

$$\begin{aligned}\operatorname{cap}_h(U) &= \|h - (h - h_U)\|_{Q^{(N)}}^2 \\ &\geq \inf\{\|h - f\|_{Q^{(N)}}^2 \mid f \in D(Q^{(N)}) \text{ with } f 1_U = 0\}.\end{aligned}$$

Let now $\varepsilon > 0$ and choose $g \in D(Q^{(N)})$ with $g 1_U = 0$ such that

$$\|h - g\|_{Q^{(N)}}^2 \leq \inf\{\|h - f\|_{Q^{(N)}}^2 \mid f \in D(Q^{(N)}) \text{ with } f 1_U = 0\} + \varepsilon.$$

Since $h - g \geq h 1_U$, this inequality and the definition of $\operatorname{cap}_h(U)$ imply

$$\operatorname{cap}_h(U) \leq \inf\{\|h - f\|_{Q^{(N)}}^2 \mid f \in D(Q^{(N)}) \text{ with } f 1_U = 0\} + \varepsilon.$$

This finishes the proof. □

Lemma 11.6.12 *Let (X, b) be a connected locally finite graph and let ρ be a strongly intrinsic path metric. Let $h \in D(Q^{(N)})$ be a 1-excessive function. Then*

$$\operatorname{cap}_h(\partial X) = \operatorname{cap}_h(\partial_\rho X).$$

Proof. Since ρ is a metric, points are closed. Hence, for any finite $K \subseteq X$ the set $\overline{X}^\rho \setminus K$ is an open neighbourhood of $\partial_\rho X$. This implies $\operatorname{cap}_h(\partial X) \geq \operatorname{cap}_h(\partial_\rho X)$.

Let O be an open neighbourhood of $\partial_\rho X$ and let $h_{X\cap O}$ be the corresponding equilibrium potential. Since $h_{X\cap O} = h$ on $X \cap O$, the function $h - h_{X\cap O}$ is supported in $X \setminus O$. Moreover, $h - h_{X\cap O} \in D(Q^{(N)})$ and so the restriction of $h - h_{X\cap O}$ to $X \setminus O$ belongs to $D(Q_{X\setminus O}^{(N)})$. The complement $X \setminus O$ is closed in the completion and therefore complete itself. Lemma 11.6.10 shows

$Q^0_{X\setminus O} = Q^{(N)}_{X\setminus O}$. Since $h - h_{X\cap O}$ is supported in $X \setminus O$, this yields that $h - h_{X\cap O}$ can be approximated with respect to $\|\cdot\|_{Q^{(N)}}$ by functions of finite support in $X \setminus O$.

Let now $\varepsilon > 0$ and let ψ be a function with finite support $K \subseteq X \setminus O$ that satisfies $\|h - h_{X\cap O} - \psi\|_{Q^{(N)}} < \varepsilon$. Using Lemma 11.6.11 and that $\psi 1_{X\setminus K} = 0$ we obtain

$$\begin{aligned}\operatorname{cap}_h(X \cap O)^{1/2} =& \|h - (h - h_{X\cap O})\|_{Q^{(N)}} \\ &\geq \|h - \psi\|_{Q^{(N)}} - \varepsilon \geq \operatorname{cap}_h(X \setminus K)^{1/2} - \varepsilon.\end{aligned}$$

This yields $\operatorname{cap}_h(X \cap O) \geq \operatorname{cap}_h(\partial X)$ and the claim is proven. □

Lemma 11.6.13 *Let h be a strictly positive 1-excessive function. If $h \in D(Q^0)$, then $Q^0 = Q^{(N)}$.*

Proof. Since $Q^{(N)}$ is a Dirichlet form, it suffices to prove that each non-negative $f \in D(Q^{(N)})$ can be approximated by finitely supported functions with respect to the form norm. Let (φ_n) be a sequence in $C_c(X)$ that converges to h with respect to the form norm. Since f is non-negative, the functions $f_n := f \wedge \varphi_n$ have finite support and they converge in $\ell^2(X,\mu)$ to $f \wedge h$. The lower semi-continuity of Q^0 and that Q^0 and $Q^{(N)}$ agree on $C_c(X)$ yield

$$\begin{aligned}\left(Q^0(f \wedge h) + \|f \wedge h\|_2^2\right)^{1/2} &\leq \liminf_{n\to\infty} \|f_n\|_{Q^0} = \liminf_{n\to\infty} \|f_n\|_{Q^{(N)}} \\ &\leq \|f\|_{Q^{(N)}} + \|h\|_{Q^{(N)}} < \infty.\end{aligned}$$

For the last inequality we use Lemma A.5. This shows $f \wedge h \in D(Q^0)$. For any $n \in \mathbb{N}$ the function nh is also 1-excessive and belongs to $D(Q^0)$. Hence, $f \wedge (nh) \in D(Q^0)$ for any $n \in \mathbb{N}$. With this at hand, the lower semi-continuity of Q^0 and that $Q^{(N)}$ and Q^0 agree on $D(Q^0)$ imply

$$\begin{aligned}Q^0(f) + \|f\|_2^2 &\leq \liminf_{n\to\infty}\left(Q^0(f \wedge (nh)) + \|f \wedge (nh)\|_2^2\right) \\ &= \liminf_{n\to\infty}\left(Q^{(N)}(f \wedge (nh)) + \|f \wedge (nh)\|_2^2\right) \\ &\leq Q^{(N)}(f) + \|f\|_2^2.\end{aligned}$$

For the last inequality we used that nh is 1-excessive and Lemma 11.6.8. This shows $f \in D(Q^0)$ and finishes the proof. □

Lemma 11.6.14 *Let $h \in D(Q^{(N)})$ be a 1-excessive function. Then $h \in D(Q^0)$ if and only if $\operatorname{cap}_h(\partial X) = 0$.*

Proof. Let $h \in D(Q^{(N)})$ be 1-excessive. Using Lemma 11.6.11 we obtain

$$\begin{aligned}\operatorname{cap}_h(\partial X) &= \inf_{K \subseteq X \text{ finite}} \operatorname{cap}_h(X \setminus K) \\ &= \inf_{K \subseteq X \text{ finite}} \inf\{\|h - f\|^2_{Q^{(N)}} \mid f 1_{X \setminus K} = 0\} \\ &= \inf\{\|h - f\|^2_{Q^{(N)}} \mid f \in C_c(X)\}.\end{aligned}$$

This shows the claim. □

Proof of Theorem 11.6.9. According to Corollary 11.6.6 for proving the equivalence of (i)–(iii) it suffices to show that the assertions (ii) and (iii) are equivalent to $Q^0 = Q^{(N)}$.

(i) ⇒ (ii): Since $Q^0 = Q^{(N)}$, we have $h \in D(Q^0)$ and Lemma 11.6.14 shows $\operatorname{cap}_h(\partial X) = 0$.

(ii) ⇒ (iii): This follows from the existence of a strictly positive 1-excessive function (see Lemma 11.6.8).

(iii) ⇒ (i): Let $h \in D(Q^{(N)})$ be 1-excessive and strictly positive. According to Lemma 11.6.14 assertion (iii) implies $h \in D(Q^0)$. With this at hand $Q^0 = Q^{(N)}$ follows from Lemma 11.6.13.

Assume now additionally that (X, b) is connected and locally finite. The equivalence of the assertions (ii) and (iv) and of the assertions (iii) and (v) follows from Lemma 11.6.12 and the existence of strongly intrinsic path metrics, which was discussed in Section 11.5.2. □

Remark 1. Under the condition $1 \in D(Q^{(N)})$, for locally finite graphs the equivalence of $Q^0 = Q^{(N)}$ and $\operatorname{cap}_1(\partial_\rho X) = 0$ for some strongly intrinsic path metric ρ is contained in [21]. The idea for the proof of Lemma 11.6.10 is also taken from [21]. The proofs of Lemma 11.6.11, 11.6.14, and 11.6.13 are taken from [49], which contains abstract versions of these results and of the equivalence of (i)–(iii) in Theorem 11.6.9. Lemma 11.6.12 seems to be a new observation.

2. If $1 \in D(Q^{(N)})$, the condition $\operatorname{cap}_1(\partial_\rho X) = 0$ can be inferred from estimates on the Minkowski dimension of $\partial_\rho X$ with respect to the given measure μ (see [21, Theorem 4]).
3. For proving $\operatorname{cap}_h(\partial X) = \operatorname{cap}_h(\partial_\rho X)$ it was essential that (X, b) is locally finite and ρ is a strongly intrinsic path metric. If ρ is only an intrinsic metric that induces the discrete topology on X, we still have $\operatorname{cap}_h(\partial X) \geq \operatorname{cap}_h(\partial_\rho X)$ but we doubt that the converse inequality holds. Nevertheless, it could still happen that $\operatorname{cap}_h(\partial_\rho X) = 0$ implies $\operatorname{cap}_h(\partial X)$. If this were the case, in the theorem we could drop the assumption that ρ is a strongly intrinsic path metric.

In the case of finite measure it is possible to relate Markov uniqueness of $\mathcal{H}$ to uniqueness of standard realizations of $\mathcal{H}$ and the criteria that we gave in Section 11.5.2. This is discussed next. We denote by $\mathcal{D}^0 = \mathcal{D}^0_V$ the functions of finite energy $f \in \mathcal{D} = \mathcal{D}_V$ for which there exists a sequence (φ_n) in $C_c(X)$ such that $\varphi_n \to f$ pointwise and $\mathcal{Q}(f - \varphi_n) \to 0$, as $n \to \infty$. In the following theorem we assume $V = 0$ for convenience. It would also be true with $(V \cdot \mu)(X) < \infty$.

Theorem 11.6.15 *Suppose that $V = 0$. The following assertions are equivalent.*

(i) There exists a finite measure μ such that $\mathcal{H}_{\mu,0}$ has a unique Markovian realization.
(ii) For all finite measures μ the operator $\mathcal{H}_{\mu,0}$ has a unique Markovian realization.
(iii) There exists a finite measure μ such that $\mathcal{H}_{\mu,0}$ has a unique standard realization.
(iv) There exists a finite measure μ and an intrinsic metric ρ with respect to b and μ that has finite distance balls.
(v) $1 \in \mathcal{D}^0$.
(vi) There exists a function $f \in \mathcal{D}$ with $\lim\limits_{x\to\infty} f(x) = \infty$.

If, additionally, (FC) holds, then these are equivalent to the following.

(vii) There exists a finite measure μ such that $H^{\min}_{\mu,0}$ is essentially self-adjoint.

To shed some light on the measures that are mentioned in (iii) and (vii) we single out the following lemma before proving the theorem.

Lemma 11.6.16 *Let $f \in \mathcal{D}$ injective with $\lim\limits_{x\to\infty} f(x) = \infty$. Then the metric $d_f(x, y) : X \times X \to [0, \infty)$, $d_f(x, y) = |f(x) - f(y)|$ has finite distance balls. The measure μ_f that is given by*

$$\mu_f(x) = \sum_{y\in X} b(x, y)|f(x) - f(y)|^2$$

satisfies $\mu_f(X) \leq 2\mathcal{Q}(f) < \infty$ and d_f is intrinsic with respect to any measure $\mu \geq \mu_f$.

Proof. According to Proposition 11.5.8 the metric d_f has finite distance balls. The other statements follow straightforward from the definitions. □

Proof. We first prove the equivalence of (i), (ii), and (v).

(ii) ⇒ (i): This is trivial.

(i) $\Rightarrow$ (v): Markov uniqueness, $V = 0$ and $\mu(X) < \infty$ imply $1 \in D(Q^{(N)}) = D(Q^0)$. Moreover, it follows from the definitions that $D(Q^0) \subseteq \mathcal{D}^0 \cap \ell^2(X,\mu)$. This shows $1 \in \mathcal{D}^0$.

(v) $\Rightarrow$ (ii): Let μ be a finite measure on X. Since the constant function 1 is 1-excessive and belongs to $D(Q^{(N)})$, by Lemma 11.6.13 it suffices to prove $1 \in D(Q^0)$. Let (φ_n) be a sequence in $C_c(X)$ that converges pointwise to 1 and satisfies $\mathcal{Q}(\varphi_n - 1) = \mathcal{Q}(\varphi_n) = \mathcal{Q}^c(\varphi_n) \to 0$, as $n \to \infty$. Consider the sequence $\tilde{\varphi}_n := (\varphi_n \wedge 1) \vee 0$ in $C_c(X)$. It satisfies $\mathcal{Q}^c(\tilde{\varphi}_n) \leq \mathcal{Q}^c(\varphi_n)$ and, since μ is finite, it converges in $\ell^2(X,\mu)$ to 1. The ℓ^2-lower semi-continuity of Q^0 and that it is the closure of $\mathcal{Q}^c$ yields

$$Q^0(1) \leq \liminf_{n\to\infty} Q^0(\tilde{\varphi}_n) = \liminf_{n\to\infty} \mathcal{Q}^c(\tilde{\varphi}_n) \leq \liminf_{n\to\infty} \mathcal{Q}^c(\varphi_n) = 0 < \infty.$$

This shows $1 \in D(Q^0)$.

(iv) $\Rightarrow$ (iii)/(vii): This follows from Corollary 11.5.5.

(iii)/(vii) $\Rightarrow$ (i): This is trivial.

(v) $\Rightarrow$ (vi): Let μ be a finite measure. Let (φ_n) be a sequence in $C_c(X)$ that converges pointwise to 1 and satisfies $\mathcal{Q}(1 - \varphi_n) = \mathcal{Q}(\varphi_n) \to 0$, as $n \to \infty$. As seen in the proof of (v) $\Rightarrow$ (ii), we can assume $0 \leq \varphi_n \leq 1$ so that $\varphi_n \to 1$ in $\ell^2(X,\mu)$. We can further assume (after choosing a subsequence) that

$$\sum_{n=1}^{\infty} \|1 - \varphi_n\|_{Q^{(N)}_{\mu,0}} < \infty.$$

Hence, $f := \sum_{n=1}^{\infty}(1 - \varphi_n)$ exists in the Hilbert space $(D(Q^{(N)}_{\mu,0}), \|\cdot\|_{Q^{(N)}_{\mu,0}})$. For $K \in \mathbb{N}$ it satisfies $|f(x)| \geq K$ whenever $x \in X \setminus \cup_{n=1}^{K} \operatorname{supp}(\varphi_n)$. This shows $\lim_{x\to\infty} f(x) = \infty$.

(vi) $\Rightarrow$ (iv): Let $f \in \mathcal{D}$ with $\lim_{x\to\infty} f(x) = \infty$. By modifying the values of f a little bit it can be chosen to be injective. With this at hand, the statement follows from Lemma 11.6.16. □

Remark 1. Let $V = 0$. Weighted graphs (X, b) that satisfy $1 \in \mathcal{D}^0$ are called *recurrent*. There is a vast amount of literature on abstract and geometric conditions ensuring recurrence. We refer to the textbooks [51, 55] as well as [50] for further details and references.

That recurrence ($1 \in \mathcal{D}^0$) is equivalent to $Q^0_{\mu,0} = Q^{(N)}_{\mu,0}$ for finite measures μ is well known (see, e.g., [50] for the discrete setting or [32, 17] in the context of general Dirichlet forms). Indeed, recurrence always implies Markov uniqueness for all measures.

The equivalence of (iv), (v), and (vi) is taken from [45].

2. Corollary 11.5.5 shows that the existence of intrinsic metrics with finite distance balls yields uniqueness of standard realizations. It is quite remarkable that the previous theorem gives some kind of converse. Uniqueness of Markovian realizations yields the existence of intrinsic metrics with finite distance balls and hence also uniqueness of standard realizations. However, we warn the readers that this is only true for particular measures. In the proof we construct finite measures and metrics with finite distance balls that depend on functions as in (vi) (cf. Lemma 11.6.16). See also Example 11.6.17.

We finish this section with an example of a graph where the previous theorem can be applied. Moreover, illustrate the construction of the finite measures that appear in assertion (iii)/(vii) of the theorem.

Example 11.6.17 Consider the graph $(\mathbb{Z}, b)$ of Example 11.5.10 and the measure $\nu_\alpha : \mathbb{Z} \to (0, \infty)$ with $\nu_\alpha(n) = |n|^{-\alpha}$ if $n \neq 0$ and $\nu(0) = 1$. With the same arguments as in Example 11.5.10 it is possible to prove that if $\alpha > 3$, the operator $H^{\min}_{\nu_\alpha,0}$ is not essentially self-adjoint on $\ell^2(X, \nu_\alpha)$. Hence, according to Corollary 11.5.5, for $\alpha > 3$ there cannot exist an intrinsic metric with finite distance balls with respect to b and the measure ν_α.

It is well known that the graph $(\mathbb{Z}, b)$ is recurrent, i.e., $1 \in \mathcal{D}^0$. Indeed, for $1/2 < \alpha \leq 1$ the function $f_\alpha : \mathbb{Z} \to \mathbb{R}$ with $f_\alpha(0) = 1$ and

$$f_\alpha(n) = 1 + \operatorname{sgn}(n) \sum_{k=1}^{|n|} \frac{1}{|k|^\alpha}, \qquad n \neq 0,$$

satisfies $f_\alpha \in \mathcal{D}$ and $\lim_{x\to\infty} f_\alpha(x) = \infty$. Therefore, the previous theorem shows $1 \in \mathcal{D}^0$ and that for any finite measure we have Markov uniqueness. In particular, for all $\alpha > 1$ the operator $\mathcal{H}_{\nu_\alpha,0}$ has a unique Markovian realization. For $\alpha \leq 1$ essential self-adjointness and Markov uniqueness can be inferred from Corollary 11.5.3.

We discuss which finite measures and intrinsic metrics with finite balls are induced by f_α (cf. Lemma 11.6.16). Let $g_\alpha = f_{\alpha/2}/\sqrt{2}$ and consider the measure

$$\mu_{g_\alpha}(n) = \frac{1}{2} \sum_{k\in\mathbb{Z}} b(n,k)(f_{\alpha/2}(n) - f_{\alpha/2}(k))^2 = \begin{cases} \frac{1}{2(|n|+1)^\alpha} + \frac{1}{2|n|^\alpha} & \text{if } |n| \geq 1 \\ 1 & \text{if } n = 0 \end{cases}.$$

It satisfies $\mu_{g_\alpha} \leq \nu_\alpha$. Hence, the metric d_{g_α} that is given by

$$d_{g_\alpha}(n, m) = |g_\alpha(n) - g_\alpha(m)| = \frac{1}{\sqrt{2}} \left| \operatorname{sgn}(n) \sum_{k=1}^{|n|} \frac{1}{k^{\alpha/2}} - \operatorname{sgn}(m) \sum_{k=1}^{|m|} \frac{1}{k^{\alpha/2}} \right|$$

is intrinsic with respect to ν_α. For $\alpha \leq 2$ we have $\lim_{x\to\infty} g_\alpha = \infty$ so that balls with respect to d_{g_α} are finite (see Proposition 11.5.8). This shows that for $\alpha \leq 2$ the operator $\mathcal{H}_{\nu_\alpha,0}$ is essentially self-adjoint on $\ell^2(X, \nu_\alpha)$, while for $\alpha > 1$ the measure ν_α is finite.

In the previous discussion we saw that the operator $H^{\min}_{\nu_\alpha,0}$ is essentially self-adjoint for $\alpha \leq 2$ and not essentially self-adjoint for $\alpha > 3$. It would be interesting to know what happens for $2 < \alpha \leq 3$. To the best of our knowledge, this is open. Note that $\mathcal{H}_{\nu_\alpha,0}$ has a unique Markovian realization for all $\alpha \in \mathbb{R}$.

11.7 Open Problems

In this section we collect some of the open problems that arose in the text. For some problems we also comment on expected answers, turning them into conjectures.

Problem 1 *Is $M^{\max}$ a closed (resp. closable) operator on $\ell^2(X, \mu; E)$?*

If (FC) is satisfied, the operator $M^{\max}$ is the adjoint of $M^{\min}$ and hence it is closed. For general graphs and finite measures however we expect the answer to be negative for two reasons. If we take a graph that is not locally finite and equip $C(X)$ with the locally convex topology of pointwise convergence, the operator $\mathcal{H}_{\mu,0} : \mathcal{F} \to C(X)$ is not closed (nor closable) as an unbounded operator on $C(X)$. For finite measures the spaces $\ell^2(X, \mu)$ and $C(X)$ and their topologies are not that different.

Problem 2 *Does $W \in \mathcal{A}_{\mu,\Phi;E}$ for all unitary connections Φ on E imply $W_{\min} \in \mathcal{A}_\mu$?*

At a first glance it seems unlikely that this is true, since by Proposition 11.3.6 $W_{\min} \in \mathcal{A}_\mu$ implies a lower bound for $\mathcal{Q}^c_{\Phi,W;E}$ that is uniform in the connection Φ. On the other hand, it is not totally unlikely that some uniform boundedness principle yields a positive answer to the question.

Problem 3 *Let M be a standard realization of $\mathcal{M}$. Is the associated quadratic form an extension of $\mathcal{Q}^c_E$? In particular, does the existence of a standard realization of $\mathcal{H}_{\mu,V}$ imply $V \in \mathcal{A}_\mu$?*

We resolved this problem in the scalar case when the associated form satisfies the first Beurling–Deny criterion (see Lemma 11.3.10). Moreover, by Proposition 11.3.17 it is satisfied whenever (FC) holds. An answer to the 'in particular'-part of this problem would show or disprove the optimality of Theorem 11.3.13.

Problem 4 *Let (X, b) be a graph and let μ be a weight such that (FC) is not satisfied. Is there $W \in \mathcal{A}_{\mu,\Phi;E}$ with $W_{\min} \notin \mathcal{A}_\mu$ such that $\mathcal{M}_{\mu,\Phi,W;E}$ has/does not have a standard realization?*

This is a case that cannot be treated with domination arguments (cf. also Section 11.3.4).

Problem 5 *Is there a graph (X, b) and a weight μ with (FC) and $W \in \mathcal{A}_{\mu,\Phi;E}$ such that $M^{\min}$ is not essentially self-adjoint but $\mathcal{M}_{\mu,\Phi,W;E}$ has a unique standard realization?*

Realizations in the sense of Definition 11.3.1 are always semi-bounded. Hence, uniqueness of standard realizations asks for uniqueness in the class of semi-bounded operators while essential self-adjointness asks for uniqueness in the class of all self-adjoint operators. For a general symmetric operator on some Hilbert space it can of course happen that it has a unique semi-bounded extension while it is not essentially self-adjoint. Therefore, the previous questions asks whether such abstract examples can be realized within the class of magnetic Schrödinger operators.

Problem 6 *Let ρ be an intrinsic pseudo-metric such that X is closed in $\overline{X}^\rho$ and such that for all $\varepsilon > 0$ bounded subsets of $\{x \in X \mid D_\rho(x) \geq \varepsilon\}$ are finite. What is the optimal constant $C \geq 0$ such that for all W with $W_{\min} \geq C D_\rho^{-2} + V$ for some $V \in \mathcal{A}_\mu$ the operator $\mathcal{M}$ has a unique standard realization?*

Theorem 11.5.4 shows that the optimal constant is smaller than or equal to $\frac{1}{2}$. For essential self-adjointness of classical Schrödinger operators $-\Delta + V$ on open subsets of Euclidean space (with D_ρ replaced by the distance to the topological boundary of the domain), the optimal constant is $\frac{3}{4}$ (see, e.g., [44] for the result in any dimension and [47, Theorem X.10] for the optimality of the constant in one dimension). Even though this may make our result seem stronger than in the Euclidean case, it is not. Our discrete operator $\mathcal{H}_{\mu,0}$ plays the same role as $-\frac{1}{2}\Delta$ (the generator of Brownian motion) in the Euclidean setting. For $\frac{1}{2}\Delta$ the optimal constant scales down to $\frac{3}{8}$, which is smaller than $\frac{1}{2}$.

Problem 7 *For which $2 < \alpha \leq 3$ is the operator $H^{\min}_{\nu_\alpha,0}$ of Example 11.6.17 essentially self-adjoint?*

It was the general theme of the uniqueness sections that small measures tend to make discrete magnetic Schrödinger operators less unique. Even for the simplest infinite graph $\mathbb{Z}$ the role of the measure for essential self-adjointness is not fully understood. The previous problem could be a starting point for further investigations.

A Quadratic Forms on Hilbert Spaces

In this appendix we collect basic facts about quadratic forms on Hilbert spaces. In particular, we treat their monotone limits and the Beurling–Deny criteria. The material presented here is standard. For the claims where we do not give detailed references, we refer the readers to the textbooks [10, 46].

A.1 Basics

Let $(H, \langle \cdot, \cdot \rangle)$ be a complex Hilbert space with induced norm $\|\cdot\|$. A functional $q : H \to (-\infty, \infty]$ is called *quadratic form* if it is homogeneous and satisfies the parallelogram identity, i.e., if for all $f, g, h \in H$ and $\lambda \in \mathbb{C}$ it satisfies

$$q(\lambda f) = |\lambda|^2 q(f)$$

and

$$q(f + g) + q(f - g) = 2q(f) + 2q(g).$$

The *domain* of a quadratic form q on H is $D(q) = \{f \in H \mid q(f) < \infty\}$. It is called densely defined if the closure of $D(q)$ in H equals H.

By a theorem of Jordan and von Neumann [22] any quadratic form induces a sesquilinear form on its domain via polarization, i.e.,

$$q : D(q) \times D(q) \to \mathbb{C},\ (f, g) \mapsto q(f, g) := \frac{1}{4} \sum_{k=1}^{4} i^k q(f + i^k g)$$

is sesquilinear. Here we abuse notation and write q for the quadratic form on H and the induced sesquilinear form on $D(q)$. In this sense, we have $q(f, f) = q(f)$ for $f \in D(q)$.

A quadratic form q' is called an *extension* of a quadratic form q if $D(q) \subseteq D(q')$ and $q(f) = q'(f)$ for $f \in D(q)$. Moreover, we say that two quadratic forms q, q' satisfy $q \leq q'$ if $D(q) \subseteq D(q')$ and $q(f) \geq q'(f)$ for all $f \in D(q)$. Note that the order relation $\leq$ on quadratic forms compares the size of their domains.

A quadratic form q on H is called *lower semi-bounded* or *semi-bounded from below* if there exists a constant $C \in \mathbb{R}$ such that

$$C\|f\|^2 \leq q(f) \text{ for all } f \in H. \qquad (\spadesuit)$$

By the definition of $D(q)$ this can be replaced by the validity of the above inequality for all $f \in D(q)$. If the quadratic form q is semi-bounded from below, the largest possible constant $C \in \mathbb{R}$ for which Inequality ($\spadesuit$) holds is denoted by $\lambda_0(q)$. The quadratic form $q - \lambda_0(q)\|\cdot\|^2$ is a non-negative quadratic form. Therefore, the induced sesquilinear form satisfies the Cauchy–Schwarz inequality.

For a lower semi-bounded quadratic form q and $\alpha > 0$ the inner product

$$\langle \cdot, \cdot \rangle_q : D(q) \times D(q) \to \mathbb{C},\ (f, g) \mapsto \langle f, g \rangle_q := q(f, g) + (\alpha - \lambda_0(q))\langle f, g \rangle$$

is called *form inner product*. The induced norm on $D(q)$ is called the *form norm* and denoted by $\| \cdot \|_q$. Even though this definition depends on $\alpha > 0$, different α yield equivalent norms and this suffices for our purposes. Note that if $f_n \to f$ with respect to the form norm, then also $f_n \to f$ in H, $q(f_n - f) \to 0$ and $q(f_n) \to q(f)$.

A densely defined lower semi-bounded quadratic form q on H is called *closed* if $(D(q), \langle \cdot, \cdot \rangle_q)$ is a Hilbert space; it is called *closable* if it possesses a closed extension. Closability and closedness can be characterized by lower semi-continuity as follows (see, e.g., [46, Theorem S.18]).

Lemma A.1 *Let q be a densely defined lower semi-bounded quadratic form on H. The following assertions are equivalent.*

(i) q is closed.
(ii) q is lower semi-continuous, i.e., $f_n \to f$ in H implies

$$q(f) \leq \liminf_{n \to \infty} q(f_n).$$

Lemma A.2 *Let q be a densely defined lower semi-bounded quadratic form on H. The following assertions are equivalent.*

(i) q is closable.
(ii) q is lower semi-continuous on its domain, i.e., for all $f \in D(q)$ the convergence $f_n \to f$ in H implies

$$q(f) \leq \liminf_{n \to \infty} q(f_n).$$

Let q be a densely defined lower semi-bounded closed quadratic form on H. Then there is a lower semi-bounded self-adjoint operator L that is associated with q. It has the domain

$$D(L) = \{f \in D(q) \mid \text{ex. } g \in H \text{ s.t. } q(f, h) = \langle g, h \rangle \text{ for all } h \in D(q)\},$$

on which it acts by

$$Lf = g.$$

The value $\lambda_0(q)$ is the infimum of the spectrum of L, which is why we also write $\lambda_0(L)$ for it. The domain of q satisfies $D(q) = D((L - \lambda_0(q))^{1/2})$. For

$\alpha > -\lambda_0(q)$ we denote by $G_\alpha := (L+\alpha)^{-1}$ the resolvent of L. Moreover, for $\alpha > -\lambda_0(q)$ we define the *approximating form* $q^{(\alpha)} : H \to \mathbb{R}$ by

$$q^{(\alpha)}(f) := \alpha\langle f - \alpha G_\alpha f, f\rangle.$$

It follows from the spectral theorem that they are monotone increasing in the parameter α. The resolvent and the approximating forms have the following properties. For details we refer to [10, Section 1.3], where non-negative forms are treated. The proofs given there work also for lower semi-bounded forms.

(a) For all $\alpha > -\lambda_0(q)$ the operator G_α is bounded and self-adjoint.
(b) The family $(G_\alpha)_{\alpha > -\lambda_0(q)}$ satisfies the *resolvent identity*, i.e., for $\alpha, \beta > -\lambda_0(q)$ we have

$$G_\alpha - G_\beta = (\beta - \alpha)G_\beta G_\alpha.$$

(c) For all $\alpha > -\lambda_0(q)$ we have $\|G_\alpha\| \leq (\alpha + \lambda_0(q))^{-1}$ and the family $(G_\alpha)_{\alpha > -\lambda_0(q)}$ is *strongly continuous*, i.e., for all $f \in H$ we have $\alpha G_\alpha f \to f$, as $\alpha \to \infty$.
(d) For all $f \in H$ we have $q(f) = \lim_{\alpha\to\infty} q^{(\alpha)}(f)$, where it is possible that the limit takes the value ∞. In particular,

$$D(q) = \{f \in H \mid \lim_{\alpha\to\infty} q^{(\alpha)}(f) < \infty\}.$$

Conversely, suppose that $(\widetilde{G}_\alpha)_{\alpha > C}$ is a family of operators that satisfies properties (a)–(c) as above with $-\lambda_0(q)$ replaced by the constant $C \in \mathbb{R}$. Such a family is called a *strongly continuous self-adjoint resolvent family* or simply *strongly continuous resolvent*. In this case the functional $\tilde{q} : H \to (-\infty, \infty]$ that is given by

$$\tilde{q}(f) := \lim_{\alpha\to\infty} \alpha\langle f - \alpha\widetilde{G}_\alpha f, f\rangle$$

is a densely defined lower semi-bounded closed quadratic form on H with $C > -\lambda_0(\tilde{q})$. For $\alpha > C$ the operator $\widetilde{G}_\alpha$ coincides with the α-resolvent of $\tilde{q}$.

A.2 Monotone Convergence of Quadratic Forms

Lemma A.3 (Monotone convergence of quadratic forms) *Let (q_n) be a sequence of densely defined lower semi-bounded closed quadratic forms on H with associated resolvents (G^n_α) and assume that there exists $C \in \mathbb{R}$ such that $C \leq \lambda_0(q_n)$ for all $n \in \mathbb{N}$. If (q_n) is monotone decreasing, i.e., $q_n(f) \geq q_{n+1}(f)$ for all $f \in H$ and $n \in \mathbb{N}$, then there exists a densely defined lower semi-bounded closed quadratic form q with resolvent (G_α) and with the following properties.*

(a) $\lambda_0(q) \geq C$.
(b) For all $\alpha > 0$ *we have* $G_\alpha^n \to G_\alpha$ *strongly, as* $n \to \infty$.
(c) The convergence $f_n \to f$ *weakly in* H *implies*

$$q(f) \leq \liminf_{n\to\infty} q_n(f_n).$$

(d) For every $f \in D(q)$ *there exists a sequence* (f_n) *with* $f_n \in D(q_n)$, $f_n \to f$ *in* H *and*

$$\limsup_{n\to\infty} q_n(f_n) \leq q(f).$$

Proof. Properties (c) and (d) say that the sequence (q_n) Mosco converges towards q. This is well known to be equivalent to (b) (see, e.g., [4, Theorem 8.3]), where a proof for non-negative forms is given. The modifications for a sequence of forms that are lower semi-bounded with a uniform lower bound is straightforward.

(b) is given in [46, Theorem S.16] and (a) follows from (d) and $\lambda_0(q_n) \geq C$. □

Remark

- Properties (c) and (d) in the previous lemma mean that the sequence (q_n) Mosco converges towards q.
- Note that q is not the pointwise limit of (q_n). Even if $f \in D(q_n)$ for all $n \in \mathbb{N}$, the equation $q(f) = \lim_{n\to\infty} q_n(f)$ may fail. It is possible to prove that q is the largest closed densely defined quadratic form that satisfies $q(f) \leq q_n(f)$ for all $f \in H$ and $n \in \mathbb{N}$ (see, e.g., [46, Theorem S.16]).

A.3 The Beurling–Deny Criteria

In this section we consider quadratic forms on the Hilbert space $\ell^2(X, \mu)$, where X is a countable set and μ is a weight on X. We discuss the compatibility of their resolvents with the order structure of this Hilbert space.

A quadratic form q on $\ell^2(X, \mu)$ is called *real* if for any real-valued $f, g \in D(q)$ it satisfies $q(f+ig) = q(f)+q(g)$. For studying a real form q it suffices to consider its restriction to $D(q)_r := \{f : X \to \mathbb{R} \mid f \in D(q)\}$ on the real Hilbert space $\ell^2_r(X, \mu) := \{f : X \to \mathbb{R} \mid f \in \ell^2(X, \mu)\}$.

We say that a lower semi-bounded closed quadratic form on $\ell^2(X, \mu)$ satisfies the *first Beurling–Deny criterion* if $f \in D(q)$ implies $|f| \in D(q)$ and $q(|f|) \leq q(f)$.

Proposition A.4 *Let* q *be a densely defined closed lower semi-bounded quadratic form on* $\ell^2(X, \mu)$. *The following assertions are equivalent.*

(i) q satisfies the first Beurling–Deny criterion.
(ii) The resolvent of q is positivity preserving, i.e., for $\alpha > -\lambda_0(q)$ and $f \in \ell^2(X, \mu)$ the inequality $f \geq 0$ implies $G_\alpha f \geq 0$.

In particular, forms satisfying the first Beurling–Deny criterion are real.

Proof. For non-negative forms this is contained in [48, Theorem XIII.50]. The proof given there can be carried out verbatim for lower semi-bounded forms.

Assertion (ii) implies that the resolvent maps $\ell^2_r(X, \mu)$ to $\ell^2_r(X, \mu)$. Using approximating forms this yields that forms satisfying the first Beurling–Deny criterion are real. □

Lemma A.5 *Let q be a densely defined closed lower semi-bounded quadratic that satisfies the* first Beurling–Deny criterion. *For $f, g \in \ell^2_r(X, \mu)$ we have*

$$\|f \wedge g\|_q \leq \|f\|_q + \|g\|_q \text{ and } \|f \vee g\|_q \leq \|f\|_q + \|g\|_q.$$

Proof. We have $f \wedge g = \frac{1}{2}(f + g - |f - g|)$ and $f \vee g = \frac{1}{2}(f + g + |f - g|)$. Hence, the inequalities follow from the fact that $\| \cdot \|_q$ is a norm with $\| |h| \|_q \leq \|h\|_q$ for all $h \in D(q)$. □

A function $C : \mathbb{C} \to \mathbb{C}$ is called a *normal contraction* if $C(0) = 0$ and $|C(x) - C(y)| \leq |x - y|$ for each $x, y \in \mathbb{C}$. A non-negative quadratic form satisfies the *second Beurling–Deny criterion* if for any normal contraction C and $f \in D(q)$ we have $C \circ f \in D(q)$ and $q(C \circ f) \leq q(f)$. Forms satisfying the second Beurling–Deny criterion are also called *Dirichlet forms*. They can be characterized as follows.

Proposition A.6 *Let q be a densely defined closed non-negative quadratic form on $\ell^2(X, \mu)$. The following assertions are equivalent.*

(i) q satisfies the second Beurling–Deny criterion.
(ii) q satisfies the first Beurling–Deny criterion and for all non-negative $f \in D(q)$ we have $(f \wedge 1) \in D(q)$ and $q((f \wedge 1)) \leq q(f)$.
(iii) The resolvent of q is Markovian, i.e., for $\alpha > 0$ and $f \in \ell^2(X, \mu)$ the inequality $0 \leq f \leq 1$ implies $0 \leq \alpha G_\alpha f \leq 1$.

Proof. This is given in [48, Theorem XIII.51]. □

The following lemma applies to Dirichlet forms but it can also be used for forms that are not closed.

Lemma A.7 *Let q be a non-negative quadratic form on $\ell^2(X, \mu)$ such that for each non-negative $f \in D(q)$ we have $f \wedge 1 \in D(q)$ and $q(f \wedge 1) \leq q(f)$. Let $f, g \in D(q)$ be non-negative with $g \leq 1$ and $g = 1$ on $\{f > 0\}$. Then $q(f, g) \geq 0$.*

Proof. Let $\varepsilon > 0$. The properties of f and g imply

$$q(g) = q((g + \varepsilon f) \wedge 1) \leq q(g + \varepsilon f) = q(g) + 2\varepsilon(f, g) + \varepsilon^2(g).$$

Dividing by ε and letting $\varepsilon \to 0+$ yields the claim. □

Acknowledgements

The author is indebted to Matthias Keller for asking him several of the questions answered in this chapter and for encouraging him to write it. Moreover, he expresses his gratitude to Simon Puchert for sharing new ideas on Markov uniqueness and for allowing him to use some of the results of the Master's thesis [45]. Furthermore, several discussions with Melchior Wirth on domination of quadratic forms and with Daniel Lenz and Radoslaw Wojciechowski on related topics resulted in various insights that improved this chapter.

References

[1] Itai Benjamini and Oded Schramm. Harmonic functions on planar and almost planar graphs and manifolds, via circle packings. *Invent. Math.*, 126(3):565–587, 1996.

[2] Xin Chen, Takashi Kumagai, and Jian Wang. Random conductance models with stable-like jumps I: Quenched invariance principle. *Preprint. arXiv:1805.04344.*

[3] Zhen-Qing Chen and Masatoshi Fukushima. *Symmetric Markov processes, time change, and boundary theory*, volume 35 of *London Mathematical Society Monographs Series*. Princeton University Press, Princeton, NJ, 2012.

[4] Zhen-Qing Chen, Panki Kim, and Takashi Kumagai. Discrete approximation of symmetric jump processes on metric measure spaces. *Probab. Theory Related Fields*, 155(3-4):703–749, 2013.

[5] Yves Colin de Verdière, Nabila Torki-Hamza, and Françoise Truc. Essential self-adjointness for combinatorial Schrödinger operators II – metrically non complete graphs. *Preprint. arXiv:1006.5778v3.*

[6] Yves Colin de Verdière, Nabila Torki-Hamza, and Françoise Truc. Essential self-adjointness for combinatorial Schrödinger operators II – metrically non complete graphs. *Math. Phys. Anal. Geom.*, 14(1):21–38, 2011.

[7] Yves Colin de Verdière, Nabila Torki-Hamza, and Françoise Truc. Essential self-adjointness for combinatorial Schrödinger operators III – Magnetic fields. *Ann. Fac. Sci. Toulouse Math. (6)*, 20(3):599–611, 2011.

[8] Matthew Folz. Gaussian upper bounds for heat kernels of continuous time simple random walks. *Electron. J. Probab.*, 16(62): 1693–1722, 2011.

[9] Rupert L. Frank, Daniel Lenz, and Daniel Wingert. Intrinsic metrics for non-local symmetric Dirichlet forms and applications to spectral theory. *J. Funct. Anal.*, 266(8):4765–4808, 2014.

[10] Masatoshi Fukushima, Yoichi Oshima, and Masayoshi Takeda. *Dirichlet forms and symmetric Markov processes*, volume 19 of *de Gruyter Studies in Mathematics*. Walter de Gruyter & Co., Berlin, extended edition, 2011.

[11] Sylvain Golénia. Unboundedness of adjacency matrices of locally finite graphs. *Lett. Math. Phys.*, 93(2):127–140, 2010.

[12] Sylvain Golénia. Hardy inequality and asymptotic eigenvalue distribution for discrete Laplacians. *J. Funct. Anal.*, 266(5):2662–2688, 2014.

[13] Sylvain Golénia and Christoph Schumacher. Comment on 'The problem of deficiency indices for discrete Schrödinger operators on locally finite graphs' [J. Math. Phys. 52, 063512 (2011)] [mr2841768]. *J. Math. Phys.*, 54(6):064101, 4, 2013.

[14] Batu Güneysu, Matthias Keller, and Marcel Schmidt. A Feynman-Kac-Itô formula for magnetic Schrödinger operators on graphs. *Probab. Theory Related Fields*, 165(1–2):365–399, 2016.

[15] Batu Güneysu, Ognjen Milatovic, and Françoise Truc. Generalized Schrödinger semigroups on infinite graphs. *Potential Anal.*, 41(2):517–541, 2014.

[16] Sebastian Haeseler and Matthias Keller. Generalized solutions and spectrum for Dirichlet forms on graphs. In *Random walks, boundaries and spectra*, volume 64 of *Progr. Probab.*, pages 181–199. Birkhäuser/Springer Basel AG, Basel, 2011.

[17] Sebastian Haeseler, Matthias Keller, Daniel Lenz, Jun Masamune, and Marcel Schmidt. Global properties of Dirichlet forms in terms of Green's formula. *Calc. Var. Partial Differential Equations*, 56(5):Art. 124, 43, 2017.

[18] Sebastian Haeseler, Matthias Keller, Daniel Lenz, and Radosław Wojciechowski. Laplacians on infinite graphs: Dirichlet and Neumann boundary conditions. *J. Spectr. Theory*, 2(4):397–432, 2012.

[19] Michael Hinz and Michael Schwarz. A note on Neumann problems on graphs. *Preprint, arXiv:1803.08559*, 2018.

[20] Xueping Huang. On stochastic completeness of weighted graphs. Ph.D. thesis, 2011.

[21] Xueping Huang, Matthias Keller, Jun Masamune, and Radosław K. Wojciechowski. A note on self-adjoint extensions of the Laplacian on weighted graphs. *J. Funct. Anal.*, 265(8):1556–1578, 2013.

[22] Pascual Jordan and John Von Neumann. On inner products in linear, metric spaces. *Ann. of Math. (2)*, 36(3):719–723, 1935.

[23] Palle E. T. Jorgensen. Essential self-adjointness of the graph-Laplacian. *J. Math. Phys.*, 49(7):073510, 33, 2008.

[24] Palle E. T. Jorgensen and Erin P. J. Pearse. Spectral reciprocity and matrix representations of unbounded operators. *J. Funct. Anal.*, 261(3):749–776, 2011.

[25] Naotaka Kajino. Equivalence of recurrence and Liouville property for symmetric Dirichlet forms. *Mat. Fiz. Komp′yut. Model.*, 3(40):89–98, 2017.

[26] Matthias Keller. Intrinsic metrics on graphs: a survey. In *Mathematical technology of networks*, volume 128 of *Springer Proc. Math. Stat.*, pages 81–119. Springer, Cham, 2015.

[27] Matthias Keller and Daniel Lenz. Unbounded Laplacians on graphs: basic spectral properties and the heat equation. *Math. Model. Nat. Phenom.*, 5(4):198–224, 2010.

[28] Matthias Keller and Daniel Lenz. Dirichlet forms and stochastic completeness of graphs and subgraphs. *J. Reine Angew. Math.*, 666:189–223, 2012.

[29] Matthias Keller, Daniel Lenz, Marcel Schmidt, and Michael Schwarz. Boundary representation of dirichlet forms on discrete spaces. *Preprint, arXiv:1711.08304*, 2017.

[30] Matthias Keller, Daniel Lenz, Hendrik Vogt, and Radosław Wojciechowski. Note on basic features of large time behaviour of heat kernels. *J. Reine Angew. Math.*, 708:73–95, 2015.

[31] Matthias Keller, Yehuda Pinchover, and Felix Pogorzelski. Criticality theory for Schrödinger operators on graphs. *preprint, arXiv:1708.09664*.

[32] Kazuhiro Kuwae. Reflected Dirichlet forms and the uniqueness of Silverstein's extension. *Potential Anal.*, 16(3):221–247, 2002.

[33] Kazuhiro Kuwae and Yuichi Shiozawa. A remark on the uniqueness of Silverstein extensions of symmetric Dirichlet forms. *Math. Nachr.*, 288(4):389–401, 2015.

[34] Daniel Lenz, Marcel Schmidt, and Melchior Wirth. Domination of quadratic forms. *Preprint, arXiv:1711.07225*, 2017.

[35] Daniel Lenz, Marcel Schmidt, and Melchiort Wirth. Uniqueness of form extensions and domination of semigroups. *preprint, arXiv:1608.06798v2*, 2017.

[36] Zhi Ming Ma and Michael Röckner. *Introduction to the theory of (nonsymmetric) Dirichlet forms*. Universitext. Springer-Verlag, Berlin, 1992.

[37] Amir Manavi, Hendrik Vogt, and Jürgen Voigt. Domination of semigroups associated with sectorial forms. *J. Operator Theory*, 54(1):9–25, 2005.

[38] Jun Masamune. A Liouville property and its application to the Laplacian of an infinite graph. In *Spectral analysis in geometry and number theory*, volume 484 of *Contemp. Math.*, pages 103–115. Amer. Math. Soc., Providence, RI, 2009.

[39] Jun Masamune and Toshihiro Uemura. Conservation property of symmetric jump processes. *Ann. Inst. Henri Poincaré Probab. Stat.*, 47(3):650–662, 2011.

[40] Ognjen Milatovic. Essential self-adjointness of magnetic Schrödinger operators on locally finite graphs. *Integral Equations Operator Theory*, 71(1):13–27, 2011.

[41] Ognjen Milatovic. A Sears-type self-adjointness result for discrete magnetic Schrödinger operators. *J. Math. Anal. Appl.*, 396(2):801–809, 2012.

[42] Ognjen Milatovic and Françoise Truc. Self-adjoint extensions of discrete magnetic Schrödinger operators. *Ann. Henri Poincaré*, 15(5):917–936, 2014.

[43] Ognjen Milatovic and Françoise Truc. Maximal accretive extensions of Schrödinger operators on vector bundles over infinite graphs. *Integral Equations Operator Theory*, 81(1):35–52, 2015.

[44] Gheorghe Nenciu and Irina Nenciu. On confining potentials and essential self-adjointness for Schrödinger operators on bounded domains in $\mathbb{R}^n$. *Ann. Henri Poincaré*, 10(2):377–394, 2009.

[45] Simon Puchert. Masterarbeit, in preparation.

[46] Michael Reed and Barry Simon. *Methods of modern mathematical physics. I. Functional analysis*. Academic Press, New York-London, 1972.

[47] Michael Reed and Barry Simon. *Methods of modern mathematical physics. II. Fourier analysis, self-adjointness*. Academic Press [Harcourt Brace Jovanovich, Publishers], New York-London, 1975.

[48] Michael Reed and Barry Simon. *Methods of modern mathematical physics. IV. Analysis of operators*. Academic Press [Harcourt Brace Jovanovich, Publishers], New York-London, 1978.

[49] Marcel Schmidt. Energy forms. *Disseration, arXiv:1703.04883*, 2017.

[50] Marcel Schmidt. Global properties of Dirichlet forms on discrete spaces. *Dissertationes Math. (Rozprawy Mat.)*, 522:43, 2017.

[51] Paolo M. Soardi. *Potential theory on infinite networks*, volume 1590 of *Lecture Notes in Mathematics*. Springer-Verlag, Berlin, 1994.

[52] Robert S. Strichartz. Analysis of the Laplacian on the complete Riemannian manifold. *J. Funct. Anal.*, 52(1):48–79, 1983.

[53] Nabila Torki-Hamza. Laplaciens de graphes infinis (I-graphes) métriquement complets. *Confluentes Math.*, 2(3):333–350, 2010.

[54] Andreas Weber. Analysis of the physical Laplacian and the heat flow on a locally finite graph. *J. Math. Anal. Appl.*, 370(1):146–158, 2010.

[55] Wolfgang Woess. *Random walks on infinite graphs and groups*, volume 138 of *Cambridge Tracts in Mathematics*. Cambridge University Press, Cambridge, 2000.

[56] Radosław K. Wojciechowski. Heat kernel and essential spectrum of infinite graphs. *Indiana Univ. Math. J.*, 58(3):1419–1441, 2009.

[57] Radoslaw Krzysztof Wojciechowski. *Stochastic completeness of graphs*. ProQuest LLC, Ann Arbor, MI, 2008. Thesis (Ph.D.)–City University of New York.

12

Box Spaces: Geometry of Finite Quotients

Ana Khukhro and Alain Valette

Abstract

This is a survey chapter trying to explain why box spaces, originating in geometric group theory and coarse geometry, are relevant for the spectral theory of towers of covers of closed Riemannian manifolds. Most of the results are borrowed from the paper [KV17] by the same authors.

12.1 Motivation

Box spaces produce, out of a group G, a family of Cayley graphs of finite quotients of G. If G is the fundamental group of a closed Riemannian manifold M, a box space yields a tower of finite covers $(M_i)_{i>0}$ of M, and the large-scale geometry of the M_i's mirrors the one of the box space. In particular the first non-zero eigenvalue $\lambda_1(M_i)$ (of the Riemannian Laplace operator on M_i) is bounded away from 0, if and only if the box space is an expander. The present note is an introductory survey on box spaces and their connections with Riemannian and spectral geometry.

A finitely generated group G is *residually finite* if homomorphisms to finite groups separate elements of the group. Informally, this means that the group possesses a plethora of finite quotients. Equivalently, there exists a decreasing sequence $(N_i)_{i>0}$ of finite index normal subgroups of G, with trivial intersection (i.e., $\cap_{i>0} N_i = \{1\}$). For brevity, we call such a sequence a *filtration* of G.

Example 1 *Take $G = SL_d(\mathbb{Z})$; let $(m_i)_{i>0}$ be a sequence of positive integers such that m_i divides m_{i+1} for every i; define N_i, a principal congruence subgroup in G, as the kernel of reduction modulo m_i. The sequence $(N_i)_{i>0}$ is a filtration of G.*

Given a filtration $(N_i)_{i>0}$ in a finitely generated residually finite group, we are interested in understanding the sequence of finite quotients $(G/N_i)_{i>0}$. One way is to observe that the G/N_i's form a projective system

$$\cdots \to G/N_3 \to G/N_2 \to G/N_1,$$

so that algebraic properties of the sequence $(G/N_i)_{i>0}$ are encoded as algebraic/topological properties of the projective limit, $\text{proj}\lim_i G/N_i$, a profinite (i.e., compact, totally disconnected) group.

Another approach, more inductive (as opposed to projective), is to consider the quotients G/N_i in a row:

$$G/N_1\,, \quad G/N_2\,, \quad G/N_3\,, \quad \cdots$$

Originally defined by J. Roe [Ro03], the *box space* of G with respect to the filtration $(N_i)_{i>0}$, as a set, is the disjoint union

$$\square_{(N_i)}G = \coprod_{i>0} G/N_i.$$

The G/N_is are called the *components* of $\square_{(N_i)}G$. Geometry returns when we choose a finite, symmetric, generating set S in G. Slightly abusing the notation, we denote by $Cay(G/N_i, S)$ the Cayley graph of G/N_i with respect to S (more precisely: with respect to the image of S in G/N_i under the quotient map). So, once S has been chosen, each component of $\square_{(N_i)}G$ becomes a graph, hence a metric space with the graph metric.

Example 2 *Take $G = \mathbb{Z}$, the additive group of integers, and $S = \{\pm 1\}$. If $(m_i)_{i>0}$ is a sequence of positive integers such that m_i divides m_{i+1} for every i, and $N_i = m_i\mathbb{Z}$, then the box space $\square_{(N_i)}\mathbb{Z}$ is the disjoint union of the cycles $C_{m_1}, C_{m_2}, \ldots$*

Our goal in the next section will be to equip a box space $\square_{(N_i)}G$ with a metric which restricts to the graph metric on each component, such that this metric is unique up to coarse equivalence, and does not depend on the choice of the finite generating set S in G. Doing this will allow us to study the geometric properties that the finite quotients possess uniformly.

12.2 Elements of Coarse Geometry

Let (X, d_X), (Y, d_Y) be metric spaces, and $f : X \to Y$ a map (not necessarily continuous). W say that f is *almost surjective* if there exists $C > 0$ such that Y

is the C-neighbourhood of $f(X)$. Recall that f is a *quasi-isometric embedding* if there exists $A > 0$ such that

$$\frac{1}{A}d_X(x, x') - A \leq d_Y(f(x), f(x')) \leq A d_X(x, x') + A,$$

for every $x, x' \in X$, and that f is a *quasi-isometry* if f is a quasi-isometric embedding which is almost surjective. A weaker condition is provided by coarse embeddings: f is a *coarse embedding* if there exist functions $\rho_+, \rho_- : \mathbb{R}^+ \to \mathbb{R}^+$ (called *control functions*) such that $\lim_{t\to\infty} \rho_\pm(t) = \infty$ and

$$\rho_-(d_X(x, x')) \leq d_Y(f(x), f(x')) \leq \rho_+(d_X(x, x'))$$

for every $x, x' \in X$. Finally, f is a *coarse equivalence* if f is a coarse embedding which is almost surjective.

Example 3 *A space X is quasi-geodesic if there exists $L > 0$ such that, for every $x, x' \in X$, there is a quasi-isometric embedding $\gamma : [0, d_X(x, x')] \to X$ with constant L such that $\gamma(0) = x$ and $\gamma(d_X(x, x')) = x'$. Examples are provided by connected Riemannian manifolds and by connected graphs. If X is quasi-geodesic and $f : X \to Y$ is a coarse embedding, then we may take $\rho_+(t) = Kt$ for some constant $K > 0$; see Theorem 1.4.13 of [NY12].*

Example 4 *A map $f : X \to Y$ is a coarse embedding if and only if, for any two sequences $(x_n)_{n>0}, (x'_n)_{n>0}$ in X:*

$$\lim_{n\to\infty} d_X(x_n, x'_n) = \infty \iff \lim_{n\to\infty} d_Y(f(x_n), f(x'_n)) = \infty\ ;$$

this provides a definition of coarse embeddings depending purely on the metrics.

Classification of metric spaces can be achieved at the following different levels (at least), in order of increasing roughness:

1. isometric;
2. bi-Lipschitz;
3. quasi-isometric;
4. coarse.

Observe that there is a definite change of paradigm when moving from the second to the third level: every bounded metric space is quasi-isometric to the one-point space. But even if bounded spaces are in some sense trivial, it is possible to get something interesting out of sequences of bounded spaces with increasing diameter.

Definition 5 *Let $(X_i, d_i)_{i>0}$ be a family of bounded metric spaces with the property that $\lim_{i\to\infty} \mathrm{diam}(X_i) = \infty$. The coarse disjoint union of the X_is is the set $\coprod_{i>0} X_i$ endowed with a metric d such that:*

- *on every X_i, the metric d restricts to the metric d_i;*
- *for $i \neq j$, we have $\lim_{i,j\to\infty} d(X_i, X_j) = \infty$.*

The uniqueness of such a metric, up to coarse equivalence, follows from Example 4. For the existence, consider the metric d on $\coprod_{i>0} X_i$ defined by

$$d(x, x') = \begin{cases} d_i(x, x') & if \quad x, x' \in X_i \\ \mathrm{diam}(X_i) + \mathrm{diam}(X_j) & if \quad x \in X_i, x' \in X_j, i \neq j. \end{cases}$$

Coming back to a finitely generated, residually finite group G with a filtration $(N_i)_{i>0}$: we can now complete the definition of the box space $\square_{(N_i)} G$.

Definition 6 *The box space $\square_{(N_i)} G$ is the coarse disjoint union of the Cayley graphs $Cay(G/N_i, S)$, for some finite symmetric generating set S of G.*

Replacing S by another finite symmetric generating set T provides a family of quasi-isometries with uniform constants $Cay(G/N_i, S) \to Cay(G/N_i, T)$. From this, it follows that up to coarse equivalence the box space $\square_{(N_i)} G$ does not depend on the choice of the finite generating set S. We note, however, that the choice of filtration can have a big influence on the geometric properties of the box space. We will elaborate on this in the next section.

12.3 Box Spaces and Expanders

For a finite, connected, d-regular graph $X = (V, E)$, we denote by

$$\Delta_X : \mathbb{C}V \to \mathbb{C}V : f \mapsto (x \mapsto d \cdot f(x) - \sum_{y:d(x,y)=1} f(y))$$

the combinatorial Laplace operator on X. It is semi-definite positive for the standard scalar product on $\mathbb{C}V$, so its spectrum consists of n non-negative eigenvalues $\lambda_0, \ldots, \lambda_{n-1}$ (counted with multiplicities):

$$0 = \lambda_0 < \lambda_1 \leq \lambda_2 \leq \ldots \leq \lambda_{n-1} \leq 2d.$$

Now let $(X_i)_{i>0}$ be a sequence of finite, connected, d-regular graphs with $\lim_{i\to\infty} |V_i| = \infty$. We say that the coarse disjoint union $\coprod_{i>0} X_i$ is an *expander* if, for some $\varepsilon > 0$, we have $\lambda_1(X_i) \geq \varepsilon$ for every $i > 0$. Informally, a sequence of graphs is an expander if the graphs are uniformly highly connected, but of the same degree. This intuitive description is made precise by the well-known Cheeger–Buser inequality, which gives a combinatorial characterization of expanders by giving a two-way inequality linking λ_1 to a combinatorial invariant called the *Cheeger constant*. The Cheeger constant $h(X)$ of a finite graph X is defined by

$$h(X) := \inf_{A \subset X} \frac{|\partial A|}{\min\{|A|, |X \backslash A|\}},$$

where ∂A denotes the edge boundary of A, i.e., the set of edges with exactly one end-vertex in A. It is a measure of how difficult it is to disconnect a subset of vertices of X from the rest of the graph. A sequence of graphs X_i as above is thus an expander if and only if there exists some $\varepsilon > 0$ such that $h(X_i) \geq \varepsilon$ for every $i > 0$. It is known that being an expander is an invariant of coarse equivalence for coarse disjoint unions of d-regular graphs (see, e.g., Lemma 15 in [KV17]).

Expanders are desirable objects in many branches outside of mathematics, such as computer science, thanks to their high connectivity. For example, expanders are ideal models for communication networks since high connectivity ensures stability and efficiency of the network, while the bounded degree makes it possible to construct a network of any size such that the number of connections in the network is linear in the number of nodes (which ensures that the network is cost-effective). While the first proof of the existence of expanders was probabilistic and non-constructive [Pin], explicit examples were needed for applications. The first constructive examples were provided by Margulis [Ma75]. In the language of box spaces, his result is as follows.[1]

Theorem 7 *Let G be an infinite, residually finite group with Kazhdan's property (T). Then any box space of G is an expander.*

12.4 Analytic Properties of Groups vs. Metric Properties of Box Spaces

Theorem 7 is an example of how properties of the group are connected to geometric properties of the box space. There are, in fact, many results of this flavour, making it possible to exploit well-studied aspects of groups in order to construct metric spaces with interesting properties. On the group side, we consider certain representation-theoretic properties; recall that, for a unitary representation π of a countable group G, a *coefficient* of π is a function on G

[1] Property (T) will be defined in Definition 8.

of the form $g \mapsto \langle \pi(g)\xi, \xi \rangle$ with $\|\xi\| = 1$. The representation π is C_0 if every coefficient vanishes at infinity on G; the representation π *almost has invariant vectors* if it admits a sequence of coefficients converging to 1 pointwise on G.

Definition 8 *Let G be a countable group.*

(1) G has Kazhdan's property (T) if every unitary representation of G almost having invariant vectors, has non-zero invariant vectors (see [BHV08]).
(2) If $(N_i)_{i>0}$ is a filtration of G, the pair $(G, (N_i))$ has property (τ) if the representation $\oplus_{i>0}\lambda^0_{G/N_i}$ does not almost have invariant vectors (here λ^0_{G/N_i} denotes the quasi-regular representation of G on the orthogonal of constants in $\ell^2(G/N_i)$) (see [LZ03]).
(3) G is amenable if the left regular representation λ_G on $\ell^2(G)$ almost has invariant vectors (see, e.g., Appendix G in [BHV08]).
(4) G has the Haagerup property, or is a-(T)-menable, if G admits a C_0-representation almost having invariant vectors (see [CCJJV01]; note that λ_G is a C_0-representation, so that every amenable group has the Haagerup property).

On the metric side, one property we will be interested in, is as follows.

Definition 9 *A discrete metric space (X, d) has property A if for all $R, \varepsilon > 0$, there is a family $\{A_x\}_{x \in X}$ of non-empty finite subsets of $X \times \mathbb{N}$ such that the following two conditions are satisfied:*

1. for all $x, y \in X$ such that $d(x, y) \leq R$,

$$\frac{|A_x \Delta A_y|}{|A_x \cap A_y|} < \varepsilon ;$$

2. there exists $S > 0$ such that for all $x \in X$, $(y, n) \in A_x$ implies $d(x, y) \leq S$.

Property A was originally defined by G. Yu in [Yu] as a way of proving that a metric space coarsely embeds into a Hilbert space. In the same paper, Yu proved that for a discrete metric space, the existence of a coarse embedding into a Hilbert space implies the coarse Baum–Connes conjecture, and in the case of a finitely generated group, the existence of such an embedding of its Cayley graph implies the Novikov conjecture. These deep consequences for topology and analysis motivated further study of property A. In particular, the following characterization of box spaces with property A was given.

Theorem 10 ([Ro03], attributed to Guentner) *A box space of a finitely generated, residually finite group G has property A if and only if G is amenable.*

A similar result holds for coarse embeddability into a Hilbert space.

Theorem 11 ([Ro03]) *A finitely generated, residually finite group is a-(T)-menable if one of its box spaces coarsely embeds into a Hilbert space.*

Note however that the converse does not hold. Indeed, there exist a-(T)-menable groups such that every box space is an expander (expanders being typical metric spaces that do not admit coarse embeddings into a Hilbert space), such as $SL_2(\mathbb{Z}[\frac{1}{p}])$.

Box spaces were used to differentiate between property A and the property of admitting a coarse embedding into a Hilbert space, first by Nowak in [Now] (for non-bounded geometry metric spaces) and then by Arzhantseva, Guentner and Špakula in [AGS], who constructed a box space of the free group on two generators that coarsely embeds into a Hilbert space (but does not have property A since the group is non-amenable). It is still unknown whether there is a group-theoretic property which captures the property of admitting a box space that coarsely embeds into a Hilbert space, but in [CWW13], a geometric property called *fibred coarse embeddability* into a Hilbert space was used to characterize a-(T)-menability for finitely generated, residually finite groups.

Given a residually finite group G and a filtration (N_i) of G, we can summarize the various connections between analytic properties of G and geometric properties of the box space as follows:

$$
\begin{aligned}
G \text{ has property (T)} &\implies \Box_{(N_i)}G \text{ is an expander [Ma75];}\\
G \text{ has property (T)} &\iff \Box_{(N_i)}G \text{ has geometric property (T) [WY14];}\\
(G,(N_i)) \text{ has property } (\tau) &\iff \Box_{(N_i)}G \text{ is an expander [LZ03];}\\
G \text{ is amenable} &\iff \Box_{(N_i)}G \text{ has property A (Guentner, [Ro03]);}\\
G \text{ is a-(T)-menable} &\Longleftarrow \Box_{(N_i)}G \text{ coarsely embeds into } \ell^2 \text{ [Ro03];}\\
G \text{ is a-(T)-menable} &\iff \Box_{(N_i)}G \text{ fibred coarsely embeds into } \ell^2\\
&\qquad \text{[CWW13].}
\end{aligned}
$$

As mentioned, expanders do not admit coarse embeddings into ℓ^2, and containing an expander was until recently the only known obstruction to such embeddings (for bounded geometry spaces; for non-bounded geometry spaces, ℓ^p for $p > 2$ is a space that does not contain expanders but does not embed coarsely into ℓ^2). Tessera succeeded in characterizing spaces that do not embed coarsely into ℓ^2 in [Tes], in terms of *generalized expanders*: a sequence of finite, graphs $(X_i)_{i>0}$ is a generalized expander if there exist a sequence r_i with $\lim_{i\to\infty} r_i = \infty$, a sequence of symmetric probability measures

$\mu_i \in \operatorname{Prob}(X_i \times X_i)$ supported on $\{(x, y) \in X_i \times X_i : d(x, y) \geq r_i\}$, and a constant $K > 0$ such that for every 1-Lipschitz map $f_i : X_i \to \ell^2$, we have

$$\sum_{x,y \in X_i} \|f(x) - f(y)\|^2 \mu_i(x, y) \leq K$$

for all i. One can see that expanders are in particular generalized expanders by first considering the uniform measures on $X_i \times X_i$ and then using the fact that at least half the mass of these measures is supported sufficiently far away from the diagonal to restrict and renormalize these measures so that they satisfy the definition (see section 2.2 in [Tes] for details).

A special case of generalized expansion called *relative expansion* was recently used by Arzhantseva and Tessera in [AT15] to give examples of bounded geometry metric spaces which do not contain expanders but do not embed into a Hilbert space. These examples also take the form of box spaces, in particular, of groups with *relative property (T)*.

In general, the coarse properties of a box space of a given group can depend on the chosen filtration. Typical examples illustrating this fact are box spaces of a non-abelian free group: there are some such box spaces which are expanders, some which admit coarse embeddings into ℓ^2, and also some which do not embed into ℓ^2, but do not contain expanders ([DK16]).

Box spaces thus often lead to examples of spaces with interesting coarse geometric properties. Another useful property of box spaces is their uniformly bounded degree, as this allows them to have a chance of being embedded (either coarsely or as a subgraph) in Cayley graphs of finitely generated groups. The motivation behind such constructions is that embedding a box space with certain properties into a group can force the group itself to behave in an interesting way. Gromov was the first to give an example of such a *monster group* containing (weakly) embedded expanders in its Cayley graph ([Grom], see also [AD08]). The resulting properties of Gromov's monster were exploited by Higson, Lafforgue, and Skandalis in [HLS] to give the first counterexample to the Baum–Connes conjecture with coefficients.

Later, Osajda gave an example of a group which contains expanders isometrically [Osaj]. All of these examples are created using small cancellation theory, and are of a non-constructive nature, since they rely on a probabilistic argument to prove the existence of a suitable labelling of graphs used in the proof. It thus remains an open problem to give explicit constructions of such groups.

Box space methods have also proved to be a powerful tool in coarsely distinguishing spaces of interest, such as expanders. With only a limited number of strategies known to produce expanders, one of which being via box spaces, it is interesting to determine whether the examples produced are actually distinct. The relevant results stem from the question, what can one deduce about groups if one knows that their box spaces are geometrically similar in some sense? In a later section, we mention in more detail a result of this type, namely, Proposition 21, which states that coarsely equivalent box spaces must come from groups with quasi-isometric Cayley graphs. This has the following corollary for expander graphs, answering a question of Mendel and Naor ([MN14], see also [Hu14] for a different method which treats graphs of large girth).

Corollary 12 *Expanders occurring as box spaces of $SL_n(\mathbb{Z})$ for varying n are pairwise not coarsely equivalent. In particular, there are infinitely many coarse equivalence classes of expanders coming from groups with property (T).*

This corollary uses a result of Eskin [Es98], which states that the Cayley graphs of $SL_n(\mathbb{Z})$ and $SL_m(\mathbb{Z})$ are not quasi-isometric when $n \neq m$. In [DK17], it is shown that for finitely presented, residually finite groups, one can detect the subgroups used to construct the box space using a coarse notion of the fundamental group, leading to the following result (compare with Proposition 21).

Theorem 13 *Given two finitely presented, residually finite groups G and H with respective filtrations (N_i) and (M_i) such that the box spaces $\square_{(N_i)}G$ and $\square_{(M_i)}H$ are coarsely equivalent, the groups G and H must be commensurable.*

This rigidity result for box spaces can be used to coarsely distinguish many examples, in particular resulting in the following corollary for Ramanujan graphs, which are expanders with asymptotically optimal spectral gap (see [Lu94], [LPS88] for more on Ramanujan graphs).

Corollary 14 *There are infinitely many coarse equivalence classes containing Ramanujan graphs.*

12.5 Box Spaces and Towers of Covers

We recall a basic result in geometric group theory (see, e.g., Theorem 7.5 in [CM17] for a proof).

Theorem 15 (The Milnor–Schwarz lemma) *Let (X, d) be a proper metric space (i.e., closed balls are compact). Let G be a group acting isometrically,* *properly discontinuously, and co-compactly on X. Then G is finitely generated and moreover X and G are quasi-isometric.*

Let G be a residually finite group with a filtration $(N_i)_{i>0}$. Assume that G is the fundamental group of some compact Riemannian manifold M, i.e., $G = \pi_1(M)$. To each N_i corresponds a finite-sheeted cover M_i of M. We are interested in the geometry of the tower $(M_i)_{i>0}$ of covers of M that we incorporate in the coarse disjoint union $\coprod_{i>0} M_i$.

Lemma 16 *The coarse disjoint union $\coprod_{i>0} M_i$ is coarsely equivalent to the box space $\square_{(N_i)}G$.*

Proof. Let $\widetilde{M}$ be the universal cover of M, equipped with the Riemannian metric lifted from M. Then G acts isometrically, properly discontinuously, co-compactly on $\widetilde{M}$; so by the Milnor-Schwarz lemma $\widetilde{M}$ is quasi-isometric to G. Actually the proof of Theorem 15 provides an explicit quasi-isometry $f : G \to \widetilde{M}$, namely any orbital map $f(g) = g \cdot x_0$, where x_0 is any point in $\widetilde{M}$ (intuitively, this map allows one to 'draw' the Cayley graph of G as the 1-skeleton of some tiling of $\widetilde{M}$). An orbital map is clearly G-equivariant, so it descends to a quasi-isometry $f_i : G/N_i \to N_i\backslash\widetilde{M} = M_i$. This family $(f_i)_{i>0}$ has uniform quasi-isometry constants (namely the quasi-isometry constant of f), so it induces a coarse equivalence $\square_{(N_i)}G \to \coprod_{i>0} M_i$. □

Note that Lemma 16, like the following Theorem 17, does not need the condition $\cap_{i>0}N_i = \{1\}$ from the definition of a filtration.

Denote by $\lambda_1(M)$ the first non-zero eigenvalue of the Riemannian Laplace operator $\Delta_M = d^*d$. The first connection between the coarse geometry of the tower of covers $(M_i)_{i>0}$ and the geometry of the corresponding box space, implicitly appealing to Lemma 16, seems to be due to Brooks (Theorem 1 in [Br86]; see also Theorem 4.3.2 in [Lu94]).

Theorem 17 *In the above situation: $\square_{(N_i)}G$ is an expander if and only if there exists $\varepsilon > 0$ such that $\lambda_1(M_i) \geq \varepsilon$ for every $i > 0$.*

Observe that this condition only depends on the fundamental group, not on the chosen metric on M. As explained in Section 4.4 of [Lu94], this allows one to construct box spaces which are expanders even if the group G does not have property (T), e.g., when G is an arithmetic group of isometries of the hyperbolic plane and the N_is are congruence subgroups, expansion being guaranteed by Selberg's $\lambda_1 \geq \frac{3}{16}$ theorem (see Theorems 4.4.2 and 4.4.3 in [Lu94]).

The following result is Corollary 5 in [Br86].

Proposition 18 *Let M be compact Riemannian manifold. Assume that there exists a surjective homomorphism $\alpha : \pi_1(M) \to G$, with G infinite, amenable, and residually finite. Let $(N_i)_{i>0}$ be a filtration of G; set $M_i = \alpha^{-1}(N_i)\backslash\widetilde{M}$, the quotient of $\widetilde{M}$ by $\alpha^{-1}(N_i)$. Then $\lim_{i\to\infty} \lambda_1(M_i) = 0$.*

Proof. By Theorem 17, $\lim_{i\to\infty} \lambda_1(M_i) = 0$ if and only if the coarse disjoint union $\coprod_{i>0} \pi_1(M)/\alpha^{-1}(N_i)$ is not an expander. Because of the third isomorphism theorem of group theory, the latter coarse disjoint union is coarsely equivalent to the box space $\Box_{(N_i)}G$. As a consequence of the amenability of G, it is not an expander (see Proposition 3.3.7 in [Lu94]). □

In view of Proposition 18, it may be asked whether a tower of covers $(M_i)_{i>0}$ with $\lim_{i\to\infty} \lambda_1(M_i) = 0$ is coarsely equivalent to a tower obtained by mapping $\pi_1(M)$ onto some infinite, amenable, residually finite group. The negative answer was first obtained in [AG12], where it is proved that if $\pi_1(M)$ surjects onto the free group $\mathbb{F}_2$, then M admits a tower of covers $(M_i)_{i>0}$ with $\lim_{i\to\infty} \lambda_1(M_i) = 0$ which is not coarsely equivalent to any tower obtained by mapping $\pi_1(M)$ onto an infinite, amenable, residually finite group. In Corollary 26 of [KV17], we generalize this result as follows.

Theorem 19 *Assume that $\pi_1(M)$ maps onto a non-amenable, residually finite group with infinite abelianization. Then M admits a tower $(M_i)_{i>0}$ of finite covers with $\lim_{i\to\infty} \lambda_1(M_i) = 0$, which is not coarsely equivalent to any tower of covers obtained by mapping $\pi_1(M)$ onto an infinite, amenable, residually finite group. Moreover, for every $\alpha \in (0, 1)$, the tower $(M_i)_{i>0}$ can be chosen so that $\lambda_1(M_i) = O(vol(M_i)^{-\alpha})$.*

Sketch of Proof The proof of Proposition 18, together with Lemma 16, shows that, if the tower of covers $(M_i)_{i>0}$ is obtained by mapping $\pi_1(M)$ onto some residually finite group H (with a given filtration $(N_i)_{i>0}$), then the coarse disjoint union $\coprod_{i>0} M_i$ is coarsely equivalent to the box space $\Box_{(N_i)}H$. The latter has property A if and only if H is amenable (see Theorem 10), and property A is an invariant of coarse equivalence for box spaces. This allows one to distinguish, up to coarse equivalence, towers obtained by mapping $\pi_1(M)$ to an amenable group, from those obtained by mapping $\pi_1(M)$ onto a non-amenable group.

Assuming that $\pi_1(M)$ maps onto G, a non-amenable residually finite group with infinite abelianization, in view of Theorem 17 it is enough to construct some box space of G which is not an expander. It is classical (see Proposition 7.3.11 in [Lu94]) that the components in an expander have a diameter which is logarithmic in the number of vertices. So to ensure non-expansion, it is enough to construct a box space of G such that the diameter of the components grows faster than logarithmically; actually we will realize $|G/N_i|^\alpha$ as the diameter, where α is any fixed number with $0 < \alpha < 1$.

Since G has infinite abelianization, it maps onto $\mathbb{Z}$, and therefore can be written as a semi-direct product $G = H \rtimes \mathbb{Z}$. Let t be a generator of the $\mathbb{Z}$-factor. Choose a finite generating set S of G of the form $S = T \cup \{t^{\pm 1}\}$, with $T \subset H$. Start with any filtration of G. Intersecting with H, we get a filtration $(K_i)_{i>0}$ of H. We are going to choose the desired filtration $(N_i)_{i>0}$ of G in the form:

$$N_i = \langle K_i, t^{m_i} \rangle$$

with m_i a suitably chosen integer (which is where the freedom lies). Observe that N_i is a finite index subgroup in G; taking for m_i a multiple of the order of t acting by conjugation on the finite group H/K_i we may assume that N_i is normal in G. Intuitively the Cayley graph $Cay(G/N_i, S)$ can be thought as a 'fat tyre', i.e., a thickening of the cycle C_{m_i}. So, $\text{diam}(Cay(G/N_i, S)) \geq \text{diam}(C_{m_i}) = \frac{m_i}{2}$. So if $m_i \geq 2^{\frac{1}{1-\alpha}}.|H/K_i|^{\frac{\alpha}{1-\alpha}}$, then (using $|G/N_i| = |H/K_i|.m_i$) we get

$$\text{diam}(Cay(G/N_i, S)) \geq |G/N_i|^\alpha \tag{12.5.1}$$

It remains to take m_i dividing m_{i+1} to ensure that $(N_i)_{i>0}$ is a filtration of G.

Now by inequality (1.3) in [BT15]:

$$\lambda_1(Cay(G/N_i, S)) \leq \frac{8|S|^2.\log|G/N_i|}{\text{diam}(Cay(G/N_i, S))} \leq \frac{8|S|^2.\log|G/N_i|}{|G/N_i|^\alpha}$$

where the second inequality follows from inequality (12.5.1). So for any $\alpha' < \alpha$ we get: $\lambda_1(Cay(G/N_i, S)) = O(|G/N_i|^{-\alpha'})$. This proof is taken from the proof of Proposition 25 in [KV17].

12.6 Rigidity for Towers of Covers

Even if coarse equivalence looks like a very loose equivalence, as mentioned, it sometimes entails surprising rigidity results. The following result was proved as Corollary 3 of [KV17].

Theorem 20 *Let $\widetilde{M}, \widetilde{N}$ be irreducible Riemannian symmetric spaces of non-compact type. Let Γ, Λ be torsion-free co-compact lattices on $\widetilde{M}, \widetilde{N}$ respectively, and let $(\Gamma_i)_{i>0}$ (respectively $(\Lambda_i)_{i>0}$) be a filtration in Γ (respectively Λ). Set $M_i = \Gamma_i\backslash\widetilde{M}$, and $N_i = \Lambda_i\backslash\widetilde{N}$, so that $(M_i)_{i>0}$ (respectively $(N_i)_{i>0}$) is a tower of finite covers of $\Gamma\backslash\widetilde{M}$ (resp. $\Lambda\backslash\widetilde{N}$). If the coarse disjoint* *union $\coprod_{i>0} M_i$ is coarsely equivalent to the coarse disjoint union $\coprod_{i>0} N_i$, then $\widetilde{M}$ is isometric to $\widetilde{N}$, after a possible scaling of the Riemannian metrics.*

The proof is in three steps:

- By Lemma 16, under the assumption of Theorem 20 the box spaces $\square_{(\Gamma_i)}\Gamma$ and $\square_{(\Lambda_i)}\Lambda$ are coarsely equivalent.
- We prove that if two groups admit coarsely equivalent box spaces, then the groups themselves are quasi-isometric.
- Then we apply results about quasi-isometric rigidity of lattices, due to Mostow [Mo73] in rank 1 and Kleiner-Leeb [KL7] in higher rank (see, e.g., Corollary 1.1.4 in [KL7]): if co-compact lattices Γ, Λ are quasi-isometric, then $\widetilde{M}$ is quasi-isometric to $\widetilde{N}$, so $\widetilde{M}$ and $\widetilde{N}$ become isometric once their metrics are suitably scaled.

Let us elaborate on the second step. The precise statement is (see Theorem 7 in [KV17]):

Proposition 21 *Let Γ, Λ be finitely generated, residually finite groups, with corresponding filtrations $(\Gamma_i)_{i>0}$, $(\Lambda_i)_{i>0}$. If $\square_{(\Gamma_i)}\Gamma$ is coarsely equivalent to $\square_{(\Lambda_i)}\Lambda$, then Γ is quasi-isometric to Λ.*

Sketch of Proof Set $X_i = \Gamma/\Gamma_i$, $X = \square_{(\Gamma_i)}\Gamma$; and $Y_j = \Lambda/\Lambda_j$, $Y = \square_{(\Lambda_i)}\Lambda$; let $f : X \to Y$ be a coarse equivalence with control functions $\rho_\pm$. The proof follows the following steps.

(a) For k large enough, there exists a unique index $\ell = \ell(k)$ such that $f(X_k) \subset Y_\ell$. Indeed, for x, x' at distance 1 in X_k, we have $f(x) \in Y_\ell$ (say), and $d(f(x), f(x')) \leq \rho_+(1)$. But for ℓ large, the distance from Y_ℓ to any other Y_j is larger than $\rho_+(1)$, so $f(x')$ belongs to the same component Y_ℓ as $f(x)$. By connectedness of X_k, we have $f(X_k) \subset Y_\ell$.

(b) f sets up a bijection between a co-finite set of components in X and a co-finite set of components in Y. Remember that f is almost surjective, i.e., there is $C > 0$ such that Y is the C-neighbourhood of $f(X)$. But for ℓ large enough, we have $\text{diam}(Y_\ell) > C$, so the component Y_ℓ intersects $f(X)$. On the other hand, if two components X_i, X_j of X are a distance d apart, their images are at least $\rho_-(d)$ apart. So if $f(X_i) \subset Y_\ell$, for j large enough d will become larger than $\text{diam}(Y_\ell)$, so that $f(X_j)$ does not meet Y_ℓ.

(c) There exists a constant $A > 0$ such that, for k large enough and $x, x' \in X_k$:

$$\frac{1}{A} d_X(x, x') - A \leq d_Y(f(x), f(x')) \leq A\, d_X(x, x').$$

Indeed, for $x_0 = x, x_1, \ldots, x_n = x'$ a geodesic path in X_k from x to x', we have by the triangle inequality: $d_Y(f(x), f(x')) \leq \sum_{i=0}^{n-1} d_Y(f(x_i), f(x_{i+1})) \leq n\rho_+(1) = \rho_+(1)d_X(x, x')$. The lower bound is obtained using a quasi-inverse for f.

At this point we have shown that, after discarding finitely many components in X and Y, the coarse equivalence f is given by a family $f_k : X_k \to Y_{\ell(k)}$ of quasi-isometries with *uniform* constants. Using vertex-transitivity, we may further assume that f_k maps the unit of X_k to the unit of $Y_{\ell(k)}$.

Now we proceed to construct a quasi-isometry $\Gamma \to \Lambda$. Fix some integer radius $R > 0$. For G a finitely generated group with a fixed word metric, we denote by $B_G(R)$ the ball of radius R centred at the unit of G. Let $p_i : \Gamma \to X_i = \Gamma/\Gamma_i$ and $q_j : \Lambda \to Y_j = \Lambda/\Lambda_j$ be the quotient homomorphisms. By residual finiteness of Γ and Λ, we may take k large enough so that:

- p_k maps $B_\Gamma(R)$ isometrically into X_k;
- $q_{\ell(k)}$ maps $B_\Lambda(AR)$ isometrically into $Y_{\ell(k)}$.

By (c) above, the map f_k maps $B_{X_k}(R)$ quasi-isometrically into $B_{Y_{\ell(k)}}(AR)$. So the map

$$\tilde{f}_{k,R} = q_{\ell(k)}^{-1}|_{B_{Y_{\ell(k)}}(AR)} \circ f_k \circ p_k$$

is a quasi-isometric embedding from $B_\Gamma(R)$ to $B_\Lambda(AR)$. Note that all these maps have uniform quasi-isometry constants. They seem to depend on k but, since $B_\Gamma(R)$ and $B_\Lambda(AR)$ are finite sets, there are finitely many maps $B_\Gamma(R) \to B_\Lambda(AR)$. Varying R in $\mathbb{N}$ and using a diagonal process, we set up a quasi-isometric embedding $\tilde{f} : \Gamma \to \Lambda$ which is easily seen to be almost surjective.

References

[AD08] G. ARZHANTSEVA and T. DELZANT, *Examples of random groups*, preprint available on authors' websites (2008)

[AGS] G. ARZHANTSEVA, E. GUENTNER and J. ŠPAKULA, *Coarse non-amenability and coarse embeddings*, Geom. Funct. Anal. 22 (2012), no. 1, 2236

[AG12] G. ARZHANTSEVA and E. GUENTNER, *Coarse non-amenability and covers with small eigenvalues*, Math. Ann. 354 (2012), no. 3, 863–870

[AT15] G. ARZHANTSEVA and R. TESSERA, *Relative expanders*, Geom. Funct. Anal. 25 (2015), no. 2, 317341

[BHV08] B. BEKKA, P. DE LA HARPE, A. VALETTE. *Kazhdan's property (T)*. Cambridge University Press, 2008

[BT15] E. BREUILLARD and M. TOINTON, *Nilprogressions and groups with moderate growth*, Adv. Math. 289 (2016), 10081055

[Br86] R. BROOKS, *The spectral geometry of a tower of coverings*, J. Differential Geometry 23 (1986), 97–107

[CWW13] X. CHEN, Q. WANG and X. WANG, *Characterization of the Haagerup property by fibred coarse embedding into Hilbert space*, Bull. Lond. Math. Soc. 45 (2013), no. 5, 1091–1099

[CCJJV01] P.-A. CHERIX, M. COWLING, P. JOLISSAINT, P. JULG, and A. VALETTE. *Groups with the Haagerup property*. Birkhäuser, Progress in Math. 197, 2001.

[CM17] M. CLAY and D. MARGALIT (eds.), *Office hours with a geometric group theorist*, Princeton University Press, 2017

[DK16] T. DELABIE and A. KHUKHRO, *Box spaces of the free group that neither contain expanders nor embed into a Hilbert space*, Adv. Math. 336 (2018), 70–96.

[DK17] T. DELABIE and A. KHUKHRO, *Coarse fundamental group and box spaces*, Proceedings of the Royal Society of Edinburgh: Section A Mathematics, 1–16. doi:10.1017/prm.2018.102

[Es98] A. ESKIN, *Quasi-isometric rigidity of nonuniform lattices in higher rank symmetric spaces*, J. Amer. Math. Soc. 11 (1998), 321–361

[Grom] M. GROMOV, *Random walk in random groups*, Geom. Funct. Anal. 13 (2003), no. 1, 73–146

[HLS] N. HIGSON, V. LAFFORGUE and G. SKANDALIS, *Counterexamples to the Baum–Connes conjecture*, Geom. Funct. Anal. 12 (2002), 330–354

[Hu14] D. HUME, *A continuum of expanders*, Preprint, arXiv:1410.0246, to appear in Fundamenta Mathematicae

[KV17] A. KHUKHRO and A. VALETTE, *Expanders and box spaces*, Advances in Math. 314 (2017), 806–834

[KL7] B. KLEINER and B. LEEB, *Rigidity of quasi-isometries for symmetric spaces and Euclidean buildings*, Publications Math. IHES, Volume 86, Issue 1, (1997) 115–197

[Lu94] A. LUBOTZKY, *Discrete groups, expanding graphs and invariant measures*, Birkhäuser, Progress in Math. 125, 1994

[LPS88] A. LUBOTZKY, R. PHILLIPS and P. SARNAK, *Ramanujan graphs*, Combinatorica 8 (1988), 261-277

[LZ03] A. LUBOTZKY, A. ZUK, On property (τ), Pre-book available on www.ma.huji.ac.il/~alexlub/

[Ma75] G.A. MARGULIS, *Explicit constructions of concentrators*, Probl. of Inform. Transm. 10 (1975), 325–332

[MN14] M. MENDEL and A. NAOR, *Nonlinear spectral calculus and super-expanders*, Publ. Math. Inst. Hautes Études Sci., 119 (2014), 1–95

[Mo73] G.D. MOSTOW, *Strong rigidity of locally symmetric spaces*, Annals of math. studies 78, Princeton University Press, 1973

[Now] P. NOWAK, *Coarsely embeddable metric spaces without Property A*, J. Funct. Anal. 252 (2007), no. 1, 126136

[NY12] P. NOWAK and G. YU, *Large scale geometry*, EMS Textbooks in Math., European Math. Soc. 2012

[Osaj] D. OSAJDA, *Small cancellation labellings of some infinite graphs and applications*, Preprint, arXiv:1406.5015

[Pin] M.S. PINSKER, *On the complexity of a concentrator*, 7th International Teletraffic Conference, Stockholm (1973)

[Ro03] J. ROE, *Lectures on Coarse Geometry*, vol. 31 of University Lecture Series. American Mathematical Society, Providence, RI, 2003.

[Tes] R. TESSERA, *Coarse embeddings into a Hilbert space, Haagerup property and Poincaré inequalities*, J. Topol. Anal. 1 (2009), no. 1, 87100

[WY14] R. WILLETT, G. YU, *Geometric property (T)*, Chinese Annals of Mathematics, Ser. B 35 (2014), no. 5, 761–800

[Yu] G. YU, *The coarse Baum–Connes conjecture for spaces which admit a uniform embedding into Hilbert space*, Inventiones 139 (2000), 201–240

13

Ramanujan Graphs and Digraphs

Ori Parzanchevski

Abstract

Ramanujan graphs have fascinating properties and history. In this chapter we explore a parallel notion of Ramanujan *digraphs*, collecting relevant results from old and recent papers, and proving some new ones. *Almost-normal* Ramanujan digraphs are shown to be of special interest, as they are extreme in the sense of an Alon-Boppana theorem, and they have remarkable combinatorial features, such as small diameter, Chernoff bound for sampling, optimal covering time, and sharp cutoff. Other topics explored are the connection to Cayley graphs and digraphs, the spectral radius of universal covers, Alon's conjecture for random digraphs, and explicit constructions of almost-normal Ramanujan digraphs.

13.1 Introduction

A connected k-regular graph is called a *Ramanujan graph* if every eigenvalue λ of its adjacency matrix (see definitions below) satisfies either

$$|\lambda| = k \,, \quad \text{or} \quad |\lambda| \leq 2\sqrt{k-1}. \qquad \textit{(Ramanujan graph)}$$

While the generalized Ramanujan conjecture appears in the first constructions of such graphs [LPS88, Mar88], the reason that lead Lubotzky, Phillips, and Sarnak to coin the term Ramanujan graphs is that by their very definition, they present the phenomenon of extremal spectral behaviour, which Ramanujan observed in a rather different setting.

Supported by ISF grant 1031/17.

In the case of graphs, this can be stated in two ways: Ramanujan graphs spectrally mimic their universal cover, the infinite k-regular tree $\mathcal{T}_k$, whose spectrum is the interval

$$\mathrm{Spec}\,(\mathcal{T}_k) = \left[-2\sqrt{k-1}, 2\sqrt{k-1}\right]$$

[Kes59]; and, they are asymptotically optimal: the Alon–Boppana theorem (cf. [LPS88, Nil91, HLW06]) states that for any $\varepsilon > 0$, there is no infinite family of k-regular graphs for which all non-trivial adjacency eigenvalues satisfy $|\lambda| \leq 2\sqrt{k-1} - \varepsilon$ (the trivial eigenvalues are by definition $\pm k$). These two observations are closely related - in fact, any infinite family of quotients of a common covering graph $\widetilde{\mathcal{G}}$ cannot 'do better' than $\widetilde{\mathcal{G}}$ (see [Gre95, GŻ99] for precise statements).

A major interest in the adjacency spectrum of a graph comes from the notion of *expanders* – graphs of bounded degree whose non-trivial adjacency spectrum is of small magnitude. Such graphs have strong connectedness properties which are extremely useful: see [Lub94, HLW06, Lub12] for extensive surveys on properties of expanders in mathematics and computer science.

Ramanujan graphs, which stand out as the optimal expanders (from the spectral point of view), have a rich theory and history. The purpose of this chapter is to suggest that a parallel theory should be developed for directed graphs (*digraphs*, for short), where by a *Ramanujan digraph* we mean a k-regular digraph whose adjacency eigenvalues satisfy either

$$|\lambda| = k\,, \quad \text{or} \quad |\lambda| \leq \sqrt{k}\,, \qquad (\textit{Ramanujan digraph}) \qquad (13.1.1)$$

where the reasons for this definition will be made clear in the chapter.

The idea of 'Ramanujan digraphs' arose during the work on the papers [PS18, LLP19]. While we believe that the term itself is new, several classic results can be interpreted as saying something about these graphs (see, e.g., Sections 13.3.5 and 13.5.2). We survey here both classic results and ones from the mentioned papers, and prove several new ones. We remark that for the most part of this chapter we focus on finite graphs and digraphs, with infinite ones appearing mainly as universal covers. Without doubt, they merit further study in their own right (see also Section 13.6).

The chapter unfolds as follows: After giving the definitions in Section 13.2 and various examples in Section 13.3, we prove that there are very few normal Ramanujan digraphs in Section 13.3.7. We then turn to almost-normal digraphs in Section 13.4, proving an Alon–Boppana type theorem, and surveying their spectral and combinatorial features, such as optimal covering, sharp cutoff, small diameter and Chernoff sampling bound. We then explore Ramanujan digraphs from the perspective of universal covers (Section 13.5.1) and infinite

Cayley graphs (Section 13.5.2). In Section 13.5.3 we discuss an explicit construction of Ramanujan digraphs as Cayley graphs of finite groups, which is similar to the LPS construction [LPS88], but applies to any PGL_d and not only to PGL_2. In Section 13.5.4 we touch upon zeta functions and the Riemann Hypothesis, in Section 13.5.5 we discuss Alon's second eigenvalue conjecture for digraphs, and finally in Section 13.6 we present some questions.

13.2 Definitions

Throughout the chapter we denote by $\mathcal{G} = \left(V_{\mathcal{G}}, E_{\mathcal{G}}\right)$ a connected k-regular graph on n vertices, where by a *graph* we always mean an undirected one. Its adjacency matrix $A = A_{\mathcal{G}}$, indexed by V, is defined by $A_{v,w} = 1$ if $v \sim w$ (v and w are neighbours in $\mathcal{G}$), and 0 otherwise.(†) Since $\sim$ is symmetric, so is A, hence it is self-adjoint with real spectrum. The constant function $\mathbb{1}$ is an eigenvector of A with eigenvalue k, and when $\mathcal{G}$ is bipartite, say $V = L \amalg R$, the function $\mathbb{1}_L - \mathbb{1}_R$ is an eigenvector with eigenvalue $-k$. We call these eigenvalues and eigenvectors *trivial*, and denote by $L_0^2 = L_0^2(V)$ their orthogonal complement in $L^2(V)$, namely

$$L_0^2(V) = \begin{cases} \mathbb{1}^{\perp} & \mathcal{G} \text{ is not bipartite} \\ \langle \mathbb{1}_L, \mathbb{1}_R \rangle^{\perp} & \mathcal{G} \text{ is bipartite.} \end{cases}$$

Observe that A restricts to a self-adjoint operator on $L_0^2(V)$, and recall that for self-adjoint (and even normal) operators, the spectral radius

$$\rho(M) = \max\left\{|\lambda| \;\middle|\; \lambda \in \operatorname{Spec}(M)\right\}$$

coincides with the operator norm

$$\|M\| = \max_{v \neq 0} \frac{\|Mv\|}{\|v\|}.$$

Definition 13.2.1 ([LPS88]) A k-regular graph $\mathcal{G}$ is a *Ramanujan graph* if

$$\rho(\mathcal{G}) \overset{def}{=} \rho\left(A_{\mathcal{G}}\big|_{L_0^2}\right) = \left\|A_{\mathcal{G}}\big|_{L_0^2}\right\| \leq 2\sqrt{k-1}.$$

Moving on to digraphs, we denote by $\mathcal{D}$ a finite connected k-regular directed graph, by which we mean that each vertex has k incoming and k outgoing edges. Now, $A_{v,w} = 1$ whenever $v \to w$ (namely, there is an edge from v to

(†) On occasions we allow loops and multiple edges, in which case $A_{v,w}$ is the number of edges between v and w.

w) and since A is no longer symmetric, its spectrum is not necessarily real. However, by regularity we still have

$$\rho(A) = \|A\|_1 = \|A\|_2 = \|A\|_\infty = k,$$

as any square matrix satisfies

$$\|A\|_2^2 = \rho\left(A^*A\right) \leq \left\|A^*A\right\|_\infty \leq \left\|A^*\right\|_\infty \|A\|_\infty = \|A\|_1 \|A\|_\infty, \quad (13.2.1)$$

and $\mathbb{1}$ is still a k-eigenfunction. If $\mathcal{D}$ is m-periodic, namely $V_{\mathcal{D}} = \coprod_{j=0}^{m-1} V_j$ with every edge starting in V_j terminating in $V_{(j+1 \bmod m)}$, then $e^{2\pi ti/m}k$ are also eigenvalues (with $t = 1, \ldots, m-1$), with corresponding eigenfunctions $\sum_{j=0}^{m-1} e^{2\pi jti/m} \mathbb{1}_{V_j}$. By Perron–Frobenius theory, all eigenvalues of absolute value k arise in this manner. We call these eigenvalues (including k) trivial, and denote by L_0^2 the orthogonal complement of their eigenfunctions in $L^2(V_{\mathcal{D}})$. Even though A is not self-adjoint or normal, the regularity assumptions ensures that it still restricts to L_0^2, and we make the following definition:

Definition 13.2.2 A k-regular digraph $\mathcal{D}$ is a *Ramanujan digraph* if

$$\rho(\mathcal{D}) \stackrel{def}{=} \rho\left(A_{\mathcal{D}}\big|_{L_0^2}\right) \leq \sqrt{k}.$$

The bound $\left\|A\big|_{L_0^2}\right\| \leq \sqrt{k}$ does not have to hold anymore; indeed, we will see that there are Ramanujan digraphs for which $\left\|A\big|_{L_0^2}\right\| = k$, which is as bad as one can have for a k-regular adjacency operator (in the undirected settings, this would mean that the graph is disconnected). For spectral analysis, the operator norm is much more important than the spectral radius, and this is what makes digraphs harder to study than graphs.

We say that a digraph $\mathcal{D}$ is self-adjoint, or normal, if its adjacency matrix is. In these cases we do have $\left\|A_{\mathcal{D}}\big|_{L_0^2}\right\| = \rho(\mathcal{D})$, and much of the theory of expanders remains as it is for graphs (see for example [Vu08]). However, we will see in Proposition 13.3.1 that there are very few normal Ramanujan digraphs. A main novelty of [LP16], which was developed further in [LLP19], is the idea of *almost-normal* digraphs:

Definition 13.2.3 A matrix is r-*normal* if it is unitarily equivalent to a block-diagonal matrix with blocks of size at most $r \times r$. A digraph is called r-normal if its adjacency matrix is r-normal, and a family of matrices (or digraphs) is said to be almost-normal if its members are r-normal for some fixed $r < \infty$.

We shall see in Section 13.4 that for many applications, almost-normal digraphs are almost as good as normal ones.

13.3 Examples

13.3.1 Complete Digraphs

For $m, k \in \mathbb{N}$, we define the complete k-regular m-periodic digraph $\mathcal{K}_{k,m}$ by

$$V_{\mathcal{K}_{k,m}} = \{(x, y) \mid x \in \mathbb{Z}/m\mathbb{Z},\ y \in [k]\}$$
$$E_{\mathcal{K}_{k,m}} = \{(x, y) \to (x+1, z) \mid x \in \mathbb{Z}/m\mathbb{Z},\ y, z \in [k]\}.$$

This is a normal Ramanujan digraph on $n = km$ vertices, with m trivial eigenvalues coming from periodicity, and $(k-1)\,m$ times the eigenvalue zero. This shows that one should focus on the case of bounded degree and periodicity, for otherwise infinite families of trivial examples arise.

13.3.2 Projective Planes and Hyperplanes

The *Projective plane over* $\mathbb{F}_p$ is the undirected bipartite graph whose vertices represent the lines and planes in $\mathbb{F}_p^3$, and whose edges correspond to the relation of inclusion. It is k-regular for $k = p+1$, and has $n = 2\left(p^2 + p + 1\right)$ vertices. Its non-trivial spectrum is $\pm\sqrt{k-1}$ (each repeating $p^2 + p$ times), so it is Ramanujan. In fact, it is twice better than Ramanujan, which only requires $|\lambda| \leq 2\sqrt{k-1}$. We can therefore consider it as a digraph, with each edge appearing with both directions, and obtain a k-regular self-adjoint Ramanujan digraph, since the adjacency matrix remains the same.

More generally, the bipartite graph of lines against d-spaces in $\mathbb{F}_p^{d+1}$ (with respect to inclusion) has $n = 2\cdot\frac{p^{d+1}-1}{p-1}$ vertices and is k-regular with $k = \frac{p^d-1}{p-1}$. Its non-trivial eigenvalues are $\pm\sqrt{p^{d-1}} = \pm\sqrt{k - \frac{p^{d-1}-1}{p-1}}$, so we obtain a self-adjoint Ramanujan digraph for every d.

13.3.3 Paley Digraphs

For a prime p with $p \equiv 3 \pmod 4$, the Paley digraph $\mathcal{PD}(p)$ [GS71] has $V = \mathbb{F}_p$ and

$$E = \left\{a \to b \,\middle|\, \left(\tfrac{b-a}{p}\right) = 1\right\},$$

where $\left(\frac{\cdot}{\cdot}\right)$ is the Legendre symbol. It is a $k = \frac{p-1}{2}$-regular normal digraph, with non-trivial eigenvalues $\frac{-1\pm i\sqrt{p}}{2}$ (this is a nice exercise in Legendre symbols). These are of absolute value $\sqrt{\frac{k+1}{2}}$, so $\mathcal{PD}(p)$ is a normal Ramanujan digraph.

It turns out that examples as in Sections 13.3.1–13.3.3 are limited. In Section 13.3.7 we will prove:

Proposition 13.3.1 *For any fixed $k \geq 2$ and $m \geq 1$ there are only finitely many k-regular m-periodic normal (and in particular, self-adjoint) Ramanujan digraphs.*

Thus, if we wish to fix the regularity k and periodicity m, and yet take $|V| = n$ to infinity we must move on to non-normal graphs.

13.3.4 Extremal Directed Expanders

The De Bruijn graph $\mathcal{DB}(k, s)$ is a k-regular aperiodic Ramanujan digraph with

$$V_{\mathcal{DB}(k,s)} = [k]^s \qquad ([k] = \{1, \ldots, k\}, \text{so } n = k^s)$$
$$E_{\mathcal{DB}(k,s)} = \{(a_1, \ldots, a_s) \to (a_2, \ldots, a_s, t) \,|\, a_i, t \in [k]\}.$$

Just as complete digraphs, the non-trivial spectrum of $\mathcal{DB}(k, s)$ consists entirely of zeros. However, its adjacency matrix is not diagonalizable, and it has Jordan blocks of size s, so in particular, these do not form an almost-normal family even for a fixed k. The Kautz digraph is another example with similar properties.(†)

In [FL92] Feng and Li show that k-regular r-periodic *diagonalizable* digraphs must have $\rho(\mathcal{D}) \geq 1$ once $n > kr$. Furthermore, for any n which is co-prime to k they give an explicit construction of a k-regular r-periodic digraph on nr vertices with $\rho(\mathcal{D}) = 1$.

Remark 13.3.2 De Bruijn graphs show that a direct analogue of the Alon–Boppana theorem (with respect to any positive ε) does not hold for digraphs in general. In Section 13.4 we will see that in the settings of almost-normal digraphs, an Alon–Boppana theorem does hold, with the bound $\sqrt{k}$.

13.3.5 Directed Line Graphs

In this section, we assume that $\mathcal{G}$ is a $(\boldsymbol{k+1})$**-regular** graph, and we define its $\boldsymbol{k}$**-regular** *line-digraph* $\mathcal{D}_L(\mathcal{G})$ as follows:

$$V_{\mathcal{D}_L(\mathcal{G})} = \left\{(v, w) \,\middle|\, v, w \in V_{\mathcal{G}},\ v \sim w\right\}$$
$$E_{\mathcal{D}_L(\mathcal{G})} = \{(v, w) \to (w, u) \,|\, u \neq v\}.$$

(†) For the spectrum of the symmetrization of De Bruin and Kautz digraphs, see [DT98].

Namely, the vertices correspond to edges in $\mathcal{G}$ with a chosen direction, and a $\mathcal{G}$-edge is connected to another one in $\mathcal{D}_L(\mathcal{G})$ if they form a non-backtracking path of length 2 in $\mathcal{G}$. The importance of this construction is that non-backtracking walks on $\mathcal{G}$ are encoded precisely by regular (memory-less) walks on $\mathcal{D}_L(\mathcal{G})$ (see Figure 13.1).

By Hashimoto's interpretation of the Ihara–Bass formula (cf. [Iha66; Sun86; Has89; Bas92; ST96; FZ99; KS00; LP16 ...]), the spectra of $\mathcal{G}$ and $\mathcal{D}_L(\mathcal{G})$ are related:

$$\operatorname{Spec}(\mathcal{D}_L(\mathcal{G})) = \left\{ \frac{\lambda \pm \sqrt{\lambda^2 - 4k}}{2} \,\middle|\, \lambda \in \operatorname{Spec} \mathcal{G} \right\} \cup \{ \underbrace{\pm 1, \ldots, \pm 1}_{|E_\mathcal{G}|-|V_\mathcal{G}| \text{ times}} \}. \tag{13.3.1}$$

One can easily check that

$$|\lambda| \leq 2\sqrt{(k+1)-1} \qquad \Longleftrightarrow \qquad \left| \tfrac{1}{2}\left(\lambda \pm \sqrt{\lambda^2 - 4k}\right) \right| = \sqrt{k}, \tag{13.3.2}$$

so that $\mathcal{G}$ is a Ramanujan graph if and only if $\mathcal{D}_L(\mathcal{G})$ is a Ramanujan digraph. Therefore, any construction of Ramanujan graphs (e.g., [LPS88, Mor94, MSS15]) can be used to construct Ramanujan digraphs.

The digraph $\mathcal{D}_L(\mathcal{G})$ is not normal (as one can easily verify by applying AA^T and $A^T A$ to some $\ominus$ in Figure 13.1), and it turns out that the singular values of A are as bad as can be: the trivial singular value k repeats $|V_\mathcal{G}|$ times. This reflects the fact that the walk described by $A^T A$ is highly disconnected: the edges entering a fixed vertex form a connected component, since

$$A^T A (v \to w) = A^T \left\{ (w \to u) \,\middle|\, \begin{matrix} w \sim u \\ u \neq v \end{matrix} \right\} = \left\{ (u' \to w) \,\middle|\, w \sim u' \right\}$$

(this is easier to see in Figure 13.1 than algebraically). In particular, this shows that $\left\| A\big|_{L_0^2(V_{\mathcal{D}_L(\mathcal{G})})} \right\| = k$. The breakthrough in [LP16] is the understanding that A is always 2-normal, and that this is good enough for the analysis of the random walk on $\mathcal{D}_L(\mathcal{G})$ (see Section 13.4).

13.3.6 Collision-free Walks on Affine Buildings

In the previous example, a certain walk on the *directed* edges of an *undirected* graph $\mathcal{G}$ gave rise to a digraph $\mathcal{D}_L(\mathcal{G})$, which was a Ramanujan digraph whenever $\mathcal{G}$ was a Ramanujan graph. In [LLP19] this is generalized to higher dimension: considering some walk W on the cells of a simplicial complex $\mathcal{X}$ (possibly oriented or ordered cells), one asks when is the digraph $\mathcal{D}_W(\mathcal{X})$ which represents this walk a Ramanujan digraph.

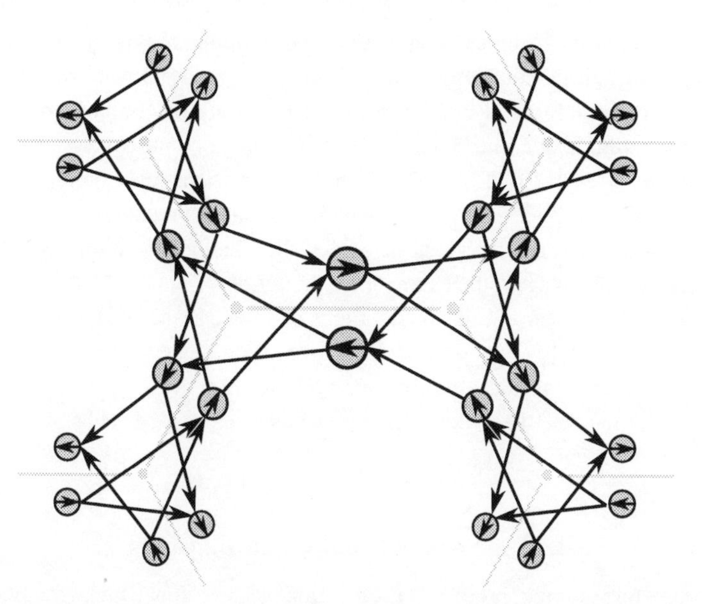

Figure 13.1 The local view of the line-digraph of a three-regular graph (the original graph is shown in the background)

It turns out that the key is the following property: We say that a digraph is *collision-free* if it has at most one (directed) path from any vertex v to any vertex w. The digraph $\mathcal{D}_L(\mathcal{G})$ from Section 13.3.5 is not collision-free – indeed, a regular graph with this property must be infinite – but the line-digraph of the universal cover of $\mathcal{G}$, namely $\mathcal{D}_L(\mathcal{T}_{k+1})$, is indeed collision-free: Two non-backtracking walkers which start on the same directed edge on the tree will never reunite, once separated. The main theorem in [LLP19] is this:

Theorem 13.3.3 ([LLP19]) *Let $\mathcal{X}$ be a complex whose universal cover is the affine Bruhat–Tits building $\mathcal{B}$, and let W be a geometric regular random walk operator. If W is collision-free on $\mathcal{B}$ (namely, $\mathcal{D}_W(\mathcal{B})$ is collision-free), and $\mathcal{X}$ is a Ramanujan complex, then $\mathcal{D}_W(\mathcal{X})$ is a Ramanujan digraph.*

Here *geometric* means that the random walk commutes with the symmetries of $\mathcal{B}$. Properly defining the other terms in the theorem will take us too far afield, and we refer the interested reader to [LSV05a, Lub14, LLP19].

Let us give one concrete example: the *geodesic edge walk* on a complex goes from a directed edge (v, w) to the directed edge (w, u) if $u \neq v$ (no backtracking), and in addition $\{v, w, u\}$ is **not** a triangle in the complex (so the path $v \to w \to u$ is not 'homotopic' to the shorter path $v \to u$).

The edges of the d-dimensional Bruhat–Tits building of type $\widetilde{A}_d$ are colored by $\{1, \ldots, d\}$ [LSV05a, Lub14, LLP19], and the geodesic walk restricted to edges of color 1 forms a regular collision-free walk on the building. Thus, by the theorem above, the same walk on Ramanujan complexes of type $\widetilde{A}_d$, as constructed in [Li04, LSV05b, Fir16], gives a Ramanujan digraph. In the case $d = 1$, the building $\widetilde{A}_1$ is a regular tree, its Ramanujan quotients are Ramanujan graphs, and the geodesic edge walk is simply the non-backtracking walk, so we obtain again the example from Section 13.3.5.

Finally, all geometric walks on quotients of a fixed building $\mathcal{B}$ form a family of almost-normal digraphs [LLP19, Prop. 4.5]. For the geodesic edge walk on $\widetilde{A}_d$-Ramanujan complexes, the corresponding Ramanujan digraphs are sharply $(d+1)$-normal [LLP19, Prop. 5.3, 5.4], and they can be made to be m-periodic for any m dividing $(d+1)$.

13.3.7 Normal Ramanujan Digraphs

We now turn to the proof of Proposition 13.3.1, for which we need a quantitative version of the Alon–Boppana theorem. We use the following:

Theorem 13.3.4 ([Nil04, Theorem 1 with $s = 2$]) *The second largest eigenvalue of a k-regular graph $\mathcal{G}$ is at least $2\sqrt{k-1}\cos\left(\frac{2\pi}{\operatorname{diam}\mathcal{G}}\right)$.*

Proof of Proposition 13.3.1. Let $\mathcal{D}$ be a k-regular normal Ramanujan digraph on n vertices, and let $\mathcal{G}$ be its symmetrization, namely, $A_\mathcal{G} = A_\mathcal{D} + A_\mathcal{D}^T$. Assume for now that $\mathcal{D}$ is aperiodic. From normality of $A_\mathcal{D}$ we obtain

$$\rho(\mathcal{G}) = \max\left\{\lambda + \overline{\lambda} \,\middle|\, \lambda \in \operatorname{Spec} A_\mathcal{D}\big|_{L_0^2}\right\} \le 2\sqrt{k}, \tag{13.3.3}$$

and we would like to combine this with Theorem 13.3.4. For a $k_\mathcal{G}$-regular graph with $k_\mathcal{G} \ge 4$, Moore's bound [HS60] gives

$$n \le 1 + k_\mathcal{G}\sum_{j=1}^{\operatorname{diam}\mathcal{G}} (k_\mathcal{G}-1)^{j-1} \le 2\left(k_\mathcal{G}-1\right)^{\operatorname{diam}\mathcal{G}},$$

so that Theorem 13.3.4 implies (for $k_\mathcal{G} \ge 4$)

$$\rho(\mathcal{G}) \ge 2\sqrt{k_\mathcal{G}-1}\cos\left(\frac{2\pi}{\log_{k_\mathcal{G}-1}(n/2)}\right) \ge 2\sqrt{k_\mathcal{G}-1}\left(1 - \frac{2\pi^2}{\log^2_{k_\mathcal{G}-1}(n/2)}\right) \tag{13.3.4}$$

Our $\mathcal{G}$ is $2k$-regular, so that (13.3.3) and (13.3.4) combine to

$$1 - \frac{2\pi^2}{\log^2_{2k-1}(n/2)} \le \sqrt{\frac{k}{2k-1}} \le \sqrt{\frac{2}{3}},$$

which gives

$$n \leq 2\,(2k-1)^{10.4}\,. \tag{13.3.5}$$

Assume now that $\mathcal{D}$ is m-periodic, and observe the k^m-regular digraph $\mathcal{D}'$ whose vertices are those of $\mathcal{D}$ and whose edges are the paths of length m in $\mathcal{D}$. Since $A_{\mathcal{D}'} = A_{\mathcal{D}}^m$, the trivial eigenvalues $e^{2\pi ji/m}k$ of $\mathcal{D}$ become the eigenvalue k^m in $\mathcal{D}'$, which has no other trivial eigenvalues. This reflects the fact that $\mathcal{D}'$ is a disconnected digraph with m aperiodic connected components. As $\mathcal{D}'$ is also normal and Ramanujan, (13.3.5) bounds the size of each component by $2\,(2k^m-1)^{10.4}$. All together, we get

$$n \leq 2m\left(2k^m-1\right)^{10.4}, \tag{13.3.6}$$

so there are only finitely many such graphs.[(†)] □

Remark 13.3.5 (*a*) In Section 13.3.2 we saw examples for 2-periodic normal Ramanujan digraphs with $n \approx 2k^2$, which is quite far from the bound (13.3.6) with $m = 2$. It seems interesting to ask what is the optimal bound.
(*b*) In [LLP19, §5.1] it is shown that for any $i \geq 1$ there is a walk W_i on cells of dimension i of a complex, such that if $\mathcal{X}$ is a Ramanujan complex of dimension d then $\mathcal{D}_{W_i}(\mathcal{X})$ are Ramanujan digraphs for $1 \leq i \leq d$. However, no such walk on vertices (i.e., for $i = 0$) is exhibited. Proposition 13.3.1 explains why: it is well known that all geometric operators on vertices commute with each other (these are 'Hecke operators' – cf. [LSV05a]). In particular, such an operator commutes with its own transpose, and therefore induces normal digraphs, which cannot be Ramanujan for an infinite family by the proposition.

13.4 Almost-normal Digraphs

In this section we explore almost-normal digraphs, and in particular almost-normal Ramanujan digraphs. Their main feature, which goes back to [LP16, LLP19] is the behaviour of powers of their adjacency matrix:

Proposition 13.4.1 *Let $\mathcal{D}$ be an r-normal, k-regular digraph with $\rho(\mathcal{D}) = \lambda$. For any $\ell \in \mathbb{N}$,*

$$\left\| A_{\mathcal{D}}^{\ell}\big|_{L_0^2} \right\| \leq \binom{\ell+r-1}{r-1} k^{r-1}\lambda^{\ell-r+1} = O\left(\ell^{r-1}\lambda^{\ell}\right).$$

(†) An alternate way to handle periodicity is to use [Nil04, Theorem 1] with $s = m + 1$.

Note that for normal digraphs $r = 1$, which gives $\left\| A_{\mathcal{D}}^{\ell}\big|_{L_0^2} \right\| \leq \lambda^{\ell}$ as should be. The upshot here is that as long as the 'failure of normality' is bounded, only a polynomial price is incurred. This shows why random walk on almost-normal digraphs is susceptible to spectral analysis: Let p_ℓ denote the probability distribution of the walk at time ℓ. Assuming for simplicity that $\mathcal{D}$ is aperiodic, so that $L_0^2 = \langle \mathbb{1} \rangle$, the distance from equilibrium is

$$\left\| p_\ell - \tfrac{\mathbb{1}}{n} \right\| = \left\| \left(\tfrac{A}{k}\right)^{\ell} p_0 - \tfrac{\mathbb{1}}{n} \right\| = \left\| \left(\tfrac{A}{k}\right)^{\ell} \left(p_0 - \tfrac{\mathbb{1}}{n}\right) \right\| \leq \tfrac{1}{k^\ell} \left\| A^{\ell}\big|_{L_0^2} \right\|$$
$$= O\left(\ell^{r-1} \left(\tfrac{\lambda}{k}\right)^{\ell}\right),$$

where we have used $p_0 - \frac{\mathbb{1}}{n} \in L_0^2$. In the case of Ramanujan digraphs $\lambda = \sqrt{k}$, and this gives an almost-optimal L^1-cutoff, at time $\log_k n + O\left(\log_k \log n\right)$ (see [LP16, Theorem 3.5] and [LLP19, Prop. 3.1], and [ABLS07] for related results).

An interesting corollary [LLP19, Theorem 2] is that in an r-normal Ramanujan digraph the sphere of radius $\ell_0 = \log_k n + (2r - 1) \log_k \log n$ around any vertex v covers almost all of the graph. Indeed, if the walk described by p_ℓ starts at v_0 then $\operatorname{supp}(p_\ell) = S_\ell(v_0)$, so that

$$\frac{n - |S_\ell(v_0)|}{n^2} = \sum_{v \notin S_\ell(v_0)} \frac{1}{n^2} = \left\| \left(p_\ell - \tfrac{\mathbb{1}}{n}\right)\big|_{V \setminus S_\ell(v_0)} \right\|^2 \leq \left\| p_\ell - \tfrac{\mathbb{1}}{n} \right\|^2$$
$$= O\left(\frac{\ell^{2r-2}}{k^\ell}\right),$$

and $\ell = \ell_0$ yields $\left|S_{\ell_0}(v_0)\right| \geq n(1 - o(1))$. This in turn implies a bound of $(2 + o(1)) \log_k(n)$ on the diameter, since the ℓ_0-spheres around any two vertices must intersect.

Yet another consequence of almost-normality is a Chernoff bound for sampling: in [PR20] we show that if f is a function from the vertices to $[-1, 1]$ with sum zero, and $v_1, \ldots, v_\ell$ are the vertices visited in a random walk on an almost-normal directed expander, then

$$\operatorname{Prob}\left[\tfrac{1}{\ell}\sum_{i=1}^{\ell} f(v_\ell) > \gamma\right] \leq e^{-C\gamma^2 \ell}$$

for small enough γ, where C depends on the expansion and normality. Using Section 13.3.5, this also gives a similar result for non-backtracking random walk on non-directed expanders, and via Section 13.3.6 to geodesic walks on high-dimensional expanders.

Proof of Proposition 13.4.1. By definition, A is unitarily equivalent to a block-diagonal matrix with blocks of size $r \times r$. The periodic functions on $\mathcal{D}$ correspond to 'trivial' blocks of size one, and the singular values of $A^{\ell}\big|_{L_0^2}$ are the union of the singular values of the ℓ-th powers of the remaining, 'nontrivial' blocks. Let B be a non-trivial block of size $s \times s$. By Schur decomposition, we can assume that B is upper triangular, in which case the absolute values of its diagonal entries are bounded by λ. In addition, since B is unitarily equivalent to the restriction of A to some invariant subspace, all entries of B are bounded by $\|B\|_2 \leq \|A\|_2 = k$, so that B is entry-wise majorized by

$$M_{s,\lambda,k} \stackrel{def}{=} \left.\begin{pmatrix} \lambda & k & \cdots & k \\ 0 & \lambda & \ddots & \vdots \\ \vdots & \ddots & \ddots & k \\ 0 & \cdots & 0 & \lambda \end{pmatrix}\right\} s.$$

It follows that B^{ℓ} is majorized by $M^{\ell}_{s,\lambda,k}$, hence using (13.2.1) we have

$$\left\|B^{\ell}\right\|_2 \leq \sqrt{\|B^{\ell}\|_1 \|B^{\ell}\|_\infty} \leq \sqrt{\left\|M^{\ell}_{s,\lambda,k}\right\|_1 \left\|M^{\ell}_{s,\lambda,k}\right\|_\infty} = \left\|M^{\ell}_{s,\lambda,k}\right\|_1,$$

and the latter is just the sum of the first row in $M^{\ell}_{s,\lambda,k}$. This is maximized for $s = r$, and equals

$$\sum_{t=0}^{r-1} \binom{r-1}{t}\binom{\ell}{t} k^t \lambda^{\ell-t} \leq \binom{\ell+r-1}{r-1} k^{r-1}\lambda^{\ell-r+1},$$

which gives the bound in the proposition. □

It is natural to ask whether symmetrization turns directed expanders into expanders, and we suspect that this is true for almost-normal aperiodic expanders in general. We can show that this is so for the symmetrization of a high enough power:

Proposition 13.4.2 *Let $\mathcal{D}$ be an aperiodic r-normal digraph with $\rho(\mathcal{D}) = \lambda$. If $\mathcal{G}_\ell$ is the symmetrization of the ℓ-th power of $\mathcal{D}$, namely $A_{\mathcal{G}_\ell} = A^{\ell}_{\mathcal{D}} + \left(A^{\ell}_{\mathcal{D}}\right)^T$, then*

$$\frac{\rho(\mathcal{G}_{r-1})}{\deg \mathcal{G}_{r-1}} = \frac{1}{2} + \frac{(r-1)^2}{2} \cdot \frac{\lambda}{k} + O\left(\left(\tfrac{\lambda}{k}\right)^2\right), \qquad \textit{and}$$

$$\frac{\rho(\mathcal{G}_r)}{\deg \mathcal{G}_r} = \frac{r\lambda}{k} + O\left(\left(\tfrac{\lambda}{k}\right)^2\right).$$

Proof. Observe that $\deg \mathcal{G}_\ell = 2k^\ell$. Maintaining the notations of the previous proof, we have by the same reasoning

$$\frac{1}{\deg \mathcal{G}_{r-1}} \left\| B^{r-1} + B^{*^{r-1}} \right\|_2 \leq \frac{1}{2k^{r-1}} \left\| M^{r-1}_{s,\lambda,k} + M^{*^{r-1}}_{s,\lambda,k} \right\|_1$$
$$= \frac{1}{2k^{r-1}} \left[\lambda^{r-1} + \sum_{t=0}^{r-1} \binom{r-1}{t}^2 k^t \lambda^{r-1-t} \right] = \frac{1}{2} + \frac{(r-1)^2}{2} \cdot \frac{\lambda}{k} + O\left(\left(\tfrac{\lambda}{k}\right)^2 \right).$$

and the computations for $\mathcal{G}_r$ are similar. □

We now prove an Alon–Boppana theorem for almost-normal digraphs:

Theorem 13.4.3 *Let $k \geq 2$ and $m \geq 1$. For any $\varepsilon > 0$, there is no infinite almost-normal family of k-regular m-periodic digraphs $\mathcal{D}$ with $\rho(\mathcal{D}) \leq \sqrt{k} - \varepsilon$.*

Proof. Let $\mathcal{D}$ be an r-normal, aperiodic k-regular digraph on n vertices and denote $\lambda = \rho(\mathcal{D})$ and $A = A_{\mathcal{D}}$. Let $\mathcal{G}$ be the graph whose adjacency matrix is $A^{*\ell} A^\ell$, for $\ell \geq r$ which will be determined later on. Namely, $V_{\mathcal{G}} = V_{\mathcal{D}}$, and each edge in $\mathcal{G}$ corresponds to a 2ℓ-path in $\mathcal{D}$ whose first ℓ steps are in accordance with the directions of the edges of $\mathcal{D}$, and the next ℓ steps are in discordance with them[(†)]. Since $\mathcal{G}$ is $k^{2\ell}$-regular, (13.3.4) gives

$$\rho(\mathcal{G}) \geq 2\sqrt{k^{2\ell} - 1} \left(1 - \frac{2\pi^2}{\log^2_{k^{2\ell}-1}(n/2)} \right).$$

On the other hand, Proposition 13.4.1 gives

$$\rho(\mathcal{G}) = \rho\left(A^{*\ell} A^\ell |_{L^2_0} \right) = \left\| A^\ell |_{L^2_0} \right\|^2 \leq \binom{\ell + r - 1}{r-1}^2 k^{2r-2} \lambda^{2(\ell - r + 1)},$$

and together we obtain for some $C_{k,r} > 0$

$$\lambda^{2(\ell-r+1)} \geq \frac{2\sqrt{k^{2\ell}-1}}{\binom{\ell+r-1}{r-1}^2 k^{2r-2}} \left(1 - \frac{2\pi^2}{\log^2_{k^{2\ell}-1}(n/2)} \right)$$
$$\geq \frac{C_{k,r} k^{\ell-r+1}}{\ell^{2r-2}} \left(1 - \frac{8(\pi \ell \ln k)^2}{\ln^2(n/2)} \right)$$
$$\implies \quad \lambda \geq \sqrt{k} \cdot \sqrt[2(\ell-r+1)]{\frac{C_{k,r}}{\ell^{2r-2}} \left(1 - \frac{8(\pi \ell \ln k)^2}{\ln^2(n/2)} \right)}.$$

(†) In particular, there are k^ℓ such closed path consisting of taking some ℓ-path and then retracing it backwards, so that one can even take the graph whose adjacency matrix is $A^{*\ell} A^\ell - k^\ell I$.

We finally choose $\ell = \sqrt{\ln(n/2)}$, obtaining

$$\lambda \geq \sqrt{k} \cdot \sqrt[2(\ell-r+1)]{\frac{C_{k,r}}{\ell^{2r-2}}\left(1 - \frac{8(\pi \ln k)^2}{\ln(n/2)}\right)} \xrightarrow{n\to\infty} \sqrt{k}.$$

This concludes the aperiodic case, and we leave the general one to the reader. □

13.5 Further Exploration

13.5.1 Universal Objects

The universal cover of all k-regular graphs is the k-regular tree $\mathcal{T}_k$; Ramanujan graphs are those which, save for the trivial eigenvalues, confine their spectrum to that of their forefather. It is possible to give an analogous interpretation for Ramanujan digraphs: consider the k-regular *directed* tree $\mathcal{T}_k^{\rightleftharpoons}$, which is obtained by choosing directions for the edges in $\mathcal{T}_{2k}$ to create a k-regular digraph. The spectrum of $\mathcal{T}_k^{\rightleftharpoons}$ was computed in [dlHRV93]:

$$\operatorname{Spec}\left(\mathcal{T}_k^{\rightleftharpoons}\right) = \left\{z \in \mathbb{C} \,\middle|\, |z| \leq \sqrt{k}\right\}, \tag{13.5.1}$$

so indeed a k-regular digraph is Ramanujan iff its non-trivial spectrum is contained in that of its 'universal directed cover' $\mathcal{T}_k^{\rightleftharpoons}$. However, one can also consider other universal objects: for example, the line digraph $\mathcal{D}_L(\mathcal{T}_{k+1})$ of the $k+1$-regular tree is a k-regular collision-free digraph which covers all of the digraphs obtained as line graphs of $(k+1)$-regular graphs (see Figure 13.2 for $k = 2$). Its spectrum is

$$\operatorname{Spec}(\mathcal{D}_L(\mathcal{T}_{k+1})) = \{\pm 1\} \cup \left\{z \in \mathbb{C} \,\middle|\, |z| = \sqrt{k}\right\},$$

and it contains the spectrum of all Ramanujan digraphs of the form $\mathcal{D}_L(\mathcal{G})$. It is also 2-normal: $L^2\left(V_{\mathcal{D}_L(\mathcal{T}_{k+1})}\right)$ decomposes as an orthogonal direct integral of one and two-dimensional spaces, each stable under the adjacency operator. Similarly, the digraph which describes the geodesic walk on the two-dimensional buildings of type $\widetilde{A}_2$ is 3-normal, and by computations in [KLW10] its spectrum is

$$\operatorname{Spec}\left(\mathcal{D}_W\left(\widetilde{A}_2\right)\right) = \left\{z \in \mathbb{C} \,\middle|\, |z| = \sqrt[4]{k}\right\} \cup \left\{z \in \mathbb{C} \,\middle|\, |z| = \sqrt{k}\right\} \tag{13.5.2}$$

(see Figure 13.3 (right) for a Ramanujan quotient of this digraph). One can continue to higher dimensions in this manner – see [Kan16, LLP19] for more details.

13.5.2 Universal Cayley Graphs

For even k, the k-regular tree $\mathcal{T}_k$ is the Cayley graph of $\mathbf{F}_{k/2}$, the free group on $S = \{x_1, \ldots, x_{k/2}\}$, with respect to the generating set $S \cup S^{-1} = \left\{x_1, \ldots, x_{k/2}, x_1^{-1}, \ldots, x_{k/2}^{-1}\right\}$ (see Figure 13.2). In fact, for any subset S of size $k/2$ in a group G, the following are tautologically equivalent:

(1) G is a free group and S is a free generating set.
(2) The Cayley graph $Cay\left(G, S \sqcup S^{-1}\right)$ is a tree.(†)

The following, however, is far from a tautology:

Theorem 13.5.1 ([Kes59]) *For* $\frac{k}{2} > 1$, (1) *and* (2) *above are equivalent to:*

(3) $\rho\left(A_{Cay(G,S \sqcup S^{-1})}\right) = 2\sqrt{k-1}$.

This does not say that $\mathcal{T}_k$ is the only k-regular graph with spectral radius $2\sqrt{k-1}$, but rather that among Cayley graphs it is the only one. In a sense, Keten's result says that the Ramanujan spectrum characterizes the free group. The analogue for directed graphs was revealed to be more complex in [dlHRV93]. First, observe that $\mathcal{T}_k^{\rightleftharpoons}$ is the Cayley digraph of the free group with respect to the *positive* generating letters:

$$\mathcal{T}_k^{\rightleftharpoons} = Cay\left(\mathbf{F}_k, \{x_1, \ldots, x_k\}\right).$$

As we have said, the spectral radius of $\mathcal{T}_k^{\rightleftharpoons}$ is $\sqrt{k}$, but it turns out that it is enough that S generate a free *semigroup* in order for this to happen:

Theorem 13.5.2 ([dlHRV93]) *Let S be a subset of size $k \geq 2$ in a group G. If S generates a free sub-semigroup of G, then*

$$\rho\left(A_{Cay(G,S)}\right) = \sqrt{k},$$

and if G has property (RD)(‡) *then the converse holds as well.*

For example, small cancellation theory shows that in the surface group of genus $g \geq 2$

$$S_g = \left\langle a_1, b_1, \ldots, a_g, b_g \,\middle|\, [a_1, b_1] \cdot \cdots \cdot [a_g, b_g]\right\rangle,$$

the elements $\{a_1, b_1, \ldots, a_g, b_g\}$ generate a free semigroup. Thus, the corresponding Cayley digraph of S_g has spectral radius $\sqrt{k}$ even though S_g is not free.

(†) Here $\sqcup$ indicates disjoint union, so that this is always a k-regular graph.

(‡) Property (RD), which stands for rapid decay, is satisfied both by hyperbolic groups and by groups of polynomial growth. For its definition we refer the reader to [Jol90, dlHRV93].

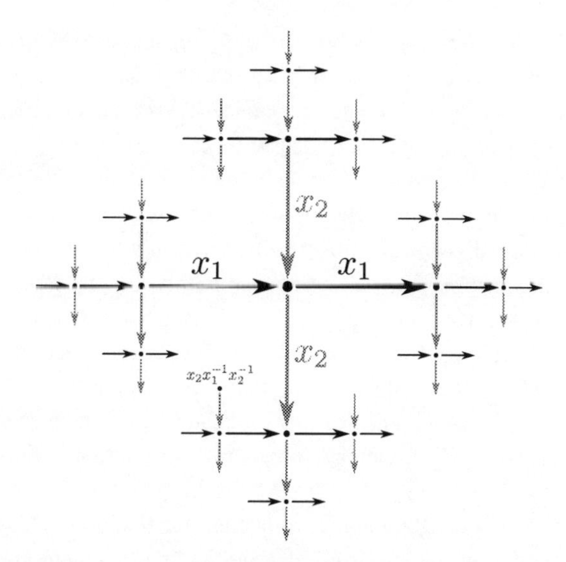

Figure 13.2 The directed tree $\mathcal{T}_2^{\rightleftharpoons}$ as $Cay\,(\mathbf{F}_2, \{x_1, x_2\})$

13.5.3 Explicit Constructions

For any k, [MSS15] shows the existence of infinitely many k-regular bipartite Ramanujan graphs, and thus there exist infinitely many k-regular, 2-periodic, 2-normal Ramanujan digraphs, namely their line-digraphs defined in Section 13.3.5. For any prime power k, the papers [LPS88, Mor94] give both aperiodic and 2-periodic k-regular Ramanujan digraphs, as line digraphs of explicit Cayley graphs.

Nevertheless, it is interesting to ask whether Ramanujan digraphs can be obtained as Cayley digraphs in themselves, and also which groups G has a generating set S such that $Cay\,(G, S)$ is an almost-normal Ramanujan digraph, as this gives the extremal results on random walk and diameter mentioned after the statement of Proposition 13.4.1.

For $k \in \{2, 3, 4, 5, 7, 11, 23, 59\}$, an infinite family of k-regular, 2-normal Ramanujan digraphs is constructed in [PS18, §5.2] as Cayley digraphs of $PSL_2\left(\mathbb{F}_q\right)$ and $PGL_2\left(\mathbb{F}_q\right)$. Each such family arises from a special arithmetic lattice in the projective unitary group $PU\,(2)$, which acts simply transitively on the directed edges of the tree $\mathcal{T}_{k+1}$, and whose torsion subgroup is a group of symmetries of a platonic solid. An example with $k = 4$ is shown in Figure 13.3.

In [Par20] we go much further, showing that for any prime power q and any $d \geq 2$ there is an explicit family of Cayley Ramanujan digraphs on $PSL_d\left(\mathbb{F}_{q^\ell}\right)$

and $PGL_d\left(\mathbb{F}_{q^\ell}\right)$ ($\ell \to \infty$), which are $k = q^{d-1}$-regular and sharply d-normal. As explained in Section 13.4, this implies that they have sharp L^1-cutoff at time $\log_k n$, and that their diameter is bounded by $(2 + o(1)) \log_k(n)$. This is quite different from the symmetric case: we have no reason to suspect that $PSL_d\left(\mathbb{F}_{q^\ell}\right)$ can be endowed with a structure of a Ramanujan Cayley graph, for $d \geq 3$. Let us sketch the main ideas: In [CS98, LSV05b] appears an arithmetic lattice Γ in a certain division algebra, which acts simply-transitively on the vertices of the building of type $\widetilde{A}_{d-1}$ associated with the group $PSL_d\left(\mathbb{F}_q((t))\right)$. This lattice can be enlarged to a lattice $\Gamma < \Gamma'$, which acts simply-transitively on the edges of color 1 in the same building. Recall from Section 13.3.6 that the geodesic walk on these edges is k-regular and collision-free. We take a set of generators $S \subseteq \Gamma'$ which induces this walk, and regard them as elements in the finite group $PSL_d\left(\mathbb{F}_{q^\ell}\right)$, which is obtained as a congruence quotient of Γ' via strong approximation. We then invoke the Jacquet–Langlands correspondence of [BR17] and the Ramanujan conjecture for function fields [Laf02] to deduce that the non-trivial spectrum of S on the finite quotient group is contained in the spectrum of S acting on the building, thus obtaining a Ramanujan digraph. Finally, sharp d-normality follows from [LLP19, Prop. 5.3, 5.4]. An example with $d = 3$ is shown in Figure 13.3, agreeing with the spectrum of geodesic walk on the $\widetilde{A}_2$ building shown in (13.5.2).

13.5.4 Riemann Hypothesis

We briefly mention the perspective of zeta functions – for a lengthier discussion see [LLP19, §6] and [KLW10, Kan16, Kam16]. Ihara [Iha66] associated with a graph $\mathcal{G}$ a zeta function $\zeta_{\mathcal{G}}(u)$ which counts closed cycles in $\mathcal{G}$, in analogy with the Selberg zeta function of a hyperbolic surface. If $\mathcal{G}$ is $(\boldsymbol{k+1})$-regular, it is Ramanujan if and only if $\zeta_{\mathcal{G}}(u)$ satisfies the following 'Riemann hypothesis': every pole at $\zeta_{\mathcal{G}}\left(k^{-s}\right)$ with $0 < \Re s < 1$ satisfies $\Re s = \frac{1}{2}$. Indeed, Hashimoto [Has89] proved that $\zeta_{\mathcal{G}}(u) = \det\left(I - u \cdot A_{\mathcal{D}_L(\mathcal{G})}\right)^{-1}$, so that (13.3.1) and (13.3.2) show this equivalence (note that the trivial eigenvalues $\pm k$ of $\mathcal{D}_L(\mathcal{G})$ and the eigenvalues ± 1 in (13.3.1) correspond to $s = 1$ and $s = 0$, respectively). For digraphs the story is simpler: the zeta function $Z_{\mathcal{D}}(u)$ of a digraph $\mathcal{D}$ (following [BL70, Has89, KS00]) is $Z_{\mathcal{D}}(u) = \prod_{[\gamma]}\left(1 - u^{\ell(\gamma)}\right)^{-1}$, where γ is a primitive directed cycle of length $\ell(\gamma)$ in $\mathcal{D}$, and $[\gamma]$ is the equivalence class of its cyclic rotations. One then has $Z_{\mathcal{D}}(u) = \det(I - u \cdot A_{\mathcal{D}})^{-1}$, so that by (13.1.1) a k-regular digraph $\mathcal{D}$ is Ramanujan if and only if every pole

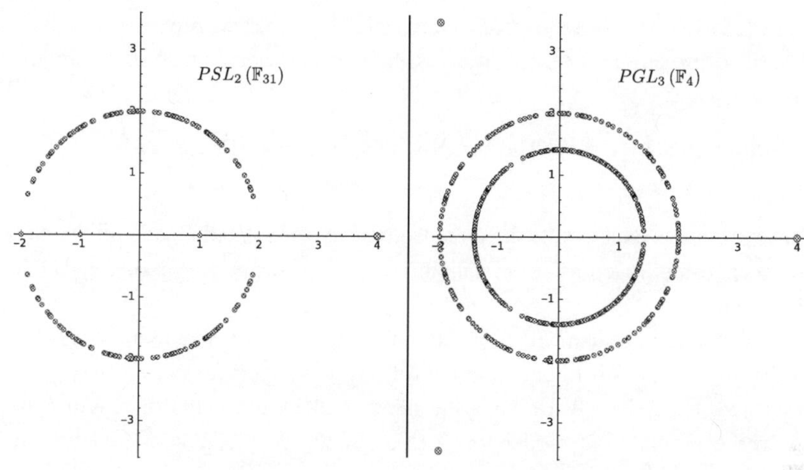

Figure 13.3 Examples of spectra of Ramanujan Cayley digraphs: $PSL_2(\mathbb{F}_{31})$ with generators $\left\{\left(\begin{smallmatrix}28 & 4\\ 12 & 4\end{smallmatrix}\right), \left(\begin{smallmatrix}15 & 13\\ 10 & 15\end{smallmatrix}\right), \left(\begin{smallmatrix}6 & 18\\ 18 & 13\end{smallmatrix}\right), \left(\begin{smallmatrix}7 & 3\\ 11 & 5\end{smallmatrix}\right)\right\}$ from [PS18], and $PGL_3(\mathbb{F}_4)$ with generators $\left\{\left(\begin{smallmatrix}0 & 0 & 1\\ 0 & x & 0\\ x & x+1 & x\end{smallmatrix}\right), \left(\begin{smallmatrix}1 & 1 & 1\\ x+1 & x+1 & 1\\ x & 0 & x\end{smallmatrix}\right), \left(\begin{smallmatrix}1 & 1 & 0\\ x+1 & 1 & 1\\ 0 & x+1 & 0\end{smallmatrix}\right), \left(\begin{smallmatrix}0 & 1 & x\\ x & x+1 & 0\\ 1 & x & x\end{smallmatrix}\right)\right\}$ from [Par20]

at $Z_{\mathcal{D}}\left(k^{-s}\right)$ satisfies $\Re s = 1$ or $0 \leq \Re s \leq \frac{1}{2}$. The fact that we cannot rule out s with $0 < \Re s < \frac{1}{2}$ is demonstrated by (13.5.2), for example.

13.5.5 Alon's Conjecture

One of the earliest results on graph expansion is that random regular graphs are expanders [KB67, Pin73]. In [Alo86], Alon conjectured that they are in fact almost Ramanujan. Namely, for any $\varepsilon > 0$

$$\operatorname{Prob}\left[\rho(\mathcal{G}) \leq 2\sqrt{k-1}+\varepsilon\right] \overset{n\to\infty}{\longrightarrow} 1 \qquad \left(\begin{smallmatrix}\text{where } \mathcal{G} \text{ is a random}\\ k\text{-regular graph on } n \text{ vertices}\end{smallmatrix}\right). \tag{13.5.3}$$

Alon's conjecture was eventually proved by Friedman [Fri08], and other proofs followed [FK14, Bor15]. While working on the paper [LLP19], the author conjectured that random regular digraphs are almost Ramanujan as well, in the sense that

$$\operatorname{Prob}\left[\rho(\mathcal{D}) \leq \sqrt{k}+\varepsilon\right] \overset{n\to\infty}{\longrightarrow} 1 \qquad \left(\begin{smallmatrix}\text{where } \mathcal{D} \text{ is a random}\\ k\text{-regular digraph on } n \text{ vertices}\end{smallmatrix}\right), \tag{13.5.4}$$

for any $\varepsilon > 0$; and furthermore, that they behave as almost-normal digraphs, in the sense that the operator norm of their powers is well behaved as in Proposition 13.4.1. In joint work with Doron Puder we tried to extend the methods from [Pud15] to prove this conjecture, and made partial progress which is described further. This project was disrupted by the appearance of a solution on the arXiv:

Theorem 13.5.3 ([Cos17, Theorem 1 with $\delta = \Delta = k$]) *Statement (13.5.4) is true.*

Since our methods are quite different from the ones in [Cos17], and might lead to other results (such as understanding of the adjacency-powers), we sketch them here.

In the seminal paper [BS87], the value of $\rho(\mathcal{G})$ for a random k-regular graph on n vertices is bounded in the following manner: for even ℓ, $\rho(\mathcal{G})^{\ell} \leq \operatorname{tr}\left(A^{\ell}\right) - k^{\ell}$, and $\operatorname{tr}\left(A^{\ell}\right)$ equals the number of closed paths of length ℓ in $\mathcal{G}$. In the permutation model for $\mathcal{G}$ (see [BS87, Wor99, Pud15]) each path of length ℓ is determined by a starting vertex, and a word ω of length ℓ in $S = \left\{x_1^{\pm 1}, \ldots, x_{k/2}^{\pm 1}\right\}$. If ω is trivial as an element of $\mathbf{F}_{k/2}$, this path is completely backtracking in every instance of $\mathcal{G}$, and in particular closed. Denoting $p_\omega = \operatorname{Prob}\left(\begin{smallmatrix}\text{a path in } \mathcal{G} \text{ which starts at } v \\ \text{and is labelled by } \omega \text{ ends at } v\end{smallmatrix}\right) - \frac{1}{n}$, one obtains

$$\mathbb{E}\left(\rho(\mathcal{G})^{\ell}\right) \leq \mathbb{E}\left(\operatorname{tr}\left(A^{\ell}\right)\right) - k^{\ell} = n \sum_{\omega \in S^{\ell}} p_\omega,$$

and each trivial ω contributes $p_\omega = 1 - \frac{1}{n}$. In [BS87] it is shown that p_ω is small for words which are not trivial or proper powers in $\mathbf{F}_{k/2}$, and the number of trivial and power words is bounded, giving a bound on $\mathbb{E}(\rho(\mathcal{G})^{\ell})$. An appropriate choice of ℓ then implies $\rho(\mathcal{G}) \leq 3k^{3/4}$ a.a.s. as $n \to \infty$. In [Pud14, PP15] it is shown that p_ω depends on the so-called *primitivity rank* of ω, and in [Pud15] this is made qualitatively precise, and words of each primitivity rank are counted, leading to $\rho(\mathcal{G}) \leq 2\sqrt{k-1} + 1$ a.a.s.

Now, let $\mathcal{D}$ be a k-regular digraph on n vertices, so that $A_{\mathcal{D}}$ is simply the sum of k independent $n \times n$ permutation matrices. We cannot use $\operatorname{tr}\left(A^{\ell}\right)$ directly to bound $\rho(\mathcal{G})$, since A is not normal anymore. Instead, denoting $A_0 = A\big|_{L_0^2}$ we use Gelfand's formula:

$$\sqrt[2\ell]{\rho\left(A_0^{*\ell} A_0^{\ell}\right)} = \sqrt[\ell]{\left\|A_0^{\ell}\right\|} \underset{\ell \to \infty}{\searrow} \rho(A_0),$$

and to bound $\rho\left(A_0^{*\ell} A_0^{\ell}\right)$ we study $\operatorname{tr}\left((A^{*\ell} A^{\ell})^t\right)$. The entries of $\left(A^{*\ell} A^{\ell}\right)^t$ correspond to 'ℓ-alternating words' of length $2\ell t$: words in $S = \left\{x_1^{\pm}, \ldots, x_k^{\pm}\right\}$ which are composed of alternating sequences of ℓ negative letters followed by ℓ positive ones. Given a starting vertex, each such word translates to a path in $\mathcal{D}$, where negative letters indicate crossing a directed edge in the 'wrong' direction. Again p_ω is the probability that this path is closed, so that

$$\rho\left(\left(A_0^{*\ell}A_0^{\ell}\right)^t\right) \le \operatorname{tr}\left(\left(A^{*\ell}A^{\ell}\right)^t\right) - k^{2\ell t} = n \cdot \sum_{\omega \in (S_+^{\ell} \times S_-^{\ell})^t} p_w.$$

Now, $\mathbb{E}\left(\operatorname{tr}\left(\left(A^{*\ell}A^{\ell}\right)^t\right)\right)$ can be bounded similarly to [Pud15], this time by counting ℓ-alternating words of each primitivity rank, and choosing both ℓ and t carefully. We discovered that already from $\ell = 2$ one obtains the bound $\rho(\mathcal{D}) \le \sqrt{2k} + \varepsilon$ a.a.s., and we expect that as ℓ goes to infinity one should recover (13.5.4) up to an additive constant. As remarked above, this analysis goes through the spectral norm of A^{ℓ}, so it might lead to other interesting results on $\mathcal{D}$.

13.6 Questions

(1) A *non-regular* graph $\mathcal{G}$ is said to be Ramanujan if its non-trivial spectrum is contained in the L^2-spectrum of its universal cover (which is a non-regular tree). This definition is justified both by the extended Alon–Boppana theorem [Gre95, GŻ99] and by the behaviour of random covers [Fri03, Pud15, FK14, BDH18]. What is the appropriate definition of a non-regular Ramanujan digraph?

(2) Can standard results on expanders (such as the Cheeger inequalities and the expander mixing lemma) be extended to almost-normal directed expanders?

(3) Does symmetrization turn a family of almost-normal directed expanders into a family of expanders?

(4) Are there infinite almost-normal families of non-periodic k-regular Ramanujan digraphs for k which is not a prime power?

(5) Is there an almost-normal family of k-regular digraphs with $n \to \infty$ whose non-trivial spectrum is contained in the circle $\{z \mid |z| = \sqrt{k}\}$?

(6) Almost-normality is an 'algebraic' phenomenon: it originates from representation theory in [LLP19], and from the special structure of line-digraphs in [LP16]. There seems to be no reason that random models will have this property, or that it will be stable under perturbations. Is there a more flexible definition of almost-normality, which still gives a theorem in the spirit of Proposition 13.4.1?

(7) How important is almost-normality? Is there a family of k-regular Ramanujan digraphs which behave like amenable graphs in terms of expansion?

(8) In the infinite case, we can even ask whether a *single* digraph is almost-normal, meaning that its adjacency operator is unitarily equivalent

to a direct integral of linear operators of bounded dimension. Such digraphs can be obtained by taking line-digraphs of infinite graphs, or more generally $\mathcal{D}_W(\mathcal{B})$ where W is some walk on a building $\mathcal{B}$ (or an infinite quotient of a building). To what extent does the spectral theory of infinite symmetric graphs carries over to almost-normal digraphs?

Acknowledgements

We would like to express our gratitude to Noga Alon, Amitay Kamber, Alex Lubotzky, Doron Puder, and Alain Valette for various helpful remarks and suggestions.

References

[ABLS07] N. Alon, I. Benjamini, E. Lubetzky, and S. Sodin, *Non-backtracking random walks mix faster*, Communications in Contemporary Mathematics **9** (2007), no. 04, 585–603.

[Alo86] N. Alon, *Eigenvalues and expanders*, Combinatorica **6** (1986), no. 2, 83–96.

[Bas92] H. Bass, *The Ihara-Selberg zeta function of a tree lattice*, International Journal of Mathematics **3** (1992), no. 06, 717–797.

[BDH18] G. Brito, I. Dumitriu, and K. D. Harris, *Spectral gap in random bipartite biregular graphs and its applications*, arXiv:1804.07808 (2018).

[BL70] R. Bowen and O Lanford, *Zeta functions of restrictions of the shift transformation*, Proc. Symp. Pure Math, 1970, pp. 43–50.

[Bor15] C. Bordenave, *A new proof of Friedman's second eigenvalue theorem and its extension to random lifts*, arXiv:1502.04482 (2015).

[BR17] A. I. Badulescu and P. Roche, *Global Jacquet-Langlands correspondence for division algebras in characteristic p*, International Mathematics Research Notices **2017** (2017), no. 7, 2172–2206.

[BS87] A. Broder and E. Shamir, *On the second eigenvalue of random regular graphs*, 28th Annual Symposium on Foundations of Computer Science, 1987, pp. 286–294.

[Cos17] S. Coste, *The spectral gap of sparse random digraphs*, arXiv:1708.00530 (2017).

[CS98] D.I. Cartwright and T. Steger, *A family of $\tilde{A}_n$-groups*, Israel J. Math. **103** (1998), no. 1, 125–140.

[dlHRV93] P. de la Harpe, A. G. Robertson, and A. Valette, *On the spectrum of the sum of generators of a finitely generated group, II*, Colloquium Mathematicum **65** (1993), no. 1, 87–102.

[DT98] C. Delorme and J.P. Tillich, *The spectrum of de Bruijn and Kautz graphs*, European Journal of Combinatorics **19** (1998), no. 3, 307–319.

[Fir16] U.A. First, *The Ramanujan property for simplicial complexes*, arXiv:1605.02664 (2016).

[FK14] J. Friedman and D.-E. Kohler, *The relativized second eigenvalue conjecture of Alon*, arXiv:1403.3462 (2014).

[FL92] K. Q. Feng and W.-C. W. Li, Appendix to *Character sums and abelian Ramanujan graphs*, J. Number Theory **41** (1992), no. 2, 199–217.

[Fri03] J. Friedman, *Relative expanders or weakly relatively Ramanujan graphs*, Duke Mathematical Journal **118** (2003), no. 1, 19–35.

[Fri08] ———, *A proof of Alon's second eigenvalue conjecture and related problems*, Mem. Amer. Math. Soc. **195** (2008), no. 910, viii+100.

[FZ99] D. Foata and D. Zeilberger, *A combinatorial proof of Bass's evaluations of the Ihara-Selberg zeta function for graphs*, Transactions of the A.M.S. **351** (1999), no. 6, 2257–2274.

[Gre95] Y. Greenberg, *On the spectrum of graphs and their universal covering*, Ph.D. Thesis, 1995.

[GS71] R. L Graham and J. H Spencer, *A constructive solution to a tournament problem*, Canad. Math. Bull **14** (1971), no. 1, 45–48.

[GŻ99] R.I. Grigorchuk and A. Żuk, *On the asymptotic spectrum of random walks on infinite families of graphs*, Random walks and discrete potential theory (Cortona, 1997), Sympos. Math **39** (1999), 188–204.

[Has89] K. Hashimoto, *Zeta functions of finite graphs and representations of p-adic groups*, Automorphic Forms and Geometry of Arithmetic Varieties, 1989, pp. 211–280.

[HLW06] S. Hoory, N. Linial, and A. Wigderson, *Expander graphs and their applications*, Bulletin of the American Mathematical Society **43** (2006), no. 4, 439–562.

[HS60] A. J. Hoffman and R. R. Singleton, *On Moore graphs with diameters 2 and 3*, IBM J. Res. Dev. **4** (1960), no. 5, 497–504.

[Iha66] Y. Ihara, *On discrete subgroups of the two by two projective linear group over p-adic fields*, Journal of the Mathematical Society of Japan **18** (1966), no. 3, 219–235.

[Jol90] P. Jolissaint, *Rapidly decreasing functions in reduced C*-algebras of groups*, Transactions of the A.M.S. **317** (1990), no. 1, 167–196.

[Kam16] A. Kamber, *L_p-expander complexes*, arXiv:1701.00154 (2016).

[Kan16] M.H. Kang, *Riemann Hypothesis and strongly Ramanujan complexes from GL_n*, Journal of Number Theory **161** (2016), 281–297.

[KB67] A.N. Kolmogorov and Ya. M. Barzdin, *On the realization of nets in 3-dimensional space*, Problems in Cybernetics **8** (1967), 261–268.

[Kes59] H. Kesten, *Symmetric random walks on groups*, Trans. Amer. Math. Soc. **92** (1959), 336–354.

[KLW10] M.H. Kang, W.C.W. Li, and C.J. Wang, *The zeta functions of complexes from* PGL(3)*: a representation-theoretic approach*, Israel J. Math. **177** (2010), no. 1, 335–348.

[KS00] M. Kotani and T. Sunada, *Zeta functions of finite graphs*, J. Math. Sci. Univ. Tokyo **7** (2000), 7–25.

[Laf02] L. Lafforgue, *Chtoucas de Drinfeld et correspondance de Langlands*, Inventiones mathematicae **147** (2002), no. 1, 1–241.

[Li04] W.C.W. Li, *Ramanujan hypergraphs*, Geometric and Functional Analysis **14** (2004), no. 2, 380–399.

[LLP19] E. Lubetzky, A. Lubotzky, and O. Parzanchevski, *Random walks on Ramanujan complexes and digraphs*, Journal of the European Mathematical Society, to appear (2019), arXiv:1702.05452.

[LP16] E. Lubetzky and Y. Peres, *Cutoff on all Ramanujan graphs*, Geometric and Functional Analysis **26** (2016), no. 4, 1190–1216.

[LPS88] A. Lubotzky, R. Phillips, and P. Sarnak, *Ramanujan graphs*, Combinatorica **8** (1988), no. 3, 261–277.

[LSV05a] A. Lubotzky, B. Samuels, and U. Vishne, *Ramanujan complexes of type* $\tilde{A}_d$, Israel J. Math. **149** (2005), no. 1, 267–299.

[LSV05b] *Explicit constructions of Ramanujan complexes of type* $\tilde{A}_d$, Eur. J. Comb. **26** (2005), no. 6, 965–993.

[Lub12] A. Lubotzky, *Expander graphs in pure and applied mathematics*, Bull. Amer. Math. Soc **49** (2012), 113–162.

[Lub14] *Ramanujan complexes and high dimensional expanders*, Japanese Journal of Mathematics **9** (2014), no. 2, 137–169.

[Lub94] *Discrete groups, expanding graphs and invariant measures*, Modern Birkhäuser Classics, Birkhäuser Verlag, Basel, 1994, With an appendix by Jonathan D. Rogawski.

[Mar88] G. A. Margulis, *Explicit group-theoretical constructions of combinatorial schemes and their application to the design of expanders and concentrators*, Problemy Peredachi Informatsii **24** (1988), no. 1, 51–60.

[Mor94] M. Morgenstern, *Existence and explicit constructions of $q+1$ regular Ramanujan graphs for every prime power q*, Journal of Combinatorial Theory, Series B **62** (1994), no. 1, 44–62.

[MSS15] A. Marcus, D.A. Spielman, and N. Srivastava, *Interlacing families I: Bipartite Ramanujan graphs of all degrees*, Annals of Mathematics **182** (2015), 307–325.

[Nil04] A. Nilli, *Tight estimates for eigenvalues of regular graphs*, The Electronic Journal of Combinatorics **11** (2004), no. 1, 9.

[Nil91] *On the second eigenvalue of a graph*, Discrete Mathematics **91** (1991), no. 2, 207–210.

[Par20] O. Parzanchevski, *Optimal generators for matrix groups* (2019). In preparation.

[Pin73] M. S Pinsker, *On the complexity of a concentrator*, 7th international telegraffic conference, 1973, pp. 1–318.

[PP15] D. Puder and O. Parzanchevski, *Measure preserving words are primitive*, Journal of the American Mathematical Society **28** (2015), no. 1, 63–97.

[PR20] O. Parzanchevski and R. Rosenthal, *Chernoff bound for non-symmetric walks* (2019). In preparation.

[PS18] O. Parzanchevski and P. Sarnak, *Super-Golden-Gates for PU(2)*, Advances in Mathematics **327** (2018), 869–901, Special volume honoring David Kazhdan.

[Pud14] D. Puder, *Primitive words, free factors and measure preservation*, Israel Journal of Mathematics **201** (2014), no. 1, 25–73.

[Pud15] ———, *Expansion of random graphs: new proofs, new results*, Inventiones Mathematicae **201** (2015), no. 3, 845–908.

[ST96] H. M Stark and A. A Terras, *Zeta functions of finite graphs and coverings*, Advances in Mathematics **121** (1996), no. 1, 124–165.

[Sun86] T. Sunada, *L-functions in geometry and some applications*, Curvature and Topology of Riemannian Manifolds, 1986, pp. 266–284.

[Vu08] V. H. Vu, *Sum-product estimates via directed expanders*, Math. Res. Lett. **15** (2008), no. 2, 375–388.

[Wor99] N. C. Wormald, *Models of random regular graphs*, Surveys in Combinatorics, 1999, pp. pp. 239–298.

14

From Partial Differential Equations to Groups

Andrzej Zuk

Abstract

We discuss a construction which associates to a KdV equation the lamplighter group. In order to establish this relation, one uses automata and random walks on ultra-discrete limits. We present it in a more general context.

14.1 Introduction

The connection between discrete and continuous objects is a fascinating topic of studies in mathematics. One of the motivations of seeking such a relation comes from physics. The partial differential equations describing the evolution of some systems are obtained as the scaling limits (dilations both in spatial and temporal variables) of some difference equations.

In the group theory setting, the scaling limits can be applied to virtually nilpotent countable groups to end up with nilpotent Lie groups.

Another example of the relation between discrete and continuous groups is the theory of lattices in semi-simple Lie groups of higher rank. The link is provided by the ergodic theory. A spectacular application of this theory is Margulis classification of such lattices (arithmeticity theorem).

In this chapter we are dealing with ultra-discrete versions of partial differential equations and we consider the case of the KdV equation which was worked out in our paper [2]. We provide here a complementary information and the reader is encouraged to look at [2] while studying this chapter.

First of all we present a detailed proof of the fact that the Ball Box System is an ultra-discrete version of the KdV equation (this is done in Section 14.2).

An important part of the construction consists of relating the ultra-discrete equation to automata. In our case the basic object of the study is a two-state automaton. In Section 14.3 we present groups which arise as groups generated by two-state automata. Among them is the lamplighter group which we related with the BBS and which is presented in detail in Section 14.4. One of its remarkable features is that it can be embedded into a finitely presented group or equivalently a fundamental group of some closed manifold. This fact is very important as our computation can be related to the approximation result concerning the L^2 Betti numbers of closed manifolds (see [2] and [10]).

In case of the Box Ball System (BBS), this construction would only lead to a semigroup. In order to obtain a group (in our case the lamplighter group), one needs to prove that matrices obtained by recurrence rules from the automaton are bi-stochastic. In Section 14.5 we present other three-state automata which have this property. We call them BBS-V and BBS-S [6, 7].

Among different features of ultra-discrete limits which make them very appealing is the role played by time. In PDEs time is just another variable denoted usually by t. In the ultra-discrete version, time corresponds to the number of iterations and relates to deep questions from the complexity theory.

14.2 Ultradiscretization and BBS

In the discrete version of partial differential equations, time and spatial variables take integer values, however, the functions still take real values. The derivative is replaced by the difference $f(x+1) - f(x)$. Formally, one obtains a PDE from the discrete version by replacing 1 by ε, dividing the expressions by suitable powers of ε and taking the limit when ε tends to 0.

In the ultra-discrete version of a PDE not only time and space variables take integer values but also a function and actually only a finite number of values, say 0 and 1. It is obtained from a discrete version by taking not a scaling limit but a limit in the sense of tropical geometry as we will see in this section. Taking such a limit changes also algebraic operations. We give here a detailed proof that the BBS is an ultra-discrete version of the KdV equation.

BBS [5, 9, 8] is one of the fundamental ultradiscrete integrable sytems derived from the discrete KdV equation:

$$\frac{1}{u_{n+1}^{(s+1)}} - \frac{1}{u_n^{(s)}} = \frac{\delta}{1+\delta}\left(u_{n+1}^{(s)} - u_n^{(s+1)}\right). \tag{14.2.1}$$

By taking the contiuum limit $\epsilon \to 0$ of (14.2.1) with $\delta = \epsilon^3, t = -\epsilon^6 s/3, x = \epsilon(n-s) - 2\epsilon^4 s, u_n^{(s)} = 1 + \epsilon^2 u(x,t)/w$, we obatin the KdV equation

$$\frac{\partial u}{\partial t} = 6u\frac{\partial u}{\partial x} + \frac{\partial^3 u}{\partial x^3}. \tag{14.2.2}$$

Here we review the procedure to derive an extended BBS from the discrete integrable systems. Let us start from the Hirota–Miwa equation [1, 3]

$$\begin{aligned} &\tau(k_1+1,k_2+1,k_3)\tau(k_1,k_2,k_3+1) \\ &-\tau(k_1+1,k_2,k_3+1)\tau(k_1,k_2+1,k_3) \\ &+\tau(k_1,k_2+1,k_3+1)\tau(k_1+1,k_2,k_3) = 0, \end{aligned} \tag{14.2.3}$$

where $k_1, k_2, k_3 \in \mathbb{Z}$ denote the independent discrete variables. Choosing the dependent variables by

$$u_{12} = \frac{\tau^{+k_1+k_2}\tau}{\tau^{+k_1}\tau^{+k_2}}, \quad u_{13} = \frac{\tau^{+k_1+k_3}\tau}{\tau^{+k_1}\tau^{+k_3}}, \quad u_{23} = \frac{\tau^{+k_2+k_3}\tau}{\tau^{+k_2}\tau^{+k_3}}, \tag{14.2.4}$$

where $\tau^{+k_1} = \tau(k_1+1,k_2,k_3)$, $\tau^{+k_1+k_2} = \tau(k_1+1,k_2+1,k_3)$ and ..., then we obtain the discrete KP lattice

$$u_{12} - u_{13} + u_{23} = 0, \quad \frac{u_{12}^{+k_3}}{u_{12}} = \frac{u_{13}^{+k_2}}{u_{13}} = \frac{u_{23}^{+k_1}}{u_{23}}, \tag{14.2.5}$$

which admits the casorati determinant solutions exhibiting the multi-soliton interactions. For later convenience, we introduce the gauge transformation

$$\tau \to \rho\tau \tag{14.2.6}$$

where $\rho = \rho(k_1, k_2, k_3)$ is some solution of the Hirota–Miwa equation (14.2.3). Note that the gauge factor ρ amounts to the 'vacuum' solution of the discrete KP lattice,

$$\widetilde{u}_{12} = \frac{\rho^{+k_1+k_2}\rho}{\rho^{+k_1}\rho^{+k_2}}, \quad \widetilde{u}_{13} = \frac{\rho^{+k_1+k_3}\rho}{\rho^{+k_1}\rho^{+k_3}}, \quad \widetilde{u}_{23} = \frac{\rho^{+k_2+k_3}\rho}{\rho^{+k_2}\rho^{+k_3}}, \tag{14.2.7}$$

and the Hirota–Miwa equation can be expressed as

$$\begin{aligned} &\widetilde{u}_{12}(k_1,k_2,k_3)\,\tau(k_1+1,k_2+1,k_3)\tau(k_1,k_2,k_3+1) \\ &-\widetilde{u}_{13}(k_1,k_2,k_3)\,\tau(k_1+1,k_2,k_3+1)\tau(k_1,k_2+1,k_3) \\ &+\widetilde{u}_{23}(k_1,k_2,k_3)\,\tau(k_1,k_2+1,k_3+1)\tau(k_1+1,k_2,k_3) = 0, \end{aligned} \tag{14.2.8}$$

which is called as 'non-autonoumous' Hirota–Miwa equation. These gauge factors normalize the vaccuum solution of (14.2.8) as $\tau = 1$.

In order to derive the discrete integrable systems related to the BBS with carrier capacity K, we take

$$\widetilde{u}_{12} = a_1(k_1) - a_2(k_2), \quad \widetilde{u}_{13} = a_1(k_1) - a_3(k_3), \quad \widetilde{u}_{23} = a_2(k_2) - a_3(k_3), \tag{14.2.9}$$

where $a_j(k_j)$ is a function of k_j $(j = 1, 2, 3)$, and impose the reduction conditions

$$\tau(k_1, k_2 + 2, k_3) \simeq \tau(k_1, k_2, k_3) \tag{14.2.10}$$

where $f \simeq g$ means that f and g differ only in trivial factor and satisfy the same equation. Under the reduction conditions (14.2.10), (14.2.8) can be divided into the system of equations:

$$\begin{aligned}
&(a_1^{(s)} - a_2^{(0)})\,\sigma_n^{(s+1)}\tau_{n+1}^{(s)} - (a_1^{(s)} - a_3^{(n)})\,\tau_{n+1}^{(s+1)}\sigma_n^{(s)} \\
&\qquad + (a_2^{(0)} - a_3^{(n)})\,\sigma_{n+1}^{(s)}\tau_n^{(s+1)} = 0, \\
&(a_1^{(s)} - a_2^{(1)})\,\tau_n^{(s+1)}\sigma_{n+1}^{(s)} - (a_1^{(s)} - a_3^{(n)})\,\sigma_{n+1}^{(s+1)}\tau_n^{(s)} \\
&\qquad + (a_2^{(1)} - a_3^{(n)})\,\tau_{n+1}^{(s)}\sigma_n^{(s+1)} = 0,
\end{aligned} \tag{14.2.11}$$

where $\tau_n^{(s)} = \tau(k_1 + s, 0, k_3 + n)$, $\sigma_n^{(s)} = \tau(k_1 + s, 1, k_3 + n)$, $a_1^{(s)} = a_1(k_1 + s)$, $a_2^{(j)} = a_2(k_2 + j)$, $a_3^{(n)} = a_3(k_3 + n)$. Then, by employing the dependent variables $u_{12} = (a_1 - a_2)v$, $u_{23} = (a_2 - a_3)w$, one can obtain the discrete sytems

$$\frac{v_{n+1}^{(s)}}{v_n^{(s)}} = \frac{(a_1^{(s)} - a_2^{(1)})/v_n^{(s)} + (a_2^{(1)} - a_3^{(n)})/w_n^{(s)}}{(a_1^{(s)} - a_2^{(0)})v_n^{(s)} + (a_2^{(0)} - a_3^{(n)})w_n^{(s)}} = \frac{w_n^{(s+1)}}{w_n^{(s)}} \tag{14.2.12}$$

from the discrete KP lattice (14.2.5). We call (14.2.12) discrete modified KdV lattice. Under the boundary conditions $\lim_{n\to\infty} v_n^{(s)} = 1$, the discrete modified KdV lattice (14.2.12) can be written as

$$\log \frac{w_n^{(s+1)}}{w_n^{(s)}} = \log \frac{(a_1^{(s)} - a_2^{(1)}) \displaystyle\prod_{j=-\infty}^{n-1} \frac{w_j^{(s)}}{w_j^{(s+1)}} + \frac{a_2^{(1)} - a_3^{(n)}}{w_n^{(s)}}}{(a_1^{(s)} - a_2^{(0)}) \displaystyle\prod_{j=-\infty}^{n-1} \frac{w_j^{(s+1)}}{w_j^{(s)}} + (a_2^{(0)} - a_3^{(n)})w_n^{(s)}}, \tag{14.2.13}$$

where we have used the relation

$$v_n^{(s)} = \frac{w_{n-1}^{(s+1)}}{w_{n-1}^{(s)}} v_{n-1}^{(s)} = \frac{w_{n-1}^{(s+1)}}{w_{n-1}^{(s)}} \frac{w_{n-2}^{(s+1)}}{w_{n-2}^{(s)}} v_{n-2}^{(s)} = \cdots = \prod_{j=-\infty}^{n-1} \frac{w_j^{(s+1)}}{w_j^{(s)}}. \tag{14.2.14}$$

Now we come to the last stage of the ultra-discretization procedure. Let us assume that $w_n^{(s)} > 0$, $a_2^{(1)} - a_2^{(0)} = 1$ and $a_1^{(s)} - a_2^{(1)}, a_2^{(0)} - a_3^{(n)} \in (0, 1)$, which allow us to substitute

$$w_n^{(s)} = \exp(B_n^{(s)}/\epsilon), \quad a_1^{(s)} - a_2^{(1)} = \exp(-K/\epsilon), \quad a_2^{(0)} - a_3^{(n)} = \exp(-1/\epsilon) \tag{14.2.15}$$

into (14.2.13). Then taking $\epsilon \to +0$, we obtain the BBS with carrier capacity K[4]

$$B_n^{(s+1)} - B_n^{(s)} = \max\left(-K + \sum_{j=-\infty}^{n-1} (B_j^{(s)} - B_j^{(s+1)}), -B_n^{(s)}\right) \quad (14.2.16)$$

$$- \max\left(\sum_{j=-\infty}^{n-1} (B_j^{(s+1)} - B_j^{(s)}), B_n^{(s)} - 1\right) \quad (14.2.17)$$

by using the simple formula[9]

$$\lim_{\epsilon \to +0} \log\left(\exp(A/\epsilon) + \exp(B/\epsilon)\right) = \max(A, B) \quad (14.2.18)$$

for $A, B \in \mathbb{R}$. If we apply the formula $\max(A, B) = -\min(-A, -B)$, (14.2.17) can be expressed as

$$B_n^{(s+1)} = \min\left(1 - B_n^{(s)}, \sum_{j=-\infty}^{n-1} (B_j^{(s)} - B_j^{(s+1)})\right) \quad (14.2.19)$$

$$+ \max\left(0, \sum_{j=-\infty}^{n} (B_j^{(s)} - B_{j-1}^{(s+1)}) - K\right). \quad (14.2.20)$$

14.3 Automata Groups

The class of automata groups contains several remarkable countable groups. Their study has led to the solution of a number of important problems in group theory. Its recent applications have extended to the fields of algebra, geometry, analysis, and probability. Together with arithmetic and hyperbolic groups, automata groups dominate the modern landscape of theory of infinite groups.

A quick introduction to this subject is given in [10], which presents also highlights of this theory, like constructions of infinite torsion groups, together with elementary proofs.

14.3.1 Definition of Groups Generated by Automata

The automata which we consider are finite, reversible, and have the same input and output alphabets, say $D = \{0, 1, \ldots, d-1\}$ for a certain integer $d > 1$. To such an automaton A are associated a finite set of states Q, a transition function $\phi : Q \times D \to Q$, and the exit function $\psi : Q \times D \to D$. The automaton A is characterized by a quadruple (D, Q, ϕ, ψ).

The automaton A is invertible if, for every $q \in Q$, the function $\psi(q, \cdot) : D \to D$ is a bijection.

In this case, $\psi(q, \cdot)$ can be identified with an element σ_q of the symmetric group S_d on $d = |D|$ symbols.

There is a convenient way to represent a finite automaton by a marked graph $\Gamma(A)$ where vertices correspond to elements of Q.

Two states $q, s \in Q$ are connected by an arrow labelled by $i \in D$ if $\phi(q, i) = s$; each vertex $q \in Q$ is labelled by a corresponding element σ_q of the symmetric group.

The automata we just defined are non-initial. To make them initial we need to mark some state $q \in Q$ as the initial state. The initial automaton $A_q = (D, Q, \phi, \psi, q)$ acts on the right on the finite and infinite sequences over D in the following way. For every symbol $x \in D$ the automaton immediately gives $y = \psi(q, x)$ and changes its initial state to $\phi(q, x)$.

By joining the exit of A_q to the input of another automaton $B_s = (S, \alpha, \beta, s)$, we get an application which corresponds to the automaton called the composition of A_q and B_s and is denoted by $A_q \star B_s$.

This automaton is formally described as the automaton with the set of the states $Q \times S$ and the transition and exit functions Φ, Ψ defined by

$$\Phi((x, y), i) = (\phi(x, i), \alpha(y, \psi(x, i))),$$

$$\Psi((x, y), i) = \beta(y, \psi(x, i))$$

and the initial state (q, s).

The composition $A \star B$ of two non-initial automata is defined by the same formulas for input and output functions but without indicating the initial state.

Two initial automata are equivalent if they define the same application. There is an algorithm to minimize the number of states.

The automaton which produces the identity map on the set of sequences is called trivial. If A is invertible then for every state q the automaton A_q admits an inverse automaton A_q^{-1} such that $A_q \star A_q^{-1}$, $A_q^{-1} \star A_q$ are equivalent to the trivial one. The inverse automaton can be formally described as the automaton $(Q, \widetilde{\phi}, \widetilde{\psi}, q)$ were $\widetilde{\phi}(s, i) = \phi(s, \sigma_s(i))$, $\widetilde{\psi}(s, i) = \sigma_s^{-1}(i)$ for $s \in Q$. The equivalence classes of finite invertible automata over the alphabet D constitute a group called a group of finite automata which depends on D. Every set of finite automata generates a subgroup of this group.

Now let A be an invertible automaton. Let $Q = \{q_1, \ldots, q_t\}$ be the set of states of A and let $A_{q_1}, \ldots, A_{q_t}$ be the set of initial automata which can be obtained from A. The group $G(A) = \langle A_{q_1}, \ldots, A_{q_t} \rangle$ is called the group generated or determined by A.

14.4 The Lamplighter Group

In our work [2] we associated to the BBS a two-state automaton. However, this automaton is not invertible (see Section 14.3 for definitions) and one could obtain from it only a semigroup.

In order to relate it to a group, we presented a construction which in the context of automata in some cases can associate to a semigroup or a group. This construction is recalled in Section 14.5. In our case it leads to the lamplighter group.

14.4.1 The Lamplighter Group as an Automaton Group

The automaton group from Figure 14.1 generates the lamplighter group [10]. In this example, the alphabet consists of two letters 0 and 1. The elements of the symmetric group S_2 are denoted by Id and e (the non-trivial element).

This group can be defined as the wreath product $\mathbb{Z}/2\mathbb{Z} \wr \mathbb{Z}$ or as a semi-direct product $(\oplus_{\mathbb{Z}}\mathbb{Z}/2\mathbb{Z}) \rtimes \mathbb{Z}$ with the action of $\mathbb{Z}$ on $\oplus_{\mathbb{Z}}(\mathbb{Z}/2\mathbb{Z})$ by translation.

The proof that the group generated by the automaton from Figure 14.1 is isomorphic to the lamplighter group is not trivial [10]. The states of the automaton do not correspond to the standard generators of the lamplighter group (see below). However, they provide a very symmetric set of generators, namely there is an automorphism of the lamplighter group which exchanges the elements corresponding to the states a and b.

Let a and b be the generators of the lamplighter group $(\oplus_{\mathbb{Z}}\mathbb{Z}/2\mathbb{Z}) \rtimes \mathbb{Z}$ such that $a = (f_a, g_a)$, $b = (f_b, g_b)$, where $g_a = g_b \in \mathbb{Z}$ is a generator of $\mathbb{Z}$, $f_a \in \oplus_{\mathbb{Z}}(\mathbb{Z}/2\mathbb{Z})$ is the identity and $f_b = (\dots, 0, 0, 1, 0, 0, \dots) \in \oplus_{\mathbb{Z}}(\mathbb{Z}/2\mathbb{Z})$ is such that 1 is in a position 1. There is an isomorphism between this group and the group generated by the automaton from Figure 14.1, where a and b correspond to the initial states of the automaton.

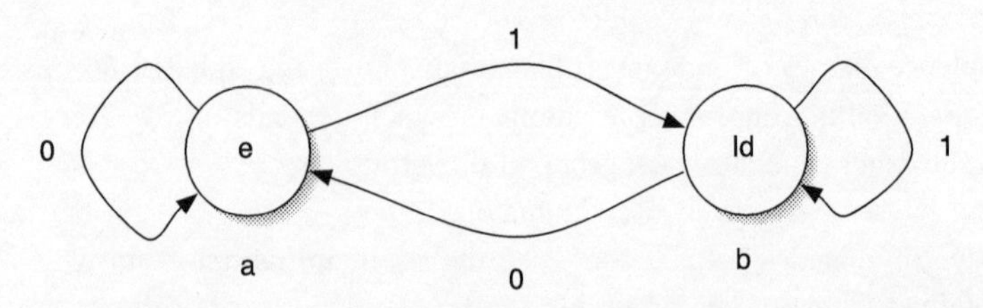

Figure 14.1 The automaton which generates the lamplighter group

14.4.2 Classification of the Automata Groups on Two States with the Alphabet $\{0, 1\}$

For the alphabet on two letters the automata with just one state produce only the trivial group or the group of order two.

We are going to analyse all groups generated by the automata on two states with the alphabet on two letters.

In addition to the lamplighter group from Figure 14.1, there are five other groups generated by the automata from Figure 14.2 (we denote Id the identity of S_2 and e the non-trivial element of S_2).

The first two automata generate the trivial group and the group of order two. The group given by the third automaton is isomorphic to the Klein group

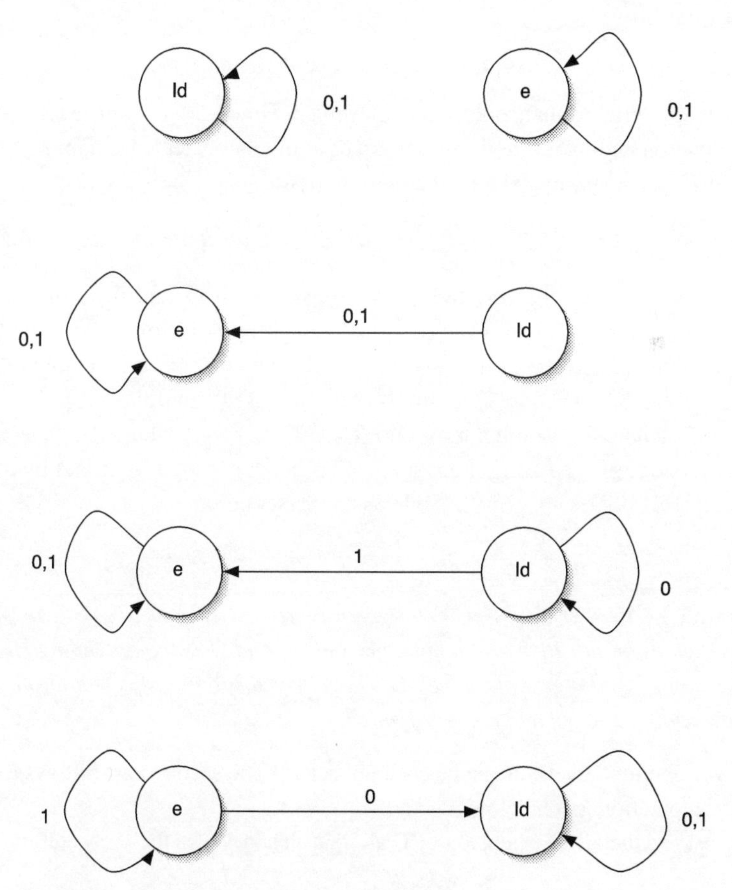

Figure 14.2 The automata generating the trivial group, the group of order two, the Klein group, the dihedral group and the infinite cyclic group

$(\mathbb{Z}/2\mathbb{Z}) \oplus (\mathbb{Z}/2\mathbb{Z})$. The fourth automaton defines the dihedral group $\mathbb{D}_\infty$. The last automaton defines the infinite cyclic group.

These are the only possibilities [10].

Theorem 14.4.1 *The only groups generated by the automata on two states over the alphabet on two letters are:*

- *the trivial group;*
- *the group of order two* $\mathbb{Z}/2\mathbb{Z}$*;*
- *the Klein group* $(\mathbb{Z}/2\mathbb{Z}) \oplus (\mathbb{Z}/2\mathbb{Z})$*;*
- *the infinite cyclic group* $\mathbb{Z}$*;*
- *the infinite dihedral group* $\mathbb{D}_\infty$*;*
- *the lamplighter group* $(\oplus_{\mathbb{Z}}\mathbb{Z}/2\mathbb{Z}) \rtimes \mathbb{Z}$.

14.4.3 Finite Presentations

The lamplighter group is not finitely presented. However, it admits a recursive presentation and therefore is a subgroup of a finitely presented group.

Let G be the group given by the presentation

$$G = \langle a, t, s \mid a^2 = 1, [t, s] = 1, [t^{-1}at, a] = 1, s^{-1}as = at^{-1}at\rangle. \quad (14.4.1)$$

We will see that the lamplighter is a subgroup of G and therefore some computations from [2] are valid for G. As the group G is finitely presented, it is a fundamental group of a closed manifold. Thus the spectral computations from [2] can be related to L^2 Betti numbers of closed manifolds [10].

Let Γ denote the lamplighter group $(\oplus_{i\in\mathbb{Z}}\mathbb{Z}/2\mathbb{Z}) \rtimes \mathbb{Z}$, where the generator of $\mathbb{Z}$ acts on $\oplus_{i\in\mathbb{Z}}\mathbb{Z}/2\mathbb{Z}$ by translation. Γ is generated by $t \in \mathbb{Z}$ and by $a = (\dots, 0, 1, 0, \dots) \in \oplus_{i\in\mathbb{Z}}\mathbb{Z}/2\mathbb{Z}$ and has the presentation

$$\Gamma = \langle a, t \mid a^2 = 1, [t^{-k}at^k, t^{-n}at^n] = 1 \ \forall k, n \in \mathbb{Z}\rangle.$$

Lemma 14.4.2 *Let* $\alpha\colon \Gamma \to \Gamma$ *be given by* $\alpha(t) = t$ *and* $\alpha(a) = at^{-1}at$*. This defines an injective group homomorphism, and* G *is the ascending HNN-extension of* Γ *along* α*. Moreover,* G' *(the derived subgroup) is isomorphic to a countable direct sum of copies of* $\mathbb{Z}/2\mathbb{Z}$*.*

Proof. The first assertion can be easily checked. The second part follows from the computation given below.

Let V be the HNN extension of Γ along α. Then V has the presentation

$$\begin{aligned} V \;=\; & \langle a, t, s \mid a^2 = 1, [s, t] = 1, s^{-1}as = at^{-1}at = [a, t], \\ & [t^{-k}at^k, t^{-n}at^n] = 1 \ \forall k, n \in \mathbb{Z}\rangle. \end{aligned}$$

Obviously, we have an epimorphism G onto V mapping a to a, s to s, and t to t. It only remains to show that every relation in the given presentation of V follows from the relations of G. Observe first in G that by conjugation with t^{-n}, $[t^{-k+n}at^{-n+k}, a] = 1$ implies $[t^{-k}at^{k}, t^{-n}at^{n}] = 1$. Moreover, commutativity is commutative, i.e., $[t^{-k}at^{k}, t^{-n}at^{n}] = 1$ implies $[t^{-n}at^{n}, t^{-k}at^{k}] = 1$. Hence, it remains to prove $[t^{-n}at^{n}, a] = 1$ in G for $n > 1$. We will do this by induction on n. Assume therefore $t^{-j}at^{j}$ commutes with $t^{-l}at^{l}$ for $0 \leq j \leq l < n$. Conjugate the relation $[t^{-(n-1)}at^{n-1}, a] = 1$ with s. We obtain

$$1 = [t^{-(n-1)}at^{-1}att^{n-1}, at^{-1}at] = [(t^{-(n-1)}at^{n-1})(t^{-n}at^{n}), a(t^{-1}at)]. \tag{14.4.2}$$

Now observe that by induction a commutes with $a_1 := t^{-1}at$ and with $a_{n-1} := t^{-(n-1)}at^{n-1}$. This second relation also implies (by conjugation with t^{-1}) that moreover, $t^{-1}at$ commutes with $a_n := t^{-n}at^{n}$. Therefore, we can simplify the commutator in (14.4.2) to the desired

$$\begin{aligned} 1 &= (a_n^{-1}a_{n-1}^{-1})(a_1^{-1}a^{-1})(a_{n-1}a_n)(aa_1) \\ &= a_n^{-1}(a_{n-1}^{-1}a_{n-1})(a_1^{-1}a_1)a^{-1}a_n a = [t^{-n}at^{n}, a]. \end{aligned}$$

By induction, we therefore see that $V = G$.

Using the presentation, we next check that the abelianization of V is isomorphic to $\mathbb{Z} \times \mathbb{Z}$, and s, t are mapped to two free generators, whereas a is mapped to zero. Therefore, G' is equal to the normal subgroup generated by a, which is generated by $s^{-l}t^{-k}at^{k}s^{l}$, $k, l \in \mathbb{Z}$, $l < 0$. All these elements are of order 2, and by conjugation with sufficiently high powers of s we see that they all commute. Therefore, G' is a vector space over $\mathbb{Z}/2\mathbb{Z}$ with countably many generators, and therefore isomorphic to a countable direct sum of copies of $\mathbb{Z}/2\mathbb{Z}$. Observe, however, that G' is quite different from the base of the HNN-extension Γ. The element sas^{-1} is a typical example which is not contained in Γ but in G'. □

14.5 Stochastic Matrices

An important part of our work [2] consisted in relating a non-invertible automaton to an invertible one, in other words, in some cases one can associate a group to a semigroup.

For each state of the automaton, one can consider its action on a finite sequence of 0 and 1s of length n. It can be represented by a finite matrix of size 2^n with 0 and 1 entries. If one orders 2^n sequences of 0 and 1s of length n

in a lexicographic oder then one puts 1 on (i, j) entry if i-th sequence is sent to j-th sequence. Let $a_k^{(n)}$ be a matrix associated with the state a_k for its action on a sequence of length n.

One obtains recurrence relations for such matrices. For instance, in case of the automaton from Figure 14.1, we have $a_0^{(0)} = a_1^{(0)} = 1$ and

$$a_0^{(n+1)} = \begin{pmatrix} 0 & a_0^{(n)} \\ a_1^{(n)} & 0 \end{pmatrix}$$

$$a_1^{(n+1)} = \begin{pmatrix} a_0^{(n)} & 0 \\ 0 & a_1^{(n)} \end{pmatrix}.$$

These matrices are invertible, i.e., there is only one 1 in each line and each column. If we consider the sum of these matrices and their transposes (in this case this corresponds to their inverses), the sum of columns and lines is constant. We refer to such a matrix as bistochastic, although in order to use the probabilistic language one should normalize it (and then we might also call it a Markov or a random walk operator).

For a non-invertible automaton, the sum of matrices and their transposes might not result in a matrix with property of being bistochastic. In case of the automaton associated to BBS, which is non-invertible we have the following recursion [2]:

$$a_0^{(n+1)} = \begin{pmatrix} a_0^{(n)} & a_1^{(n)} \\ 0 & 0 \end{pmatrix}$$

$$a_1^{(n+1)} = \begin{pmatrix} 0 & 0 \\ a_0^{(n)} & a_1^{(n)} \end{pmatrix}$$

and we proved that we do get in this case bi-stochastic matrices. Actually, we proved that they are conjugated to the matrices associated to the automaton generating the lamplighter group.

In this section, we look at other examples of non-invertible automata which lead to bistochastic matrices.

14.5.1 BBS-V and S

The following two automata BBS-V and S were originally introduced in [6, 7]. We define a sequence of three matrices $(a_0^{(n)}, \ldots, a_2^{(n)})$ of dimension 2^n, for

$n = 0, 1, \ldots$ by the following matrix recursion (0 represents here $2^n \times 2^n$ nul matrix).

For BBS-V we have

$$a_0^{(n+1)} = \begin{pmatrix} a_0^{(n)} & a_2^{(n)} \\ 0 & 0 \end{pmatrix}$$

$$a_1^{(n+1)} = \begin{pmatrix} 0 & 0 \\ a_0^{(n)} & a_1^{(n)} \end{pmatrix}$$

and

$$a_2^{(n+1)} = \begin{pmatrix} a_1^{(n)} & 0 \\ 0 & a_2^{(n)} \end{pmatrix}$$

with the initial data $a_i^{(0)} = 1$ for all $i = 0, 1, 2$ and for BBS-S we have

$$a_0^{(n+1)} = \begin{pmatrix} a_0^{(n)} & a_1^{(n)} \\ 0 & 0 \end{pmatrix}$$

$$a_1^{(n+1)} = \begin{pmatrix} a_2^{(n)} & 0 \\ 0 & a_2^{(n)} \end{pmatrix}$$

and

$$a_2^{(n+1)} = \begin{pmatrix} 0 & 0 \\ a_0^{(n)} & a_1^{(n)} \end{pmatrix}$$

We consider the following $2^n \times 2^n$ matrix $M_k^{(n)}$

$$M^{(n)} = \frac{1}{6}(a_0^{(n)} + (a_0^{(n)})^* + \ldots + a_2^{(n)} + (a_2^{(n)})^*).$$

Proposition 14.5.1 *The matrix $M^{(n)}$ is double stochastic for all $n \geq 0$, i.e., the sum of each row and each column is equal to* 1.

Proof. The matrix $M^{(n)}$ is symmetric and therefore it suffices to prove that the sum of columns is constant.

Clearly, the recursive relations for $a_0^{(n+1)}, \ldots, a_2^{(n+1)}$ show that the matrix we obtain from each of them is the matrix with constant sum of columns (equal to 1 as it has only one 1 in each column).

Thus, it is enough to show that $(a_0^{(n)})^* + \ldots + (a_2^{(n)})^*$ has constant column sum. Let us prove this by induction. It is clear for $n = 0$. Then using recursion formula for BBS-V

$$(a_0^{(n+1)})^* + \cdots + (a_2^{(n+1)})^* = \begin{pmatrix} (a_0^{(n)})^* + (a_1^{(n)})^* & (a_0^{(n)})^* \\ (a_2^{(n)})^* & (a_2^{(n)})^* + (a_1^{(n)})^* \end{pmatrix}$$

and for BBS-S

$$(a_0^{(n+1)})^* + \cdots + (a_2^{(n+1)})^* = \begin{pmatrix} (a_0^{(n)})^* + (a_2^{(n)})^* & (a_0^{(n)})^* \\ (a_1^{(n)})^* & (a_1^{(n)})^* + (a_2^{(n)})^* \end{pmatrix}.$$

Thus, the sum of the left matrix blocks and right matrix blocks (both for BBS-V and S) is equal to

$$(a_0^{(n)})^* + \ldots + (a_2^{(n)})^*.$$

Therefore, the statement follows by induction. □

Acknowledgement

I would like to thank Satoshi Tsujimoto for writing the computation from Section 14.2 and explaining his work [6, 7].

References

[1] R. HIROTA, *Discrete analogue of a generalized Toda equation*, J. Phys. Soc. Japan, **50** pp. 3785–3791 (1981).

[2] T. KATO, S. TSUJIMOTO, A. ZUK, *Spectral analysis of transition operators, automata groups and translation in BBS*, Commun. Math. Phys., **350** pp. 205–229 (2017).

[3] T. MIWA, *On Hirota's difference equations*, Proc. Japan Acad. A, **58** pp. 9–12 (1982).

[4] D. TAKAHASHI AND J. MATSUKIDAIRA, *Box and ball system with a carrier and ultradiscrete modified KdV equation*, J. Phys. A: Math. Gen., **30** pp. 733–739 (1997).

[5] D. TAKAHASHI AND J. SATSUMA, *A soliton cellular automaton*, J. Phys. Soc. Japan, **59** pp. 3514–3519 (1990).

[6] S. TSUJIMOTO *Rules of new box-ball systems and their analysis*, Reports of RIAM Symposium, **28AO-S6** pp.13–18 (2017) in Japanese.

[7] S. TSUJIMOTO *On soliton automata*, RIMS Kôkyûroku, **2071** pp.134–140 (2018) in Japanese.

[8] S. Tsujimoto and R. Hirota *Ultradiscrete KdV equation*, J. Phys. Soc. Japan, **67** pp.1809–1810 (1998).

[9] T. Tokihiro, D. Takahashi, J. Matsukidaira and J. Satsuma, *From soliton equations to integrable cellular automata through a limiting procedure*, Phys. Rev. Lett., **76** pp. 3247–3250 (1996).

[10] A. Zuk, *Groupes engendrés par les automates*, Séminaire Bourbaki, Astérisque 311, 2008, p. 141–174.

15

Spectral Properties of Limit-Periodic Operators

David Damanik and Jake Fillman

Abstract

We survey results concerning the spectral properties of limit-periodic operators. The main focus is on discrete one-dimensional Schrödinger operators, but other classes of operators, such as Jacobi and CMV matrices, continuum Schrödinger operators and multi-dimensional Schrödinger operators, are discussed as well.

We explain that each basic spectral type occurs, and it does so for a dense set of limit-periodic potentials. The spectrum has a strong tendency to be a Cantor set, but there are also cases where the spectrum has no gaps at all. The possible regularity properties of the integrated density of states range from extremely irregular to extremely regular. Additionally, we present background about periodic Schrödinger operators and almost-periodic sequences.

In many cases we outline the proofs of the results we present.

15.1 Introduction

We will talk about Schrödinger operators H acting in $\ell^2(\mathbb{Z})$ via

$$[Hu](n) = u(n-1) + u(n+1) + V(n)u(n)$$

with periodic and limit-periodic potentials. We say that V (or, interchangeably, H) is *periodic* if there is some $q \in \mathbb{Z}_+$ with $V(n+q) = V(n)$ for every n. We say that V (or H) is *limit-periodic* if V lies in the closure of the space of periodic potentials, i.e., if there exist $V_1, V_2, \ldots$ periodic with

$$\lim_{n\to\infty} \|V - V_n\|_\infty = 0.$$

D.D. was supported in part by NSF grants DMS–1361625 and DMS–1700131.
J.F. was supported in part by an AMS-Simons travel grant, 2016–2018

One might think that such operators would be rather similar to periodic operators, and that many characteristics and results can be 'pushed through' to the limit using uniform convergence. However, one would be very wrong! Periodic operators always exhibit purely absolutely continuous spectrum of multiplicity two supported on a spectrum which is a finite union of non-degenerate closed intervals, and the associated quantum dynamics are ballistic (i.e., a particle subjected to a periodic potential behaves like a free particle). Every single one of these characteristics can be broken in the limit-periodic regime. Concretely, in the limit-periodic regime:

- The spectrum can be a Cantor set (a dense set of gaps). In fact, the spectrum may have zero Hausdorff dimension.
- The spectral type may be purely singular continuous or even pure point.
- The quantum dynamics may be localized in a very strong sense.

The study of these operators was begun in the 1980s by work of Avron-Simon [4], Moser [43], and the Soviet school [8, 9, 21, 41, 48, 49]. These classical results focused on the presence of Cantor spectrum and purely absolutely continuous spectral measures. A notable exception is the work [53] by Pöschel from that era, which exhibited examples that are uniformly localized. There has been a recent uptick in activity, ushered in by Avila [3], which has led to the realization that new phenomena are possible, such as the genericity of purely singular continuous spectrum [3, 16] and the denseness of pure point spectrum [18] in the space of all limit-periodic potentials. Another new phenomenon established in a recent paper is the Lipschitz continuity of the integrated density of states in Pöschel's examples [12].

The chapter is structured as follows. We provide a quick introduction to the periodic case in Section 15.2. Almost periodic potentials and their hulls are discussed in Section 15.3. Every limit-periodic sequence is almost-periodic, and those almost-periodic sequences that are limit-periodic can be characterized via a topological property of their hulls: they need to be totally disconnected. Moreover, this discussion also shows that the study of almost-periodic operators fits in the broader class of ergodic operators, which are discussed in Section 15.4, along with the standard quantities associated with them: the Lyapunov exponent and the integrated density of states. Section 15.5 contains results about limit-periodic operators with purely absolutely continuous spectral measures, supported on thick (Carleson homogeneous) Cantor sets. The genericity of zero-measure Cantor spectrum and purely singular continuous spectral measures is the subject of Section 15.6. Limit-periodic operators with pure point spectrum are presented in Section 15.7, both those that arise in the large coupling regime and those that arise due to local randomness.

Section 15.8 addresses regularity results for the integrated density of states. While we focus on discrete Schrödinger operators in this chapter, much of what we say has analogues for related families of operators in mathematical physics, namely: Jacobi matrices, CMV matrices/quantum walks, and continuum Schrödinger operators. We describe these operators and some of the known results in Section 15.9. Motivated by the known results, we present some interesting open problems in Section 15.10. Finally, since the hull of a limit-periodic potential is a totally disconnected group, we discuss some characteristics of totally disconnected groups in Appendix Appendix A.

15.2 A Crash Course on Periodic Operators

In the present section, we will review some aspects of the spectral analysis of periodic operators in 1D. Throughout, assume that V is q-periodic and that $q \in \mathbb{Z}_+$ is chosen to be minimal with respect to this property. We will at least sketch most of the arguments here. With the exception of the discussion of ballistic motion in Section 15.2.4, the results in this section are textbook material; lucid references include [61, Chapter 5] and [67, Chapter 7].

The eigenvalue equation for H reads

$$u(n-1) + u(n+1) + V(n)u(n) = Eu(n) \tag{15.2.1}$$

where E is a scalar. We shall see presently that this equation has no (nontrivial) solution $u \in \ell^2(\mathbb{Z})$ when V is periodic, so we will need to be somewhat more generous in our interpretation of (15.2.1). Namely, we will consider (15.2.1) for arbitrary $u \in \mathbb{C}^{\mathbb{Z}}$, not just $u \in \ell^2(\mathbb{Z})$. Through a standard abuse of notation, we will also write Hu for $u \in \mathbb{C}^{\mathbb{Z}}$, where $[Hu](n)$ is simply defined by the left-hand side of (15.2.1).

15.2.1 Floquet Theory

The difference equation (15.2.1) can be rewritten as a 2×2 matrix recursion:

$$\begin{bmatrix} u(n+1) \\ u(n) \end{bmatrix} = \begin{bmatrix} E - V(n) & -1 \\ 1 & 0 \end{bmatrix} \begin{bmatrix} u(n) \\ u(n-1) \end{bmatrix}. \tag{15.2.2}$$

Clearly then, the asymptotic characteristics of u can be encoded by the *monodromy matrix*

$$\Phi(E) = \begin{bmatrix} E - V(q) & -1 \\ 1 & 0 \end{bmatrix} \begin{bmatrix} E - V(q-1) & -1 \\ 1 & 0 \end{bmatrix} \cdots \begin{bmatrix} E - V(1) & -1 \\ 1 & 0 \end{bmatrix}$$

and the relationship between $(u(1), u(0))^\top$ and the eigenvector(s) of $\Phi(E)$. Since $\Phi \in \mathrm{SL}(2, \mathbb{C})$, the character of its eigenvectors is entirely encoded by its trace, called the *discriminant*, and defined by $D(E) := \mathrm{Tr}\, \Phi(E)$. We may characterize the spectrum of H as follows:

Theorem 15.2.1 *If V is q-periodic with discriminant D, then*

$$\sigma(H_V) = \{E \in \mathbb{C} : D(E) \in [-2, 2]\} = \{E \in \mathbb{R} : |D(E)| \leq 2\}.$$

For all $c \in (-2, 2)$, all solutions of $D(E) = c$ are real and simple. For $c = \pm 2$, all solutions of $D(E) = c$ are real and at most doubly degenerate.

Proof Sketch. Since H_V is self-adjoint, we focus on real E. For each $E \in \mathbb{R}$, there are four possibilities for $\Phi(E)$:

(1) $|\mathrm{Tr}\, \Phi(E)| < 2$: Then, the powers of $\Phi(E)$ are uniformly bounded, so all solutions of (15.2.1) are bounded.

(2) $|\mathrm{Tr}\, \Phi(E)| = 2$ and $\Phi(E) \notin \{\pm I\}$: There is a Jordan anomaly. One solution of (15.2.1) is bounded, and any linearly independent solution grows linearly.

(3) $\Phi(E) = \pm I$. Clearly, all solutions of (15.2.1) are bounded.

(4) $|\mathrm{Tr}\, \Phi(E)| > 2$. There are linearly independent solutions u_+ and u_- of (15.2.1) with the property that $u_\pm$ decays exponentially at $\pm\infty$ and grows exponentially at $\mp\infty$.

In cases (1), (2), and (3), one can use a bounded solution to construct a Weyl sequence, and hence $E \in \sigma(H)$. In case (4), we have two special solutions $u_\pm$, defined as follows: let $(u_+(1), u_+(0))^\top$ be a contracting eigenvector of $\Phi(E)$ and let $(u_-(1), u_-(0))^\top$ be an expanding eigenvector of $\Phi(E)$, and use (15.2.2) to extend these initial data to produce solutions of (15.2.1). Then, $u_\pm$ decays exponentially at $\pm\infty$ and one can use these solutions to construct a Green function for H. Concretely, one can check that

$$\langle \delta_n, R\delta_m \rangle = \frac{u_-(n \wedge m)u_+(n \vee m)}{u_-(0)u_+(1) - u_+(0)u_-(1)}$$

($n \wedge m = \min(n, m)$ and $n \vee m = \max(n, m)$) defines a bounded operator with $R(H - E) = (H - E)R = I$ by Schur's criterion.

In view of the discussion above, it follows that any solution of $D(E) = c$ with $c \in [-2, 2]$ must be real, since such an E belongs to the spectrum of H and H is self-adjoint. The multiplicity statements follow from Rouché's theorem. Concretely, if $D(E) - c$ has a double (or higher) root at $E \in \mathbb{R}$ for some $c \in (-2, 2)$, then Rouché's theorem implies the existence of non-real z near E such that $D(z) \in (-2, 2)$, contradicting reality of such solutions. Similarly, if $D(E) - c$ vanishes to third order or higher for some choice of $c \in \{\pm 2\}$, then Rouché's theorem implies the existence of nearby non-real z with $D(z) \in [-2, 2]$. □

Thus, we see that the spectrum of H consists of q non-degenerate closed intervals, which we call the *bands*:

$$\sigma(H) = \bigcup_{j=1}^{q} [\alpha_j, \beta_j].$$

Here, $\alpha_1 < \beta_1 \le \alpha_2 < \cdots \le \alpha_q < \beta_q$ denotes an enumeration of the solutions of $D(E) = \pm 2$, counted with multiplicity. Each of the closed intervals $[\alpha_j, \beta_j]$ may be obtained as the closure of a connected component of $D^{-1}((-2, 2))$. In the event that $\beta_j < \alpha_{j+1}$, we say that the jth *gap* is open. Otherwise, we say that we have a *closed gap* at the point $\alpha_j = \beta_{j+1}$. Notice that one has a closed gap at c if and only if $D^2 - 4$ vanishes to second order at c, i.e., $D(c) = \pm 2$ and $D'(c) = 0$.

15.2.2 Bloch Waves

If $E \in \sigma(H)$, then $D(E) \in [-2, 2]$. Writing $D(E) = 2\cos\theta$ for some $\theta \in [0, \pi]$, we see that $\Phi(E)$ has eigenvalues $e^{\pm i\theta}$. Using an eigenvector corresponding to the eigenvalue $e^{i\theta}$ as an initial condition, we may generate a solution u_+ to $Hu = Eu$ so that

$$u_+(n + q) = e^{i\theta} u_+(n). \tag{15.2.3}$$

Similarly, using the eigenvector for $e^{-i\theta}$ as an initial condition, one obtains a solution u_- satisfying $u_-(n + q) = e^{-i\theta} u_-(n)$. On can encode solutions of $Hu = Eu$ satisfying a condition like (15.2.3) in terms of $q \times q$ matrices with suitable boundary conditions. Concretely, define

$$H_\theta = H_{\theta,q} = \begin{bmatrix} V_1 & 1 & & & e^{-i\theta} \\ 1 & V_2 & 1 & & \\ & \ddots & \ddots & \ddots & \\ & & 1 & V_{q-1} & 1 \\ e^{i\theta} & & & 1 & V_q \end{bmatrix}.$$

Theorem 15.2.2 *If V is q-periodic, we have*

$$\sigma(H) = \bigcup_{\theta \in [0,\pi]} \sigma(H_\theta). \tag{15.2.4}$$

Moreover, the discriminant and the characteristic polynomials are related via

$$\det(E - H_\theta) = D(E) - 2\cos\theta. \tag{15.2.5}$$

Proof Sketch. Given $E \in \sigma(H)$, write $D(E) = 2\cos\theta$ as before, and define $u_\pm$ as above. One can then check that $(u_+(1), u_+(2), \ldots, u_+(q))^\top$ is an eigenvector of H_θ with eigenvalue E. On the other hand, if E is an eigenvalue of H_θ, an eigenvector of H_θ can be extended to a solution of $Hu = Eu$ satisfying (15.2.3). Thus, $e^{i\theta}$ is an eigenvalue of $\Phi(E)$, which implies $D(E) = 2\cos\theta$. This discussion proves (15.2.4).

Thus, for fixed θ, $D(E) - 2\cos(\theta)$ and $\det(E - H_\theta)$ are monic polynomials in E with the same degree and the same roots. For $\theta \in (0, \pi)$, all roots of $D(E) - 2\cos\theta$ are simple by Theorem 15.2.1, so (15.2.5) follows for all $E \in \mathbb{C}$ and all $\theta \in (0, \pi)$. Then, fixing E, both sides of (15.2.5) define entire functions of θ that agree for $\theta \in (0, \pi)$ and hence must agree everywhere. □

Let us note a couple of pleasant consequences of the foregoing result. First of all, since $D(E) - c$ can only vanish to first order for $c \in (-2, 2)$, it follows from (15.2.5) that H_θ has simple spectrum for $\theta \in (0, \pi)$. One can also prove this directly (cf. [61, Section 5.3]). We can also read off from (15.2.5) that the spectrum of H has a closed gap at energy E_0 with $D(E_0) = 2$ (resp. $D(E_0) = -2$) if and only if all solutions to $Hu = E_0 u$ are periodic (resp. anti-periodic).

One can formulate Theorem 15.2.2 in terms of direct integrals in a natural manner. Concretely, consider the space

$$\begin{aligned} \mathcal{H}_q &:= L^2([0, 2\pi), \mathbb{C}^q) \\ &= \left\{ \vec{f} : [0, 2\pi) \to \mathbb{C}^q : \int_0^{2\pi} \|\vec{f}(\theta)\|_{\mathbb{C}^q}^2 < \infty \right\}. \end{aligned}$$

Naturally, $\mathcal{H}_q$ is a Hilbert space with respect to the inner product

$$\langle f, g \rangle_{\mathcal{H}_q} = \int_0^{2\pi} \langle f(\theta), g(\theta) \rangle_{\mathbb{C}^q} \frac{d\theta}{2\pi}.$$

This space is sometimes written as a direct integral, especially in the physics literature:

$$\mathcal{H}_q = \int_{[0,2\pi)}^{\oplus} \mathbb{C}^q \frac{d\theta}{2\pi}.$$

We may define a linear operator $\mathcal{F} : \ell^2(\mathbb{Z}) \to \mathcal{H}_q$ by $\mathcal{F} : u \mapsto \widehat{u}$, where

$$\widehat{u}_j(\theta) = \sum_{\ell \in \mathbb{Z}} u_{j+\ell q} e^{-i\ell\theta}, \quad \theta \in [0, 2\pi), \ 1 \le j \le q. \tag{15.2.6}$$

Of course, $\mathcal{F}$ is initially only defined on $\ell^1(\mathbb{Z})$, but it admits a unique extension to a unitary operator on $\ell^2(\mathbb{Z})$ by Parseval's formula.

Lemma 15.2.3 *The map $\mathcal{F}$ extends to a unitary operator from $\ell^2(\mathbb{Z})$ to $\mathcal{H}_q$; its inverse maps $f \in \mathcal{H}_q$ to $\check{f} \in \ell^2(\mathbb{Z})$, defined by*

$$\check{f}_{j+\ell q} = \int_0^{2\pi} e^{i\ell\theta} f_j(\theta) \frac{d\theta}{2\pi}, \quad 1 \le j \le q, \ \ell \in \mathbb{Z}.$$

Proof. For each $1 \le j \le q$ and each $\ell \in \mathbb{Z}$, we define $\varphi_{j,\ell} \in \mathcal{H}_q$ by $\varphi_{j,\ell} = \widehat{\delta}_{j+\ell q}$, i.e.:

$$\varphi_{j,\ell}(\theta) = e^{-i\ell\theta} \vec{e}_j,$$

where $\vec{e}_1, \ldots, \vec{e}_q$ denotes the standard basis of $\mathbb{C}^q$. It is easy to see that $\{\varphi_{j,\ell} : 1 \le j \le q, \ \ell \in \mathbb{Z}\}$ is an orthonormal basis of $\mathcal{H}_q$. Armed with this knowledge, all claims in the lemma are straightforward. □

For any operator B on $\ell^2(\mathbb{Z})$, we denote by $\widehat{B}$ its action on Fourier space, i.e., $\widehat{B} = \mathcal{F} B \mathcal{F}^{-1}$, where $\mathcal{F}$ is defined by (15.2.6). Using the definition of $\mathcal{F}$, we see that $\widehat{H}$ acts as a multiplication operator on $\mathcal{H}_q$. That is, for any $f \in \mathcal{H}_q$, one has

$$\left(\widehat{H} f\right)(\theta) = H_\theta f(\theta) \text{ for Lebesgue almost every } \theta \in [0, 2\pi).$$

Thus, even though H has no eigenvalues and no point spectrum, we still colloquially say that $\mathcal{F}$ 'diagonalizes' H. If we denote the eigenvalues of H_θ as $\lambda_1(\theta) \le \cdots \le \lambda_q(\theta)$, then,

$$\sigma(H) = \bigcup_{q-j \text{ even}} [\lambda_j(\pi), \lambda_j(0)] \cup \bigcup_{q-j \text{ odd}} [\lambda_j(0), \lambda_j(\pi)].$$

This point of view takes the problem of determining the spectrum and replaces a numerically unstable problem (polynomial root-finding) by a very stable procedure (computing eigenvalues of Hermitian matrices). Concretely, one can completely determine the spectrum by diagonalizing H_0 and H_π. There are fast, stable numerical algorithms for computing eigenvalues of finite tridiagonal Hermitian matrices; however, H_0 and H_π are not tridiagonal, but rather 'tridiagonal with corner entries' which can lead to numerical bottlenecks with a naïve implementation leading to needlessly performing $O(q^3)$ floating-point operations. However, conjugating H_0 and H_π by a breadth-first permutation matrix creates penta-diagonal matrices which can be reduced to tridiagonal form very efficiently (in $O(q^2)$ operations). See [54] for more details on this point of view.

15.2.3 The Density of States

Define a sequence of measures by

$$\int g\,dk_N = \frac{1}{N}\mathrm{Tr}(P_N\, g(H)\, P_N^*), \tag{15.2.7}$$

for bounded Borel functions g, where P_N denotes projection onto coordinates $\{0, 1, \ldots, N-1\}$. We will prove that dk_N has a weak limit, dk, as $N \to \infty$, which we will call dk the *density of states measure* (DOS) for H. The accumulation function of the DOS is called the *integrated density of states* (IDS) and is denoted by

$$k(E) = \int \chi_{(-\infty,E]}\,dk.$$

Theorem 15.2.4 *The DOS of H exists. That is, $dk = \lim dk_N$ exists, where the limit is taken with respect to the weak topology. Denoting $\Sigma = \sigma(H)$, the IDS of H is differentiable away from the edges of open gaps of Σ (i.e., on $\mathbb{R}\setminus\partial\Sigma$); more explicitly, if V is q-periodic, we have*

$$\frac{dk}{dE}(E_0) = \frac{|D'(E_0)|}{\pi q\sqrt{4 - D(E_0)^2}} = \frac{1}{\pi q}\left|\frac{d\theta}{dE}(E_0)\right|, \tag{15.2.8}$$

whenever $D(E_0) \in (-2, 2)$, where $\theta = \theta(E)$ is chosen continuously so that

$$2\cos(\theta) = D(E), \quad E \in \Sigma. \tag{15.2.9}$$

If E_0 is a closed gap of Σ, then

$$\frac{dk}{dE}(E_0) = \lim_{E\to E_0}\frac{|D'(E)|}{\pi q\sqrt{4 - D(E)^2}}; \tag{15.2.10}$$

in particular, the limit on the right-hand side exists and is finite. Otherwise, when $E_0 \in \partial\Sigma$, then E_0 borders an open gap of Σ, and dk/dE diverges at the rate $|E - E_0|^{-1/2}$ as $E \in \Sigma$ approaches E_0. Moreover, if $B = [\alpha, \beta]$ is any band of the spectrum,

$$\int_B dk = k(\beta) - k(\alpha) = \frac{1}{q}. \tag{15.2.11}$$

Proof. For $m \in \mathbb{Z}_+$, consider H_m^{per}, the restriction of H to $[1, mq]$ with periodic boundary conditions, i.e.,

$$H_m^{\text{per}} = \begin{bmatrix} V(1) & 1 & & & 1 \\ 1 & V(2) & 1 & & \\ & \ddots & \ddots & \ddots & \\ & & 1 & V(mq-1) & 1 \\ 1 & & & 1 & V(mq) \end{bmatrix},$$

and denote the corresponding eigenvalues by $E_{m,1}^{\text{per}} \leq \cdots \leq E_{m,mq}^{\text{per}}$ (notice that eigenvalues of H_m^{per} may occur with multiplicity two).

For $m \in \mathbb{Z}_+$, consider the probability measure

$$dk_m^{\text{per}} = \frac{1}{mq} \sum_{j=1}^{mq} \delta_{E_{m,j}^{\text{per}}}$$

defined by uniformly distributing point masses on the eigenvalues of H_m^{per}. One can easily check that E is an eigenvalue of H_m^{per} if and only if 1 is an eigenvalue of $[\Phi(E)]^m$, which holds if and only if the monodromy matrix $\Phi(E)$ has an mth root of unity as an eigenvalue. Thus, E is an eigenvalue of H_m^{per} if and only if

$$D(E) = 2\cos\left(\frac{2\pi j}{m}\right)$$

for some integer $0 \leq j \leq m/2$. Notice that any such eigenvalue with $0 < j < m/2$ is necessarily of multiplicity two, while the eigenvalues corresponding to $j = 0$ and $j = m/2$ (for even m) can have multiplicity one or two.

Let B_n denote the nth band of H and D_n the restriction of D to B_n so that D_n^{-1} is a continuous map $[-2, 2] \to B_n$. Given a continuous compactly supported function g, the previous discussion implies

$$\lim_{m\to\infty} \int_{B_n} g(E)\, dk_m^{\text{per}}(E) = \frac{1}{\pi q} \int_0^{\pi} g\left(D_n^{-1}(2\cos(\theta))\right) d\theta.$$

From the existence of the weak limit of dk_m^{per}, one may deduce that dk_N converges weakly and to the same limit. From this and the foregoing work, the first equality in (15.2.8) follows by using the change of variables suggested in (15.2.9) and keeping track of the sign of D'; the second is a consequence of the definition of θ. This also shows that dk/dE diverges as the inverse square root at borders of open gaps, since $D'(E_0) \neq 0$ whenever E_0 borders an open gap of Σ (which may be seen from (15.2.5)). Now, if E_0 is a closed gap of Σ, then the limit in (15.2.10) exists since D' vanishes linearly at E_0 and $4 - D^2$ vanishes quadratically at E_0. $\square$

15.2.4 Ballistic Motion

Let us consider the position operator:

$$[X\psi]_n = n\psi_n,$$

which is self-adjoint on $D(X) = \left\{\psi \in \ell^2(\mathbb{Z}) : X\psi \in \ell^2(\mathbb{Z})\right\}$. We are interested in the Heisenberg evolution when H is periodic:

$$X(t) = e^{itH} X e^{-itH}, \quad t \in \mathbb{R}.$$

One can check that e^{itH} preserves $D(X)$ and that $X(t)$ is self-adjoint on $D(X)$ for all t.

Theorem 15.2.5 *If H is periodic, the strong limit*

$$Q = \operatorname*{s\text{-}lim}_{t\to\infty} t^{-1} X(t)$$

exists and $\ker(Q) \cap D(X) = \{0\}$.

In fact, the conclusion of this theorem applies to any operator family with a Bloch-like decomposition. It was proved first for continuum operators by Asch and Knauf [2], and then for block Jacobi operators [19] (which include discrete Schrödinger operators as a special case), CMV matrices, and quantum walks [1, 15].

We will sketch the main strokes of the proof and refer the reader to [19] for details in the present setting. Formally (we are avoiding some niceties regarding domains, since X is unbounded), differentiating $X(t)$ with respect to time, one obtains:

$$\frac{dX(t)}{dt} = ie^{itH}(HX - XH)e^{-itH} = A(t),$$

where

$$A := i[H, X] = i(HX - XH).$$

Then, one can identify

$$\frac{1}{t}(X(t) - X) = \frac{1}{t}\int_0^t A(s)\, ds. \tag{15.2.12}$$

If H and A were finite matrices, the existence of the limit on the right-hand side of (15.2.12) would be trivial. To wit, if H_0 and A_0 are $m \times m$ matrices with H_0 Hermitian-symmetric, then, writing the spectral decomposition of H_0 as

$$H_0 = \sum_{\lambda\in\sigma(H_0)} \lambda P_\lambda,$$

one can readily compute that

$$\lim_{t \to \infty} \frac{1}{t} \int_0^t e^{isH_0} A_0 e^{-isH_0} \, ds = \sum_{\lambda \in \sigma(H_0)} P_\lambda A_0 P_\lambda, \tag{15.2.13}$$

the 'diagonal part' of A_0 with respect to H_0. More generally, this argument obtains for operators having pure point spectrum. However, as H has purely absolutely continuous spectrum, one has to work a little harder. The main idea is to apply (15.2.13) on the Floquet fibers to obtain the existence of the limit defining A and compute the diagonal part explicitly (fiberwise) to verify that $\ker(Q)$ is trivial.

Thus, the Bloch decomposition from Section 15.2.2 is the key ingredient in the proof of Theorem 15.2.5. Recall that we defined the direct integral space $\mathcal{H}_q$ by

$$\begin{aligned} \mathcal{H}_q &:= L^2\left([0, 2\pi), \mathbb{C}^q; \frac{d\theta}{2\pi}\right) \\ &= \left\{ f : [0, 2\pi) \to \mathbb{C}^q : \int_0^{2\pi} \|f(\theta)\|_{\mathbb{C}^q}^2 \frac{d\theta}{2\pi} < \infty \right\}. \end{aligned}$$

Let $\mathcal{F}$ be the Fourier transform defined by (15.2.6). It is not hard to see that $\mathcal{F}$ also diagonalizes the commutator A into a multiplication operator. Specifically, we have $\left(\widehat{A} f\right)(\theta) = A_\theta f(\theta)$ for $f \in \mathcal{H}_q$ and Lebesgue a.e. $\theta \in [0, 2\pi)$, where

$$A_\theta = i \begin{bmatrix} 0 & 1 & & & & -e^{-i\theta} \\ -1 & 0 & 1 & & & \\ & \ddots & \ddots & \ddots & & \\ & & & -1 & 0 & 1 \\ e^{i\theta} & & & & -1 & 0 \end{bmatrix}.$$

Now let $\Lambda_\theta \in \mathbb{C}^{q \times q}$ denote the diagonal matrix with diagonal entries $\langle e_k, \Lambda_\theta e_k \rangle = e^{ik\theta/q}$, i.e.,

$$\Lambda_\theta = \begin{bmatrix} e^{i\theta/q} & & & & \\ & e^{2i\theta/q} & & & \\ & & \ddots & & \\ & & & e^{i(q-1)\theta/q} & \\ & & & & e^{i\theta} \end{bmatrix}.$$

Define

$$\widetilde{H}_\theta = \Lambda_\theta^* H_\theta \Lambda_\theta, \quad \widetilde{A}_\theta = \Lambda_\theta^* A_\theta \Lambda_\theta.$$

By a direct calculation, one can verify that

$$\frac{\partial}{\partial\theta}\widetilde{H}_\theta = \frac{1}{q}\widetilde{A}_\theta. \tag{15.2.14}$$

The key calculation now is to compute the limit of the Heisenberg evolution of the truncated operators.

Lemma 15.2.6 *For each $\theta \in [0, 2\pi)$, let $\lambda_1(\theta) \le \cdots \le \lambda_q(\theta)$ denote the eigenvalues of H_θ (enumerated with multiplicities). Then, for every $\theta \in [0, 2\pi) \setminus \{0, \pi\}$, we have*

$$\lim_{t\to\infty} \frac{1}{t}\int_0^t e^{isH_\theta} A_\theta e^{-isH_\theta}\, ds = q\sum_{j=1}^{q} \frac{\partial\lambda_j}{\partial\theta}(\theta) P_j(\theta), \tag{15.2.15}$$

where $P_j(\theta)$ denotes projection onto the eigenspace corresponding to $\lambda_j(\theta)$.

Proof Sketch. Since H_θ is a finite Hermitian-symmetric matrix, the limit on the left-hand side of (15.2.15) exists and is simply equal to the diagonal part of A_θ with respect to the eigenspaces of H_θ. Since H_θ has simple spectrum for $\theta \ne 0, \pi$, it then suffices to show

$$\langle v_j(\theta), A_\theta v_j(\theta)\rangle = q\frac{\partial\lambda_j}{\partial\theta}(\theta)$$

for $1 \le j \le q$ and $\theta \ne 0, \pi$. This follows from (15.2.14). To see this, let $v_j(\theta)$ denote a continuous choice of a unit vector from the $\lambda_j(\theta)$ eigenspace of H_θ, put $\widetilde{v}_j(\theta) := \Lambda_\theta^* v_j(\theta)$ and observe that

$$\begin{aligned}
\frac{\partial\lambda_j}{\partial\theta}(\theta) &= \frac{\partial}{\partial\theta}\langle v_j(\theta), H_\theta v_j(\theta)\rangle \\
&= \frac{\partial}{\partial\theta}\langle \widetilde{v}_j(\theta), \widetilde{H}_\theta \widetilde{v}_j(\theta)\rangle \\
&= \frac{1}{q}\langle \widetilde{v}_j(\theta), \widetilde{A}_\theta \widetilde{v}_j(\theta)\rangle \\
&= \frac{1}{q}\langle v_j(\theta), A_\theta v_j(\theta)\rangle
\end{aligned}$$

for $\theta \ne 0, \pi$. □

Proof of Theorem 15.2.5. Once one has (15.2.15), we get the conclusions of Theorem 15.2.5 with

$$Q = \mathcal{F}^{-1}\left(\int_{[0,2\pi)}^{\oplus} \sum_{j=1}^{q} q\frac{\partial\lambda_j}{\partial\theta}(\theta) P_j(\theta)\, \frac{d\theta}{2\pi}\right)\mathcal{F}.$$

In particular, the limit exists by (15.2.15) and dominated convergence. The kernel of Q is trivial, since $\partial_\theta \lambda_j(\theta) \neq 0$ for $\theta \neq 0, \pi$ (which may be proved using (15.2.5) and implicit differentiation). □

15.3 Almost-Periodic Hulls

15.3.1 Almost-Periodic Sequences

The shift acts on $\ell^\infty(\mathbb{Z})$ via

$$[SV]_n = V_{n+1}.$$

Given V, the *orbit* of V is the set of its translates:

$$\mathrm{orb}(V) = \left\{ S^k V : k \in \mathbb{Z} \right\};$$

the *hull* of V is the closure of the orbit in ℓ^∞:

$$\mathrm{hull}(V) = \overline{\mathrm{orb}(V)}^{\ell^\infty(\mathbb{Z})}.$$

Naturally, it is easy to see that V is periodic if and only if hull(V) is finite. We say that V is *Bochner almost-periodic* if hull(V) is compact (in the $\ell^\infty(\mathbb{Z})$ topology). We say that V is *Bohr almost-periodic* if, for every $\varepsilon > 0$, the ε-almost periods of V are relatively dense in $\mathbb{Z}$. That is, V is Bohr almost-periodic if for every $\varepsilon > 0$, there is an $R > 0$ so that for any $k \in \mathbb{Z}$, there exists $\ell \in \mathbb{Z}$ with $|k - \ell| \leq R$ and

$$\|S^\ell V - V\|_\infty < \varepsilon.$$

Proposition 15.3.1 *An element $V \in \ell^\infty(\mathbb{Z})$ is almost-periodic in the Bohr sense if and only if it is almost-periodic in the Bochner sense.*

Thus, going forward, we only speak of potentials being *almost-periodic* and freely use both characterizations.

Proof Sketch. If V is Bochner-almost periodic, let $\varepsilon > 0$ and cover hull(V) by finitely many ε-balls centred at elements of orb(V), say

$$\mathrm{hull}(V) \subseteq \bigcup_{j=1}^{n} B_\varepsilon(S^{m_j} V).$$

Take $R = \max |m_j|$. Then, given k, choose m_j with $S^k V \in B_\varepsilon(S^{m_j} V)$ to get $\|S^{k-m_j} V - V\|_\infty < \varepsilon$ and hence V is Bohr almost-periodic.

Conversely, if V is Bohr almost-periodic, then one can cover hull(V) by finitely many ε-balls centred at elements of orb(V). Thus, hull(V) is complete and totally bounded, hence compact. □

The operation

$$(S^j V) \circledast (S^k V) = S^{j+k} V$$

makes orb(V) into a group. For abelian groups, one usually uses '+' or '*'; we have eschewed these symbols to avoid confusion with addition and convolution of functions.

In the event that V is periodic of minimal period q, it is clear that $\mathrm{orb}(V) \cong \mathbb{Z}_q$. Otherwise, when V is aperiodic, $\mathrm{orb}(V) \cong \mathbb{Z}$. We should note that this isomorphism is purely as groups without topology; in particular, the reader is invited to verify that orb(V) *never* has the discrete topology when V is almost-periodic and aperiodic.

Proposition 15.3.2 *The operation* $\circledast$ *extends uniquely to a continuous operation* $\circledast : \mathrm{hull}(V)^2 \to \mathrm{hull}(V)$. *With respect to this operation,* hull(V) *is a compact abelian topological group.*

Proof. Since S is an isometry, we have

$$\begin{aligned} \|S^{k+\ell} V - S^{k'+\ell'} V\|_\infty &= \|S^{k-k'} V - S^{\ell'-\ell} V\|_\infty \qquad (15.3.1) \\ &\leq \|S^{k-k'} V - V\|_\infty + \|S^{\ell'-\ell} V - V\|_\infty \\ &= \|S^k V - S^{k'} V\|_\infty + \|S^{\ell'} V - S^{\ell} V\|_\infty . \end{aligned}$$

This calculation shows that $\circledast$ is uniformly continuous on $\mathrm{orb}(V) \times \mathrm{orb}(V)$ and hence extends uniquely to a continuous operation

$$\circledast : \mathrm{hull}(V) \times \mathrm{hull}(V) \to \mathrm{hull}(V).$$

The abelian group properties of (hull(V), $\circledast$) follow from those of (orb(V), $\circledast$) and continuity. □

Clearly, if V is almost-periodic, then $\Omega = \mathrm{hull}(V)$ is also a *monothetic* group, i.e., Ω contains a dense cyclic subgroup (namely orb(V)). In view of this observation and Proposition 15.3.2, we see that any almost-periodic $V \in \ell^\infty(\mathbb{Z})$ can be written in the form

$$V(n) = f(n\alpha), \qquad (15.3.2)$$

where G is a compact, monothetic group whose topology is generated by a translation-invariant metric; $\alpha \in G$ generates a dense cyclic subgroup; and $f : G \to \mathbb{R}$ is continuous. The converse is also true, i.e., if V is of the form

(15.3.2) with f, G, and α as just described, then V is almost periodic. We leave the proof of this direction to the reader. In light of this connection, the theory of compact topological groups naturally plays an important role in the analysis of Schrödinger operators with almost-periodic potentials.

15.3.2 Limit-Periodic Hulls

One can characterize exactly what compact abelian groups are hulls of limit-periodic sequences.

Proposition 15.3.3 *Every limit-periodic V is almost-periodic. Moreover, an almost-periodic V is limit-periodic if and only if* $\mathrm{hull}(V)$ *is totally disconnected. Finally, an almost-periodic V is aperiodic if and only if no point of* $\mathrm{hull}(V)$ *is isolated.*

Proof. Suppose V is limit-periodic. For each $j \in \mathbb{Z}_+$, choose $V_j \in \ell^\infty(\mathbb{Z})$ and $q_j \in \mathbb{Z}_+$ such that $S^{q_j} V_j = V_j$ and

$$\lim_{j \to \infty} \|V - V_j\|_\infty = 0.$$

As before, it suffices to prove that $\mathrm{orb}(V)$ is totally bounded (since this also implies that $\mathrm{hull}(V)$ is totally bounded). Given $\varepsilon > 0$, choose j such that

$$\|V - V_j\|_\infty < \varepsilon.$$

One can then check that $\mathrm{orb}(V)$ is contained in the following finite union of ε-balls:

$$\bigcup_{k=0}^{q_j - 1} B(S^k V_j, \varepsilon).$$

Thus, $\mathrm{hull}(V)$ is complete and totally bounded, hence compact, so V is almost-periodic. Let us show that $\mathrm{hull}(V)$ is totally disconnected. Since the hull of V is a topological group, it suffices to show that there are arbitrarily small neighbourhoods of the identity (i.e., V) that are simultaneously open and closed. Therefore, given $\varepsilon > 0$ we will show that $B(V, \varepsilon) \cap \mathrm{hull}(V)$ contains a non-empty set that is both closed and open. Choose j so that $\|V_j - V\|_\infty \le \frac{\varepsilon}{2}$. Then,

$$\mathrm{hull}^{q_j}(V) := \overline{\{S^{kq_j} V : k \in \mathbb{Z}\}}$$

is a compact subgroup of $\mathrm{hull}(V)$ of index at most q_j. Clearly, $\mathrm{hull}^{q_j}(V)$ is closed, but it is also open since it is the complement of the union of no more than $q_j - 1$ other closed cosets. Moreover, since $S^{q_j} V_j = V_j$ and

$\|V_j - V\|_\infty \le \frac{\varepsilon}{2}$, every element $W \in \mathrm{hull}^{q_j}(V)$ satisfies $\|W - V_j\|_\infty \le \varepsilon/2$ and hence $\mathrm{hull}^{q_j}(V)$ is contained in the ε-ball centred at V. Consequently, $\mathrm{hull}(V)$ is totally disconnected.

Conversely, let V be almost-periodic and suppose its hull is totally disconnected. We have to show that V is limit-periodic. Given $\varepsilon > 0$, we have to find $W \in \ell^\infty(\mathbb{Z})$ and $p \in \mathbb{Z}_+$ with $S^p W = W$ and $\|W - V\|_\infty < \varepsilon$. By Proposition A.6, we may choose a compact open neighbourhood N of V in $\mathrm{hull}(V)$, small enough so that

$$\|(W_1 \circledast W_2) - W_1\|_\infty < \varepsilon/2 \quad \text{for all } W_1 \in \mathrm{hull}(V),\ W_2 \in N. \tag{15.3.3}$$

This is possible since $\circledast$ is uniformly continuous, V is the identity of $\mathrm{hull}(V)$ with respect to $\circledast$, and $\mathrm{hull}(V)$ is totally disconnected. Since the sets N and $\mathrm{hull}(V) \setminus N$ are compact and disjoint, there exists $\delta > 0$ so that $\|X - Y\| \ge \delta$ for all $X \in N$ and all $Y \in \mathrm{hull}(V) \setminus N$. By almost-periodicity of V, we can choose $p \ge 1$ so that

$$\|S^p V - V\|_\infty < \delta,$$

hence $S^p V \in N$ by our choice of δ. But then we find inductively that $\{S^{kp} V : k \in \mathbb{Z}\} \subseteq N$, by isometry of S and the choice of δ. Now consider the p-periodic W that coincides with V on $[0, p-1]$. Given $n \in \mathbb{Z}$, we write $n = r + \ell p$ with $\ell \in \mathbb{Z}$ and $0 \le r \le p - 1$. Then, it follows from (15.3.3) that

$$\begin{aligned} |V(n) - W(n)| &= |V(r + \ell p) - V(r)| \\ &= \left|(S^r V \circledast S^{\ell p} V)(0) - (S^r V)(0)\right| \\ &< \varepsilon/2, \end{aligned}$$

since $S^{\ell p} V \in N$. This shows that the p-periodic W obeys $\|W - V\|_\infty \le \varepsilon/2 < \varepsilon$, concluding the proof of the first two claims.

For the final claim, note first that $\mathrm{hull}(V)$ consists only of isolated points when V is periodic. Conversely, if V is aperiodic (and almost-periodic), let ε_j be any sequence converging to zero and pick ε_j almost-periods $0 < q_1 < q_2 < \cdots$. Then $S^{q_j} V \to V$ (in ℓ^∞), which implies that V is not isolated. We leave it to the reader to confirm that this implies that no point of $\mathrm{hull}(V)$ is isolated. □

In light of Proposition 15.3.3, the following definition is natural.

Definition A *Cantor group* is a compact, totally disconnected group that has no isolated points.

Example 15.3.4 The cyclic group $\mathbb{Z}_q$ is not a Cantor group, nor is the circle group $\mathbb{T} = \mathbb{R}/\mathbb{Z}$. Given a prime p, the group $\mathbb{J}_p$ of p-adic integers is a Cantor group. The group $\mathbb{J}_p$ admits a minimal translation, namely $T\omega = \omega + 1$. More generally, if $\mathfrak{q} = (q_1, q_2, \ldots)$ is a sequence of positive integers so that $q_j | q_{j+1}$ for every j, the inverse limit

$$\mathbb{J}_{\mathfrak{q}} = \varprojlim \mathbb{Z}_{q_j}$$

is a Cantor group; (the map from $\mathbb{Z}_{q_{j+1}}$ to $\mathbb{Z}_{q_j}$ is the canonical projection). For more about the p-adic integers and inverse limits, see the appendix.

In the course of the proof of Proposition 15.3.3, we encountered a device that is quite useful in the study of limit-periodic operators and hulls, in the guise of the closed subgroup generated by $S^q V$, which we denoted $\text{hull}^q(V)$. Namely, a Cantor group will have many compact finite-index subgroups. These compact subgroups are useful, as they provide a means of producing precisely those periodic potentials that can be represented via continuous functions on the hull. To be more specific, fix a monothetic Cantor group Ω and let us define $\mathcal{P}$ to be the subset of $f \in C(\Omega, \mathbb{R})$ with the property that $f \circ T^q = f$ for some $q \in \mathbb{Z}_+$. Clearly, if f satisfies such an identify, one has

$$V_\omega(n+q) = f(T^{n+q}\omega) = f(T^n\omega) = V_\omega(n),$$

so that the associated potentials are periodic.

Theorem 15.3.5 *$\mathcal{P}$ is dense in $C(\Omega, \mathbb{R})$.*

Proof. Since $\mathcal{P}$ is an algebra and contains all constant functions, it suffices (by Stone–Weierstrass) to verify that $\mathcal{P}$ separates points of ω. To that end, let $\omega \neq \omega'$ be given.

By Theorem A.7, the open subgroups of Ω comprise a neighbourhood basis for the topology of Ω at the identity. In light of this, choose $\Omega_0 \subseteq \Omega$ to be an open subgroup with $\omega - \omega' \notin \Omega_0$. Define a function $g : \Omega/\Omega_0 \to \mathbb{R}$ such that $g(\omega + \Omega_0) \neq g(\omega' + \Omega_0)$ (which is possible precisely because ω and ω' are inequivalent mod Ω_0). Then,

$$f(\omega) = g(\omega + \Omega_0)$$

yields a continuous[1] function with $f(\omega) \neq f(\omega')$. Moreover, since Ω_0 is open and Ω is compact, Ω_0 is a finite-index subgroup; denoting $q = \text{index}(\Omega_0)$, one necessarily has $q\alpha \in \Omega_0$, which implies

$$f(T^q x) = f(q\alpha + x) = f(x)$$

[1] Since Ω_0 is open, Ω/Ω_0 has the discrete topology, so continuity of f is free.

for all $x \in \Omega$, whence $f \in \mathcal{P}$. Thus, $\mathcal{P}$ is dense in $C(\Omega, \mathbb{R})$ by Stone–Weierstrass. $\square$

15.4 Ergodic Schrödinger Operators

We have seen that limit-periodic Schrödinger operators are those having potentials of the form

$$V(n) = V_\omega(n) = f(T^n\omega),$$

where ω is an element of Ω a monothetic metrizable Cantor group, $f : \Omega \to \mathbb{R}$ is continuous, and $T : \Omega \to \Omega$ is a translation of Ω by a generator of a dense cyclic subgroup. Thus, limit-periodic operators naturally fall into the category of operators with ergodic, dynamically defined potentials. The present subsection will describe some tools, techniques, and objects that this dynamical formalism enables us to use in the study of ergodic Schrödinger operators.

Definition Let $(\Omega, \mathcal{B}, \mu)$ be a probability measure space. That is, Ω is a non-empty set, $\mathcal{B}$ is a σ-algebra of subsets of Ω, and μ is a (positive) probability measure defined on $\mathcal{B}$. A *μ-ergodic transformation* $T : \Omega \to \Omega$ is a measurable transformation which is *μ-preserving* in the sense that

$$\mu(T^{-1}E) = \mu(E) \text{ for every } E \in \mathcal{B}$$

and which has the property that $\mu(E) \in \{0, 1\}$ whenever $E \subseteq \Omega$ satisfies $T^{-1}E = E$. We will also interchangeably say that μ is a *T-ergodic measure*.

Suppose T is invertible and that T^{-1} is also measurable, and let $f : \Omega \to \mathbb{R}$ denote a bounded, measurable function. The associated family of *ergodic Schrödinger operators* is defined by $H_\omega = \Delta + V_\omega$, where

$$V_\omega(n) = f(T^n\omega), \quad \omega \in \Omega, \ n \in \mathbb{Z}.$$

In view of Proposition 15.3.3, limit-periodic operators fall into this categorization by taking $\Omega = \text{hull}(V)$, $T\omega = \omega \circledast SV$, and $f(\omega) = \omega(0)$. The notation becomes somewhat redundant in that case, since one has $V_\omega = \omega$ for $\omega \in \text{hull}(V)$, but we will continue to write V_ω to better match current notational conventions in the literature. In this setting, one can check that the normalized Haar measure is the unique T-ergodic measure on Ω.

Throughout the remainder of the present section, we fix a measure space $(\Omega, \mathcal{B}, \mu)$, an invertible μ-ergodic transformation $T : \Omega \to \Omega$, a bounded measurable $f : \Omega \to \mathbb{R}$, and we let $\{H_\omega\}_{\omega \in \Omega}$ denote the associated family of ergodic Schrödinger operators. In this setting, one can leverage tools, techniques, and ideas from dynamical systems to prove results that hold μ-almost surely. First, one can alternatively characterize ergodicity in the following manner: if $f : \Omega \to \mathbb{R}$ is measurable and T-*invariant* in the sense that $f \circ T = f$, then f is almost-surely constant; i.e., there exists $c \in \mathbb{R}$ such that $f(\omega) = c$ for μ-a.e. $\omega \in \Omega$. In the present formalism, if $S : \ell^2(\mathbb{Z}) \to \ell^2(\mathbb{Z})$ denotes the left shift $\delta_n \mapsto \delta_{n-1}$, it is easy to verify that

$$H_{T\omega} = SH_\omega S^*, \tag{15.4.1}$$

so $H_{T\omega}$ is unitarily equivalent to H_ω via the shift. By an induction followed by a limiting argument, it follows that (15.4.1) holds for sufficiently nice functions of the operator, e.g.,

$$g(H_{T\omega}) = Sg(H_\omega)S^*, \quad g \in C(\mathbb{R}).$$

Consequently, any spectral data of H_ω is T-invariant so one should expect that if one can prove suitable measurability statements, then any spectral data of H_ω to be almost-surely constant with respect to the ergodic measure μ. As a sample result, this holds for the spectrum as a set as well as the spectral decomposition with respect to Lebesgue measure.

Theorem 15.4.1 *There exist compact sets $\Sigma, \Sigma_{\mathrm{ac}}, \Sigma_{\mathrm{sc}}, \Sigma_{\mathrm{pp}} \subseteq \mathbb{R}$ with the property that*

$$\begin{aligned}
\sigma(H_\omega) &= \Sigma, \ \mu\text{-almost surely}\\
\sigma_{\mathrm{ac}}(H_\omega) &= \Sigma_{\mathrm{ac}}, \ \mu\text{-almost surely}\\
\sigma_{\mathrm{sc}}(H_\omega) &= \Sigma_{\mathrm{sc}}, \ \mu\text{-almost surely}\\
\sigma_{\mathrm{pp}}(H_\omega) &= \Sigma_{\mathrm{pp}}, \ \mu\text{-almost surely}.
\end{aligned}$$

Moreover, one also has

$$\sigma_{\mathrm{disc}}(H_\omega) = \emptyset, \ \mu\text{-almost surely,}$$

and, for any $E \in \mathbb{R}$,

$$\mu\{\omega \in \Omega : E \text{ is an eigenvalue of } H_\omega\} = 0.$$

The previous theorem summarizes results of Kunz–Souillard [38] and Pastur [46]. Consult [11, Chapter 9] for proofs. See also [7, 13, 47]. Imposing additional assumptions on the base dynamics (Ω, T) and the sampling function f can allow one to draw stronger conclusions. For example, in the case where Ω is the hull of an almost periodic sequence and T is the shift map, the dynamical system (Ω, T) is *minimal* in the sense that $\{T^n\omega : n \in \mathbb{Z}\}$ is dense in Ω for all $\omega \in \Omega$ and the sampling function f is continuous. Under the assumptions of minimality of (Ω, T) and continuity of f, one can use strong operator approximation to upgrade the almost-sure spectrum to a completely-sure spectrum. That is, one has

$$\sigma(H_\omega) = \Sigma \text{ for all } \omega \in \Omega.$$

One can prove this by using very general results, e.g., [33, Theorem VIII-1.14] or [55, Theorem VIII.24]. It is a deep result of Last and Simon that this also holds true for the absolutely continuous spectrum. That is, if (Ω, T) is minimal and f is continuous, then $\sigma_{\mathrm{ac}}(H_\omega) = \Sigma_{\mathrm{ac}}$ for *all* $\omega \in \Omega$ [40].

15.4.1 The Lyapunov Exponent

Motivated by (15.2.2), define

$$A_z(\omega) = \begin{bmatrix} z - f(T\omega) & -1 \\ 1 & 0 \end{bmatrix}, \qquad \omega \in \Omega,\ z \in \mathbb{C}.$$

For $n \in \mathbb{Z}$, let

$$A_z^n(\omega) = \begin{cases} A_z(T^{n-1}\omega) A_z(T^{n-2}\omega) \cdots A_z(\omega) & n > 0, \\ I & n = 0, \\ A_z^{-n}(T^n\omega)^{-1} & n < 0. \end{cases}$$

Thus, if $u \in \mathbb{C}^{\mathbb{Z}}$ solves $H_\omega u = zu$ in the difference equation sense, one has

$$\begin{bmatrix} u(n+1) \\ u(n) \end{bmatrix} = A_z^n(\omega) \begin{bmatrix} u(1) \\ u(0) \end{bmatrix}$$

for all $n \in \mathbb{Z}$.

Definition The *Lyapunov exponent* of the ergodic family is defined by

$$L(z) = \lim_{n\to\infty} \frac{1}{n} \int \log \|A_z^n(\omega)\| \, d\mu(\omega), \quad z \in \mathbb{C}.$$

By Kingman's subadditive ergodic theorem [34], for each fixed $z \in \mathbb{C}$,

$$L(z) = \lim_{n\to\infty} \frac{1}{n} \log \|A_z^n(\omega)\| \tag{15.4.2}$$

for μ-a.e. $\omega \in \Omega$. In general, one cannot reverse the quantifiers. That is to say, it is not the case in general that there exists a z-independent full-measure set of $\Omega_{\mathrm{L}} \subseteq \Omega$ with the property that (15.4.2) holds simultaneously for all $z \in \mathbb{C}$ and every $\omega \in \Omega_{\mathrm{L}}$. In general, one can partially reverse the quantifiers by applying Fubini's theorem on a suitable product space. For example, one can say that, for μ-a.e. ω, one has (15.4.2) for Lebesgue a.e. $E \in \mathbb{R}$.

15.4.2 The Density of States

Next, we consider the density of states (DOS) in the ergodic setting. For each $\omega \in \Omega$ and each $N \in \mathbb{Z}_+$, we may define $dk_{\omega,N}$ as in (15.2.7), i.e.,

$$\int g(E)\, dk_{\omega,N} = \frac{1}{N} \mathrm{Tr}(P_N\, g(H_\omega)\, P_N^*),$$

for continuous g, where P_N is projection onto coordinates in $[0, N)$. In this setting, the weak limit of $dk_{\omega,N}$ (i.e., the DOS) exists μ-almost surely and is deterministic.

Theorem 15.4.2 *There is a deterministic probability measure dk and full-measure subset $\Omega_{\mathrm{DOS}} \subseteq \Omega$ with the property that $dk_{\omega,N}$ converges weakly to dk for all $\omega \in \Omega_{\mathrm{DOS}}$. Moreover,*

$$\int g\, dk = \int_\Omega \langle \delta_0, g(H_\omega)\delta_0 \rangle\, d\mu(\omega) \tag{15.4.3}$$

for $g \in C(\mathbb{R})$.

Proof. Let us consider a continuous $g : \mathbb{R} \to \mathbb{R}$ and define $\widetilde{g} : \Omega \to \mathbb{R}$ by

$$\widetilde{g}(\omega) = \langle \delta_0, g(H_\omega)\delta_0 \rangle.$$

By definition of $dk_{\omega,N}$, we then observe that

$$\begin{aligned} \int g\, dk_{\omega,N} &= \frac{1}{N} \mathrm{Tr}(P_N\, g(H_\omega)\, P_N^*) \\ &= \frac{1}{N} \sum_{n=0}^{N-1} \langle \delta_n, g(H_\omega)\delta_n \rangle. \end{aligned}$$

Applying (15.4.1), we obtain

$$\int g\, dk_{\omega,N} = \frac{1}{N} \sum_{n=0}^{N-1} \widetilde{g}(T^n \omega). \tag{15.4.4}$$

By Birkhoff's ergodic theorem, we know that the right-hand side converges to the integral of $\widetilde{g}$ μ-almost surely, and hence (15.4.3) holds by definition of $\widetilde{g}$. This provides a g-dependent full measure set Ω_g on which (15.4.3) holds; to obtain Ω_{DOS}, note that boundedness of the sampling function f implies that there is a uniform compact set K that contains the support of $dk_{\omega,N}$ for all ω and N; then, choose a countable dense set $\{g_n : n \geq 1\} \subseteq C(K, \mathbb{R})$, take

$$\Omega_{\mathrm{DOS}} = \bigcap_n \Omega_{g_n},$$

and apply an $\varepsilon/3$ argument. $\square$

For almost-periodic (and in particular limit-periodic) operators, the situation is even better. In this setting, normalized Haar measure is the unique invariant measure on the hull, so one says that (Ω, T) is *uniquely ergodic*. Then, an argument using unique ergodicity implies that $dk_{\omega,N}$ converges weakly to dk for *all* ω in the hull. Specifically, unique ergodicity implies that the limit of the quantity on the right-hand side of (15.4.4) exists for all ω, not just μ-a.e. $\omega \in \Omega$. Consequently, if V is limit-periodic, then $dk_{V,N}$ converges weakly to dk, and we may speak of 'the density of states of V' without worry.

We will be concerned with the regularity of the DOS and IDS. In full generality, the IDS is continuous [20].

Theorem 15.4.3 *For any ergodic family, the integrated density of states is a continuous function of E. Equivalently, the DOS is an atomless measure.*

Proof. Fix $E_0 \in \mathbb{R}$ and suppose g_n are continuous compactly supported functions with $g_n(E_0) = 1$ and $g_n(E) \downarrow 0$ for $E \neq E_0$. Then, by dominated convergence,

$$\int g_n \, dk \to \int \chi_{\{E_0\}} \, dk. \tag{15.4.5}$$

Denoting $A_\omega = \chi_{\{E_0\}}(H_\omega)$, we then also have $\langle \delta_0, g_n(H_\omega)\delta_0 \rangle \to \langle \delta_0, A_\omega \delta_0 \rangle$ for all ω, and hence, again by dominated convergence,

$$\begin{aligned} \int g_n \, dk &= \int \langle \delta_0, g_n(H_\omega)\delta_0 \rangle \, d\mu(\omega) \\ &\to \int \langle \delta_0, A_\omega \delta_0 \rangle \, d\mu(\omega). \end{aligned}$$

For ωs from a set of full μ-measure,[2]

$$\int \langle \delta_0, A_\omega \delta_0 \rangle \, d\mu(\omega) = \lim_{N \to \infty} \frac{1}{N} \operatorname{Tr}(P_N A_\omega P_N^*).$$

Since $\operatorname{Tr}(P_N A_\omega P_N^*)$ is bounded by 1, the limit is zero. Thus, $dk(\{E_0\}) = 0$, and dk is a continuous measure. Since k is the accumulation function of dk, continuity of k follows immediately. □

Theorem 15.4.4 *The almost sure spectrum is given by the points of increase of k, i.e.,* $\operatorname{supp}(dk) = \Sigma$.

[2] To get this full-measure set, apply the argument from Theorem 15.4.2 to the function $g = \chi_{\{E_0\}}$.

Proof. If $E_0 \notin \Sigma$, there is an open interval I containing E_0 with $I \cap \Sigma = \emptyset$. We then have $\chi_I(H_\omega) = 0$ for a.e. ω and hence

$$\int \chi_I \, dk = \int_\Omega \langle \delta_0, \chi_I(H_\omega)\delta_0 \rangle \, d\mu(\omega) = 0.$$

Thus, $E_0 \notin \operatorname{supp}(dk)$.

Conversely, if $E_0 \notin \operatorname{supp}(dk)$, there is an interval I containing E_0 such that $I \cap \operatorname{supp}(dk) = \emptyset$. Then,

$$\begin{aligned}\int \chi_I \, dk &= \int \langle \delta_0, \chi_I(H_\omega)\delta_0 \rangle \, d\mu(\omega) \\ &= \int \langle \delta_0, \chi_I(H_{T^n\omega})\delta_0 \rangle \, d\mu(\omega) \\ &= \int \langle \delta_n, \chi_I(H_\omega)\delta_n \rangle \, d\mu(\omega).\end{aligned}$$

By positivity, we obtain

$$\dim \operatorname{ran} \chi_I(H_\omega) = \operatorname{Tr} \chi_I(H_\omega) = 0$$

for μ-a.e. $\omega \in \Omega$ and hence $E_0 \notin \Sigma$. □

The density of states and the Lyapunov exponent are related via the *Thouless formula*:

$$L(z) = \int \log|E - z| \, dk(E), \quad z \in \mathbb{C}.$$

This was discovered on an intuitive basis by Thouless in the early 1970s [68], with the first rigorous proof due to Avron–Simon [5].

15.5 Absolutely Continuous Spectrum

In this section, we shall consider the limit-periodic potentials in the perturbative regime. Concretely, we will consider potentials V with q_n-periodic approximants such that

$$\lim_{n\to\infty} e^{bq_{n+1}} \|V_n - V\|_\infty = 0 \text{ for every } b > 0. \qquad (15.5.1)$$

In this case, we will write $V \in \mathrm{PT}$, after the contributions of [48, 49]; the inverse spectral problem for Jacobi matrices satisfying a condition like (15.5.1) was studied by Egorova [21]. See also [8, 9, 41]. The main point of the present section is that potentials in PT behave extremely similarly to periodic potentials.

We consider first the spectrum as a set. In the case when V is q-periodic, we have seen that the spectrum consists of q non-degenerate closed intervals, each of which is given as the closure of a connected component of $D^{-1}((-2, 2))$, where D is the Floquet discriminant. The intervals may touch at the endpoints but do not overlap otherwise. It should not be surprising that this behaviour does not persist for genuine aperiodic limit-periodic models. The number of spectral bands grows with the period, and the spectrum must lie within the interval

$$I = [-2 - \|V\|_\infty, 2 + \|V\|_\infty],$$

and so we cannot expect $\sigma(H)$ to consist of non-trivial closed intervals when V is aperiodic. Indeed, one can get Cantor sets for $V \in \mathrm{PT}$:

Theorem 15.5.1 *For a dense set $\mathcal{C} \subseteq \mathrm{PT}$, $\sigma(H_V)$ is a Cantor set for all $V \in \mathcal{C}$.*

However, for $V \in \mathrm{PT}$, the spectrum of H is thick in a precise sense that is nice from the point of view of inverse spectral theory. Concretely, one says that a set $\Sigma \subseteq \mathbb{R}$ is τ-*homogeneous* if there exists $\delta_0 > 0$ with the property that

$$\mathrm{Leb}(\Sigma \cap (x - \delta, x + \delta)) \geq \delta\tau$$

for all $0 < \delta \leq \delta_0$ and all $x \in \Sigma$. For instance, it is easy to see that $\sigma(H)$ is 1-homogenous if V is periodic. The property of homogeneity persists for all $V \in \mathrm{PT}$, which is due to Fillman and Lukic [24].

Theorem 15.5.2 *If $V \in \mathrm{PT}$, then $\sigma(H_V)$ is τ-homogeneous for every $\tau < 1$.*

Proof. We will describe how to prove homogeneity with $\tau = 1/2$; modifying the proof for $1/2 < \tau < 1$ involves fiddling with constants. If $V \in \mathrm{PT}$, let $V_j \to V$ be such that (15.5.1) holds, abbreviate $H_j = \Delta + V_j$, and denote $\Sigma_j = \sigma(H_j)$. There is a constant that only depends on $\|V\|_\infty$ with the property that each band of Σ_j has length at least K^{-q_j}. Using the PT condition, we may remove finitely many terms of the sequence $\{V_n\}_{n=1}^\infty$ and renumber to ensure that

$$\sum_{n=1}^{\infty} K^{q_{n+1}} \|V_n - V_{n+1}\|_\infty < \frac{1}{10}. \tag{15.5.2}$$

Put $\delta_0 = K^{-q_1}$. We will prove the following estimate:

$$|B_\delta(x) \cap \Sigma_N| \geq \delta/2 \text{ for all } x \in \Sigma_N \text{ and every } 0 < \delta \leq \delta_0 \tag{15.5.3}$$

for all $N \in \mathbb{Z}_+$. To that end, fix $N \in \mathbb{Z}_+$, $x \in \Sigma_N$, and $0 < \delta \leq \delta_0$. If $\delta \leq K^{-q_N}$, (15.5.3) is an obvious consequence of our choice of K, since δ is less than the length of the band of Σ_N which contains x in this case. Otherwise, $\delta > K^{-q_N}$, and there is a unique integer n with $1 \le n \le N-1$ such that

$$K^{-q_{n+1}} < \delta \le K^{-q_n}. \tag{15.5.4}$$

This integer n is relevant, as it determines the periodic approximant corresponding to the length scale δ. More precisely, by our choice of K, any band of Σ_n has length at least δ. On the other hand, by standard perturbation theory (e.g., Lemma 15.6.4), there exists $x_0 \in \Sigma_n$ with

$$|x - x_0| \le \sum_{\ell=n}^{N-1} \|V_\ell - V_{\ell+1}\|_\infty. \tag{15.5.5}$$

Using (15.5.4), we deduce

$$\begin{aligned} |x - x_0| &\le \sum_{\ell=n}^{N-1} \|V_\ell - V_{\ell+1}\|_\infty \\ &< \delta K^{q_{n+1}} \sum_{\ell=n}^{N-1} \|V_\ell - V_{\ell+1}\|_\infty \\ &< \delta \sum_{\ell=n}^{N-1} K^{q_{\ell+1}} \|V_\ell - V_{\ell+1}\|_\infty \\ &< \frac{\delta}{10}. \end{aligned} \tag{15.5.6}$$

Thus, there exists an interval I_0 with $x_0 \in I_0 \subseteq B_\delta(x) \cap \Sigma_n$ such that

$$\mathrm{Leb}(I_0) = \delta - \frac{\delta}{10} = \frac{9\delta}{10}.$$

By standard measure theory, we have

$$\mathrm{Leb}(B_\delta(x) \cap \Sigma_N) \ge \mathrm{Leb}(I_0 \cap \Sigma_n) - \sum_{\ell=n}^{N-1} \mathrm{Leb}(I_0 \cap (\Sigma_\ell \setminus \Sigma_{\ell+1})).$$

Our choice of K implies that the interval I_0 completely contains at most $\delta K^{q_{\ell+1}}$ bands of $\Sigma_{\ell+1}$ for each $\ell \ge n$. Consequently, perturbation theory (cf. Lemma 15.6.4) yields

$$\begin{aligned} \mathrm{Leb}(I_0 \cap (\Sigma_\ell \setminus \Sigma_{\ell+1})) &\le 2(\delta K^{q_{\ell+1}} + 1)\|V_\ell - V_{\ell+1}\|_\infty \\ &\le 4\delta K^{q_{\ell+1}} \|V_\ell - V_{\ell+1}\|_\infty. \end{aligned}$$

Notice that the extra term in the parentheses on the first line is needed to account for possible boundary effects. Summing this over ℓ and estimating the result with (15.5.2), we obtain

$$\sum_{\ell=n}^{N-1} \mathrm{Leb}(I_0 \cap (\Sigma_\ell \setminus \Sigma_{\ell+1})) \le \sum_{\ell=n}^{N-1} 4\delta K^{q_{\ell+1}} \|V_\ell - V_{\ell+1}\|_\infty < \frac{2\delta}{5}.$$

Putting all of this together, we have

$$|B_\delta(x) \cap \Sigma_N| \ge |I_0 \cap \Sigma_n| - \sum_{\ell=n}^{N-1} |I_0 \cap (\Sigma_\ell \setminus \Sigma_{\ell+1})| > \frac{9\delta}{10} - \frac{2\delta}{5} = \frac{\delta}{2}.$$

This proves (15.5.3) for arbitrary $N \in \mathbb{Z}_+$. Since Lebesgue measure is upper semi-continuous with respect to the topology induced by the Hausdorff metric, we obtain

$$|B_\delta(x) \cap \Sigma| \ge \delta/2 \text{ for all } x \in \Sigma, \text{ and } 0 < \delta \le \delta_0,$$

so Σ is homogeneous, as promised. □

The motivation to study homogeneity of the spectrum of a Schrödinger operator (or Jacobi matrix) arises from inverse spectral theory. Loosely speaking, if $\Sigma \subseteq \mathbb{R}$ is homogeneous, then one has a very nice description of the set of Jacobi matrices with spectrum Σ and which are reflectionless thereupon, due to M. Sodin and Yuditskii; all such operators are almost-periodic, and they comprise a torus whose dimension coincides with the number of gaps in Σ [66]. On homogeneous sets, one has a potent generalization of a theorem of Kotani for ergodic Schrödinger operators. Concretely, let $\{H_\omega\}$ be an ergodic family with Lyapunov exponent L. Kotani showed that if L vanishes in an interval, then (almost surely with respect to the underlying ergodic measure) the spectrum of H_ω is purely absolutely continuous thereon (compare [35, Theorem 4.2]). By a result of Poltoratski and Remling [51], one may replace 'interval' by 'homogeneous set' and deduce the pure absolute continuity of the spectrum (in fact, one can assume a 'weak' version of homogeneity in which some constants are allowed to vary over the spectrum). There are also results for reflectionless continuum Schrödinger operators [27, 65] and CMV matrices [28].

Theorem 15.5.3 *If $V \in \mathrm{PT}$, then H has purely a.c. spectrum.*

Proof Sketch. Let $\Omega = \mathrm{hull}(V)$, denote by Σ the common spectrum of H_ω for all $\omega \in \Omega$, and let $\Sigma_\omega^{(q)}$ denote the spectrum of the q-periodic operator obtained by repeating the string $V_\omega(1), \ldots, V_\omega(q)$. From the definition of PT and perturbation theory, one can see that

$$\lim_{n\to\infty} \mathrm{Leb}\left(\Sigma \setminus \Sigma_\omega^{(q_n)}\right) = 0 \tag{15.5.7}$$

for all ω. Defining $\mathcal{Z} = \{z \in \mathbb{C} : L(z) = 0\}$ where L denotes the Lyapunov exponent, one can use (15.5.7) and a result of Last to show that

$\mathrm{Leb}(\Sigma \setminus \mathcal{Z}) = 0$ [39]. From this, Remling's theorem implies that H is reflectionless on Σ [56]. Consequently, the spectral type is purely absolutely continuous by homogeneity and Poltoratski–Remling [51]. □

At the level of quantum dynamics, the potentials in PT also exhibit strong ballistic motion in the sense of Asch–Knauf, which is due to Fillman [23].

Theorem 15.5.4 *If* $V \in \mathrm{PT}$*, then the Schrödinger group generated by* H_V *exhibits strong ballistic motion in the sense that*

$$Q = \operatorname*{s-lim}_{t \to \infty} t^{-1} X(t)$$

exists and $\ker(Q) \cap D(X) = \{0\}$.

The key ingredient in the proof of Theorem 15.5.4 is to get precise quantitative estimates on the rate at which the strong convergence in Theorem 15.2.5 occurs. Naïvely thinking about how the convergence was originally proved, one can see easily that the convergence rate is roughly proportional to t^{-1}. However, the constant of proportionality depends on the length of the smallest gap, and thus the naïve estimates completely break when a gap degenerates; moreover, it is difficult to divine the length of the smallest gap from coefficient data alone; finally, even in the generic ('all gaps open') scenario, since we are interested in the aperiodic case, one absolutely needs estimates that are independent of the period and the gap size. Consequently, one needs a more robust estimate that averages out resonances. One also needs to be sure that the estimate one obtains must have constants that do not depend on the period (since one would eventually need to send the period to ∞) or at least some quantitative control on the constants as functions of the period.

Theorem 15.5.5 *For every q-periodic Schrödinger operator H, and every $\varphi \in D(X)$, one has*

$$\left\| Q\varphi - \frac{1}{t} X(t)\varphi \right\| \le t^{-1} \|X\varphi\| + C_1^q \|\varphi\|_1 t^{-1/5}, \tag{15.5.8}$$

where C_1 denotes a constant that depends solely on $\|V\|_\infty$ and $\|\varphi\|_1$ denotes the ℓ^1 norm of φ.

15.6 Singular Continuous Spectrum

Fix a monothetic Cantor group Ω and an $\alpha \in \Omega$ so that $\{n\alpha : n \in \mathbb{Z}\}$ is dense in Ω. If $f \in C(\Omega, \mathbb{R})$ and $\omega \in \Omega$, then the potential given by

$$V_{\omega,f}(n) = f(\omega + n\alpha), \quad n \in \mathbb{Z} \tag{15.6.1}$$

is limit-periodic. The associated Schrödinger operator is denoted by $H_{\omega,f}$ and the spectrum of $H_{\omega,f}$, which is independent of ω by minimality, is denoted by $\Sigma(f)$.

Theorem 15.6.1 *There exists a dense G_δ set $\mathcal{S} \subseteq C(\Omega, \mathbb{R})$ such that for every $f \in \mathcal{S}$ and every $\omega \in \Omega$, the spectrum of $H_{\omega,f}$ is a Cantor set of zero Lebesgue measure and $H_{\omega,f}$ has purely singular continuous spectrum.*

This theorem follows from two separate observations that lead to generic spectral results, the combination of which is the conclusion in Theorem 15.6.1. The first observation is that one can carry out periodic approximation with control on the measure of the spectrum of the approximant. This leads to generic zero-measure spectrum, which as a byproduct also precludes any absolutely continuous spectral measures. The second observation shows that quantitative approximation with periodic potentials allows one to generically conclude that the limit-operators share an important property with the approximants: the absence of square-summable (and in fact decaying) solutions, which in turn shows that the spectral measures will also have no atoms. The next two sub-sections explain these two parts in more detail.

15.6.1 Zero-Measure Spectrum

Theorem 15.6.2 *There exists a dense G_δ subset $\mathcal{C} \subseteq C(\Omega, \mathbb{R})$ so that $\Sigma(f)$ is a Cantor set of zero Lebesgue measure for all $f \in \mathcal{C}$.*

Remark 15.6.3 In fact, one may pass from the generic setting to the dense setting and obtain even finer control on the size of the spectrum. Namely, there is a dense family whose spectra are of Hausdorff dimension zero. Both Theorem 15.6.2 and the strengthening described in this remark are due to Avila [3].

The proof of Theorem 15.6.2 is based on two key perturbative lemmas. Recall that the *Hausdorff distance* between two compact non-empty sets $F, K \subset \mathbb{R}$ is defined by

$$d_{\mathrm{H}}(F, K) = \inf\{\varepsilon > 0 : F \subset B_\varepsilon(K) \text{ and } K \subset B_\varepsilon(F)\},$$

where $B_\varepsilon(S)$ denotes the ε-neighbourhood of the set S. Then the following estimate is not difficult to prove (exercise).

Lemma 15.6.4 *If A and B are bounded self-adjoint operators on a Hilbert space $\mathcal{H}$, then*

$$d_{\mathrm{H}}(\sigma(A), \sigma(B)) \leq \|A - B\|.$$

The second key lemma shows that any sampling function may be uniformly approximated by periodic sampling functions whose spectra are exponentially small (with respect to the period). To formulate the lemma, we need to precisely say what periods a function in $\mathcal{P}$ may have. To that end, fix a sequence $\Omega = \Omega_0 \supset \Omega_1 \supset \Omega_2 \supset \cdots$ of open subgroups of Ω with

$$\bigcap_{n=1}^{\infty} \Omega_n = \{0\},$$

where 0 denotes the identity element of Ω. Such a sequence exists by Proposition A.7 and metrizability of Ω. Each Ω_n has finite index q_n by compactness, and one clearly has $q_n \to \infty$ as $n \to \infty$. One can use Ω_n to define periodic sampling functions of period q_n by the construction in the proof of Theorem 15.3.5. That is, if $\tilde{f} : \Omega/\Omega_n \to \mathbb{R}$ is any function, then

$$f(\omega) = \tilde{f}(\omega + \Omega_n)$$

defines a q_n-periodic sampling function in $\mathcal{P}$.

Lemma 15.6.5 *Fix $f \in C(\Omega, \mathbb{R})$. For all $\varepsilon > 0$, there exist $c = c(f, \varepsilon) > 0$ and $n_0 = n_0(f, \varepsilon) \in \mathbb{Z}_+$ such that the following holds true. For all $n \geq n_0$, there exists $g = g_n \in C(\Omega, \mathbb{R})$ of period q_n such that*

$$\mathrm{Leb}(\Sigma(g)) \leq e^{-cq_n}$$

and $\|f - g\|_\infty < \varepsilon$.

Proof Sketch. From Theorem 15.3.5, $\mathcal{P}$ is dense in $C(\Omega, \mathbb{R})$, so it suffices to prove the lemma for periodic sampling functions. Let $f \in C(\Omega, \mathbb{R})$ be a given q-periodic sampling function. The construction works by first perturbing f to produce a family of sampling functions which are very close to f, and whose resolvent sets cover the line. Thus, for every $E \in \mathbb{R}$, one of these new potentials will have $L(E) > 0$. We then form a new potential by concatenating these finite families over long blocks. Positive exponents over sub-blocks enable us to produce growth of transfer matrices, and one can then parlay growth of transfer matrices into upper bounds on band lengths. □

Proof of Theorem 15.6.2. For each $\delta > 0$, define

$$U_\delta = \{f \in C(\Omega, \mathbb{R}) : \mathrm{Leb}(\Sigma(f)) < \delta\}.$$

To prove the theorem, we will show that U_δ is open and dense for all $\delta > 0$. To that end, suppose given $f \in U_\delta$. In essence, small perturbations of f remain in U_δ by continuity of the spectrum. The details follow.

Since $\Sigma(f)$ is a compact set with $\mathrm{Leb}(\Sigma(f)) < \delta$, we may choose finitely many open intervals $J_1, \ldots, J_n$ such that

$$\Sigma(f) \subseteq \bigcup_{k=1}^{n} J_k, \qquad \sum_{k=1}^{n} \mathrm{Leb}(J_k) < \delta.$$

Next, choose $\varepsilon > 0$ small enough that

$$B_\varepsilon(\Sigma(f)) \subseteq \bigcup_{k=1}^{n} J_k.$$

Now, if $\|f - f'\|_\infty < \varepsilon$, then Lemma 15.6.4 implies

$$\Sigma(f') \subseteq B_\varepsilon(\Sigma(f)) \subseteq \bigcup_{k=1}^{n} J_k.$$

Consequently, $\mathrm{Leb}(\Sigma(f')) < \delta$. Thus, U_δ is open.

On the other hand, Lemma 15.6.5 clearly implies that U_δ is dense for all $\delta > 0$. Consequently,

$$\mathcal{C} = \bigcap_{k=1}^{\infty} U_{1/k}$$

is a dense G_δ in $C(\Omega, \mathbb{R})$ by the Baire Category Theorem; clearly $\mathrm{Leb}(\Sigma(f)) = 0$ for every $f \in \mathcal{C}$. □

15.6.2 Continuous Spectrum

In order to prove that spectral measures are continuous, one has to exclude eigenvalues. In other words, one needs to show that for any E, the difference equation (15.2.1) does not admit any square-summable solutions u.

A classical sufficient condition is due to Gordon. This condition is applicable whenever one has very good approximation of a given potential V by periodic potentials over a (rather small) fixed number of periods around the origin. Clearly, for limit-periodic V, the number of periods for which one has the desired rate of approximation may be arbitrary, so all one needs to take care of is the size of the error relative to the size of the period. It turns out that the absence of eigenvalues can be established in this way for a generic set of sampling functions.

Let us begin by recalling the abstract Gordon criterion. Due to the reformulation (15.2.2) of (15.2.1), a sequence $u \in \mathbb{C}^{\mathbb{Z}}$ solves (15.2.1) if and only if the sequence $\mathbf{U} : \mathbb{Z} \to \mathbb{C}^2$ defined by

$$\mathbf{U}(n) = \begin{bmatrix} u(n+1) \\ u(n) \end{bmatrix}, \quad n \in \mathbb{Z} \tag{15.6.2}$$

solves

$$\mathbf{U}(n) = M_E(n)\mathbf{U}(0), \tag{15.6.3}$$

where

$$M_E(n) = \begin{cases} A_E(n)A_E(n-1)\cdots A_E(1) & n > 0, \\ I & n = 0, \\ A_E(n+1)^{-1}A_E(n+2)^{-1}\cdots A_E(0)^{-1} & n < 0 \end{cases}$$

and

$$A_E(m) = \begin{bmatrix} E - V(m) & -1 \\ 1 & 0 \end{bmatrix}.$$

Lemma 15.6.6 *Suppose* $V : \mathbb{Z} \to \mathbb{R}$ *obeys* $V(n+q) = V(n)$ *for some* $q \in \mathbb{Z}_+$ *and* $-q+1 \le n \le q$, $z \in \mathbb{C}$, *and* u *solves* (15.6.3). *Then, we have*

$$\max\left\{ \|\mathbf{U}(-q)\|, \|\mathbf{U}(q)\|, \|\mathbf{U}(2q)\| \right\} \ge \frac{1}{2}\|\mathbf{U}(0)\|. \tag{15.6.4}$$

Proof. By assumption, we have

$$\mathbf{U}(2q) = M_E(2q)\cdot\mathbf{U}(0) = M_E(q)^2\cdot\mathbf{U}(0)$$

and similarly

$$\mathbf{U}(q) = M_E(q)^2\cdot\mathbf{U}(-q).$$

Moreover, the Cayley–Hamilton theorem implies

$$M_E(q)^2 - \mathrm{Tr}(M_E(q))\cdot M_E(q) + I = 0.$$

Consequently, we have

$$\mathbf{U}(2q) - \mathrm{Tr}(M_E(q))\cdot\mathbf{U}(q) + \mathbf{U}(0) = \begin{bmatrix} 0 \\ 0 \end{bmatrix} \tag{15.6.5}$$

and

$$\mathbf{U}(q) - \mathrm{Tr}(M_E(q))\mathbf{U}(0) + \mathbf{U}(-q) = \begin{bmatrix} 0 \\ 0 \end{bmatrix}. \tag{15.6.6}$$

The assertion (15.6.4) follows from (15.6.5) when $|\mathrm{Tr}(M_E(q))| \le 1$ and it follows from (15.6.6) when $|\mathrm{Tr}(M_E(q))| > 1$. □

The estimate (15.6.4) can of course be used to exclude the existence of decaying solutions.

Theorem 15.6.7 *Suppose that there exist $q_k \to \infty$ such that*

$$V(n - q_k) = V(n) = V(n + q_k) \text{ for all } 1 \le n \le q_k.$$

Then H has purely continuous spectrum.

Proof. Let E be given, and let u denote a non-trivial solution to (15.6.3), then Lemma 15.6.6 implies that

$$\max(\|\mathbf{U}(-q_k)\|, \|\mathbf{U}(q_k)\|, \|\mathbf{U}(2q_k)\|) \ge \frac{1}{2}\|\mathbf{U}(0)\|.$$

Thus, u cannot go to zero at $\pm\infty$, so $u \notin \ell^2(\mathbb{Z})$. Consequently, z is not an eigenvalue of H. □

Notice that the energy E plays no role in the argument. As soon as the potential V has the required local periodicity for infinitely many values of q, we have the estimate (15.6.4) for infinitely many values of q, which in turn shows that no E can be an eigenvalue.

It is clear that one can perturb about this situation a little bit and still deduce useful estimates. In light of this, the following definition is natural.

Definition A bounded potential $V : \mathbb{Z} \to \mathbb{R}$ is called a *Gordon potential* if there are positive integers $q_k \to \infty$ such that

$$\max_{1 \le n \le q_k} |V(n) - V(n \pm q_k)| \le k^{-q_k} \tag{15.6.7}$$

for every $k \ge 1$. Equivalently, there are positive integers $q_k \to \infty$ such that

$$\forall C > 0 : \lim_{k\to\infty} C^{q_k} \max_{1 \le n \le q_k} |V(n) - V(n \pm q_k)| = 0. \tag{15.6.8}$$

Theorem 15.6.8 *If V is a Gordon potential, then H has purely continuous spectrum. More precisely, for every $E \in \mathbb{C}$ and every solution u of* (15.2.1), *we have*

$$\limsup_{|n|\to\infty} \|\mathbf{U}(n)\| \ge \frac{1}{2} \|\mathrm{U}(0)\|, \tag{15.6.9}$$

where $\mathbf{U}$ is defined by (15.6.2).

Proof. By assumption, there is a sequence $q_k \to \infty$ such that (15.6.8) holds. Given $E \in \mathbb{C}$, we consider a solution u of (15.6.3) and, for every k, a solution u_k of

$$u_k(n+1) + u_k(n-1) + V_k(n)u_k(n) = Eu_k(n)$$

with $u_k(1) = u(1)$ and $u_k(0) = u(0)$, where V_k is the q_k-periodic potential that coincides with V on the interval $1 \le n \le q_k$. It follows from Lemma 15.6.6 that u_k satisfies the estimate

$$\max\left\{ \|\mathbf{U}_k(-q_k)\|, \|\mathbf{U}_k(q_k)\|, \|\mathbf{U}_k(2q_k)\| \right\} \geq \frac{1}{2}\|\mathbf{U}(0)\|,$$

where $\mathbf{U}_k(n) = (u_k(n+1), u_k(n))^\top$, as usual. Since V is very close to V_k on the interval $[-q_k+1, 2q_k]$ and u and u_k have the same initial conditions, we expect that they are close throughout this interval and hence u obeys a similar estimate.

Let us make this observation explicit. Denote the transfer matrices associated with V_k by $M_{k,E}(n)$. We have

$$\begin{aligned}\max_{-q_k \leq n \leq 2q_k} \|\mathbf{U}(n) - \mathbf{U}_k(n)\| &\leq \max_{-q_k \leq n \leq 2q_k} \left\| M_E(n) - M_{k,E}(n) \right\| \|\mathbf{U}(0)\| \\ &\leq C^{q_k} \max_{-q_k+1 \leq n \leq 2q_k} |V(n) - V_k(n)| \|\mathbf{U}(0)\|,\end{aligned}$$

which goes to zero by (15.6.8). □

We are now ready to apply the Gordon criterion in the context of limit-periodic potentials; compare the paper [16] by Damanik and Gan.

Proposition 15.6.9 *There exists a dense G_δ set $\mathcal{G} \subseteq C(\Omega, \mathbb{R})$ such that the potential $V_{\omega,f}$ defined by* (15.6.1) *is a Gordon potential for all $f \in \mathcal{G}$ and all $\omega \in \Omega$. That is, $V_{\omega,f}$ satisfies* (15.6.7) *for suitable positive integers $q_k \to \infty$.*

Proof. Define $\varepsilon_n = n^{-n}$ and

$$\mathcal{O}_k = \bigcup_{q=1}^{\infty} \bigcup_{f \in P_q} B\left(f, \frac{\varepsilon_{kq}}{2}\right), \quad k \in \mathbb{Z}_+,$$

where $P_q \subset C(\Omega, \mathbb{R})$ denotes the set of q-periodic sampling functions defined over Ω. Since $\mathcal{O}_k$ is clearly open, it follows that

$$\mathcal{G} := \bigcap_{k=1}^{\infty} \mathcal{O}_k$$

is a G_δ in $C(\Omega, \mathbb{R})$. One can easily verify that $\mathcal{G}$ contains all periodic sampling functions, so, since $\mathcal{P}$ is dense by Theorem 15.3.5, $\mathcal{G}$ is dense as well.

We next show that all elements of $\mathcal{G}$ produce Gordon potentials. To that end, let $f \in \mathcal{G}$ and $\omega \in \Omega$ be given. By definition, $f \in \mathcal{O}_k$ for each k, so we may produce a sequence $f_k \in \mathcal{P}$ such that f_k is ℓ_k-periodic for some $\ell_k \in \mathbb{Z}_+$ and $\|f - f_k\|_\infty < \frac{1}{2}\varepsilon_{k \cdot \ell_k}$ for each k. Set $q_k = k \cdot \ell_k$ and note that $q_k \geq k$, whence $q_k \to \infty$. For all k, n such that $1 \leq n \leq q_k$, we use the triangle inequality and periodicity to obtain

$$\begin{aligned}|V_\omega(n) - V_\omega(n \pm q_k)| &= \left|f(T^n\omega) - f(T^{n\pm q_k}\omega)\right| \\ &\le \left|f(T^n\omega) - f_k(T^n\omega)\right| + \left|f_k(T^n\omega) - f(T^{n\pm q_k}\omega)\right| \\ &= \left|f(T^n\omega) - f_k(T^n\omega)\right| + \left|f_k(T^{n\pm k\ell_k}\omega) - f(T^{n\pm q_k}\omega)\right| \\ &< \varepsilon_{k\cdot\ell_k} \\ &\le k^{-q_k}.\end{aligned}$$

Taking the max over n with $1 \le n \le q_k$ yields

$$\max_{1\le n\le q_k} |V(n) - V(n \pm q_k)| \le k^{-q_k},$$

and hence $V_\omega = V_{\omega,f}$ is a Gordon potential for all $\omega \in \Omega$ and all $f \in \mathcal{G}$. □

Proof of Theorem 15.6.1. Let $\mathcal{C}$, $\mathcal{G} \subseteq C(\Omega, \mathbb{R})$ be the sets constructed in Theorem 15.6.2 and Proposition 15.6.9 respectively, and put $\mathcal{S} = \mathcal{C} \cap \mathcal{G}$. By the Baire Category Theorem, $\mathcal{S}$ is a dense G_δ. Since $\mathcal{S} \subseteq \mathcal{C}$, the zero-measure Cantor set statement follows.

For all $f \in \mathcal{S}$ and $\omega \in \Omega$, the Lebesgue measure of $\sigma(H_\omega)$ is zero, which precludes the presence of absolutely continuous spectrum. Similarly, for every $f \in \mathcal{S}$ and $\omega \in \Omega$, V_ω is a Gordon potential, so by Theorem 15.6.8, it follows that H_ω has no eigenvalues. Thus, H_ω has purely singular continuous spectrum, as claimed. □

15.7 Pure Point Spectrum

We have seen that purely absolutely continuous spectrum and purely singular continuous spectrum are each dense phenomena in the space of limit-periodic Schrödinger operators, with the latter being even generic. This raises the natural question of what can be said about cases with pure point spectrum. It turns out that even this spectral type is a dense phenomenon. Clearly this is surprising as in these cases the spectral type of the limit object is as different from the spectral type of the approximants as it can be.

Theorem 15.7.1 *For every limit-periodic V and every $\varepsilon > 0$, there is a limit-periodic $\widetilde{V}$ with $\|V - \widetilde{V}\|_\infty < \varepsilon$ such that the Schrödinger operator with potential $\widetilde{V}$ has pure point spectrum.*

Note that we do not fix the base dynamics and vary the sampling function only. Rather we perturb in the space of all limit-periodic potentials. It would be interesting to prove a variant of Theorem 15.7.1 with fixed base dynamics.

The proof of Theorem 15.7.1 actually does not really care about the fact that the potential V we are perturbing is limit-periodic. Indeed, the proof directly yields the following more general result, which is due to Damanik and Gorodetski [18]:

Theorem 15.7.2 *For every bounded V and every $\varepsilon > 0$, there is a limit-periodic potential V_{lp} with $\|V_{\mathrm{lp}}\|_\infty < \varepsilon$ such that the Schrödinger operator with potential $V + V_{\mathrm{lp}}$ has pure point spectrum.*

In other words, a suitable arbitrarily small limit-periodic perturbation can turn any given spectral type into one that is pure point. Such a strong dominance property was previously only known for small random perturbations. Not coincidentally, the proof of Theorem 15.7.2 is based on a generalization of the argument that yields the latter statement. Thus, let us first explain how pure point spectrum can be proved for random potentials, then for perturbations of fixed background potentials by arbitrarily small random potentials, and finally for the setting in question – perturbations of fixed background potentials by arbitrarily small limit-periodic potentials.

The way we want to explain why random potentials lead to pure point spectrum is based on the Kunz–Souillard method [38]. Suppose V_ω is a bounded random potential and H_ω is the associated Schrödinger operator. Integration with respect to the underlying probability measure μ will be denoted by $\mathbb{E}(\cdot)$. A key quantity in this approach is

$$a(n,m) = \mathbb{E}\left(\sup_{t\in\mathbb{R}} \left|\left\langle \delta_n, e^{-itH_\omega}\delta_m\right\rangle\right|\right), \quad n, m \in \mathbb{Z}.$$

The following sufficient criterion for (almost sure) pure point spectrum follows from the RAGE theorem:

Proposition 15.7.3 *Suppose that*

$$\sum_{n\in\mathbb{Z}} |a(n,m)| < \infty$$

for $m = 0, 1$. Then, for μ-almost every ω, the operator H_ω has pure point spectrum.

For $L \in \mathbb{Z}_+$, we denote by $H_\omega^{(L)}$ the restriction of H_ω to $\ell^2(\{-L, \ldots, L\})$ with Dirichlet boundary conditions. For $|n|, |m| \leq L$, we define $a_L(n,m)$ to be $a(n,m)$ with H_ω replaced by $H_\omega^{(L)}$, i.e.,

$$a_L(n,m) = \mathbb{E}\left(\sup_{t\in\mathbb{R}} \left|\left\langle \delta_n, e^{-itH_\omega^{(L)}}\delta_m\right\rangle\right|\right).$$

It is easy to see that $H_\omega^{(L)}$ has $2L + 1$ real simple eigenvalues

$$E_\omega^{L,1} < E_\omega^{L,2} < \cdots < E_\omega^{L,2L+1}.$$

By our boundedness assumption there exists a compact interval Σ_0 that contains all eigenvalues $E_\omega^{L,k}$. For each k, let $\varphi_\omega^{L,k}$ denote a normalized eigenvector corresponding to $E_\omega^{L,k}$. We now define

$$\rho_L(n,m) = \mathbb{E}\left(\sum_{k=1}^{2L+1} \left|\left\langle \delta_n, \varphi_\omega^{L,k}\right\rangle\right| \left|\left\langle \delta_m, \varphi_\omega^{L,k}\right\rangle\right|\right).$$

Lemma 15.7.4 (a) *For $n, m \in \mathbb{Z}$, we have*

$$a(n,m) \le \liminf_{L\to\infty} a_L(n,m). \tag{15.7.1}$$

(b) *If $|n|, |m| \le L$, then*

$$a_L(n,m) \le \rho_L(n,m). \tag{15.7.2}$$

Part (a) follows from strong approximation and Fatou's lemma and part (b) follows by simply expanding δ_m in the basis of eigenvectors and using that the eigenvalues are real. It follows that we can estimate $a(n,m)$ from above by proving upper bounds for $\rho_L(n,m)$ that are uniform in L.

Let us now assume, for the time being, that the potential of the random Schrödinger operator H_ω is generated by independent identically distributed random variables, each of which has a bounded compactly supported density r. Thus, we can write

$$\rho_L(n,m) = \int \cdots \int \left(\sum_{k=1}^{2L+1} \left|\left\langle \delta_n, \varphi_{v_{-L},\dots,v_L}^{L,k}\right\rangle\right| \left|\left\langle \delta_m, \varphi_{v_{-L},\dots,v_L}^{L,k}\right\rangle\right|\right) \left(\prod_{j=-L}^{L} r(v_j)\right) dv_{-L} \dots dv_L, \tag{15.7.3}$$

since $H_\omega^{(L)}$ only depends on the random variables corresponding to $|j| \le L$. Moreover, it also follows that $\rho_L(n,m) = \rho_L(m,n)$ and $\rho_L(n,m) = \rho_L(n - m, 0)$, so that it suffices to estimate $\rho_L(n,m)$ for $n \ge 0$ and $m = 0$.

Let us apply a suitable change of variables to the iterated integral expressing $\rho_L(n,0)$, $n \ge 0$. If E is $E_{v_{-L},\dots,v_L}^{L,k}$ and u is $\varphi_{v_{-L},\dots,v_L}^{L,k}$, then we have

$$u(\ell+1) + u(\ell-1) + v_\ell\, u(\ell) = Eu(\ell) \tag{15.7.4}$$

for $-L \le \ell \le L$, where $u(-L-1) = u(L+1) = 0$. We rewrite this identity as

$$v_\ell = E - \frac{u(\ell+1)}{u(\ell)} - \frac{u(\ell-1)}{u(\ell)}, \tag{15.7.5}$$

and this in turn motivates the change of variables

$$\{v_\ell\}_{\ell=-L}^{L} \quad \longleftrightarrow \quad \{x_{-L}, \ldots, x_{-1}, E, x_1, \ldots, x_L\},$$

where

$$E = E^{L,k}_{v_{-L},\ldots,v_L}$$

and

$$x_\ell = \begin{cases} \dfrac{\varphi^{L,k}_{v_{-L},\ldots,v_L}(\ell+1)}{\varphi^{L,k}_{v_{-L},\ldots,v_L}(\ell)} & \ell < 0 \\[2ex] \dfrac{\varphi^{L,k}_{v_{-L},\ldots,v_L}(\ell-1)}{\varphi^{L,k}_{v_{-L},\ldots,v_L}(\ell)} & \ell > 0, \end{cases} \tag{15.7.6}$$

so that

$$v_\ell = \begin{cases} E - x_{\ell-1}^{-1} - x_\ell & \ell < 0 \\ E - x_{-1}^{-1} - x_1^{-1} & \ell = 0 \\ E - x_{\ell+1}^{-1} - x_\ell & \ell > 0, \end{cases} \tag{15.7.7}$$

with the conventions $x_{-L-1}^{-1} = x_{L+1}^{-1} = 0$ (which are natural in view of the definitions above).

Implementing this change of variables, the $2L+1$-fold iterated integral expressing $\rho_L(n,0)$ takes the following form:

Lemma 15.7.5 *Fix $L \in \mathbb{Z}_+$ and n with $0 < n \le L$. Set*

$$\begin{aligned} \phi_E(x) &= r(E-x), \\ (U_0 f)(x) &= |x|^{-1} f\left(|x|^{-1}\right), \\ (S_E f)(x) &= \int_{\mathbb{R}} r\left(E - x - y^{-1}\right) f(y)\, dy, \\ (T_E f)(x) &= \int_{\mathbb{R}} r\left(E - x - y^{-1}\right) |y|^{-1} f(y)\, dy. \end{aligned}$$

Then, we have

$$\rho_L(n,0) = \int_{\Sigma_0} \left\langle T_E^{n-1} S_E^{L-n} \phi_E, U S_E^L \phi_E \right\rangle_{L^2(\mathbb{R})} dE. \tag{15.7.8}$$

In this representation one then proceeds by applying the Cauchy–Schwarz inequality inside the integral and applying suitable norm estimates for the operators involved that are uniform in $E \in \Sigma_0$. Specifically, the following estimates, where we denote the norm of an operator $T : L^p(\mathbb{R}) \to L^q(\mathbb{R})$ by $\|T\|_{p,q}$, can be proved and then used in this way:

Lemma 15.7.6

(a) *For every* $E \in \mathbb{R}$, *we have* $\|S_E\|_{1,1} \leq 1$.
(b) *For every* $E \in \mathbb{R}$, *we have* $\|S_E\|_{1,2} \leq \|r\|_\infty^{1/2}$.
(c) *For every* $E \in \mathbb{R}$, *we have* $\|T_E\|_{2,2} \leq 1$.
(d) *There exists* $q < 1$ *such that for every* $E \in \Sigma_0$, *we have* $\|T_E^2\|_{2,2} \leq q$.

With these estimates one can show exponential decay of $\rho_L(n, 0)$ and hence Proposition 15.7.3 is clearly applicable.

Before turning our attention to limit-periodic perturbations and the idea underlying the proof of Theorem 15.7.2, we first point out that the argument sketched above can be modified, and in fact generalized, in a number of ways.

First of all, while independence is crucially used, it is not necessary to consider identically distributed random variables. Indeed, if the density r changes from site to site, one simply replaces r by an ℓ-dependent density r_ℓ as one processes the random variable at site ℓ. For simplicity we will consider rescalings of a fixed density: $r_\ell(x) = d_\ell^{-1} r(d_\ell^{-1} x)$ for suitable $d_\ell > 0$. This results in obvious changes to (15.7.3) and Lemma 15.7.5. The necessary change to Lemma 15.7.6 is trickier. Namely, part (d) has no obvious replacement and needs to be modified via a quantitative estimate that is based on the following key lemma [58]:

Lemma 15.7.7 *For* $\lambda \geq 0$ *sufficiently small,* $\widehat{r}$, *the Fourier transform of* r, *obeys*

$$\sup_{|\eta| \geq \lambda} |\widehat{r}(\eta)|^2 \leq e^{-c|\lambda|^2} \tag{15.7.9}$$

for a suitable constant $c = c(r) > 0$. *Furthermore, there exist constants* $K_0 = K_0(r)$ *and* $\lambda = \lambda(r)$ *such that for every* ℓ *and every* $E \in \Sigma_0$, *we have*

$$\begin{aligned} \|T_E^{(\ell)} T_E^{(\ell-1)}\|_{2,2} &\leq \frac{1}{4} \left(15 + \sup_{|\eta| \geq K_0 \min\{d_\ell, d_{\ell-1}\}} |\widehat{r}(\eta)|^2 \right)^{1/2} \\ &\leq e^{-cK_0^2 \min\{d_\ell^2, d_{\ell-1}^2, \lambda\}}. \end{aligned} \tag{15.7.10}$$

With this replacement of Lemma 15.7.6.(d) one can still prove that Proposition 15.7.3 is applicable under suitable assumptions on the sequence $\{d_\ell\}$ (which obviously cannot decay too fast since the modification of

Lemma 15.7.6.(b) results in a factor $d_\ell^{-1/2}$ and the influence of the exponential in (15.7.10) becomes weaker with smaller $d_\ell, d_{\ell-1}$).

The second generalization of the basic argument outlined above is the addition of a fixed bounded background potential. Indeed, this simply results in obvious changes to (15.7.4)–(15.7.7) and then to Lemma 15.7.5. Note that this generalization produces the precursor to Theorem 15.7.2, which says that any fixed bounded potential can be turned into one whose associated Schrödinger operator has pure point spectrum by an arbitrarily small *random* perturbation.

The third generalization is the most severe. It is here that the independence of the values of the perturbation potential is given up, which is clearly necessary if we are to produce limit-periodic perturbations.

To explain what kind of correlations we wish to introduce, let us discuss the general setup of the proof of Theorem 15.7.2. The limit-periodic perturbation V_{lp} will be obtained as a convergent series

$$V_{\mathrm{lp}} = \sum_{k=1}^{\infty} V_{\mathrm{per}}^{(k)}$$

where $V_{\mathrm{per}}^{(k)}$ is $(2q_k + 1)$-periodic and we have

$$\sum_{k=1}^{\infty} \| V_{\mathrm{per}}^{(k)} \|_\infty < \varepsilon$$

in order to satisfy $\| V_{\mathrm{lp}} \|_\infty < \varepsilon$. The values of $V_{\mathrm{per}}^{(k)}$ on $\{-q_k, -q_k+1, \ldots, q_k - 1, q_k\}$ will be generated by independent identically distributed random variables with density $r_k(x) = d_k^{-1} r(d_k^{-1} x)$, where r is a fixed density and $d_k > 0$ is chosen suitably. Note that for each ℓ, there will be infinitely many random variables (with scaling factors d_k, $k \geq k_0(\ell)$) participating in determining the value of V_{lp} at site ℓ. The key idea will be to choose a decaying sequence $\{d_k\}$ and to consider the lowest-level participating random variable with scaling factor $d_{k_0(\ell)}$ as the *essential* random variable at site ℓ, and all others as *inessential* random variables that are frozen and put in the background potential for the purpose of estimating the relevant quantities as above *uniformly* in the background potential. One can then average the resulting estimate over the inessential random variables to obtain the desired estimate of the overall expectation defining $\rho_L(n, 0)$. Of course the periodicity of $V_{\mathrm{per}}^{(k)}$ results in another change to the equations (15.7.4)–(15.7.7) and then to Lemma 15.7.5.

This strategy may be implemented in a way that allows one to choose the sequences $\{q_k\}$ (determining the periods of the $V_{\mathrm{per}}^{(k)}$'s) and $\{d_k\}$ (determining the sizes of the $V_{\text{per}}^{(k)}$'s) so that Proposition 15.7.3 may be applied, showing that almost all of these limit-periodic perturbations will produce operators with pure point spectrum. Choosing one of them then establishes Theorem 15.7.2. We refer the reader to [18] for further details. Note that in this line of reasoning one only gets the pure point property for the V_{lp} as constructed (chosen from the full measure set), but not for the elements of its hull. As a consequence, as pointed out above, the formulation of the result is different from some of the earlier results about the spectral type and does not apply uniformly across the hull of the limit-periodic sequence in question.

It is however possible to prove pure point spectrum uniformly across the hull in certain scenarios (which are not dense). This follows from a combination of works of Pöschel [53] and Damanik-Gan [17].

First, we discuss Pöschel's work contained in [53]. Given a function $\Omega : [0, \infty) \to [0, \infty)$, define the quantities $\Phi_\Omega(t)$, κ, and $\Psi_\Omega(t)$ for $t > 0$ by

$$\Phi_\Omega(t) := t^{-4} \sup\{\Omega(r)\, e^{-tr} : r \geq 0\}$$

$$\kappa_t := \left\{ \{t_j\}_{j=0}^{\infty} : t \geq t_0 \geq t_1 \geq \cdots \text{ and } \sum_{j=0}^{\infty} t_j \leq t \right\}$$

$$\Psi_\Omega(t) := \inf_{(t_j) \in \kappa_t} \prod_{j=0}^{\infty} \Phi(t_j) 2^{-j-1}$$

for $t > 0$. We call Ω an *approximation function* if $\Phi_\Omega(t)$ and $\Psi_\Omega(t)$ are finite for every $t > 0$. For example, for any $\alpha \geq 0$, $\Omega(r) = r^\alpha$ defines an approximation function.

Now, suppose that $\mathcal{M}$ is a Banach sub-algebra of $\ell^\infty(\mathbb{Z})$ with respect to the operations of pointwise addition and pointwise multiplication; in particular, we assume that $\mathbb{1}$ is contained in $\mathcal{M}$ (where $\mathbb{1}(n) \equiv 1$). We say that $\lambda : \mathbb{Z} \to \mathbb{R}$ is a *distal sequence for* $\mathcal{M}$ if, for each $k \in \mathbb{Z} \setminus \{0\}$, one has $\left(\lambda - S^k \lambda\right)^{-1} \in \mathcal{M}$, and one has the bound

$$\left\| \left(\lambda - S^k \lambda\right)^{-1} \right\|_\infty \leq \Omega(|k|), \quad \text{for all } k \neq 0,$$

where Ω is an approximation function. Here, the inverse refers to the multiplicative inverse in the Banach algebra.

Pöschel proved the following theorem in [53]:

Theorem 15.7.8 *If λ is a distal sequence for the Banach algebra $\mathcal{M}$, then there exists $\varepsilon_0 > 0$ such that the following holds true. For any $0 < \varepsilon \leq \varepsilon_0$, there is a sequence V so that $\lambda - V \in \mathcal{M}$ and the Schrödinger operator $H_{\varepsilon^{-1}V} = \Delta + \varepsilon^{-1} V$ is spectrally localized with eigenvalues $\{\varepsilon^{-1}\lambda_j : j \in \mathbb{Z}\}$.*

Moreover, if ψ_k is the normalized eigenvector corresponding to the eigenvalue λ_k, then there are constants $c > 0$ and $d > 1$ such that

$$|\psi_k(n)|^2 \leq cd^{-|k-n|}$$

for all k and n.

He also supplied some examples. Here is one that is particularly interesting because it shows not only that the theorem above can be applied and yields uniformly localized limit-periodic Schrödinger operators, but also that the spectrum in this case has no gaps (which is surprising in view of our earlier results showing that Cantor spectra are typical for limit-periodic Schrödinger operators):

Pöschel's Example: A Limit-Periodic Distal Sequence. Let $\mathcal{D}_n$ denote the set of sequences in $\ell^\infty(\mathbb{Z})$ having period 2^n, and $\mathcal{D} = \bigcup_n \mathcal{D}_n$; the space

$$\mathcal{L} = \overline{\mathcal{D}}$$

is a Banach algebra and a subspace of the space of all limit-periodic sequences. Let us describe how to construct a distal sequence for $\mathcal{L}$. For $j \in \mathbb{Z}_+$, define the set B_j by

$$B_j := \begin{cases} \bigcup\limits_{N \in \mathbb{Z}} [N \cdot 2^j, N \cdot 2^j + 2^{j-1}), & \text{if } j \text{ is even;} \\ \bigcup\limits_{N \in \mathbb{Z}} [N \cdot 2^j + 2^{j-1}, (N+1) \cdot 2^j), & \text{if } j \text{ is odd.} \end{cases}$$

For example, B_1 is precisely the set of odd integers. Denoting the indicator function of B_j by $b_j = \chi_{B_j}$, define λ by

$$\lambda_n = \sum_{j=1}^{\infty} b_j(n) 2^{-j}.$$

It is immediate from the definition that $\lambda \in \mathcal{L}$; moreover, the inequality

$$\left\| \left(\lambda - S^k \lambda\right)^{-1} \right\| \leq 16|k|$$

for $k \neq 0$ means that λ is distal for $\mathcal{L}$. It is not too difficult to prove the following statement:

Lemma 15.7.9 *For any $m \in \mathbb{Z}_+$ and any integer $0 \leq j < 2^m$, there is an integer $\ell = \ell(j, m)$ so that*

$$\lambda_k \in I_{m,j} := \left[\frac{j}{2^m}, \frac{j+1}{2^m}\right) \iff k \in \ell + 2^m \mathbb{Z}.$$

We see that in this particular example, the Schrödinger operators produced by Theorem 15.7.8 have spectrum $[0, \varepsilon^{-1}]$.

Damanik and Gan showed in [17] that Pöschel's results extend to the hull, i.e., whenever Theorem 15.7.8 can be applied to produce limit-periodic potentials for which the associated Schrödinger operator is uniformly localized, then the same statement is true, with the same constants, for all elements of the hull of the potential in question; see also [29] for a generalization of this statement.

15.8 The Density of States

Recall that we defined the density of states (DOS) measure dk to be the weak limit of dk_N, where

$$\int g\, dk_N = \frac{1}{N}\mathrm{Tr}(P_N\, g(H)\, P_N^*),$$

for Borel sets B, and P_N denotes projection onto coordinates $\{0, \ldots, N-1\}$ (provided that the limit exists). In the event that V is limit-periodic, we saw that the DOS exists by unique ergodicity. The accumulation function of the DOS is called the *integrated density of states* (IDS) and is denoted by

$$k(E) = \int \chi_{(-\infty, E]}\, dk.$$

The results of the present section are concerned with the regularity of k as a function of E. In full generality, dk is continuous (Theorem 15.4.3). In light of this, it is natural to ask whether one has a quantitative modulus of continuity, e.g., α-Hölder continuity for some $\alpha > 0$. In full generality, this is too ambitious, but one can wring just a bit more continuity out of the Thouless formula, as the following theorem of Craig and Simon [10] illustrates:

Theorem 15.8.1 *For any almost-periodic potential, the integrated density of states is log-Hölder continuous. That is, there is a constant $C > 0$ with the property that*

$$|k(E_1) - k(E_2)| \le C(\log|E_1 - E_2|^{-1})^{-1}$$

for all $E_1, E_2 \in \mathbb{R}$ with $|E_1 - E_2| \le 1/2$.

Proof. Without loss of generality, assume $E_1 < E_2 \le E_1 + \frac{1}{2}$. Then, since all the transfer matrices have determinant one, we have $L(E_1) \ge 0$, which leads to

$$\begin{aligned}0 &\le L(E_1)\\ &= \int \log|E - E_1|\, dk(E)\\ &= \int_{(E_1,E_2)} \log|E - E_1|\, dk(E) + \int_{\mathbb{R}\setminus(E_1,E_2)} \log|E - E_1|\, dk(E)\end{aligned}$$

Rearranging, we get

$$-\int_{(E_1,E_2)} \log|E - E_1|\, dk(E) \le \int_{\mathbb{R}\setminus(E_1,E_2)} \log|E - E_1|\, dk(E).$$

Bounding the integrands of each side, we obtain

$$-\log|E_2 - E_1| \int_{(E_1,E_2)} dk(E) \le \log(|E_1| + \|f\|_\infty + 2) \int dk(E),$$

since $\mathrm{supp}(dk) \subseteq [-2 - \|f\|_\infty, 2 + \|f\|_\infty]$. Thus, with $C = \log(|E_1| + \|f\|_\infty + 2)$, we have

$$|k(E_2) - k(E_1)| = \int_{(E_1,E_2)} dk(E) \le C\left[\log|E_1 - E_2|^{-1}\right]^{-1}. \qquad \square$$

Within the class of limit-periodic potentials, this result is optimal; in particular, the following result of Krüger and Gan shows that there is a dense set of limit-periodic potentials whose potentials are not h-Hölder continuous for any function h that goes to zero faster than $[\log(1/\delta)]^{-1}$ [37].

Theorem 15.8.2 *Let Ω denote a Cantor group with a minimal translation T. There is a dense set $\mathcal{I} \subseteq C(\Omega, \mathbb{R})$ with the property that the density of states of $V(n) = f(T^n 0)$ is no better than log-Hölder continuous. That is, if k denotes the IDS of V, given any increasing function $h : \mathbb{R}_+ \to \mathbb{R}_+$ with*

$$\lim_{\delta \downarrow 0} h(\delta) \log(1/\delta) = 0,$$

one has

$$\limsup_{E \to E_0} \frac{|k(E) - k(E_0)|}{h(|E - E_0|)} = \infty$$

for at least one E_0. In fact, one can guarantee that this holds for all E_0 in the spectrum.

The main idea is the following: since $\mathcal{P}$ is dense, start with some $f_0 \in \mathcal{P}$ and let $\varepsilon > 0$. Then, inductively choose $f_1, \ldots \in \mathcal{P}$ using Lemma 15.6.5 so that $\|f_j - f_{j-1}\|_\infty < \varepsilon \cdot 2^{-j}$ and so that $\mathrm{Leb}(\Sigma(f_j))$ is exponentially small, i.e.,

$$\mathrm{Leb}(\Sigma(f_j)) \lesssim e^{-c_j q_j}$$

where q_j is the period of f_j. Then, by completeness, $f_\infty = \lim f_j$ exists. The density of states corresponding to the potential function f_j gives weight $1/q_j$ to an interval having length no greater than $\exp(-c_j q_j)$. By carefully tuning the rate at which $f_j \to f_\infty$, one can push a statement like this through to the DOS of f_∞.

On the other hand, for a dense set, namely, for $f \in \mathcal{P}$, one knows that the IDS is 1/2-Hölder continuous (and no better); compare Theorem 15.2.4. It is interesting to ask whether one can do better than 1/2. In the localization regime of Pöschel, the IDS can be shown to be 1-Hölder continuous, which was proved by Damanik and Fillman [12].

Theorem 15.8.3 *There exist limit-periodic V whose associated IDS is Lipschitz-continuous.*

15.9 Other Families of Operators

Let us explore some models that are closely connected with, but distinct from the discrete 1D Schrödinger operators that we have considered thus far. For each of these families of models, many of the results from previous sections have analogues. To avoid being tedious, we will be somewhat selective with our presentation in this section, focusing on results whose analogues involve interesting challenges.

15.9.1 Jacobi Matrices

A *Jacobi matrix* is an operator on $\ell^2(\mathbb{Z})$ of the form

$$[\mathcal{J}u]_n = a_{n-1}u_{n-1} + b_n u_n + a_n u_{n+1}, \quad n \in \mathbb{Z},\ u \in \ell^2(\mathbb{Z}),$$

where $a_n, b_n \in \mathbb{R}$ with

$$\sup_n |b_n| < \infty, \quad \sup_n a_n < \infty, \quad \inf_n a_n > 0.$$

Thus, discrete Schrödinger operators are obtained from Jacobi matrices as a special case (in which $a_n \equiv 1$). Most of the results discussed in the foregoing section have Jacobi analogues.

A periodic Jacobi matrix is one for which the parameters a and b satisfy $a_{n+q} = a_n$ and $b_{n+q} = b_n$ for all n. Thus, a limit-periodic Jacobi matrix $\mathcal{J}$ is one for which there are periodic Jacobi matrices $\{\mathcal{J}_j\}_{j=1}^\infty$ such that

$$\lim_{j\to\infty} \|\mathcal{J} - \mathcal{J}_j\| = 0,$$

where $\|\cdot\|$ denotes the operator norm.

Jacobi matrices are not only considered simply for the sake of a more general setting (even though it is the natural generalization that still keeps many of the essential features intact; notably self-adjointness and a 2×2 transfer matrix formalism) but also because they provide the natural setting for the study of inverse spectral problems, where it is not at all clear that a solution to the given problem at hand can be found in the class of discrete Schrödinger operators, but where it can be shown to be solvable within the class of Jacobi matrices.

The most natural instance of this principle is the association of a spectral measure to an operator, either a discrete Schrödinger operators or a Jacobi matrix, on the discrete half-line (i.e., an operator acting in $\ell^2(\mathbb{Z}_+)$) and the corresponding spectral measure (associated with the cyclic vector δ_1). This correspondence, which can be established via the spectral theorem when passing from operator to measure and via orthogonal polynomials or a continued fraction expansion of the Borel transform of the measure when passing from measure to operator, sets up a natural bijection between bounded Jacobi matrices and compactly supported probability measures on the real line, but it continues to be an open problem to characterize explicitly those measures that correspond to discrete Schrödinger operators.

More closely related to the topic of this survey, however, is the way in which Jacobi matrices arise as solutions of certain renormalization equations, which can sometimes be shown to give rise to limit-periodic coefficients. Let us describe work in this spirit by Peherstorfer, Volberg, and Yuditskii [50].

Suppose T is an expanding polynomial (we will define this term shortly) and denote its real Julia set by $\mathrm{Julia}(T)$ and its degree by d. Recall that $\mathrm{Julia}(T)$ is the compact set of real numbers that do not go to infinity under forward iterations of T. By shifting and rescaling the domain of T and rescaling the range, one may assume that

$$T^{-1}([-1,1]) \subseteq [-1,1]; \ \pm 1 \in \mathrm{Julia}(T).$$

Under this normalization, such a polynomial is well-defined by the position of its critical values

$$\mathrm{CV}(T) = \{t_i = T(c_i) : T'(c_i) = 0, \ c_i > c_j \text{ for } i > j\}.$$

T is *expanding* (or *hyperbolic*) if

$$c_i \notin \mathrm{Julia}(T) \quad \forall i.$$

Then we have the following pair of theorems from [50]:

Theorem 15.9.1 *Let $\tilde{J}$ be a Jacobi matrix with spectrum contained in* $[-1, 1]$. *With the polynomial T of degree d from above, consider the renormalization equation*

$$V^*(z - \mathcal{J})^{-1}V = (T(z) - \tilde{\mathcal{J}})^{-1}\frac{T'(z)}{d} \tag{15.9.1}$$

where

$$V\delta_k = \delta_{dk}$$

for each $k \in \mathbb{Z}$. It has a solution $\mathcal{J} = \mathcal{J}(\tilde{\mathcal{J}})$ with spectrum contained in $T^{-1}([-1, 1])$.

Moreover, if $\min_i |t_i| \geq 10$, *then*

$$\|\mathcal{J}(\tilde{\mathcal{J}}_1) - \mathcal{J}(\tilde{\mathcal{J}}_2)\| \leq \kappa \|\tilde{\mathcal{J}}_1 - \tilde{\mathcal{J}}_2\|$$

with an absolute constant $\kappa < 1$.

Theorem 15.9.2 *Let us assume that T is* sufficiently hyperbolic *in the sense that*

$$\text{dist}(\text{CV}(T), [-1, 1]) \geq 10.$$

Then the renormalization equation (15.9.1) *has a unique fixed point. That is, there is a unique Jacobi matrix $\mathcal{J}$ such that*

$$V^*(z - \mathcal{J})^{-1}V = (T(z) - \mathcal{J})^{-1}\frac{T'(z)}{d}.$$

Moreover, the coefficients of $\mathcal{J}$ are limit-periodic.

Remark 15.9.3 To be more accurate, one has to expand the notion of a Jacobi matrix slightly for the purpose of these theorems and allow the off-diagonal terms to vanish. Indeed, the central off-diagonal element a_0 of the fixed point $\mathcal{J}$ is zero, and hence J splits into the direct sum of two half line Jacobi matrices. It turns out that spectral measure of the Jacobi matrix corresponding to the right half line in this decomposition is the balanced (equilibrium) measure on Julia(T). In other words, the equilibrium measure on the Julia set of a suitable hyperbolic polynomial has orthogonal polynomials with limit-periodic recursion coefficients.

15.9.2 Continuum Schrödinger Operators

The continuum Schrödinger operator acts as a self-adjoint operator in $L^2(\mathbb{R})$ via

$$L_V y = -y'' + Vy, \tag{15.9.2}$$

where $V : \mathbb{R} \to \mathbb{R}$ is a sufficiently nice function; for most of the present section, V will be bounded and continuous. For such V, (15.9.2) defines a self-adjoint operator in $L^2(\mathbb{R})$ in a canonical fashion. These operators enjoy a transfer matrix formalism quite similar to the one described in the discrete setting: for each $z \in \mathbb{C}$ and $x \in \mathbb{R}$, there is an $\mathrm{SL}(2,\mathbb{C})$ matrix $A_z^x(V)$ such that

$$\begin{bmatrix} y'(x) \\ y(x) \end{bmatrix} = A_z^x(V) \begin{bmatrix} y'(0) \\ y(0) \end{bmatrix}$$

whenever y satisfies $L_V y = zy$. We then consider potentials that are uniform limits of continuous periodic potentials; concretely, define

$$\mathrm{P}(\mathbb{R}) = \{V \in C(\mathbb{R}) : \text{there exists } T > 0 \text{ such that } V(x+T) \equiv V(x)\}$$

and then let $\mathrm{LP}(\mathbb{R})$ denote the closure of the space of periodic potentials (with respect to the topology induced by the uniform metric):

$$\mathrm{LP}(\mathbb{R}) = \overline{\mathrm{P}(\mathbb{R})}^{\|\cdot\|_\infty}.$$

In particular, every $V \in \mathrm{LP}(\mathbb{R})$ is continuous and uniformly almost-periodic.

In the event that $V \in \mathrm{P}(\mathbb{R})$, say $V(x+T) \equiv V(x)$ for some $T > 0$, the formalism from Section 15.2.1 enables one to describe the spectrum of L_V. Concretely, the discriminant is again given by

$$D(z) = \mathrm{Tr}\, A_z^T(V),$$

and then the spectrum of L_V is again given by

$$\sigma(L_V) = \{E \in \mathbb{R} : |D(E)| \le 2\}.$$

On the other hand, one can also consider the analogue of Section 15.2.2; here, we consider $L(\theta) = L_V(\theta)$ acting via

$$L_V(\theta)y = -y'' + Vy,$$
$$D(L(\theta)) = \left\{ f \in H^2([0,T]) : f(T) = e^{i\theta} f(0) \text{ and } f'(T) = e^{i\theta} f'(0) \right\}.$$

One can then show that $L_V(\theta)$ has compact resolvent and hence a sequence of eigenvalues $\lambda_1(\theta) \le \lambda_2(\theta) \le \cdots$. As before, the eigenvalues of $L_V(0)$ and $L_V(\pi)$ provide the endpoints of the spectral bands. Fixing $T = \pi$ for concreteness, we can consider $V \equiv 0$ as a π-periodic potential[3] and explicitly compute the eigenvalues of $L_0(\theta)$ for $\theta = 0, \pi$:

$$\sigma(L_0(0)) = \left\{ (2k)^2 : k = 0, 1, 2, \ldots \right\},$$

$$\sigma(L_0(\pi)) = \left\{ (2k+1)^2 : k = 0, 1, 2, \ldots \right\}.$$

[3] Recall that the subscript in L_V refers to the potential, so L_0 refers to the free Laplacian with $V \equiv 0$.

Here, all eigenvalues are doubly degenerate with the exception of $0 \in \sigma(L_0(0))$ which is simple. Thus, (counting from $k = 1$ at the bottom of the spectrum) the kth spectral band is $[(k-1)^2, k^2]$. So, if V is then a bounded π-periodic potential, the kth spectral band of L_V satisfies

$$[(k-1)^2 + \|V\|_\infty, k^2 - \|V\|_\infty] \subseteq B_k \subseteq [(k-1)^2 - \|V\|_\infty, k^2 + \|V\|_\infty]$$

by general eigenvalue perturbation theory [33]. Then, it follows that the length of the kth band of L_V, grows approximately linearly in k. This can be viewed as one instance of a tendency for the spectrum of L_V to thicken in the high-energy region (which is not present in the discrete case). In view of this, it is quite surprising that one can beat this tendency and prove the following result, due to Damanik, Fillman, and Lukic:

Theorem 15.9.4 *There is a Baire-generic subset $Z \subseteq \mathrm{LP}(\mathbb{R})$ with the property that $\sigma(L_{\lambda V})$ is an (unbounded) Cantor set of zero Lebesgue measure for all $V \in Z$ and all $\lambda > 0$. There is a dense subset $H \subseteq \mathrm{LP}(\mathbb{R})$ with the property that $\sigma(L_{\lambda V})$ is an (unbounded) Cantor set of zero Hausdorff dimension for all $V \in H$ and all $\lambda > 0$. Moreover, for all $V \in Z$ and all $V \in H$, the spectral type of L_V is purely singular continuous.*

The proof of Theorem 15.9.4 parallels that of Theorem 15.6.2, except that one is only able to prove suitable measure estimates in compact energy windows, since, as observed above, the Lebesgue measure of the bands grows linearly in the band label. Thus, one has to expand the compact sets on which one has effective measure estimates on the spectrum in a way that misses the lengthening effect of the high-energy region. See [14] for more details.

15.9.3 CMV Matrices and Quantum Walks

CMV matrices arise naturally in the study of orthogonal polynomials on the unit circle (OPUC) and are universal within the class of unitary operators of spectral multiplicity one. Concretely, one starts with μ, a Borel probability measure supported on the unit circle which does not admit a support having only finitely many points. A *CMV matrix* is a five-diagonal semi-infinite matrix that is determined by a sequence of *Verblunsky coefficients* $\{\alpha_n\}_{n \in \mathbb{Z}_+} \subset \mathbb{D}$; these coefficients arise as the recursion coefficients of the orthogonal polynomials associated μ. In terms of α_n and the derived quantities $\rho_n = \left(1 - |\alpha_n|^2\right)^{1/2}$, the CMV matrix takes the form

$$\mathcal{C} = \begin{bmatrix} \overline{\alpha_0} & \overline{\alpha_1}\rho_0 & \rho_1\rho_0 & & & & \\ \rho_0 & -\overline{\alpha_1}\alpha_0 & -\rho_1\alpha_0 & & & & \\ & \overline{\alpha_2}\rho_1 & -\overline{\alpha_2}\alpha_1 & \overline{\alpha_3}\rho_2 & \rho_3\rho_2 & & \\ & \rho_2\rho_1 & -\rho_2\alpha_1 & -\overline{\alpha_3}\alpha_2 & -\rho_3\alpha_2 & & \\ & & & \overline{\alpha_4}\rho_3 & -\overline{\alpha_4}\alpha_3 & \overline{\alpha_5}\rho_4 & \rho_5\rho_4 \\ & & & \rho_4\rho_3 & -\rho_4\alpha_3 & -\overline{\alpha_5}\alpha_4 & -\rho_5\alpha_4 \\ & & & & \ddots & \ddots & \ddots \end{bmatrix}. \tag{15.9.3}$$

This matrix defines a unitary operator in $\ell^2(\mathbb{Z}_+)$, and the spectral measure corresponding to $\mathcal{C}$ and the vector δ_0 is given by μ. This sets up a one-to-one correspondence between measures μ and coefficient sequences $\{\alpha_n\}_{n\in\mathbb{Z}_+}$, which has been extensively studied in recent years, mainly due to the infusion of ideas from Simon's monographs [59, 60].

Similarly, an *extended CMV matrix* is a unitary operator on $\ell^2(\mathbb{Z})$ defined by a bi-infinite sequence $\{\alpha_n\}_{n\in\mathbb{Z}} \subset \mathbb{D}$ in an analogous way:

$$\mathcal{E} = \begin{bmatrix} \ddots & \ddots & \ddots & & & & & \\ \overline{\alpha_0}\rho_{-1} & -\overline{\alpha_0}\alpha_{-1} & \overline{\alpha_1}\rho_0 & \rho_1\rho_0 & & & & \\ \rho_0\rho_{-1} & -\rho_0\alpha_{-1} & -\overline{\alpha_1}\alpha_0 & -\rho_1\alpha_0 & & & & \\ & & \overline{\alpha_2}\rho_1 & -\overline{\alpha_2}\alpha_1 & \overline{\alpha_3}\rho_2 & \rho_3\rho_2 & & \\ & & \rho_2\rho_1 & -\rho_2\alpha_1 & -\overline{\alpha_3}\alpha_2 & -\rho_3\alpha_2 & & \\ & & & & \overline{\alpha_4}\rho_3 & -\overline{\alpha_4}\alpha_3 & \overline{\alpha_5}\rho_4 & \rho_5\rho_4 \\ & & & & \rho_4\rho_3 & -\rho_4\alpha_3 & -\overline{\alpha_5}\alpha_4 & -\rho_5\alpha_4 \\ & & & & & \ddots & \ddots & \ddots \end{bmatrix}. \tag{15.9.4}$$

From the point of view of orthogonal polynomials, the study of $\mathcal{C}$ is more natural; however, when the Verblunsky coefficients are generated by an invertible ergodic map (such as a minimal translation of a compact abelian group as in this chapter), the study of $\mathcal{E}$ is more natural.

CMV matrices also arise in a natural fashion in the study of 1-dimensional coined quantum walks. A 1D quantum walk on $\mathbb{Z}$ is a quantum mechanical analogue of a classical random walk, with two twists:

- The walker has an internal degree of freedom (called 'spin') that influences her probability of hopping to the left or to the right.
- The walker may exist as a superposition of pure states, rather than being fully localized.

The relevant state space is $\mathcal{H}_{\mathrm{QW}} = \ell^2(\mathbb{Z}) \otimes \mathbb{C}^2$; the $\ell^2(\mathbb{Z})$ component captures the spatial position of the walker, while the $\mathbb{C}^2$ variable captures her spin. We will denote the pure states as $\delta_n^{\pm} = \delta_n \otimes e_{\pm}$, where $\{e_+, e_-\}$ denotes the usual basis of $\mathbb{C}^2$. Each quantum coin is then a superposition of two 'classical coins', so the quantum walk is parameterized by a sequence of 2×2 unitaries:

$$Q_n = \begin{bmatrix} q_n^{11} & q_n^{12} \\ q_n^{21} & q_n^{22} \end{bmatrix}.$$

As time advances one unit forward, the quantum walk update rule acts as follows on pure states:

$$U\delta_n^+ = q_n^{11}\delta_{n+1}^+ + q_n^{21}\delta_{n-1}^-$$
$$U\delta_n^- = q_n^{12}\delta_{n+1}^+ + q_n^{22}\delta_{n-1}^-.$$

Due to a seminal paper of Cantero, Grünbaum, Moral, and Velázquez, we know that the unitary update rule U is unitarily equivalent to a CMV matrix [6]. Thus, any and all tools relevant to the study of CMV matrices enter the game and can be used to study 1D quantum walks.

Most of the results from the previous sections have CMV analogues (and hence analogues for quantum walks as well) – however, owing to the more complicated structure of the CMV matrix compared to a Schrödinger operator, the proofs are generally more involved (see, e.g., [25, 26, 44]).

However, there is one notable exception to the previous remark: it is not yet known that CMV matrices having pure point spectrum are dense in the space of all limit-periodic CMV matrices. To spell things out more carefully, the Kunz–Souillard approach to localization is an essential ingredient in the arguments that proved Theorem 15.7.1. However, there is not a satisfactory version of the Kunz–Souillard localization proof for the CMV operators. In addition, to the best of our knowledge, no one has worked out an analogue of Pöschel's KAM scheme in the CMV setting. We would regard resolutions to either of these issues as very interesting results.

15.9.4 Multi-dimensional Discrete Operators

Given a bounded potential $V : \mathbb{Z}^d \to \mathbb{R}$, the discrete Schrödinger operator H_V acts in $\ell^2(\mathbb{Z}^d)$ via

$$[H_V u]_{\mathbf{n}} = V_{\mathbf{n}} u_{\mathbf{n}} + \sum_{\|\mathbf{m}-\mathbf{n}\|_1 = 1} u_{\mathbf{m}}.$$

Given $\mathfrak{p} = (p_1, p_2, \ldots, p_d) \in \mathbb{Z}_+^d$, we say that a potential V is $\mathfrak{p}$-periodic if

$$V_{\mathbf{n}+p_j\mathbf{e}_j} = V_{\mathbf{n}} \text{ for all } \mathbf{n} \in \mathbb{Z}^d \text{ and all } 1 \le j \le d.$$

Theorem 15.9.5 *Suppose $d \ge 2$. For all $\mathfrak{p} = (p_1, p_2, \ldots, p_d) \in \mathbb{Z}_+^d$, there is a constant $C = C_{\mathfrak{p}} > 0$ such that the following holds true.*

- *If $V : \mathbb{Z}^d \to \mathbb{R}$ is $\mathfrak{p}$-periodic and $\|V\|_\infty \le C$, then $\sigma(H_V)$ consists of at most two connected components.*

- *If at least one entry of* $\mathfrak{p}$ *is odd,* V *is* $\mathfrak{p}$*-periodic, and* $\|V\|_\infty \le C$*, then* $\sigma(H_V)$ *consists of a single interval.*

Theorem 15.9.5 was proved first in the special case $d = 2$ when $\gcd(p_1, p_2) = 1$ by Krüger [36] (note that at least one of p_1 and p_2 must be odd in this case). This was generalized to all periods in $d = 2$ by Embree and Fillman [22] and to all $d \ge 2$ by Han and Jitomirskaya [30]. One should view this result as a discrete analogue of the Bethe–Sommerfeld conjecture:

Theorem 15.9.6 *If* $V : \mathbb{R}^d \to \mathbb{R}$ *is periodic, then* $\sigma(-\nabla^2 + V)$ *has only finitely many gaps.*

Theorem 15.9.6 has a rich history with contributions from many authors, including (but certainly not limited to) [31, 32, 52, 62, 63, 64, 69], and culminating in the paper of Parnovskii [45].

As a consequence of Theorem 15.9.5, one can show that small limit-periodic operators in $\ell^2(\mathbb{Z}^d)$ for $d \ge 2$ also will have spectra comprising only one or two intervals.

Corollary 15.9.7 *Let* $d \ge 2$ *and* $\mathfrak{p}_n \in \mathbb{Z}_+^d$ *be such that the* j*th coordinate of* $\mathfrak{p}_n$ *divides the* j*th coordinate of* $\mathfrak{p}_{n+1}$ *for all* $1 \le j \le d$ *and all* $n \in \mathbb{Z}_+$*. Then, there exists a sequence* $\delta_n > 0$ *with the following property: if* V_n *is* $\mathfrak{p}_n$*-periodic and* $\|V_n\|_\infty \le \delta_n$ *for each* n*, then the limit-periodic potential*

$$V_{\mathrm{lp}} = \sum_{n=1}^{\infty} V_n$$

is such that $\sigma(H_{V_{\mathrm{lp}}})$ *consists of at most two connected components. If at least one component of* $\mathfrak{p}_n$ *is odd for every* $n \in \mathbb{Z}_+$*, then* $\sigma(H_{V_{\mathrm{lp}}})$ *is a single interval.*

In particular, it is substantially harder to produce Cantor spectrum in higher dimensions than in dimension one.

15.10 Open Problems

In this section we describe a number of open problems that are suggested by the existing results. Our first set of questions can be summarized by asking whether or not any limit-periodic operator ever experiences any sort of phase transition. Concretely, whenever a result is known about a limit-periodic potential V, it is known that the spectral type of H_V is pure. Thus far, any results that give information about the hull of V show that the spectral type is constant on the hull of V. Finally, whenever one is able to prove a result about the one-parameter family $\{\Delta + \lambda V\}_{\lambda > 0}$, the spectral type of $H_{\lambda V}$ does not vary with $\lambda > 0$. This leads us to our first three questions:

Question 15.10.1 Is the spectral type of a limit-periodic operator always pure?

Question 15.10.2 Does the spectral type ever change as one passes to other elements of the hull?

Question 15.10.3 Considering a limit-periodic Schrödinger operator and replacing the potential by a non-trivial multiple of it (i.e., by varying the coupling constant), can the spectral type ever change?

Question 15.10.4 We know that the occurrence of Cantor spectra is a generic phenomenon in the limit-periodic universe. What about the failure of Cantor spectrum? We know that it is possible (certainly for periodic cases, but also for non-periodic cases due to Pöschel's work [53]). Is it a dense phenomenon among the non-periodic limit-periodic cases?

Let us discuss the general questions above, and others, in specific settings:

Question 15.10.5 Suppose V is limit-periodic and $\sigma_{\mathrm{ac}}(H_V) \neq \emptyset$.

(1) Is it true that H_V has purely a.c. spectrum?
(2) Is it true that $H_{\lambda V}$ has purely a.c. spectrum for all λ?
(3) Is it true that H_W has purely a.c. spectrum for all $W \in \mathrm{hull}(V)$?

Question 15.10.6 Consider a limit-periodic Schrödinger operator with pure point spectrum that arises via the generalized Kunz-Souillard approach presented in Section 15.7. Pass to a different element of the hull and/or vary the coupling constant. Does the spectral type remain pure point?

Question 15.10.7 In the setting of the previous problem (limit-periodic operators obtained via the generalized Kunz-Souillard approach), can one show that the eigenfunctions *cannot* decay exponentially? In other words, can one show that the Lyapunov exponent vanishes throughout the spectrum? This would be a new phenomenon within the class of ergodic Schrödinger operators: eigenvalues inside the set of energies where the Lyapunov exponent vanishes.

Question 15.10.8 Consider a limit-periodic Schrödinger operator with pure point spectrum that arises via the Pöschel approach presented in Section 15.7. Recall that this puts us in the regime of large potentials. Vary the coupling constant, and consider in particular the case where it is chosen to be small. Does the spectral type remain pure point? Does is remain constant across the hull? If so, is this due to the persistence of uniform localization?

Question 15.10.9 In the general Pöschel approach, can one find additional examples that display a variety of features? Pöschel provided two examples in [53], showing that the spectrum can be a Cantor set or it can have no gaps at all. What about other sets (sets with finitely many gaps; sets that are non-trivial unions of Cantor sets and finite unions of non-degenerate intervals; sets that are Cantorvals)?

Speaking of the Pöschel and Kunz-Souillard approaches, it would be interesting to work out their analogues for CMV matrices.

Question 15.10.10 Can the approach to limit-periodic Schrödinger operators from Pöschel's paper [53] be carried over to CMV matrices?

Question 15.10.11 Can the Kunz–Souillard approach to random Schrödinger operators from [38] be carried over to CMV matrices? If so, is there an extension of it analogous to [18], which then has similar consequences for almost-periodic CMV matrices?

Appendix A. Profinite Groups

Since the hull of a limit-periodic potential is a totally disconnected group, we will discuss some characteristics of totally disconnected groups. The goal of the appendix is to provide a short, self-contained proof of two results: that compact totally disconnected groups are profinite and that monothetic Cantor groups are procyclic. Let us recall that a *topological group* is a Hausdorff topological space G that is endowed with a group structure in such a way that the group operations (composition and inversion) are both continuous. For thorough treatments of totally disconnected groups, see [57, 70].

A.1 Inverse Limits

Definition A *partial order* on a set I is a binary relation $\preceq$ with the following properties:

(1) $i \preceq i$ for every $i \in I$.
(2) $i = j$ whenever $i \preceq j$ and $j \preceq i$.
(3) $i \preceq k$ whenever $i \preceq j$ and $j \preceq k$.

Given a non-empty set I equipped with a partial order $\preceq$, an *inverse system* of topological groups over $(I, \preceq)$ is an ordered pair $((G_i), (\varphi_i^j))$, in which G_i is a topological group for each $i \in I$ and $\varphi_i^j : G_j \to G_i$ is a continuous homomorphism whenever $i \preceq j$ such that

(1) $\varphi_i^i = \mathrm{id}_{G_i}$ for all $i \in I$.
(2) The φs are compatible in the sense that $\varphi_i^j \circ \varphi_j^k = \varphi_i^k$ whenever $i \preceq j \preceq k$. Equivalently, the following diagram commutes for every triple $i, j, k \in I$ for which $i \preceq j \preceq k$:

$$\begin{array}{ccc} G_k & \xrightarrow{\varphi_i^k} & G_i \\ & {\scriptstyle \varphi_j^k} \searrow \quad \nearrow {\scriptstyle \varphi_i^j} & \\ & G_j & \end{array}$$

An *inverse limit* (or *projective limit*) of an inverse system is a group G together with continuous homomorphisms $\psi_i : G \to G_i$ with the following properties:

(1) The maps ψ_i are compatible with the φs in the sense that $\varphi_i^j \circ \psi_j = \psi_i$ whenever $i \preceq j$, i.e., the following diagram commutes:

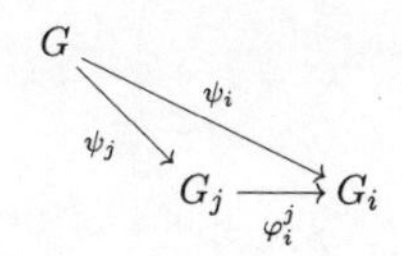

(2) The pair $(G, (\psi_i))$ is minimal with respect to property (1) in the following sense: if G' and ψ_i' are a group and a family of morphisms with $\varphi_i^j \circ \psi_j' = \psi_i'$ whenever $i \preceq j$, then there exists a unique continuous homomorphism $\theta : G' \to G$ such that $\psi_i' = \psi_i \circ \theta$ for every $i \in I$. Pictorially, the induced map θ is such that the following diagram commutes:

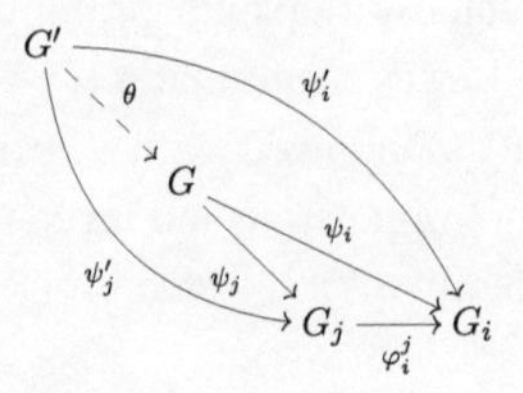

Theorem A.1 *Any inverse system of topological groups has an inverse limit which is unique up to isomorphism.*

Proof. Put

$$G = \left\{ (g_i)_{i \in I} : \varphi_i^j(g_j) = g_i \text{ whenever } i \preceq j \right\} \subseteq \prod_{i \in I} G_i, \tag{A.1}$$

and let $\psi_i : G \to G_i$ denote projection onto the ith coordinate. A simple diagram chase shows that (G, ψ_i) satisfies the definition of an inverse limit. If G_0 is another inverse limit, then the θs guaranteed by the universal mapping property furnish isomorphisms between G and G_0. □

We denote the inverse limit of the inverse system $((G_i), (\varphi_i^j))$ by

$$G = \varprojlim G_i = \varprojlim (G_i, \varphi_i^j),$$

and we typically use the first notation, even though G clearly depends on the maps φ_i^j.

Remark A.2 Of course, we have described the construction of the categorical inverse limit in the category whose objects are topological groups and whose morphisms are continuous homomorphisms. One could just as well speak of the inverse limit of rings, modules over a PID, or topological spaces, e.g., but we will not have any need for more general inverse limits.

Example A.3 Let I be any non-empty set, and equip I with the trivial partial order where $\alpha \prec \alpha$ for any $\alpha \in I$ and $\alpha \not\prec \beta$ for every $\alpha \neq \beta$. Then, one can verify that

$$\varprojlim G_i = \prod_{i \in I} G_i;$$

note that there is no need to specify maps φ_i^j in this case.

Given a prime $p \in \mathbb{Z}_+$, the additive group of p-adic integers is defined to be the set of formal sums:

$$\mathbb{J}_p = \left\{ \sum_{j=0}^{\infty} a_j p^j : a_j \in \{0, 1, \ldots, p-1\} \right\},$$

where the group operation is given by 'adding with carrying'. One can realize $\mathbb{J}_p$ as an inverse limit of cyclic groups. To see this, let $I = \mathbb{Z}_+$ with the usual order, and let $G_k = \mathbb{Z}_{p^k}$ for $k \in \mathbb{Z}_+$. Define

$$\varphi_k^\ell \left(n + p^\ell \mathbb{Z} \right) = n + p^k \mathbb{Z}$$

whenever $k \le \ell$. The inverse limit of this inverse system is canonically isomorphic to $\mathbb{J}_p$. Specifically, if we define $\psi_k : \mathbb{J}_p \to \mathbb{Z}_{p^k}$ by

$$\psi_k \left(\sum_{j=0}^{\infty} a_j p^j \right) = \left(\sum_{j=0}^{k-1} a_j p^j \right) + p^k \mathbb{Z},$$

one can check that $\mathbb{J}_p$ together with this family of maps satisfies the definition of an inverse limit.

Definition We say that a group G is *profinite* if it can be written as the inverse limit of an inverse system of finite groups and *procyclic* if it can be written as the inverse limit of an inverse system of cyclic groups.

Proposition A.4 *The inverse limit of a family of compact totally disconnected groups is itself compact and totally disconnected. In particular, profinite groups are compact, Hausdorff, and totally disconnected.*

Proof. The product of compact spaces is compact by Tychonoff's theorem. It is straightforward to check that G defined in (A.1) is closed with respect to the product topology, and hence compact. Since products and subspaces of totally disconnected (respectively Hausdorff) spaces are totally disconnected (respectively Hausdorff), the proposition follows. □

The remainder of the appendix will focus on the converse of the preceding proposition; i.e., any totally disconnected compact group is profinite. Indeed, such a group may be naturally identified with its so-called *profinite completion*.

Definition Let G be a topological group, and consider the index set

$$I = \mathcal{N} := \{N \subseteq G : N \text{ is an open normal subgroup of } G\},$$

partially ordered by reverse inclusion (i.e., $H \preceq K$ if and only if $H \supseteq K$). For $H \supseteq K$, define $\varphi_H^K : G/K \to G/H$ denote the natural projection, i.e.,

$$\varphi_H^K(gK) = gH.$$

The inverse limit of this system is called the *profinite completion* of G, denoted

$$\hat{G} = \varprojlim G/H.$$

There is a canonical map $\rho : G \to \hat{G}$ given by

$$\rho(g) = (gN)_{N \in \mathcal{N}} \in \prod_{N \in \mathcal{N}} G/N. \tag{A.2}$$

Notice that $\hat{G}$ may be trivial, and hence, ρ need not be injective. For example, the only open subgroup of $\mathbb{T}$ is $\mathbb{T}$ itself, so the profinite completion of $\mathbb{T}$ is trivial.

Theorem A.5 *A compact topological group G is profinite if and only if it is totally disconnected. Indeed, any compact totally disconnected group is isomorphic to its profinite completion.*

To prove this, we need a pair of preliminary results.

Proposition A.6 *Any totally disconnected compact Hausdorff space is of topological dimension zero in the sense that its topology enjoys a neighbourhood basis consisting of compact open sets.*

Proof. Suppose X is a totally disconnected compact Hausdorff space, let $x \in X$ be given, and denote by $C = C(x)$ the intersection of all compact open sets containing x; notice that C is necessarily closed, but it need not be open. First, we claim that $C = \{x\}$. If not, then total disconnectedness of X implies that C is disconnected, so it can be written as a disjoint union

$$C = A \cup B,$$

where A and B are disjoint, non-empty, closed subsets of C. Since C is a closed subset of X, both A and B are closed subsets of X as well. Since X is compact and Hausdorff, we may choose disjoint open sets $U, V \subseteq X$ with $A \subseteq U$ and $B \subseteq V$. Since $C \subseteq U \cup V$, the *complements* of the compact open sets containing x comprise an open cover of $X \setminus (U \cup V)$, so, by using compactness to reduce to a finite sub-cover, we see that there are finitely many compact open sets $C_1, \ldots, C_n$ containing x such that

$$P := \bigcap_{j=1}^{n} C_j \subseteq U \cup V.$$

Of course, P is itself compact and open. Notice that

$$\overline{P \cap U} \subseteq P \cap \overline{U} = (U \cup V) \cap P \cap \overline{U} = P \cap U.$$

Consequently, $P \cap U$ is compact, open, and contains no points of B. However, this contradicts the definition of C. Thus, $C(x) = \{x\}$.

Now, given an open set $V \subseteq X$ containing x, $C = \{x\}$ implies that the complements of the compact open sets containing x comprise an open cover of $X \setminus V$. In particular, we may choose a finite collection $K_1, \cdots, K_n$ of compact open subsets of X with

$$x \in \bigcap_{j=1}^{n} K_j \subseteq V.$$

As the intersection of finitely many compact open sets is itself compact and open, the proposition follows. □

Proposition A.7 *Suppose G is a totally disconnected compact group. Then the collection*

$$\mathcal{N} = \{N \subseteq G : N \text{ is an open normal subgroup of } G\}$$

is a neighbourhood basis for the topology of G at the identity, e. In particular,

$$\bigcap_{N \in \mathcal{N}} N = \{e\}. \tag{A.3}$$

Proof. By Proposition A.6, it suffices to prove that any compact open set containing e contains an open normal subgroup of G. To that end, suppose that $K \ni e$ is compact and open.

Claim. There is a symmetric neighbourhood V of e for which $K \cdot V \subseteq K$.

For each $g \in K$, openness of K and continuity of multiplication in G implies that there exists an open set $U = U_g$ containing e with $U_g^2 \subseteq g^{-1} \cdot K$. Here, we denote

$$U^2 = \{uv : u \in U \text{ and } v \in U\}.$$

By replacing U with $U \cap U^{-1}$ we may assume without loss of generality that $U_g = U_g^{-1}$ for each g. Since K is compact, one has

$$K \subseteq \bigcup_{j=1}^{m} (g_j \cdot U_{g_j})$$

for some finite collection $g_1, \dots, g_m \in K$. The open set $V = \bigcap_{j=1}^m U_{g_j}$ is a symmetric neighbourhood of the identity with $KV \subseteq K$.

Let us see how the proposition follows from the claim. Since V is symmetric, the subgroup it generates is given by

$$N = \bigcup_{n=1}^{\infty} V^n.$$

Since V is open, N is clearly open. Moreover, $K \cdot V \subseteq K$ implies that $N \subseteq K$. By compactness, N has finite index and hence there are only finitely many subgroups of G conjugate to N. The intersection of these (finitely many) conjugate subgroups is an open normal subgroup of G contained in K. □

Sketch of Proof of Theorem A.5. Let G be a totally disconnected compact group and let $\rho : G \to \hat{G}$ be the homomorphism defined in (A.2).

Step 1. The image of ρ is dense in $\hat{G}$.

Suppose $U \subseteq \hat{G}$ is open and non-empty. Without loss, we may assume that U is of the form

$$U = \hat{G} \cap \left(\prod_{N \in \mathcal{N}} U_N \right),$$

where $U_N = G/N$ for all but finitely many $N \in \mathcal{N}$. If $N_1, \dots, N_k$ denote the exceptional elements of $\mathcal{N}$, put

$$\widetilde{N} = \bigcap_{j=1}^{k} N_j$$

and choose $(g_N \cdot N)_{N \in \mathcal{N}} \in U$. By the compatibility conditions, $g_{\widetilde{N}} N_j = g_{N_j} N_j$ for each $1 \le j \le k$, whence $\rho(g_{\widetilde{N}}) \in U$.

Step 2. ρ is surjective.

By Step 1, the image of ρ is dense. On the other hand, it must be closed by compactness of G and continuity of ρ.

Step 3. ρ is injective. This is immediate from (A.3).

Thus, ρ is a continuous bijective homomorphism. To see that ρ^{-1} is continuous, simply apply the Open Mapping Theorem (see, e.g., [42, Theorem 3, Chapter 1]) □

Corollary A.8 *A compact monothetic totally disconnected group is procyclic.*

Proof. Suppose G is compact, monothetic and totally disconnected. Let α denote a generator of a dense cyclic subgroup. If N is any open subgroup of

G, then G/N is cyclic – indeed, the coset $\alpha + N$ can be seen to generate G/N by minimality. Since G is isomorphic to the inverse limit of such quotients by Theorem A.5, G is procyclic. □

Acknowledgement

Some of this material was presented in a lecture series at the Academy of Mathematics and Systems Science (Chinese Academy of Sciences), Beijing in February 2018. D.D. would like to express his gratitude to Zhe Zhou and AMSS/CAS for the kind invitation and extraordinary hospitality and all the lecture series attendees for their interest and participation.

References

[1] A. Ahlbrecht, H. Vogts, A. Werner, R. Werner, Asymptotic evolution of quantum walks with random coin, *J. Math. Phys.* **52** (2011), 042201, 36 pp.

[2] J. Asch, Joachim, A. Knauf, Motion in periodic potentials, *Nonlinearity* **11** (1998), 175–200.

[3] A. Avila, On the spectrum and Lyapunov exponent of limit periodic Schrödinger operators, *Commun. Math. Phys.* **288** (2009) 907–918.

[4] J. Avron, B. Simon, Almost periodic Schrödinger operators. I. Limit periodic potentials, *Commun. Math. Phys.* **82** (1981), 101–120.

[5] J. Avron, B. Simon, Almost periodic Schrödinger operators. II. The integrated density of states, *Duke Math. J.* **50** (1983), 369–391.

[6] M.-J. Cantero, A. Grünbaum, L. Moral, L. Velázquez, Matrix-valued Szegő polynomials and quantum random walks, *Comm. Pure Appl. Math.* **63** (2010), 464–507.

[7] R. Carmona, J. Lacroix, *Spectral Theory of Random Schrödinger operators,* Probability and its Applications, Birkhäuser, Boston, 1990.

[8] V. Chulaevskii, Perturbations of a Schrödinger operator with periodic potential (Russian). *Uspekhi Mat. Nauk* **36** (1981), 203–204.

[9] V. Chulaevskii, An inverse spectral problem for limit-periodic Schrödinger operators. (Russian) *Funktsional. Anal. i Prilozhen.* **18** (1984), no. 3, 63–66.

[10] W. Craig, B. Simon, Log Hölder continuity of the integrated density of states for stochastic Jacobi matrices, *Comm. Math. Phys.* **90** (1983), 207–218.

[11] H. L. Cycon, R. G. Froese, W. Kirsch, B. Simon, *Schrödinger Operators With Applications to Quantum Mechanics and Global Geometry*, Texts and Monographs in Physics, Springer, Berlin, 1987

[12] D. Damanik, J. Fillman, Limit-periodic Schrödinger operators with Lipschitz continuous IDS, *Proc. Amer. Math. Soc.* **147** (2019), 1531–1539.

[13] D. Damanik, J. Fillman, *Spectral Theory of Discrete One-Dimensional Ergodic Schrödinger Operators*, monograph in preparation.

[14] D. Damanik, J. Fillman, M. Lukic, Limit-periodic continuum Schrödinger operators with zero-measure Cantor spectrum, *J. Spectral Th.* **7** (2017), 1101–1118.

[15] D. Damanik, J. Fillman, D.C. Ong, Spreading estimates for quantum walks on the integer lattice via power-law bounds on transfer matrices, *J. Math. Pures Appl.* **105** (2016), 293–341.

[16] D. Damanik, Z. Gan, Spectral properties of limit-periodic Schrödinger operators, *Commun. Pure Appl. Anal.* **10** (2011), 859–871.

[17] D. Damanik, Z. Gan, Limit-periodic Schrödinger operators with uniformly localized eigenfunctions, *J. Anal. Math.* **115** (2011), 33–49.

[18] D. Damanik, A. Gorodetski, An extension of the Kunz-Souillard approach to localization in one dimension and applications to almost-periodic Schrödinger operators, *Adv. Math.* **297** (2016), 149–173.

[19] D. Damanik, M. Lukic, W. Yessen, Quantum dynamics of periodic and limit-periodic Jacobi and block Jacobi matrices with applications to some quantum many body problems, *Commun. Math. Phys.* **337** (2015), 1535–1561.

[20] F. Delyon, B. Souillard, Remark on the continuity of the density of states of ergodic finite difference operators, *Comm. Math. Phys.* **94** (1984), 289–291.

[21] I. E. Egorova, Spectral analysis of Jacobi limit-periodic matrices, *Dokl. Akad. Nauk Ukrain. SSR Ser. A* **3** (1987), 7–9. (in Russian)

[22] M. Embree, J. Fillman, Spectra of discrete two-dimensional periodic Schrödinger operators with small potentials, *J. Spectral Th.* **9** (2019), 1063–1087.

[23] J. Fillman, Ballistic transport for limit-periodic Jacobi matrices with applications to quantum many-body problems, *Comm. Math. Phys.* **350** (2017), 1275–1297.

[24] J. Fillman, M. Lukic, Spectral homogeneity of limit-periodic Schrödinger operators, *J. Spectral Th.* **7** (2017), 387–406.

[25] J. Fillman, D.C. Ong, Purely singular continuous spectrum for limit-periodic CMV operators with applications to quantum walks, *J. Funct. Anal.* **272** (2017), 5107–5143.

[26] J. Fillman, D.C. Ong, T. VandenBoom, Spectral approximation for ergodic CMV operators with an application to quantum walks, *J. Math. Anal. Appl.* **467** (2018), 132–147.

[27] F. Gesztesy, P. Yuditskii, Spectral properties of a class of reflectionless Schrödinger operators, *J. Func. Anal.* **351** (1999), 619–646.

[28] F. Gesztesy, M. Zinchenko, Local spectral properties of reflectionless Jacobi, CMV, and Schrödinger operators, *J. Diff. Eq.* **246** (2009), 78–107.

[29] R. Han, Uniform localization is always uniform, *Proc. Amer. Math. Soc.* **144** (2016), 609–612.

[30] R. Han, S. Jitomirskaya, Discrete Bethe–Sommerfeld Conjecture, *Commun. Math. Phys.* **361** (2018), 205–216.

[31] B. Helffer, A. Mohamed, Asymptotics of the density of states for the Schrödinger operator with periodic electric potential, *Duke Math. J.* **92** (1998), 1–60.

[32] Y. E. Karpeshina, Perturbation theory for the Schrödinger operator with a periodic potential, *Lecture Notes in Math.* **1663**, Springer, Berlin, 1997.

[33] T. Kato, *Pertubation Theory for Linear Operators*, Springer, Berlin (1980).

[34] J. Kingman, Subadditive ergodic theory, *Ann. Probability* **1** (1973), 883–909.

[35] S. Kotani, Ljapunov indices determine absolutely continuous spectra of stationary random one-dimensional Schrödinger operators, in *Stochastic Analysis (Kakata/Kyoto, 1982)*, North Holland, Amsterdam (1984) 225–247.

[36] H. Krüger, Periodic and limit-periodic discrete Schrödinger operators, arXiv:1108.1584.

[37] H. Krüger, Z. Gan, Optimality of log Hölder continuity of the integrated density of states, *Math. Nachr.* **284** (2011), 1919–1923.

[38] H. Kunz, B. Souillard, Sur le spectre des opérateurs aux différences finies aléatoires, *Commun. Math. Phys.* **78** (1980/81), 201–246.

[39] Y. Last, A relation between a.c. spectrum of ergodic Jacobi matrices and the spectra of periodic approximants, *Commun. Math. Phys.* **151** (1993), 183–192.

[40] Y. Last, B. Simon, Eigenfunctions, transfer matrices, and absolutely continuous spectrum of one-dimensional Schrödinger operators, *Invent. Math.* **135** (1999), 329–367.

[41] S.A. Molchanov, V. Chulaevskii, The structure of a spectrum of the lacunary-limit-periodic Schrödinger operator (Russian). *Funktsional. Anal. i Prilozhen.* **18** (1984), 90–91.

[42] S. A. Morris, *Pontryagin duality and the structure of locally compact abelian groups*, London Mathematical Society Lecture Note Series, No. 29. Cambridge University Press, Cambridge-New York-Melbourne, 1977.

[43] J. Moser, An example of a Schrödinger equation with almost periodic potential and nowhere dense spectrum, *Comment. Math. Helv.* **56** (1981), 198–224.

[44] D.C. Ong, Limit-periodic Verblunsky coefficients for orthogonal polynomials on the unit circle, *J. Math. Anal. Appl.* **394** (2012), 633–644.

[45] L. Parnovski, Bethe–Sommerfeld conjecture, *Ann. Henri Poincaré* **9** (2008), 457–508.

[46] L. Pastur, Spectral properties of disordered systems in the one-body approximation, *Commun. Math. Phys.* **75** (1980) 179–196.

[47] L. Pastur, A. Figotin, *Spectra of Random and Almost-Periodic Operators*, Grundlehren der Mathematischen Wissenschaften, *297*. Springer-Verlag, Berlin, 1992.

[48] L. Pastur, V. A. Tkachenko, On the spectral theory of the one-dimensional Schrödinger operator with limit-periodic potential (Russian), *Dokl. Akad. Nauk SSSR* **279** (1984) 1050–1053.

[49] L. Pastur, V. A. Tkachenko, Spectral theory of a class of one-dimensional Schrödinger operators with limit-periodic potentials (Russian), *Trudy Moskov. Mat. Obshch.* **51** (1988) 114–168.

[50] F. Peherstorfer, A. Volberg, P. Yuditskii, Limit periodic Jacobi matrices with a prescribed p-adic hull and a singular continuous spectrum, *Math. Res. Lett.* **13** (2006), 215–230.

[51] A. Poltoratski, C. Remling, Reflectionless Herglotz functions and Jacobi matrices, *Commun. Math. Phys.* **288** (2009), 1007–1021.

[52] V. N. Popov, M. Skriganov, A remark on the spectral structure of the two dimensional Schrödinger operator with a periodic potential, *Zap. Nauchn. Sem.* LOMI AN SSSR **109** (1981), 131–133 (in Russian).

[53] J. Pöschel, Examples of discrete Schrödinger operators with pure point spectrum, *Commun. Math. Phys.* **88** (1983), 447–463.

[54] C. Puelz, M. Embree, J. Fillman, Spectral approximation for quasiperiodic Jacobi operators, *Int. Eq. Op. Th.* **82** (2015), 533–554.

[55] M. Reed, B. Simon, *Methods of Modern Mathematical Physics, I: Functional Analysis*, Academic Press, New York, 1972.

[56] C. Remling, The absolutely continuous spectrum of Jacobi matrices, *Ann. of Math.* **174** (2011), 125–171.

[57] L. Ribes, P. Zalesskii, *Profinite Groups*, Springer-Verlag, Berlin, 2000.

[58] B. Simon, Some Jacobi matrices with decaying potential and dense point spectrum, *Commun. Math. Phys.* **87** (1982), 253–258.

[59] B. Simon, *Orthogonal Polynomials on the Unit Circle. Part 1. Classical Theory*, Colloquium Publications, 54, American Mathematical Society, Providence (2005).

[60] B. Simon, *Orthogonal Polynomials on the Unit Circle. Part 2. Spectral Theory*, Colloquium Publications, 54, American Mathematical Society, Providence (2005).

[61] B. Simon, *Szegö's Theorem and its Descendants: Spectral Theory for L^2 Perturbations of Orthogonal Polynomials*, M.B. Porter Lectures, Princeton University Press, Princeton, NJ 2011.

[62] M. Skriganov, Proof of the Bethe–Sommerfeld conjecture in dimension two, *Soviet Math. Dokl.* **20** (1979), 89–90.

[63] M. Skriganov, Geometric and arithmetic methods in the spectral theory of multidimensional periodic operators, *Proc. Steklov Math. Inst.* **171** (1984), 3–122.

[64] M. Skriganov, The spectrum band structure of the three-dimensional Schrödinger operator with periodic potential, *Inv. Math.* **80** (1985), 107–121.

[65] M. Sodin, P. Yuditskii, Almost periodic Sturm-Liouville operators with Cantor homogeneous spectrum, *Comment. Math. Helv.* **70** (1995), 639–658.

[66] M. Sodin, P. Yuditskii, Almost periodic Jacobi matrices with homogeneous spectrum, infinite-dimensional Jacobi inversion, and Hardy spaces of character-automorphic functions, *J. Geom. Anal.* **7** (1997), 387–435.

[67] G. Teschl, *Jacobi Operators and Completely Integrable Nonlinear Lattices*, Mathematical Surveys and Monographs **72**, American Mathematical Society, Providence, RI, 2000.

[68] D. Thouless, A relation between the density of states and range of localization for one-dimensional systems, *J. Phys. C* **5** (1972), 77–81.

[69] O. A. Veliev, Spectrum of multidimensional periodic operators, *Teor. Funktsiĭ Funktsional. Anal. i Prilozhen* **49** (1988), 17–34 (in Russian).

[70] J. Wilson, *Profinite Groups*, Oxford University Press, New York, NY, 1998.

16

Uniform Existence of the IDS on Lattices and Groups

C. Schumacher, F. Schwarzenberger, and I. Veselić

Abstract

We present a general framework for thermodynamic limits and its applications to a variety of models. In particular we will identify criteria such that the limits are uniform in a parameter. All results are illustrated with the example of eigenvalue counting functions converging to the integrated density of states. In this case, the convergence is uniform in the energy.

16.1 Introduction

The thermodynamic limit, i.e., taking averages over larger and larger volumes, performs the transition from microscopic to macroscopic models. In the context of this chapter, thermodynamic limits are used to define the integrated density of states (IDS) of discrete Schrödinger operators, in particular, random ones. The IDS of random Schrödinger operators has been studied in the mathematical literature at least since 1971 (see, e. g., [Pas71, Shu79, KM07, Ves08]). The IDS measures the number of quantum states per unit volume below a threshold energy E and is thus a function of E. The rigorous implementation of this notion involves a thermodynamic limit (see Section 16.2). Classical results about the existence of this macroscopic volume limit apply to fixed energy, which is to say that the limit is considered pointwise. More recently, there has been increased interest to study stronger forms of convergence, for instance, uniformly in the energy parameter. This has been studied in various geometric and stochastic contexts. For instance, there are works devoted to Hamiltonians associated to quasicrystals [LS03, LS06], to graphings [Ele07] and percolations graphs [Ves05b, Ves06, AV08, SSV14], as well as abstracts frameworks covering general classes of examples [LV09], where this is certainly a non-exhaustive list of references. We will review here in particular the results obtained in [LMV08], [LSV10, LSV12], [PS16], [SSV17], and [SSV18]. They have all in common that their findings can be formulated as Banach-valued ergodic theorems.

The convergence results in this work are not restriced to the IDS of random Schrödinger operators, but provide a general framework for thermodynamic limits with respect to a Banach space topology. To this end, we introduce almost additive fields, which, after normalization with volume, converge as the volume exhausts the physical space. The microscopic structure, like the values of the potential of the Schrödinger operator and/or whether a percolation site is open or closed, are encoded in a *coloring*. In applications, the coloring is often random, for the lack of detailed knowledge about the microscopic properties of the material.

For the thermodynamic limit to exist, one needs a certain homogeneity of the coloring. In our case, finite portions of the coloring, called *patterns*, should occur with a certain frequency. If there are only finitely many *colors*, i.e., values of the coloring, this condition is natural and allows the existence of the thermodynamic limit to be proven in any Banach space. For random Schrödinger operators, one chooses the Banach space of right continuous bounded functions with ∞-norm. Unfortunately, typical potentials have infinitely many values. In that setting, we model the colourings as random fields with some independence. This allows to employ a multivariate version of the theorem of Glivenko–Cantelli.

This chapter is structured as follows. In Section 16.2, we introduce some specific random Schrödinger operators and their IDS in more detail. They serve as examples for the abstract theorems as well as illustrations of the meaning of our assumptions. We then present a series of theorems on thermodynamic limits with increasing complexity. Section 16.3 contains the oldest of the presented results, which was in some sense the motivation for further developments. It considers fields over $\mathbb{Z}^d$ obeying the finiteness condition (16.3.1) on the set of colors. Since the results are easier to motivate and less technical than later ones, they serve well as a point of departure.

In Section 16.4, we introduce briefly finitely generated amenable groups and show how the main result generalizes to this setting. We first consider amenable groups which satisfy an additional tiling condition. To remove this tiling condition, we outline quasi tilings and discuss the result on amenable groups, still keeping the finiteness condition.

The finiteness condition (16.3.1) is finally tackled and removed in Section 16.5. For clarity, we first deal with the Euclidean lattice $\mathbb{Z}^d$ and only after that exhibit the most general formulation on finitely generated amenable groups without finiteness condition.

An outlook, a list of symbols, and a list of references conclude this chapter.

16.2 Physical Models

In this section, we introduce example systems to motivate the abstract results. For the moment we will stick to the simple geometry of $\mathbb{Z}^d$ but use a notation that later generalizes to general geometries on amenable groups in a straightforward way.

16.2.1 The Anderson Model on $\mathbb{Z}^d$

The Anderson model, introduced by P. W. Anderson in [And58], is a prototypical random Schrödinger operator. To define it, let us introduce some notation.

As physical space we choose the group $G := \mathbb{Z}^d$. To make neighbourhood relations explicit, we introduce the *Cayley graph* of G. The Cayley graph of G has G itself as vertex set, and two vertices $v, w \in G$ are connected by an undirected edge, if $\|v - w\|_1 = 1$, where $\|\cdot\|_1$ is the usual norm on $\mathbb{Z}^d$ when viewed as subset of $\ell^1\{1, \ldots, d\}$. In this case, we write $v \sim w$. Thus, the considered geometry is nothing but the usual d-dimensional lattice. However, to stay consistent with the notation required in later parts of the article, we already use the notion of a Cayley graph. The general definition of Cayley graphs is introduced in Section 16.4.

The group G acts transitively on its Cayley graph by translation:

$$\tilde{\tau}\colon G \times G \to G, \quad \tilde{\tau}_g v := (g, v) := v - g.$$

This group action lifts to a unitary group action on the square summable functions on the vertices of the Cayley graph, $\ell^2(G)$, which we denote by

$$U_g\colon \ell^2(G) \to \ell^2(G), \quad (U_g\varphi)(v) := \varphi(\tilde{\tau}_g v).$$

The *Laplace operator* $\Delta\colon \ell^2(G) \to \ell^2(G)$, given by

$$(\Delta\varphi)(v) := \sum_{w \sim v} \big(\varphi(w) - \varphi(v)\big), \tag{16.2.1}$$

mimics the sum of the second derivatives. The operator $-\Delta$ is bounded, self-adjoint, positive semi-definite and serves as the quantum mechanical observable of the kinetic energy. Its spectrum is $\sigma(-\Delta) = [0, 2d]$, as one can see with Fourier analysis. Note that the Laplace operator is equivariant with respect to G, i.e., for all $g \in G$, we have $\Delta \circ U_g = U_g \circ \Delta$.

To build a Schrödinger operator, we need a multiplication operator V on $\ell^2(G)$, which plays the role of the observable of potential energy. At this stage, the randomness enters. To this end, we choose a (measurable) set $\mathcal{A} \subseteq \mathbb{R}$, which we call the set of colors, equip it with the trace topology inherited from $\mathbb{R}$ and its Borel σ-algebra $\mathcal{B}(\mathcal{A})$, and choose a probability measure $\mathbb{P}_0$ on $(\mathcal{A}, \mathcal{B}(\mathcal{A}))$, i.e., the colors. The probability space for the random potential is $(\Omega, \mathcal{B}, \mathbb{P}) := (\mathcal{A}, \mathcal{B}(\mathcal{A}), \mathbb{P}_0)^{\otimes G} := (\mathcal{A}^G, \mathcal{B}(\mathcal{A})^{\otimes G}, \mathbb{P}_0^{\otimes G})$. The *random potential* $V : \Omega \times G \to \mathcal{A}$ returns for $v \in G$ the v-th coordinate of $\omega \in \Omega$:

$$V_\omega(v) := V(\omega, v) := \omega_v := \omega(v), \tag{16.2.2}$$

so that the random variables $\Omega \ni \omega \mapsto V_\omega(v)$, $v \in G$, are independent and identically distributed. For each $\omega \in \Omega$, the random potential V_ω operates on $\ell^2(G)$ by multiplication. If $\mathcal{A} \in \mathcal{B}(\mathbb{R})$ is bounded, the multiplication operator V_ω is bounded and self-adjoint. Analogous to above, the group action of G on the Cayley graph lifts to an ergodic group action $\tau_g : \Omega \to \Omega$ on Ω:

$$(\tau_g \omega)_v := \omega_{\tilde{\tau}_g v},$$

and the random potential is equivariant, meaning $V_{\tau_g \omega} \circ U_g = U_g \circ V_\omega$ for all $g \in G$.

For each $\omega \in \Omega$, the Schrödinger operator

$$H_\omega := -\Delta + V_\omega : \ell^2(G) \to \ell^2(G) \tag{16.2.3}$$

is well-defined, bounded, and self-adjoint. The operator family $(H_\omega)_{\omega \in \Omega}$ is the famous *Anderson Hamiltonian* and is equivariant, too: $H_{\tau_g \omega} \circ U_g = U_g \circ H_\omega$ for all $g \in G$. This equivariance shows that the spectrum $\sigma(H_\omega)$ is a shift invariant quantity. Since the group action of G on Ω is ergodic, the spectrum of H_ω is almost surely constant: there is a closed set $\Sigma \subseteq \mathbb{R}$ such that

$$\Sigma = \sigma(H_\omega)$$

for $\mathbb{P}$-almost all $\omega \in \Omega$. In quantum mechanics, the spectrum is interpreted as the set of possible energies of the particle described by H_ω. The fact that it is deterministic and does not depend on the microscopic structure of the material makes the interpretation as a homogeneous material possible. More details can be found, e.g., in [PF92, Kir08].

The main example to motivate and to illustrate the results in later chapters is the integrated density of states (IDS), also known as the spectral distribution function, of the Anderson Hamiltonian. Its definition is

$$N\colon \mathbb{R} \to \mathbb{R}, \quad N(E) := \mathbb{E}[\langle \delta_0, \mathbf{1}_{(-\infty,E]}(H_\omega)\delta_0 \rangle],$$

where $\delta_v \in \ell^2(G)$ is the Kronecker delta at $v \in G$, i.e., $\delta_0 = \mathbf{1}_{\{v\}}$, and $\mathbf{1}_{(-\infty,E]}(H_\omega)$ is the spectral projection of H_ω onto the energy interval $(-\infty, E]$. The IDS is monotone, bounded by 1. Hence it gives rise to a probability measure $\mathrm{d}N$, called the *density of states measure*. The topological support of the density of states is the almost sure spectrum of H_ω. In the following, we will describe how to approximate the IDS using only finite matrices.

For each $\Lambda \subseteq G$, we write $\ell^2(\Lambda)$ for the subspace of $\varphi \in \ell^2(G)$ with support $\operatorname{supp} \varphi \subseteq \Lambda$. The indicator function of Λ, used as multiplication operator on $\ell^2(G)$, is the self-adjoint orthogonal projection $\mathbf{1}_\Lambda : \ell^2(G) \to \ell^2(\Lambda)$. The operator $H_\omega^\Lambda : \ell^2(\Lambda) \to \ell^2(\Lambda)$ is given by

$$H_\omega^\Lambda := \mathbf{1}_\Lambda \circ H_\omega \circ (\mathbf{1}_\Lambda)^*.$$

Now let $\mathcal{F}$ be the (countable) set of finite subsets of G and assume $\Lambda \in \mathcal{F}$. The representing matrix of H_ω^Λ contains the matrix elements $\langle \delta_v, H_\omega \delta_w \rangle$ with $v, w \in \Lambda$ and is thus the $\Lambda \times \Lambda$ clipping of the representing $G \times G$ matrix of H_ω. Of course, H_ω^Λ is Hermitian and has real eigenvalues. For $\omega \in \Omega$, $\Lambda \in \mathcal{F}$, the eigenvalue counting function $n(\Lambda, \omega)\colon \mathbb{R} \to \mathbb{R}$,

$$n(\Lambda, \omega)(E) := \operatorname{tr} \mathbf{1}_{(-\infty,E]}(H_\omega^\Lambda) = \dim\{\varphi \in \ell^2(\Lambda) \mid \langle \varphi, H_\omega^\Lambda \varphi \rangle \leq E\}$$

is continuous from the right and counts the eigenvalues of H_ω^Λ below the threshold energy E according to their multiplicity. We will view the family of eigenvalue counting functions as

$$n\colon \mathcal{F} \times \Omega \to \mathbb{B},$$

where $\mathbb{B}$ is the Banach space of right continuous functions from $\mathbb{R}$ to $\mathbb{R}$.

Denote by $\Lambda_L := [0, L)^d \cap \mathbb{Z}^d \in \mathcal{F}$ a cube of side length $L \in \mathbb{N}$. The celebrated *Pastur–Shubin formula* states that, for $\mathbb{P}$-almost all $\omega \in \Omega$ and all $E \in \mathbb{R}$ where the IDS N is continuous, we have

$$N(E) = \lim_{L \to \infty} |\Lambda_L|^{-1} n(\Lambda_L, \omega)(E) \tag{16.2.4}$$

(see, e. g., [SS15]). In fact, since for this model in particular the IDS N is continuous at all energies, a straight-forward argument, using also the boundedness and the monotonicity of N, shows that the convergence does not only hold for all $E \in \mathbb{R}$, but is actually uniform in E. The Pastur–Shubin formula can be viewed as an ergodic theorem, since it states the equality of an ensemble average and a spatial average.

16.2.2 Quantum Percolation Models

16.2.2.1 Site and Edge Percolation

We first introduce *site percolation* on the Cayley graph of $G = \mathbb{Z}^d$. Let $\mathcal{A} := \{0, 1\}$, and as in Section 16.2.1 let $(\Omega, \mathcal{B}, \mathbb{P}) := (\mathcal{A}, \mathcal{B}(\mathcal{A}), \mathbb{P}_s)^{\otimes G}$ with a (non-degenerate) Bernoulli probability $\mathbb{P}_s$ on $\mathcal{A}$. In the percolation setting, the randomness does not determine the potential but configuration space itself. The *site percolation graph* $(\mathcal{V}_\omega, \mathcal{E}_\omega)$ corresponding to $\omega \in \Omega$ is induced by the Cayley graph of G on the vertex set

$$\mathcal{V}_\omega := \{v \in G \mid \omega_v = 1\}.$$

This means that the edge set is

$$\mathcal{E}_\omega := \{\{v, w\} \subseteq \mathcal{V}_\omega \mid v \sim w\},$$

that is, we keep the edges of the Cayley graph which have both end points in $\mathcal{V}_\omega$.

In *edge percolation*, one does not erase random sites from the Cayley graph but instead random edges. The color set $\mathcal{A} := \{0, 1\}^d$ is suitable to implement this strategy. Namely, for a (non-degenerate) Bernoulli probability $\mathbb{P}_e$ on $\{0, 1\}$, we define $(\Omega, \mathcal{B}, \mathbb{P}) := (\mathcal{A}, \mathcal{B}(\mathcal{A}), \mathbb{P}_e^{\otimes d})^{\otimes G}$ and define the edge set corresponding to $\omega \in \Omega$ by

$$\mathcal{E}_\omega := \bigcup_{j=1}^{d} \{\{v, v + e_j\} \mid (\omega_v)_j = 1\}$$

where e_j is the j-th standard basis vector of $\mathbb{Z}^d$. The vertex set of the edge percolation graph contains all vertices to which an edge in $\mathcal{E}_\omega$ is attached:

$$\mathcal{V}_\omega := \{v \in G \mid \exists w \in G\colon \{v, w\} \in \mathcal{E}_\omega\}. \tag{16.2.5}$$

The Laplace operator on a subgraph $(\mathcal{V}_\omega, \mathcal{E}_\omega)$ of the Cayley graph of G is

$$\Delta_\omega\colon \ell^2(\mathcal{V}_\omega) \to \ell^2(\mathcal{V}_\omega), \quad (\Delta_\omega\varphi)(v) := \sum_{w \in \mathcal{V}_\omega, \{v,w\} \in \mathcal{E}_\omega} \big(\varphi(w) - \varphi(v)\big). \tag{16.2.6}$$

Since we want to use the group action of G, we define the Hamiltonians on $\ell^2(G)$ via

$$H_\omega\colon \ell^2(G) \to \ell^2(G), \quad (H_\omega\varphi)(v) := \begin{cases} -(\Delta_\omega\varphi)(v) & \text{if } v \in \mathcal{V}_\omega \text{ and} \\ \alpha\varphi(v) & \text{if } v \in G \setminus \mathcal{V}_\omega \end{cases}$$

with a constant $\alpha \in (2d, \infty)$. The operator H_ω leaves the subspaces $\ell^2(\mathcal{V}_\omega)$ and $\ell^2(G \setminus \mathcal{V}_\omega)$ invariant. Since Δ_ω is bounded by $2d$ and $\alpha > 2d$, the spectrum of of H_ω is the disjoint union of $\sigma(-\Delta_\omega)$ and $\{\alpha\}$, and the two components can be studied separately.

In fact, the percolation Hamiltonians are equivariant, and again their spectrum is almost surely constant. As in Section 16.2.1, we define the IDS $N\colon \mathbb{R} \to \mathbb{R}$ and the eigenvalue counting function $n\colon \mathcal{F} \times \Omega \to \mathbb{B}$ of H_ω. The Pastur–Shubin formula (16.2.4) remains correct $\mathbb{P}$-almost surely for all energies $E \in \mathbb{R}$ at which N is continuous (see, for instance, [Ves05a, Ves05b, KM06, Ves06, AV08]). But in contrast to Anderson Hamiltonian, the IDS of a percolation Hamiltonian is not continuous. In fact, the set of discontinuities of N is dense in Σ. Note that since the IDS is monotone, so there can be at most countably many discontinuities.

That the IDS is discontinuous can be seen as follows. The percolation graph splits into its connected components, the so-called clusters. More precisely, the clusters of a graph are the equivalence classes of the minimal equivalence relation for which neighbours are equivalent. The event that $0 \in G$ is contained in a finite cluster has positive probability (Figure 16.1). A finite cluster of size $s \in \mathbb{N}$ supports s eigenfunctions of H_ω. In particular, the constant function with value $s^{-1/2}$ is an ℓ^2-normalized eigenfunction of H_ω with eigenvalue 0. On the event that 0 is contained in a finite cluster of size s, we have $\langle \delta_0, \mathbf{1}_{(-\infty,0]}(H_\omega)\delta_0 \rangle = s^{-1}$. Thus

$$N(0) = \mathbb{E}[\langle \delta_0, \mathbf{1}_{(-\infty,0]}(H_\omega)\delta_0 \rangle] \geq \sum_{s \in \mathbb{N}} s^{-1}\mathbb{P}(\text{the cluster of 0 has size } s) > 0,$$

and since $N(E) = 0$ for all $E < 0$, we verified that N is discontinuous.

Note that there can be compactly supported eigenfunctions of H_ω on infinite clusters. To construct an example, consider a finite symmetric cluster like $\{-e_1, 0, e_1\} \subseteq G$. The symmetry that flips the sign of e_1 commutes with the Laplace operator restricted to this cluster. Therefore, there are antisymmetric eigenfunctions like $(-1, 0, 1)/\sqrt{2}$, which have to vanish on each fixed point of the symmetry. Now connect the finite symmetric cluster to an infinite cluster with edges only touching the fixed points. For more details see [CCF85, Ves05b].

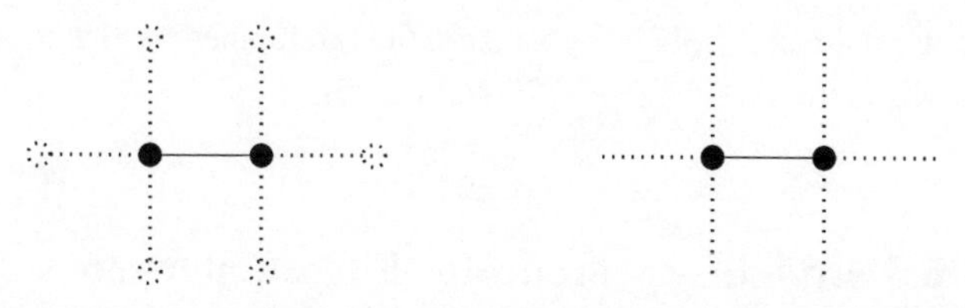

Figure 16.1 The probability that $0 \in G$ is contained in a cluster of size 2 is $2d\mathbb{P}_s(1)^2\mathbb{P}_s(0)^{4d-2}$ in site percolation and $2d\mathbb{P}_e(1)\mathbb{P}_e(0)^{4d-2}$ in edge percolation. Illustration for $d = 2$

16.2.2.2 The Anderson Model on a Percolation Graph

We will now combine the random kinetic energy of the percolation Hamiltonian with the random potential energy of the Anderson model. A suitable set of colors is $\mathcal{A} = \mathcal{A}_0 \times \{0,1\}$ for site and $\mathcal{A} = \mathcal{A}_0 \times \{0,1\}^d$ for edge percolation, where the bounded set $\mathcal{A}_0 \subseteq \mathbb{R}$ contains the values of the random potential. Accordingly, the probability space is

$$(\Omega, \mathcal{B}, \mathbb{P}) := \begin{cases} (\mathcal{A}, \mathcal{B}(\mathcal{A}), \mathbb{P}_0 \otimes \mathbb{P}_s)^{\otimes G} & \text{or} \\ (\mathcal{A}, \mathcal{B}(\mathcal{A}), \mathbb{P}_0 \otimes \mathbb{P}_e^{\otimes d})^{\otimes G}. \end{cases}$$

For each $v \in G$, the random parameter $\omega = (\omega_v)_{v \in G} \in \Omega$ has a coordinate $\omega_v = (\omega'_v, \omega''_v)$ with $\omega'_v \in \mathcal{A}_0$ and either $\omega''_v \in \{0,1\}$ or $\omega''_v \in \{0,1\}^d$. The random potential $V : G \to \mathcal{A}_0 \subseteq \mathbb{R}$ is defined by

$$V_\omega(v) := V(\omega, v) := \omega'_v. \tag{16.2.7}$$

Analogously to the construction in Section 16.2.2.1, we define the site percolation graph as the graph induced by the Caley graph of G on the vertex set

$$\mathcal{V}_\omega := \{v \in G \mid \omega''_v = 1\}.$$

The edge percolation graph has edge set

$$\mathcal{E}_\omega := \bigcup\nolimits_{j=1}^{d} \big\{\{v, v+e_j\} \bigm| (\omega''_v)_j = 1\big\}$$

and vertex set given by (16.2.5). The Laplace operator on the percolation graph is given by (16.2.6). The Hamiltonians $H_\omega \colon \ell^2(G) \to \ell^2(G)$ are defined by

$$(H_\omega \varphi)(v) := \begin{cases} -(\Delta_\omega \varphi)(v) + V_\omega(v) & \text{if } v \in \mathcal{V}_\omega \text{ and} \\ \alpha\varphi(v) & \text{if } v \in G \setminus \mathcal{V}_\omega \end{cases}$$

with $\alpha > 2d + \sup \mathcal{A}_0$. Again, the spectrum $\sigma(H_\omega)$ will have a component $\{\alpha\}$ and a disjoint remainder, which is the part we are most interested in. We have again an equivariant group action, namely for all $g \in G$,

$$H_{\tau_g \omega} \circ U_g = U_g \circ H_{\tau_g \omega}.$$

If the potential has a probability density, the randomness will smooth out at least some discontinuities of the IDS, see [Ves06].

16.3 Ergodic Theorems for Finite Colors on $\mathbb{Z}^d$

In [LMV08, Theorem 2], the authors prove a quantitative version of the following statement.

Theorem 16.3.1 *In either of the settings presented in Section 16.2, assume that the set $\mathcal{A}$ of colors is finite:*

$$|\mathcal{A}| < \infty. \tag{16.3.1}$$

Then, for $\mathbb{P}$-almost all $\omega \in \Omega$,

$$\lim_{L\to\infty} \| |\Lambda_L|^{-1} n(\Lambda_L, \omega) - N\|_\infty = 0,$$

where $N\colon \mathbb{R} \to \mathbb{R}$ is the IDS.

Condition (16.3.1) is automatically satisfied, unless a potential with infinitely many values is present.

In [LMV08], the authors do not talk about a probability space with many configurations, but rather fix one configuration and argue on some required properties. This properties turn out to be almost surely satisfied in the examples in Section 16.2. Nonetheless, it is illuminating to see a deterministic example, which we present next.

16.3.1 Visible Points

A point of $\mathbb{Z}^d$ is *visible* from the origin, if there is no other point of $\mathbb{Z}^d$ on the straight line connecting the origin and the point in question, see Figures 16.2 and 16.3. The set of visible points is thus

$$\mathcal{V} := \{x \in \mathbb{Z}^d \mid \{tx \mid t \in [0, 1]\} \cap \mathbb{Z}^d = \{0, x\}\}.$$

Equivalently, a point $x \in \mathbb{Z}^d$ is visible, if $x = 0$ or if the greatest common divisor of its coordinates is 1. See [BMP00] for a systematic exploration of $\mathcal{V}$.

The indicator function $\mathbf{1}_\mathcal{V}\colon \mathbb{Z}^d \to \mathcal{A} := \{0, 1\}$ can serve as a potential for a Schrödinger operator:

$$H_{\text{vis}}\colon \ell^2(\mathbb{Z}^d) \to \ell^2(\mathbb{Z}^d), \quad H_{\text{vis}} := -\Delta + \mathbf{1}_\mathcal{V}.$$

Analogous to Section 16.2.1, we can define the eigenvalue counting function

$$n\colon \mathcal{F} \to \mathbb{B}, \quad n(\Lambda)(E) := \operatorname{tr} \mathbf{1}_{(-\infty,E]}(H_{\text{vis}}^\Lambda).$$

The results in [LMV08] imply that the thermodynamic limit

$$\lim_{L\to\infty} |\Lambda_L|^{-1} n(\Lambda_L)(E)$$

exists uniformly for all $E \in \mathbb{R}$, but, to the best of our knowledge, there is no Pastur–Shubin formula.

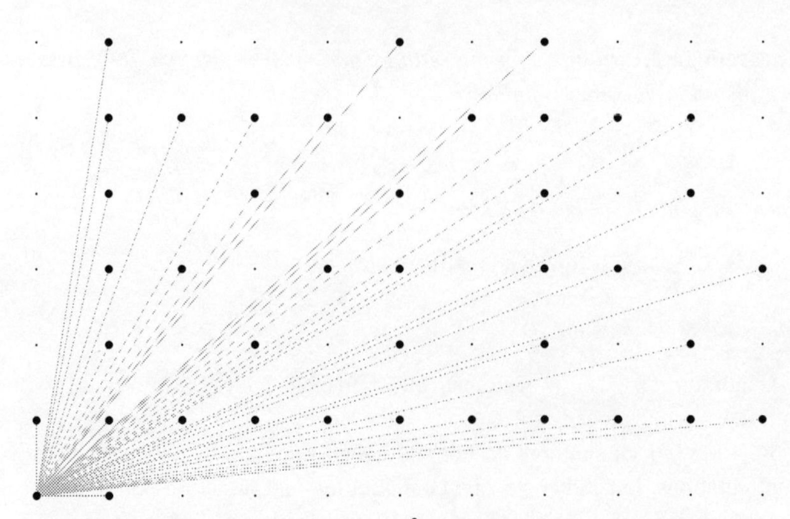

Figure 16.2 The visible points in $[0, 10]^2$ with the line connecting them to the origin

Figure 16.3 The visible points in $[0, 150] \times [0, 100]$

16.3.2 Patterns and Frequencies

To understand the mechanism behind Theorem 16.3.1 better and to motivate the abstract formulation of the next theorem, let us examine the situation of percolation without potential in more detail. We already indicated in

Section 16.2.2.1 with the example of the eigenvalue 0, how discontinuities of N arise. We now want to understand how the limit of the eigenvalue counting functions on large boxes obtains discontinuities of the same height as the IDS. Let us focus on finite clusters again. In $n(\Lambda_L, \omega)$, the eigenvalues corresponding to eigenfunctions supported on finite clusters are counted at least as often as a copy of their cluster is contained in Λ_L. Because we normalize with the volume of the box, $|\Lambda_L|^{-1}n(\Lambda_L, \omega)$, the important quantity turns out to be the relative frequency with which a finite cluster occurs in a large box Λ_L. Since the translations $(\tau_g)_{g\in G}$ act ergodically on Ω, the relative frequencies converge to the probability of their respective cluster at a fixed location. But exactly these probabilities cause the discontinuities of the IDS.

Let us formalize the counting of copies of clusters, or, in the more general setting of a finite set of colors $|\mathcal{A}| < \infty$, shifts of patterns in boxes, following [LMV08]. Recall that the set of finite subsets of G is denoted by $\mathcal{F}$. A (finite) *pattern* with *domain* $\Lambda \in \mathcal{F}$ is a map $P\colon \Lambda \to \mathcal{A}$. Since a pattern $P \in \mathcal{A}^\Lambda$ can be identified with the set $\{\omega \in \Omega \mid \omega_\Lambda = P\}$, we reuse the notation for the group action of G on Ω for the shifts of patterns:

$$\tau_g\colon \mathcal{A}^\Lambda \to \mathcal{A}^{\tilde{\tau}_g\Lambda}, \quad (\tau_g P)(v) := P(\tilde{\tau}_g v).$$

This G-action induces an equivalence relation on the set of all finite patterns. We denote the equivalence class of a pattern $P\colon \Lambda \to \mathcal{A}$ by $[P]_G := \{\tau_g P \mid g \in G\}$.

A coloring $\omega \in \Omega = \mathcal{A}^G$ and a finite subset $\Lambda \in \mathcal{F}$ define the pattern $\omega_\Lambda := \omega|_\Lambda\colon \Lambda \to \mathcal{A}$, $\omega_\Lambda(v) := \omega_v$. Similarly, for $\Lambda \subseteq \Lambda' \in \mathcal{F}$, a pattern P' with domain Λ' induces the pattern $P'_\Lambda := P'|_\Lambda\colon \Lambda \to \mathcal{A}$ via $P'_\Lambda(v) := P'(v)$. Given $\Lambda' \in \mathcal{F}$, set

$$P'|_\mathcal{F} := \{P'|_\Lambda \mid \Lambda \in \mathcal{F}, \Lambda \subseteq \Lambda'\}.$$

This set of all finite patterns induced by $P' \in \mathcal{A}^{\Lambda'}$ is useful to count how often a copy of a pattern $P \in \mathcal{A}^\Lambda$ occurs in P':

$$\sharp_P P' := |[P]_G \cap P'|_\mathcal{F}|. \tag{16.3.2}$$

See Figure 16.4 for a visualisation of this pattern counting function.

In this notation and for the probabilistic models from Section 16.2, the ergodic theorem for $\mathbb{Z}^d$-actions (see [Kel98]) states that, for all $\Lambda \in \mathcal{F}$, $P \in \mathcal{A}^\Lambda$, and $\mathbb{P}$-almost all $\omega \in \Omega$,

$$|\Lambda_L|^{-1}\sharp_P(\omega_{\Lambda_L}) \xrightarrow{L\to\infty} \mathbb{P}\{\omega \in \Omega \mid \omega_\Lambda = P\}.$$

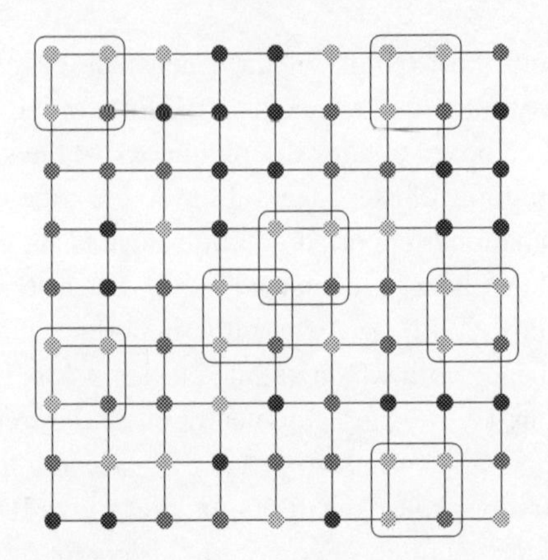

Figure 16.4 The figure shows a pattern P' defined on a square Q of side length 9. The marked 2×2 squares highlight the 7 copies of a pattern P in P'. Thus, we count $\sharp_P P' = 7$. Obviously, to obtain a meaningful proportion of the occurrences of P in P', an appropriate normalization term is the size of the domain of P', i.e., $\frac{\sharp_P P'}{|Q|} = \frac{7}{81}$.

The existence of the limit $\lim_{L\to\infty}|\Lambda_L|^{-1}\sharp_P(\mathcal{V}|_{\Lambda_L})$ for all patterns $P \in \mathcal{A}^\Lambda$ with $\Lambda \in \mathcal{F}$ in the setting of Section 16.3.1 is shown in [BMP00] with different methods. See Figure 16.3 for an optical impression.

Another feature to extract from the percolation example is the following. By restricting the Hamiltonian to finite boxes, we modify some clusters so that some part of their boundary is more 'straight'. As a consequence, the clusters with one 'straight' boundary are over-represented in the sample. Of course, as the boxes grow larger and larger, their surface increases as well. Fortunately, the even faster growth of the volume of the boxes makes the boundary negligible, or more precisely: the proportion of the surface to the volume vanishes in the limit. By this mechanism, the surplus of clusters with artificial straight boundary becomes negligible.

To formalize the notions above, let us introduce the *r-boundary* of $\Lambda \in \mathcal{F}$ for $r > 0$ as

$$\partial^r\Lambda := \{x \in G\backslash\Lambda \mid \mathrm{dist}(x,\Lambda) \le r\} \cup \{v \in \Lambda \mid \mathrm{dist}(v, G\backslash\Lambda) \le r\}. \quad (16.3.3)$$

The distance dist is the length of the shortest path in the Cayley graph, or, for $G = \mathbb{Z}^d$, the distance induced by the 1-norm. A sequence $(Q_j)_{j\in\mathbb{N}}$ of finite sets $Q_j \in \mathcal{F}$ is called a *Følner sequence*, if, for all $r > 0$,

$$\lim_{j\to\infty} \frac{|\partial^r Q_j|}{|Q_j|} = 0. \quad (16.3.4)$$

Another common name for Følner sequences is *van Hove sequence*. The finite boxes $\Lambda_L := [0, L)^d \cap \mathbb{Z}^d \in \mathcal{F}$, $L \in \mathbb{N}$, form an example of a Følner sequence, because $|\partial^r \Lambda_L| = (L+2\lfloor r \rfloor)^d - (L-2\lfloor r \rfloor)^d = 4\lfloor r \rfloor \sum_{k=0}^{d-1}(L-2\lfloor r \rfloor)^k (L - 2\lfloor r \rfloor)^{d-1+k} \leq 4dr(L+2r)^{d-1}$ is of the order of L^{d-1}, while $|\Lambda_L| = L^d$. If, for a pattern $P \in \mathcal{A}^\Lambda$, an $\omega \in \Omega$, and a Følner sequence $(Q_j)_{j \in \mathbb{N}}$, the limit

$$\nu_P := \lim_{j \to \infty} \frac{\sharp_P(\omega_{Q_j})}{|Q_j|}$$

exists, it is called the *frequency of P along $(Q_j)_{j \in \mathbb{N}}$*. See Figure 16.4 for an illustration of one element of this sequence. Note that the existence of ν_P, for all patterns P with finite domain, can be shown almost surely for all our examples from Section 16.2.

The eigenvalue counting functions are in a certain sense local enough to reflect the effect of small boundary compared to the volume. The following notions encapsulate the crucial properties.

Definition 16.3.2 A map $b \colon \mathcal{F} \to [0, \infty)$ is called *boundary term*, if

- b is invariant under G: $b(\Lambda) = b(\tilde{\tau}_g \Lambda)$ for all $g \in G$,
- $\lim_{j \to \infty} |Q_j|^{-1} b(Q_j) = 0$ for all Følner sequences $(Q_j)_{j \in \mathbb{N}}$,
- b is bounded, i.e., its *bound* $D_b := \sup\{|\Lambda|^{-1} b(\Lambda) \mid \Lambda \in \mathcal{F}, \Lambda \neq \emptyset\} < \infty$ is finite, and
- for $\Lambda, \Lambda' \in \mathcal{F}$ we have $b(\Lambda \cup \Lambda') \leq b(\Lambda) + b(\Lambda')$, $b(\Lambda \cap \Lambda') \leq b(\Lambda) + b(\Lambda')$, and $b(\Lambda \setminus \Lambda') \leq b(\Lambda) + b(\Lambda')$.

The last property is natural to require, as illustrated in Figure 16.5, but in fact only necessary when used with quasi tilings (see Section 16.4.2).

Definition 16.3.3 Consider a Banach space $(\mathbb{B}, \|\cdot\|)$, a field $F \colon \mathcal{F} \to \mathbb{B}$, and a coloring $\omega \in \mathcal{A}^G$.

- The field F is *almost additive*, if there exists a boundary term b such that, for all $n \in \mathbb{N}$, disjoint sets $\Lambda_1, \ldots \Lambda_n \in \mathcal{F}$, and $\Lambda := \bigcup_{k=1}^n \Lambda_k$, it holds true that

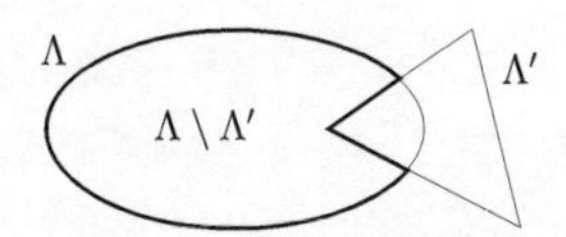

Figure 16.5 The requirement $b(\Lambda \setminus \Lambda') \leq b(\Lambda) + b(\Lambda')$ is motivated by the fact that the boundary of $\Lambda \setminus \Lambda'$ is a subset of the boundary of Λ united with the boundary of Λ'

$$\|F(\Lambda) - \sum_{k=1}^{n} F(\Lambda_k)\| \leq \sum_{k=1}^{n} b(\Lambda_k).$$

- The field F is *ω-invariant*, if for all $\Lambda, \Lambda' \in \mathcal{F}$ such that the patterns ω_Λ and $\omega_{\Lambda'}$ are G-equivalent, we have

$$F(\Lambda) = F(\Lambda').$$

To an ω-invariant field $F\colon \mathcal{F} \to \mathbb{B}$, the *pattern function* $\tilde{F}\colon \bigcup_{\Lambda\in\mathcal{F}} \mathcal{A}^\Lambda \to \mathbb{B}$ with

$$\tilde{F}(P) := \begin{cases} F(\Lambda) & \text{if there is a set } \Lambda \in \mathcal{F} \text{ with } P \in [\omega_\Lambda]_G, \text{ and} \\ 0 & \text{otherwise} \end{cases}$$

is well defined. Every almost additive field and ω-invariant f is *bounded* and has a *bound* C_F in the following sense:

$$C_F := \sup\{|\Lambda|^{-1}\|F(\Lambda)\| \mid \Lambda \in \mathcal{F} \setminus \{\emptyset\}\} < \infty.$$

Indeed, since $\mathcal{A}$ is finite,

$$C_F \leq \sup_{\Lambda\neq\emptyset} \frac{1}{|\Lambda|} \sum_{v\in\Lambda} \big(\|F(\{v\})\| + b(\{v\})\big) \leq \max_{a\in\mathcal{A}^{\{0\}}} \tilde{F}(a) + b(\{0\}) < \infty$$

The eigenvalue counting functions are good examples for these notions.

Proposition 16.3.4 ([LMV08, Proposition 2]) *For almost all $\omega \in \Omega$ and with respect to the Banach space $(\mathbb{B}, \|\cdot\|_\infty)$ of right continuous bounded $\mathbb{R}$-valued functions on $\mathbb{R}$, the eigenvalue counting functions n of the models given in Section 16.2 are ω-invariant, almost additive with boundary term $b(\Lambda) := 4|\partial^1\Lambda|$, $D_b \leq 8d + 4$, and bounded with bound $C_{n(\bullet,\omega)} = 1$.*

The following theorem thus applies to the eigenvalue counting functions of Section 16.2. We emphasize that the error estimates imply that the IDS is approximated by the eigenvalue counting functions uniformly in the energy.

Theorem 16.3.5 ([LMV08, Theorem 1]) *Let $\mathcal{A}$ be a finite set, $\omega \in \mathcal{A}^G$ a coloring, $(\mathbb{B}, \|\cdot\|)$ an arbitrary Banach space, $(Q_j)_{j\in\mathbb{N}}$ a Følner sequence, $F\colon \mathcal{F} \to \mathbb{B}$ a bounded, ω-invariant, and almost additive field with bound C_F, pattern function $\tilde{F}$, boundary term b, and bound D_b of b.*

Assume that for every finite pattern $P\colon \Lambda \to \mathcal{A}$, $\Lambda \in \mathcal{F}$, the frequency $\nu_P := \lim_{j\to\infty}|Q_j|^{-1}\sharp_P(\omega|_{Q_j})$ exists. Then, the limits

$$\overline{F} := \lim_{j\to\infty} \frac{F(Q_j)}{|Q_j|} = \lim_{L\to\infty} \sum_{P\in\mathcal{A}^{\Lambda_L}} \nu_P \frac{\tilde{F}(P)}{|\Lambda_L|} \tag{16.3.5}$$

exist and are equal. Moreover, for all $j, L \in \mathbb{N}$, the bounds

$$\left\| \overline{F} - \sum_{P\in\mathcal{A}^{\Lambda_L}} \nu_P \frac{\tilde{F}(P)}{|\Lambda_L|} \right\| \le \frac{b(\Lambda_L)}{|\Lambda_L|} \tag{16.3.6}$$

and

$$\begin{aligned} &\left\| \frac{F(Q_j)}{|Q_j|} - \sum_{P\in\mathcal{A}^{\Lambda_L}} \nu_P \frac{\tilde{F}(P)}{|\Lambda_L|} \right\| \\ &\le \frac{b(\Lambda_L)}{|\Lambda_L|} + (C_F + D_b)\frac{|\partial^L Q_j|}{|Q_j|} + C_F \sum_{P\in\mathcal{A}^{\Lambda_L}} \left| \frac{\sharp_P(\omega|_{Q_j})}{|Q_j|} - \nu_P \right| \end{aligned} \tag{16.3.7}$$

hold true.

The error estimates are the crucial part of the theorem. We note that there are two length scales, indexed by j and L, in the approximation of the limiting object $\overline{F}$. They correspond to two stages of approximation in the proof. In the first step, $\overline{F}$ is compared to the weighted average of the contributions of the patterns on Λ_L, the weights being the frequencies of the patterns. According to (16.3.6), the patterns on Λ_L capture the behaviour of the limiting function up to an error of size $|\Lambda_L|^{-1}b(\Lambda_L)$. We mentioned above, that $(\Lambda_L)_L$ is a Følner sequence, so by choosing L large enough, we can make this error term small.

Equation (16.3.7) addresses the problem that we used the limiting frequencies in (16.3.5) and (16.3.6) instead of the actual number of occurrences of patterns in the finite region Q_j. This error bound requires us to choose j so large that the empirical frequencies of all patterns on Λ_L get close to their actual frequencies. To recapitulate: First, we have to choose L large enough to make the patterns on Λ_L meaningful for the limit. Then we have to choose j large enough such that the patterns on Λ_L are actually observed in proportions that are close to the asymptotic frequencies.

In the random Schrödinger operator settings from Section 16.2, the second error is controlled by the randomness. In fact, the ergodic theorem predicts that the relative number of occurrences of a pattern converges to its frequencies almost surely. It follows from the theory of large deviations that the probability of a fixed difference between the two is exponentially small for large samples, that is, for large j.

The frequencies $(\nu_P)_{P \in \mathcal{A}^{\Lambda_L}}$ as well as their empirical counterparts $(|Q_j|^{-1} \sharp_P(\omega|_{Q_j}))_{P \in \mathcal{A}^{\Lambda_L}}$ can be interpreted as a probability mass function on $\mathcal{A}^{\Lambda_L}$. The 1-norm of their difference in (16.3.7) is also known as the *total variation norm* of the corresponding probability measures.

In the remainder of this note, we present two generalizations which correspond to the two types of errors described above. The main property of the group $G = \mathbb{Z}^d$ used in Theorem 16.3.5 is that it hosts Følner sequences. That this is actually the crucial property necessary for this proof is made explicit by generalizing the result to amenable groups, which are characterized by the existence of a Følner sequence (see Section 16.4).

The second generalization concerns the finiteness of the set of colors $\mathcal{A}$. The main obstacle to overcome this restriction is the probabilistic error. The total variation norm of the difference of the distribution and the empirical measure of a random variable does not converge to zero for continuous random variables. In order to allow infinitely many colors, we will be forced to exploit more properties of the eigenvalue counting functions (see Section 16.5).

16.4 Ergodic Theorems for Finite Colors on Amenable Groups

In this section we discuss how the above ideas generalize to less restricted geometries. In particular, we present Banach space-valued ergodic theorems for Cayley graphs generated by amenable groups. Let us emphasize that the groups considered in this chapter will always be finitely generated and therefore countable. An amenable group is by definition a group containing subsets with an arbitrary small ratio between boundary and volume. It is well known, that amenability is equivalent to the existence of a Følner sequence.

It turns out that even though amenability is the natural condition to generalize the geometry of $\mathbb{Z}^d$, there is an additional requirement needed to almost directly implement the $\mathbb{Z}^d$-methods to this setting. Here we are speaking about a so-called tiling condition, namely the condition that there exists a Følner sequence consisting of monotiles. Obviously, in $\mathbb{Z}^d$ a sequence of cubes serves as such a sequence, since for each j the group $\mathbb{Z}^d$ can be tiled with cubes of side length j. For an arbitrary amenable group it is not known whether such a sequence exists or not. Therefore, this section is structured in a first part discussing the monotile situation and a second more involved part dealing with the general amenable groups using the technique of quasi tilings.

We proceed with some definitions which generalize the notion of previous sections. Given a finitely generated group G with a finite and symmetric generating set $S \subseteq G$, i.e., $G = \langle S \rangle$, $|S| < \infty$ and $S = S^{-1} \not\ni \mathrm{id}$, the corresponding Cayley graph Γ has vertex set G and two vertices $v, w \in G$ are connected if and only if $vs = w$ for some $s \in S$. The induced graph distance, sometimes called word metric, is denoted by dist. A sequence (Q_j) of finite subsets of G is called a Følner sequence if $|\partial^r Q_j|/|Q_j| \to 0$ as $j \to \infty$ for all $r > 0$, see (16.3.4). Here the r-boundary is given as in (16.3.3).

For an amenable group G, the introduction of the physical models of Sections 16.2.1 and 16.2.2 works completely analogous. Due to the fact that for general groups the group action is usually written as a multiplication, the only difference is that the group action $\tilde{\tau}$ is defined here as

$$\tilde{\tau}\colon G \times G \to G, \quad (g, v) \mapsto \tilde{\tau}_g v := vg^{-1}. \tag{16.4.1}$$

As in (16.2.3) the Schödinger operator in the Anderson model is given by

$$H_\omega := -\Delta + V_\omega \colon \ell^2(G) \to \ell^2(G)$$

with Laplace operator $\Delta\colon \ell^2(G) \to \ell^2(G)$ as in (16.2.1) and a (random) potential $V\colon \Omega \times G \to \mathcal{A}$ as in (16.2.2). As before we consider the probability space $(\Omega, \mathcal{B}, \mathbb{P}) := (\mathcal{A}^G, \mathcal{B}(A)^{\otimes G}, \mathbb{P}_0^{\otimes G})$ with a finite set $\mathcal{A}$. An element $\omega \in \Omega$ is then interpreted as a coloring of the elements of the group using the finite set of colors $\mathcal{A}$.

Besides this, also the definition of site and edge percolation does not depend on the $\mathbb{Z}^d$-structure and generalizes straightforwardly to the case of amenable groups. Thereby the Anderson model on percolation Cayley graphs over finitely generated amenable groups is well defined.

16.4.1 Symmetrical Tiling Condition

As mentioned before, an additional condition is needed in order to implement the methods of $\mathbb{Z}^d$ to amenable groups. This so-called symmetric tiling condition is formulated as follows.

Definition 16.4.1 Let G be a group. A subset $\Lambda \subseteq G$ *symmetrically tiles* G if there exists a set T such that

(i) $T = T^{-1}$,
(ii) $\Lambda T = G$,
(iii) $\{\Lambda t \mid t \in T\}$ are pairwise disjoint.

An amenable G is said to satisfy the *symmetric tiling condition*, if there exists a Følner sequence (Λ_L) such that each Λ_L, $L \in \mathbb{N}$, symmetrically tiles G. In this situation, we call G an *ST-amenable group*.

Note that here $\Lambda T := \{xt \mid x \in \Lambda, t \in T\}$ and $\Lambda t = \{xt \mid x \in \Lambda\}$.

Let us briefly remark on this condition. Krieger proved in [Kri07] based on work of Weiss [Wei01] that an amenable group satisfies the symmetrical tiling condition if it is residually finite. For instance, each group of polynomial volume growth is nilpotent (by Gromov's theorem) and thus residually finite. Since it is also of sub-exponential growth, it is amenable, too, and hence ST-amenable.

An intensively studied and slightly more general condition than the one stated in Definition 16.4.1 can be obtained when not assuming the symmetry (i). In this situation, a set Q satisfying (ii) and (iii) is usually referred to as a *monotile* and a group G containing a Følner sequence consisting only of monotiles is called *monotileable*. Let us remark that it is still not known if there exists an amenable group which is not monotileable.

In the situation of ST-amenable groups, one can prove the following:

Theorem 16.4.2 ([LSV10, LSV12]) *Let G be a finitely generated ST-amenable group. Let (Q_j) and (Λ_L) be Følner sequences such that each Λ_L, $L \in \mathbb{N}$ symmetrically tiles G. Let $\mathcal{A}$ be finite and $\omega \in \mathcal{A}^G$ be given.*

Moreover, let $(\mathbb{B}, \|\cdot\|)$ be a Banach space and $F\colon \mathcal{F} \to \mathbb{B}$ a bounded, ω-invariant and almost additive field with bound C_F, pattern function $\tilde{F}$, boundary term b, and bound D_b of b. As before we use the notation $\mathcal{F} = \{A \subseteq G \mid A \text{ finite}\}$.

Assume that for every finite pattern $P\colon \Lambda \to \mathcal{A}$, $\Lambda \in \mathcal{F}$, the frequency $\nu_P := \lim_{j\to\infty} |Q_j|^{-1} \sharp_P(\omega|_{Q_j})$ exists. Then, the limits

$$\overline{F} := \lim_{j\to\infty} \frac{F(Q_j)}{|Q_j|} = \lim_{L\to\infty} \sum_{P\in\mathcal{A}^{\Lambda_L}} \nu_P \frac{\tilde{F}(P)}{|\Lambda_L|} \tag{16.4.2}$$

exist and are equal. Moreover, for all $j, L \in \mathbb{N}$, the bound

$$\left\| \frac{F(Q_j)}{|Q_j|} - \sum_{P\in\mathcal{A}^{\Lambda_L}} \nu_P \frac{\tilde{F}(P)}{|\Lambda_L|} \right\| \leq \frac{b(\Lambda_L)}{|\Lambda_L|} + (C_F + D_b) \frac{|\partial^{\mathrm{diam}(\Lambda_L)} Q_j|}{|Q_j|} + C_F \sum_{P\in\mathcal{A}^{\Lambda_L}} \left| \frac{\sharp_P(\omega|_{Q_j})}{|Q_j|} - \nu_P \right| \tag{16.4.3}$$

holds true. Here, $\operatorname{diam}(\Lambda_L) = \max\{\operatorname{dist}(x, y) \mid x, y \in \Lambda_L\}$ *is the diameter of the finite set* Λ_L.

When comparing Theorem 16.3.5 and Theorem 16.4.2, it turns out that the transition from $\mathbb{Z}^d$ to ST-amenable groups does not imply a quantitative difference in the strength of the result. In particular, the only difference is that in the $\mathbb{Z}^d$ setting (Λ_L) is a sequence of cubes with side length L (which naturally tiles $\mathbb{Z}^d$) and in the ST-amenable group setting (Λ_L) is a Følner sequence assumed to be symmetrically tiling. In the error bound this results in the substitution of the side length L of a cube by the diameter of Λ_L.

In the following we give a sufficient condition for the existence of the frequencies in the situation of a random coloring. Here we need the notion of a *tempered* Følner sequence, which is a Følner (Q_j) sequence with the additional property that there is some $C > 0$ such that

$$\left| \bigcup_{k<n} Q_k^{-1} Q_j \right| \leq C |Q_j|$$

for all $j \in \mathbb{N}$. Each Følner sequence has a tempered subsequence, see [Lin01].

Theorem 16.4.3 *Let* G *be a finitely generated amenable group,* $\mathcal{A}$ *some finite set and let* μ *be a probability measure on* $(\Omega, \mathcal{B}) = (\mathcal{A}^G, \mathcal{B}(A)^{\otimes G})$. *We assume that the action* $\tilde{\tau}$ *given via* (16.4.1) *of* G *on* Ω *is measure preserving and ergodic w.r.t.* μ. *Then, for any tempered Følner sequence* (Q_j), *there exists an event* $\tilde{\Omega}$ *of full measure, such that the limit*

$$\lim_{j\to\infty} \frac{\sharp_P(\omega|_{Q_j})}{|Q_j|}$$

exists for all patterns $P \in \bigcup_{Q\in\mathcal{F}} \mathcal{A}^Q$ *and all* $\omega \in \tilde{\Omega}$. *Moreover, the limit is deterministic in the sense that it is independent of the specific choice of* $\omega \in \tilde{\Omega}$.

The above result is a direct consequence (see, for instance, [Sch08]) of Lindenstrauss ergodic theorem [Lin01]. We omit the precise definitions of measure preserving and ergodic action. However, we want to emphasize that in the particular case where $\mu = \mathbb{P}_0^{\otimes G}$ is a product measure, the assumptions on $\tilde{\tau}$ are met. Thus, in the percolation setting of Section 16.2.2 one obtains that almost all configurations satisfy the assumption of well-defined frequencies.

16.4.2 General Amenable Groups

As outlined before, it is still not known if each amenable group satisfies the symmetrical tiling condition. Roughly speaking, in the previous sections this tiling condition is the crucial tool to mediate between the two Følner sequence (Λ_L) and (Q_j). Thus, when considering amenable groups without an additional tiling assumption the situation is far more challenging.

A way to overcome this lack is to apply the theory of ϵ-quasi tilings developed by Ornstein and Weiss [OW87] in 1987, see also [PS16] for the quantitative estimates used in the present setting. The key idea here is to soften the condition of a *perfect* tiling with copies of *one* set taken from a Følner sequence, and rather

- use finitely many different sets of the Følner sequence, and
- allow imperfectness in the tiling (in sense of small *overlaps* and *uncovered areas*).

More precisely we use the following definition:

Definition 16.4.4 Let G be a finitely generated group, $Q \subseteq G$ a finite set and $\epsilon > 0$. We say that $K_1, \dots, K_N \subseteq G$ with centre sets $T_1, \dots, T_N$, short $(K_i, T_i)_{i=1}^N$, are an ϵ-*quasi tiling* of Q if

(i) the sets $K_i T_i$, $i \in \{1, \dots, N\}$ are pairwise disjoint and subsets of Q;
(ii) $|Q \setminus \bigcup_{i=1}^N K_i T_i| \leq 2\epsilon |Q|$;
(iii) there are subsets $K_i \subseteq K_i$, $i \in \{1, \dots, N\}$, such that
  - for each i the sets $K_i t$, $t \in T_i$ are pairwise disjoint
  - $|K_i \setminus K_i| \leq \epsilon |K_i|$

For technical reasons, the sequence (Λ_L) which will provide the Følner sets K_i to quasi tile the group is assumed to be *nested*, i.e., for each $L \in \mathbb{N}$ we have $\mathrm{id} \in \Lambda_L \subseteq \Lambda_{L+1}$. Note that, starting from an arbitrary Følner sequence, one can construct a nested Følner sequence by translating elements of an appropriate subsequence.

In the following, it turns out to be convenient to define for given $\epsilon \in (0, 1)$ and $i \in \mathbb{N}_0$ the numbers $N(\epsilon)$ and $\eta_i(\epsilon)$ by

$$N(\epsilon) := \left\lceil \frac{\ln(\epsilon)}{\ln(1-\epsilon)} \right\rceil \quad \text{and} \quad \eta_i(\epsilon) := \epsilon(1-\epsilon)^{N(\epsilon)-i}. \tag{16.4.4}$$

As usual, we use the Gaußian bracket notation $\lceil b \rceil := \inf\{z \in \mathbb{Z} \mid z \geq b\}$. The following theorem shows that $N(\epsilon)$ is the number of required shapes K_i in order to ϵ-quasi tile a set Q. Moreover, for fixed i the $\eta_i(\epsilon)$ can be interpreted as the ratio of the points covered by copies of K_i in the ϵ-quasi tiling.

Theorem 16.4.5 *Let G be a finitely generated amenable group, (Λ_L) a nested Følner sequence, and $\epsilon \in (0, 0.1)$. Then there is a finite and strictly increasing selection of sets $K_i \in \{\Lambda_L \mid L \in \mathbb{N}\}$, $i \in \{1, \dots, N(\epsilon)\}$, with the following* *property: For each Følner sequence* (Q_j)*, there exists* $j_0(\epsilon) \in \mathbb{N}$ *satisfying that for all* $j \geq j_0(\epsilon)$ *there exists sets* $T_1^j, \ldots, T_{N(\epsilon)}^j$ *such that* $(K_i, T_i^j)_{i=1}^{N(\epsilon)}$ *is an* ϵ*-quasi tiling of* Q_j*. Moreover, for all* $j \geq j_0(\varepsilon)$ *and all* $i \in \{1, \ldots, N(\epsilon)\}$*, the proportion of* Q_j *covered by the tile* K_i *satisfies*

$$\left| \frac{|K_i T_i^j|}{|Q_j|} - \eta_i(\varepsilon) \right| \leq \frac{\varepsilon^2}{N(\varepsilon)}. \tag{16.4.5}$$

The proof of Theorem 16.4.5 is to be found in [PS16]. Although Theorem 16.4.5 provides all the elements used to formulate the desired ergodic theorem for general amenable groups (Theorem 16.4.6), a far more involved result is applied in the proof. More precisely, in the proof of Theorem 16.4.6 one does not only need *one* possibility to quasi tile a given set Q_j with $K_1, \ldots, K_{N(\epsilon)}$, but one rather needs a bunch of *different* possibilities to quasi tile Q_j with $K_1, \ldots, K_{N(\epsilon)}$. These different tilings of Q_j need to be chosen such that for (almost) all $g \in Q_j$ the frequency (over different tilings) that it is covered by one K_i is up to a small error $\eta_i(\epsilon)/|K_i|$. In this sense, this covering result is referred to as uniform ϵ-quasi tiling. We refer to [PS16] for details.

Let us formulate the ergodic theorem based on the ϵ-quasi tiling results.

Theorem 16.4.6 *Assume:*

- G *is a finitely generated amenable group.*
- $\mathcal{A}$ *is a finite set and* $\omega \in \mathcal{A}^G$.
- (Q_j) *is a Følner sequence such that the frequency* $\nu_P := \lim_{j\to\infty} |Q_j|^{-1} \sharp_P(\omega|_{Q_j})$ *exists for every finite pattern* $P\colon \Lambda \to \mathcal{A}$, $\Lambda \in \mathcal{F}$.
- (Λ_L) *is a nested Følner sequence.*
- *For given* $\epsilon \in (0, \frac{1}{10})$ *the sets* K_i, $i \in \{1, \ldots, N(\epsilon)\}$ *are chosen according to Theorem 16.4.5.*
- $(\mathbb{B}, \|\cdot\|)$ *is a Banach space and* $F\colon \mathcal{F} \to \mathbb{B}$ *is* ω*-invariant, and almost additive with bound* C_F*, pattern function* $\tilde{F}$*, boundary term* b*, and bound* D_b *of* b*.*

Then, the limits

$$\overline{F} := \lim_{j\to\infty} \frac{F(Q_j)}{|Q_j|} = \lim_{\substack{\epsilon \searrow 0 \\ \epsilon < 0.1}} \sum_{i=1}^{N(\epsilon)} \eta_i(\epsilon) \sum_{P \in \mathcal{A}^{K_i}} \nu_P \frac{\tilde{F}(P)}{|K_i|} \tag{16.4.6}$$

exist and are equal. Moreover, for given $\epsilon \in (0, \frac{1}{10})$ *there exists some* $j(\epsilon), r(\epsilon) \in \mathbb{N}$ *such that for all* $j \geq j(\epsilon)$ *the bound*

$$\left\| \frac{F(Q_j)}{|Q_j|} - \sum_{i=1}^{N(\epsilon)} \eta_i(\epsilon) \sum_{P \in \mathcal{A}^{K_i}} \nu_P \frac{\tilde{F}(P)}{|K_i|} \right\|$$
$$\leq 4 \sum_{i=1}^{N(\epsilon)} \eta_i(\epsilon) \frac{b(K_i)}{|K_i|} + (C_F + 4D_b) \frac{|\partial^{r(\epsilon)} Q_j|}{|Q_j|} \sum_{i=1}^{N(\epsilon)} |K_i|$$
$$+ C_F \sum_{i=1}^{N(\epsilon)} \eta_i(\epsilon) \sum_{P \in \mathcal{A}^{K_i}} \left| \frac{\sharp_P(\omega|_{Q_j})}{|Q_j|} - \nu_P \right| + (11C_F + 32D_b)\epsilon \tag{16.4.7}$$

holds true.

When comparing the estimate with the one in Theorem 16.4.2, note that the difference (16.4.7) gets small if one firstly executes the limit $j \to \infty$ and afterwards $\epsilon \searrow 0$. Here the limit $\epsilon \searrow 0$ corresponds to $L \to \infty$ in the previous setting. For a detailed discussion why the error terms tend to zero we refer to [PS16]. We confine ourselves to the (rough) statement that the first three terms in the estimate (16.4.7) correspond in this ordering to three terms in (16.4.3) or (16.3.7), respectively.

16.5 Glivenko–Cantelli Type Theorems

In this section we will consider the situation that the set of colors $\mathcal{A}$ is no longer finite. This means that we are leaving a combinatorial setting and relying on a probabilistic framework instead. This has been already introduced for a number of examples in Sections 16.2.1 and 16.2.2. In particular, we will need some independence and monotonicity assumptions. The monotonicity property will allow us to smooth out and regularize certain quantities which we otherwise do not know how to estimate. The independence assumption gives us a tool to describe the existence of frequencies used in Section 16.3 in a constructive and quantitative manner.

We reformulate the notions of Section 16.3 involving fields $F \colon \mathcal{F} \to \mathbb{B}$ with values in an arbitrary Banach space $\mathbb{B}$ for fields of the form $f \colon \mathcal{F} \times \Omega \to \mathbb{B}$ with a probability space Ω and the specific Banach space of right continuous functions with sup-norm. For easy distinction, we denote the latter with lowercase letters. Almost additive is assumed to hold true uniformly on Ω. The property ω-invariance for fixed $\omega \in \Omega$ is split up in equivariance and locality.

Definition 16.5.1 A field $f\colon \mathcal{F} \times \Omega \to \mathbb{B}$ is

- *almost additive*, if there is a boundary term $b\colon \mathcal{F} \to [0, \infty)$ such that for all $\omega \in \Omega$, pairwise disjoint $\Lambda_1, \ldots, \Lambda_n \in \mathcal{F}$, and $\Lambda := \bigcup_{i=1}^{n} \Lambda_i$, we have

$$\left\| f(\Lambda, \omega) - \sum_{i=1}^{n} f(\Lambda_i, \omega) \right\| \leq \sum_{i=1}^{n} b(\Lambda_i).$$

- *equivariant*, if for $\Lambda \in \mathcal{F}$, $g \in G$ and $\omega \in \Omega$ we have $f(\tilde{\tau}_g \Lambda, \omega) = f(\Lambda, \tau_g \omega)$.
- *local*, if for all $\Lambda \in \mathcal{F}$ and $\omega, \omega' \in \Omega$, $\omega_\Lambda = \omega'_\Lambda$ implies $f(\Lambda, \omega) = f(\Lambda, \omega')$.
- *bounded*, if $\sup_{\omega \in \Omega} \| f(\{\mathrm{id}\}, \omega) \| < \infty$.

16.5.1 Glivenko–Cantelli Theory

To simplify the motivation, we first consider $G = \mathbb{Z}^d$ and assume that the probability measure $\mathbb{P} = \mathbb{P}_0^{\otimes G}$ on Ω is a product measure. This implies in particular that, for every local field $f\colon \mathcal{F} \times \Omega \to \mathbb{B}$, the random variables $\omega \mapsto f(\{0\}, \tau_g \omega)$, $g \in G$, are independent. Here, we used that for local f, in particular, $f(\{0\}, \omega)$ depends only on ω_0 and not on $\omega_{G \setminus \{0\}}$.

To explain the relation of our methods to Glivenko–Cantelli theory, assume for the moment that $f\colon \mathcal{F} \times \Omega \to \mathbb{B}$ is not only almost additive but *exactly* additive. That means that for a finite subset $\Lambda \in \mathcal{F}$ of G and a realization of colors $\omega \in \Omega = \mathcal{A}^G$, we can split $f(\Lambda, \omega)$ without any errors into a sum over singleton sets

$$f(\Lambda, \omega) = \sum_{v \in \Lambda} f(\{v\}, \omega).$$

By equivariance, we can rewrite $f(\{v\}, \omega) = f(\{0\}, \tau_v^{-1} \omega)$. For the special case $\mathbb{B} = \mathbb{R}$, the law of large numbers allows us to calculate the thermodynamic limit

$$\lim_{L \to \infty} \frac{f(\Lambda_L, \omega)}{|\Lambda_L|} = \lim_{L \to \infty} \frac{1}{|\Lambda_L|} \sum_{v \in \Lambda_L} f(\{0\}, \tau_v^{-1} \omega) = \mathbb{E}[f(\{0\}, \cdot\,)]$$

$\mathbb{P}$-almost surely.

Of course, in view of our examples in Section 16.2, we are more interested in the Banach space $\mathbb{B}$ of right continuous functions from $\mathbb{R}$ to $\mathbb{R}$ with sup-norm. To head in this direction, it is advantageous to lift the point of view to the empirical probability $\widehat{\mathbb{P}}_\Lambda^\omega := |\Lambda|^{-1} \sum_{v \in \Lambda} \delta_{\omega_v}$ and to write integration as dual pair $\langle \cdot\,, \,\cdot \rangle$. In this notation, the average from above can be written as

$$|\Lambda|^{-1} f(\Lambda, \omega) = \langle f(\{0\}, \,\cdot\,), \widehat{\mathbb{P}}^{\omega}_{\Lambda} \rangle.$$

Here is a special case, where a theorem from classical probability theory helps. Assume that $\mathcal{A} \in \mathcal{B}(\mathbb{R})$, and let $f : \mathcal{F} \times \Omega \to \mathbb{B}$ count the number of random variables in Λ with value less than a given threshold $E \in \mathbb{R}$:

$$f(\Lambda, \omega)(E) := \sum_{v \in \Lambda} \mathbf{1}_{[\omega_v, \infty)}(E) = \sum_{v \in \Lambda} \mathbf{1}_{(-\infty, E]}(\omega_v) = |\Lambda| \langle \mathbf{1}_{(-\infty, E]}, \widehat{\mathbb{P}}^{\omega}_{\Lambda} \rangle.$$

This defines an additive, local, and equivariant field, which can be interpreted as the (not normalized) empirical distribution function of the sample $(\omega_v)_{v \in \Lambda}$. The thermodynamic limit in $\mathbb{B}$, i.e., uniformly, is $\mathbb{P}$-almost surely

$$\begin{aligned} \lim_{L \to \infty} |\Lambda_L|^{-1} f(\Lambda_L, \omega) &= \lim_{L \to \infty} \langle \mathbf{1}_{(-\infty, \bullet]}, \widehat{\mathbb{P}}^{\omega}_{\Lambda} \rangle \\ &= \langle \mathbf{1}_{(-\infty, \bullet]}, \mathbb{P}_0 \rangle = \mathbb{P}(\omega_0 \leq \,\cdot\,) \end{aligned}$$

by the theorem of Glivenko and Cantelli:

Theorem 16.5.2 ([Gli33, Can33]) *Let V_j, $j \in \mathbb{N}$, be real valued, independent and identically distributed random variables on $(\Omega, \mathcal{B}, \mathbb{P})$ and $\widehat{\mathbb{P}}_n := \frac{1}{n} \sum_{j=1}^{n} \delta_{V_j}$ the corresponding empirical distribution. Then there exists an event Ω_{unif} with probability $\mathbb{P}(\Omega_{\mathrm{unif}}) = 1$ such that for all $\omega \in \Omega_{\mathrm{unif}}$:*

$$\sup_{E \in \mathbb{R}} |\langle \mathbf{1}_{(-\infty, E]}, \widehat{\mathbb{P}}_n(\omega) - \mathbb{P} \rangle| \xrightarrow{n \to \infty} 0.$$

We learn that the choice of this Banach space $\mathbb{B}$ means to prove the convergence of $\widehat{\mathbb{P}}^{\omega}_{\Lambda}$ to $\mathbb{P}_0$ with respect to a supremum over appropriate test functions. The route pursued in Theorems 16.3.5, 16.4.2, and 16.4.6 for finite alphabets corresponds to the estimate

$$|\langle g, \widehat{\mathbb{P}}^{\omega}_{\Lambda} - \mathbb{P}_0 \rangle| \leq \|g\|_{\infty} \|\widehat{\mathbb{P}}^{\omega}_{\Lambda} - \mathbb{P}_0\|_{\mathrm{TV}}$$

with $g := f(\{0\}, \,\cdot\,)$. But, as the next example shows, for smooth random variables, the difference does not converge to zero in total variation.

Example 16.5.3 Let $\mathcal{A} = [0, 1]$ and X_n, $n \in \mathbb{N}$, be real valued i. i. d. random variables, distributed uniformly on $\mathcal{A}$. The empirical distribution $\widehat{\mathbb{P}}_n := n^{-1} \sum_{j=1}^{n} \delta_{X_j}$ is an atomic measure on $\mathcal{A}$, while the uniform distribution on $\mathcal{A}$ is absolutely continuous with respect to Lebesgue measure. The TV-norm of their difference does not vanish for $n \to \infty$:

$$\|\widehat{\mathbb{P}}_n - \mathbb{P}\|_{\mathrm{TV}} = \sup_{A\in\mathcal{B}(\mathcal{A})} |\widehat{\mathbb{P}}_n(A) - \mathbb{P}(A)| \geq 1,$$

as the set $A := \{X_n \mid n \in \mathbb{N}\}$ shows.

We have to follow a different path. Assume again $\mathcal{A} \in \mathcal{B}(\mathbb{R})$. Let us abbreviate the difference of the cumulative distribution functions by

$$F_\Lambda(E) := \langle \mathbf{1}_{(-\infty,E]}, \widehat{\mathbb{P}}^\omega_\Lambda - \mathbb{P}_0\rangle, \quad E \in \mathbb{R},$$

and assume that $g = f(\{0\}, \cdot)$ has bounded variation, or more specifically that it is monotone. Then, we can perform the following partial integration with Riemann-Stieltjes integrals:

$$\begin{aligned} |\langle g, \widehat{\mathbb{P}}^\omega_\Lambda - \mathbb{P}_0\rangle_{\mathcal{A}}| &= \left|\int_{\mathcal{A}} g(E)\,\mathrm{d}F_\Lambda(E)\right| = \left|-\int_{\mathcal{A}} F_\Lambda(E)\,\mathrm{d}g(E)\right| \\ &\leq \|F_\Lambda\|_\infty \int_{\mathcal{A}} \mathrm{d}|g|(E) \\ &= \|g\|_{\mathrm{TV}} \cdot \sup_{E\in\mathbb{R}} |\langle \mathbf{1}_{(-\infty,E]}, \widehat{\mathbb{P}}^\omega_\Lambda - \mathbb{P}_0\rangle_{\mathcal{A}}|. \end{aligned}$$

This calculation generalizes the theorem by Glivenko and Cantelli to bounded monotone functions: For $M > 0$, we have

$$\sup\{|\langle g, \widehat{\mathbb{P}}_{\Lambda_L} - \mathbb{P}_0\rangle| \mid g\colon \mathbb{R} \to [-M, M] \text{ monotone}\} \xrightarrow{L\to\infty} 0.$$

In order to deal with fields that are only *almost* additive, we have to treat patterns of all finite sizes and not only singletons. Each pattern corresponds to a multivariate random variable. This means that we require a multivariate version of Glivenko–Cantelli theory. To formulate this, we introduce a multivariate version of the empirical measure: For given (large) set Λ_j, smaller set Λ_L, a grid $T_{j,L}$ and a coloring $\omega \in \mathcal{A}^G$ we define the empirical measure by

$$\widehat{\mathbb{P}}^\omega_{j,L} : \mathcal{B}(\mathcal{A}^{\Lambda_j}) \to [0,1], \qquad \widehat{\mathbb{P}}^\omega_{j,L} := \frac{1}{|T_{j,L}|} \sum_{t\in T_{j,L}} \delta_{(\tau_t\omega)_{\Lambda_L}}.$$

Here, the grid $T_{j,L}$ is a set of basepoints to (almost) cover the Λ_j with translated versions of Λ_L along $T_{j,L}$. An illustration of this it to be found in Figure 16.6. Let us emphasize that the illustration serves well in the $\mathbb{Z}^d$-case or in the ST-amenable case. However, for general amenable groups there is usually no grid $T_{j,L}$ for a (perfect) covering a set with *one* set. In this case one can still use the above definition of the empirical measure, but, as in Section 16.4.2, one needs to implement the technique of ϵ-quasi tilings. Moreover, let us emphasize that counting patterns in the empirical measure along a grid is substantially different from counting patterns in the definition of frequencies

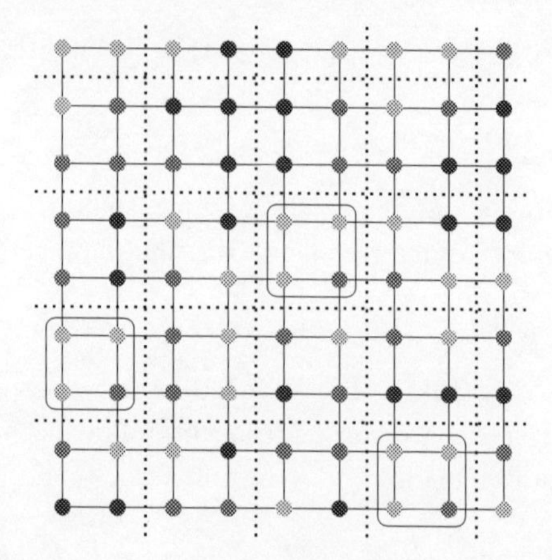

Figure 16.6 The set Λ_j is a square of side length 9 and Λ_L a square of side length 2. Λ_j is (almost) covered when translating Λ_L along the positions of the dashed grid (given by $T_{j,L}$). The marked pattern P is found at 3 (of 16 possible) positions along the grid. Thus, the empirical measure of this pattern is $\widehat{\mathbb{P}}^\omega_{j,L}(P) = \widehat{\mathbb{P}}^\omega_{9,2}(P) = \frac{3}{16}$.

in Section 16.3.2, see the definition of $\sharp_P P'$ in (16.3.2) and compare Figure 16.4 with Figure 16.6.

In order to apply the multivariate version of Glivenko–Cantelli, we aim to integrate functions mapping from $\mathcal{A}^\Lambda$ to $\mathbb{R}$. Such a function $g\colon \mathcal{A}^\Lambda \to \mathbb{R}$ is called monotone, if it is monotone in each coordinate. Besides these generalizations due to higher dimensionality, there is another fundamental difference between univariate and multivariate Glivenko–Cantelli theory: While Theorem 16.5.2 makes no assumptions on the distribution of the random variables, the following example shows that we will have to impose some restrictions on the joint distribution of the coordinates of the random vector.

Example 16.5.4 Let $X_j \sim \mathcal{N}(0,1)$, $j \in \mathbb{N}$, be i. i. d. standard normal random variables and $Y_j := -X_j$. We consider the vectors $(X_j, Y_j) \in \mathbb{R}^2$. Let $\widehat{\mathbb{P}}_n(\omega) := n^{-1}\sum_{j=1}^n \delta_{(X_j,Y_j)}$ be the empirical distribution of $(X_j, Y_j)_{j=1}^n$ on $\mathbb{R}^2$. The random test function

$$g_\omega := \mathbf{1}_{\{(x,y)\in\mathbb{R}^2 \mid x+y<0\}} + \mathbf{1}_{\{(X_j(\omega),Y_j(\omega)) \mid j\in\mathbb{N}\}} \colon \mathbb{R}^2 \to [-1,1]$$

is monotone in each coordinate, and we have

$$\sup\{|\langle g, \widehat{\mathbb{P}}^\omega_n - \mathbb{P}\rangle| \mid g\colon \mathbb{R}^2 \to [-1,1] \text{ monotone}\} \geq |\langle g_\omega, \widehat{\mathbb{P}}^\omega_n - \mathbb{P}\rangle| = 1.$$

The problem arises because the set of discontinuities of the monotone function has positive probability. A correct generalization of Theorem 16.5.2 to the multivariate case is as follows.

Theorem 16.5.5 (DeHardt [DeH71], Wright [Wri81]) *Let*

- V_j, $j \in \mathbb{N}$, *be i. i. d. random variables with values in* $\mathbb{R}^k$ *and distribution* $\mathbb{P}$,
- $\widehat{\mathbb{P}}_n := \frac{1}{n}\sum_{j=1}^n \delta_{V_j}$ *for* $n \in \mathbb{N}$ *the empirical distribution, and*
- $M > 0$ *and* $\mathcal{M} := \{g \colon \mathbb{R}^k \to [-M, M] \mid g \text{ monotone}\}$.

Then, the following are equivalent.

(i) For all $J \subseteq \{1, \dots, k\}$, $J \neq \emptyset$, *strictly monotone* $g \colon \mathbb{R}^J \to \mathbb{R}$, *and* $E \in \mathbb{R}$, *the continuous part* $\mathbb{P}_c^J$ *of the marginal* $\mathbb{P}^J$ *of* $\mathbb{P}$ *satisfies*

$$\mathbb{P}_c^J\big(\partial g^{-1}((-\infty, E])\big) = 0.$$

(ii) There exists an almost sure event Ω_{unif} *on which*

$$\sup_{g\in\mathcal{M}} |\langle g, \widehat{\mathbb{P}}_n - \mathbb{P}\rangle| \xrightarrow{n\to\infty} 0.$$

(iii) For all $\kappa > 0$, *there are* $a_\kappa, b_\kappa > 0$ *such that for all* $n \in \mathbb{N}$, *there is an event* $\Omega_{\kappa,n}$ *with* $\mathbb{P}(\Omega_{\kappa,n}) \geq 1 - b_\kappa \exp(-a_\kappa n)$ *on which*

$$\sup_{g\in\mathcal{M}} |\langle g, \widehat{\mathbb{P}}_n - \mathbb{P}\rangle| \leq \kappa.$$

Here $\partial g^{-1}\big((-\infty, E]\big)$ denotes the boundary of the sub-level set $g^{-1}\big((-\infty, E]$. Condition (i) is trivial in the classical case $k = 1$. Also, in any dimension, each product measure $\mathbb{P}$ satisfies condition (i). In fact, the following theorem holds true.

Theorem 16.5.6 *Let* $\mathbb{P}$ *be a probability measure on* $\mathbb{R}^k$ *which is absolutely continuous with respect to a product measure* $\bigotimes_{j=1}^k \mu_j$ *on* $\mathbb{R}^k$, *where* μ_j, $j \in \{1, \dots, k\}$ *are measures on* $\mathbb{R}$. *Then, condition Theorem 16.5.5(i) is satisfied.*

See [SSV17, Theorem 5.5] for a proof.

16.5.2 Uniform Limits for Monotone Fields

The theorems which follow for the case of infinitely many colors $\mathcal{A}$ all have an additional assumption, namely the monotonicity in the random parameters. This is a natural assumption, indeed: The IDS depends on the potential antitonely in our models. Also, the IDS is monotone in site percolation.

Example 16.5.7 We revisit the Anderson model on a site percolation graph from Section 16.2.2.2, this time with $\mathcal{A} \subseteq \mathbb{R}$. Fix a bounded set $\mathcal{A}_0 \in \mathcal{B}(\mathbb{R})$ for the values of the potential, and let $\mathcal{A} := \mathcal{A}_0 \cup \{\alpha\}$ with $\alpha > 2d + \sup \mathcal{A}_0$. The value α of the potential is interpreted as a closed site in the percolation graph:

$$\mathcal{V}_\omega := \{v \in G \mid V_\omega(v) \neq \alpha\}.$$

The edges of the percolation graph are as before

$$\mathcal{E}_\omega := \{\{v, w\} \subseteq \mathcal{V}_\omega \mid v \sim w\}.$$

The Hamiltonian $H_\omega \colon \ell^2(G) \to \ell^2(G)$ is given by

$$(H_\omega \varphi)(v) := \begin{cases} -\Delta_\omega \varphi(v) + V_\omega \varphi(v) & \text{if } v \in \mathcal{V}_\omega, \text{ and} \\ \alpha \varphi(v) & \text{if } v \in G \setminus \mathcal{V}_\omega. \end{cases}$$

By the min–max principle, the eigenvalues do not decrease when we increase the potential at a site $v \in G$. Particularly, when the potential reaches the value α and the site closes, the eigenfunction experiences de facto a Dirichlet boundary condition on that site, which also at most increases the kinetic energy.

The eigenvalue counting functions count less eigenvalues below a given threshold, if the eigenvalues increase. Therefore, the eigenvalue counting functions decrease when the potential is raised. The same holds true for the limit, i.e., the IDS. This is the reason why the IDS in the quantum percolation model with random potential is antitone in the randomness.

We first turn to the special case $G = \mathbb{Z}^d$. It is physically most relevant and, since the group $\mathbb{Z}^d$ satisfies the tiling property, we do not need to resort to quasi tilings in this case.

Theorem 16.5.8 ([SSV17]) *Let $\mathcal{A} \in \mathcal{B}(\mathbb{R})$, $\Omega := \mathcal{A}^{\mathbb{Z}^d}$, and let $(\Omega, \mathcal{B}(\Omega), \mathbb{P})$ be a probability space such that $\mathbb{P}$ satisfies*

- *$\mathbb{P}$ is translation invariant with respect to the $\mathbb{Z}^d$-action,*
- *for all $\Lambda \in \mathcal{F}$, the marginal $\mathbb{P}_\Lambda$ is absolutely continuous with respect to a product measure on $\mathbb{R}^\Lambda$, and*
- *for a given $r \geq 0$ and all $\Lambda_j \in \mathcal{F}$, $j \in \mathbb{N}$, with $\min\{\mathrm{dist}(\Lambda_i, \Lambda_j) \mid i \neq j\} > r$, the random variables $\omega \mapsto \omega_{\Lambda_j}$, $j \in \mathbb{N}$, are independent.*

Further, let $f : \mathcal{F} \times \Omega \to \mathbb{B}$ be a translation invariant, local, almost additive, monotone, bounded field. Then there exists an event $\tilde{\Omega}$ of full probability and a function $\overline{f} \in \mathbb{B}$ such that for every $\omega \in \tilde{\Omega}$ we have

$$\lim_{j\to\infty}\left\|\frac{f(\Lambda_j,\omega)}{|\Lambda_j|}-\overline{f}\right\|=0, \tag{16.5.1}$$

where $\Lambda_j := [0, j) \cap \mathbb{Z}^d$ *for* $j \in \mathbb{N}$. *For an estimate on the speed of convergence, denote the bound of* f *by* C_f, *the bound of the boundary term* b *of* f *by* D_b. *Then, for every* $\kappa > 0$ *and* $L \in \mathbb{N}$, $L > 2r$, *there are* $a, b > 0$, *depending on* κ, L, *and* C_f, *such that for all* $j \in \mathbb{N}$, $j > 2L$, *there is an event* $\Omega_{\kappa,j}$ *with probability* $\mathbb{P}(\Omega_{\kappa,j}) \geq 1 - b\exp(-a\lfloor j/L\rfloor^d)$, *on which*

$$\left\|\frac{f(\Lambda_j,\omega)}{|\Lambda_j|}-\overline{f}\right\| \leq 2^{2d+1}\Big(\frac{(2C_f+D_b)L^d+D_br^d}{j-2L} + \frac{2(C_f+D_b)r^d+3D_br^d}{L-2r}\Big)+\kappa.$$

holds true.

Of course, there is a version of Theorem 16.5.8 for amenable groups. We follow the strategy in Section 16.4.2 and use quasi tilings to deal with infinitely many colors on amenable groups. This brings new challenges. Theorem 16.5.5 needs as input i. i. d. samples. But, as can be seen in Definition 16.4.4, quasi tilings are allowed to overlap (in a relatively small volume). This destroys the independence of eigenvalue counting functions associated to overlapping tiles. One is tempted to excise the overlap from some of the tiles. However, this would leave us with an independent but not identically distributed sample. In this situation, Glivenko–Cantelli theory is difficult to apply. The solution is to independently resample the portions of the quasi tiles which overlap and to account for the error by a volume estimate.

The last result we present here treats fields with infinitely many colors on amenable groups.

Theorem 16.5.9 ([SSV18]) *Let* G *be a finitely generated amenable group with a Følner sequence* $(Q_j)_j$. *Furthermore, fix* $\mathcal{A} \in \mathcal{B}(\mathbb{R})$, *and let* $(\Omega = \mathcal{A}^G, \mathcal{B}(\Omega), \mathbb{P})$ *be a probability space such that* $\mathbb{P}$ *is translation invariant with respect to* G, *has finite marginals with density w.r.t. a product measure, and independence at a distance. Further, let* $\mathcal{U}$ *be a set of translation invariant, local, almost additive, monotone, bounded fields* $f : \mathcal{F} \times \Omega \to \mathbb{B}$ *with common bound, i.e.,* $C := \sup\{C_f \mid f \in \mathcal{U}\} < \infty$, *common boundary term* $b : \mathcal{F} \to \mathbb{R}$ *with bound* $D := D_b$.

(a) *Then, there exists an event* $\tilde{\Omega} \in \mathcal{B}(\Omega)$ *such that* $\mathbb{P}(\tilde{\Omega}) = 1$ *and for any* $f \in \mathcal{U}$ *there exists a function* $\overline{f} \in \mathbb{B}$, *which does not depend on the specific Følner sequence* $(Q_j)_j$, *with*

$$\forall \omega \in \tilde{\Omega}: \qquad \lim_{j\to\infty} \sup_{f \in \mathcal{U}} \left\| \frac{f(Q_j, \omega)}{|Q_j|} - \overline{f} \right\| = 0.$$

(b) *Furthermore, for each $\varepsilon \in (0, 1/10)$, there exists $j_0(\varepsilon) \in \mathbb{N}$, independent of C, such that for all $f \in \mathcal{U}$, there are $a(\varepsilon, C), b(\varepsilon, C) > 0$, such that for all $j \in \mathbb{N}$, $j \geq j_0(\varepsilon)$, there is an event $\Omega_{j,\varepsilon,C} \in \mathcal{B}(\Omega)$, with the properties*

$$\mathbb{P}(\Omega_{j,\varepsilon,C}) \geq 1 - b(\varepsilon, C) \exp\bigl(-a(\varepsilon, C)|Q_j|\bigr)$$

and

$$\left\| \frac{f(Q_j, \omega)}{|Q_j|} - \overline{f} \right\| \leq (37C_f + 47D_{\mathcal{U}} + 47)\varepsilon$$

for all $\omega \in \Omega_{j,\varepsilon,C}$ and all $f \in \mathcal{U}$.

16.6 Outlook

We have presented a number of theorems concerning convergence in sup-norm and other Banach-space norms of averaged almost additive fields. More important than the convergence itself are the corresponding quantitative error estimates. They split into two parts of two different origins: The geometric part and the probabilistic part. The probabilistic error measures how far off certain empirical measures are from their theoretical counterparts. This difference can be estimated using large deviations techniques, which is implicit in our use of the Theorem 16.5.5 of DeHardt and Wright. An aspect which is not completely satisfactory is that we are not able to specify the dependence of the positive coefficients a_κ and b_κ on the small parameter $\kappa > 0$ and the dimension of the pattern $k \in \mathbb{N}$. For this reason, we are not able to choose the two lengths scales which tend both to infinity as functions of each other.

Furthermore, the theorems in Section 16.5 assume certain monotonicity with respect to individual random parameters. While this is sufficient for a wide variety of models in statistical physics, there are examples, e.g., random hopping Hamiltonians, which do not satisfy this assumption. For this reason, it is desirable to relax the monotonicity assumption, or formulate alternative sufficient conditions. These are aims which we will pursue in a forthcoming project.

Table of Notation

G :	a finitely generated amenable group, or its Cayley graph
g :	element of G, when G is used as group
v, w :	elements of G, when G is used as Cayley graph
$\tilde{\tau}_g$:	group action of the group G on its Cayley graph G
U_g :	unitary group action of G on $\ell^2(G)$
Δ :	Laplace operator on $\ell^2(G)$, $-\Delta \geq 0$
$\mathcal{A}$:	set of colors
$\mathcal{B}(\mathcal{A})$:	Borel sets on $\mathcal{A}$
$\mathbb{P}_0$:	probability measure on $(\mathcal{A}, \mathcal{B}(\mathcal{A}))$
Ω :	set of colorings of G
ω :	coloring
$\mathbb{P}$:	probability measure on $(\Omega, \mathcal{B}(\Omega))$
V :	random potential
τ_g :	group action of G on Ω and on patterns
H_ω :	random Schrödinger operator
Σ :	almost sure spectrum of H_ω
N :	integrated density of states (IDS)
$\mathrm{d}N$:	density of states measure
Λ :	(finite) subset of G, often domain of pattern
$\ell^2(\Lambda)$:	subspace of $\ell^2(G)$
$\mathbf{1}_\Lambda$:	projection $\ell^2(G) \to \ell^2(\Lambda)$
H_ω^Λ :	restriction of H_ω to $\ell^2(\Lambda)$
$\mathcal{F}$:	set of finite subsets of G
δ_v :	Kronecker delta on v
$n(\Lambda, \omega)$:	eigenvalue counting function, not normalized
$\mathbb{B}$:	Banach space, often the right continuous $\mathbb{R} \to \mathbb{R}$-functions
Λ_L :	cube of side length L or small Følner sequence
$\mathbb{P}_s$:	Bernoulli measure for site percolation
$\mathcal{V}_\omega$:	vertices of percolation graph
$\mathcal{E}_\omega$:	edges of percolation graph
$\mathbb{P}_e$:	Bernoulli measure for edge percolation
e_j :	j-th basis vector of $\mathbb{Z}^d$
Δ_ω :	Laplace operator on percolation graph
α :	eigenvalue of H_ω on $G \setminus \mathcal{V}_\omega$
$\mathcal{A}_0$:	set of values of the random potential on a percolation graph
$\mathcal{V}$:	set of visible points in $\mathbb{Z}^d$
H_{vis} :	Schrödinger operator with $\mathbf{1}_\mathcal{V}$ as potential

Table of Notation (continued)

$\tilde{N}$:	limiting function in Theorem 16.3.1
P :	pattern, $P\colon \Lambda \to \mathcal{A}$
$[P]_G$:	equivalence class of patterns with respect to G
ω_Λ :	pattern induced by ω on Λ
$P'\|_\Lambda$:	pattern induced by $P' \in \mathcal{A}^{\Lambda'}$ on Λ
$P'\|_\mathcal{F}$:	set of all patterns induced by P'
$\sharp_P P'$:	number of occurrences of P in P'
$\partial^r \Lambda$:	(two-sided) r-boundary of Λ
dist :	graph distance on Cayley graph
$(Q_j)_j$:	Følner sequence
ν_P :	(asymptotic) frequency of P
b :	boundary term
D_b :	bound of b
F :	field (without explicit dependence on ω)
$\tilde{F}$:	pattern function of F
C_F :	bound of field F
$\overline{F}$:	limit in Theorem 16.3.5
$\widehat{\mathbb{P}}^\omega_\Lambda$:	empirical distribution

References

[And58] P.W. Anderson. Absence of diffusion in certain random lattices. *Phys. Rev.*, 109:1492, 1958.

[AV08] T. Antunović and I. Veselić. Spectral asymptotics of percolation Hamiltoninas on amenable Cayley graphs. *Operator Theory: Advances and Applications*, 186:1–29, 2008. http://arxiv.org/abs/0707.4292.

[BMP00] M. Baake, R.V. Moody, and P. A. B. Pleasants. Diffraction from visible lattice points and kth power free integers. *Discrete Math.*, 221(1–3):3–42, 2000.

[Can33] F. P. Cantelli. Sulla determinazione empirica delle leggi di probabilità. *G. Ist. Ital. Attuari*, 4:421–424, 1933.

[CCF85] J.T. Chayes, L. Chayes, and J. Fröhlich. The low-temperature behavior of disordered magnets. *Commun. Math. Phys.*, 100:399–437, 1985.

[DeH71] J. H. DeHardt. Generalizations of the Glivenko–Cantelli theorem. *The Annals of Mathematical Statistics*, 42(6):2050–2055, 1971.

[Ele07] G. Elek. On limits of finite graphs. *Combinatorica*, 27(4):503–507, 2007.

[Gli33] V.I. Glivenko. Sulla determinazione empirica delle leggi di probabilità. *G. Ist. Ital. Attuari*, 4:92–99, 1933.

[Kel98] G. Keller. *Equilibrium States in Ergodic Theory*, volume 42 of *London Mathematical Society Student Texts*. Cambridge University Press, Cambridge, 1998.

[Kir08] W. Kirsch. An invitation to random Schrödinger operators. In *Random Schrödinger operators*, volume 25 of *Panor. Synthèses*, pages 1–119. Soc. Math. France, Paris, 2008.

[KM06] W. Kirsch and P. Müller. Spectral properties of the Laplacian on bond-percolation graphs. *Math. Zeit.*, 252(4):899–916, 2006.

[KM07] W. Kirsch and B. Metzger. The integrated density of states for random Schrödinger operators. In *Spectral Theory and Mathematical Physics*, volume 76 of *Proceedings of Symposia in Pure Mathematics*, pages 649–698. AMS, 2007.

[Kri07] F Krieger. Le lemme d'Ornstein–Weiss d'aprés gromov. *Dynamics, Ergodic Theory and Geometry*, 54:99–112, 2007.

[Lin01] E. Lindenstrauss. Pointwise theorems for amenable groups. *Invent. Math.*, 146(2):259–295, 2001.

[LMV08] D. Lenz, P. Müller, and I. Veselić. Uniform existence of the integrated density of states for models on $\mathbb{Z}^d$. *Positivity*, 12(4):571–589, 2008.

[LS03] D. Lenz and P. Stollmann. Aperiodic order and quasicrystals: spectral properties. *Ann. Henri Poincaré*, 4(suppl. 2):S933–S942, 2003.

[LS06] D. H. Lenz and P. Stollmann. An ergodic theorem for Delone dynamical systems and existence of the density of states. *J. Anal. Math.*, 97:1–23, 2006. http://www.arxiv.org/math-ph/0310017.

[LSV10] D. Lenz, F. Schwarzenberger, and I. Veselić. A Banach space-valued ergodic theorem and the uniform approximation of the integrated density of states. *Geometriae Dedicata*, 150(1):1–34, 2010.

[LSV12] D. Lenz, F. Schwarzenberger, and I. Veselić. Erratum to: A Banach space-valued ergodic theorem and the uniform approximation of the integrated density of states. *Geometriae Dedicata*, 159(1):411–413, 2012.

[LV09] D. Lenz and I. Veselić. Hamiltonians on discrete structures: jumps of the integrated density of states and uniform convergence. *Math. Z.*, 263(4):813–835, 2009.

[OW87] D Ornstein and B Weiss. Entropy and isomorphism theorems for actions of amenable groups. *J. Anal. Math.*, 48(1):1–141, 1987.

[Pas71] L. A. Pastur. Selfaverageability of the number of states of the Schrödinger equation with a random potential. *Mat. Fiz. i Funkcional. Anal.*, (Vyp. 2):111–116, 1971.

[PF92] L.A. Pastur and A.L. Figotin. *Spectra of Random and Almost-Periodic Operators*. Springer Verlag, Berlin, 1992.

[PS16] F. Pogorzelski and F. Schwarzenberger. A Banach space-valued ergodic theorem for amenable groups and applications. *Journal d'Analyse Mathématique*, 130(1):19–69, 2016.

[Sch08] F. Schwarzenberger. The integrated density of states for operators on groups. http://nbn-resolving.de/urn:nbn:de:bsz:ch1-qucosa-123241. PhD Thesis, TU Chemnitz, 2013.

[Shu79] M. A. Shubin. The spectral theory and the index of elliptic operators with almost periodic coefficients. *Uspekhi Mat. Nauk*, 34:95–135,

1979. English translation: Russian Mathematical Surveys 34, 109–157, 1969.

[SS15] Christoph Schumacher and Fabian Schwarzenberger. Approximation of the integrated density of states on sofic groups. *Ann. Henri Poincaré*, 16(4):1067–1101, 2015.

[SSV14] R. Samavat, P. Stollmann, and I. Veselić. Lifshitz asymptotics for percolation Hamiltonians. *Bull. Lond. Math. Soc.*, 46(6):1113–1125, 2014.

[SSV17] C. Schumacher, F. Schwarzenberger, and I. Veselić. A Glivenko-Cantelli theorem for almost additive functions on lattices. *Stochastic Process. Appl.*, 127(1):179–208, 2017.

[SSV18] C. Schumacher, F. Schwarzenberger, and I. Veselić. Glivenko-Cantelli theory, ornstein–weiss quasi-tilings, and uniform ergodic theorems for distribution-valued fields over amenable groups. *Annals of Applied Probability*. To appear, 2018.

[Ves05a] I. Veselić. Quantum site percolation on amenable graphs. In *Proceedings of the Conference on Applied Mathematics and Scientific Computing*, pages 317–328. Springer, 2005. http://arxiv.org/math-ph/0308041.

[Ves05b] I. Veselić. Spectral analysis of percolation Hamiltonians. *Math. Ann.*, 331(4):841–865, 2005.

[Ves06] I. Veselić. Spectral properties of Anderson-percolation Hamiltonians. *Oberwolfach Rep.*, 3(1):545–547, 2006.

[Ves08] I. Veselić. *Existence and Regularity Properties of the Integrated Density of States of Random Schrödinger Operators*, volume Volume 1917 of *Lecture Notes in Mathematics*. Springer, Berlin, 2008.

[Wei01] B. Weiss. Monotileable amenable groups. *Advances in Mathematical Sciences*, 202:257–262, 2001. Topology, Ergodic Theory, Real Algebraic Geometry.

[Wri81] F. T. Wright. The empirical discrepancy over lower layers and a related law of large numbers. *The Annals of Probability*, 9(2):323–329, 1981.

Printed in the United States
By Bookmasters